Sedimentology and Stratigraphy

Second Edition

Gary Nichols

WILEY-BLACKWELL

A John Wiley & Sons, Ltd., Publication

This edition first published 2009, © 2009 by Gary Nichols
First published 1999

Blackwell Publishing was acquired by John Wiley & Sons in February 2007. Blackwell's publishing program has been merged with Wiley's global Scientific, Technical and Medical business to form Wiley-Blackwell.

Registered office
John Wiley & Sons Ltd, The Atrium, Southern Gate, Chichester, West Sussex, PO19 8SQ, UK

Editorial offices
9600 Garsington Road, Oxford, OX4 2DQ, UK
The Atrium, Southern Gate, Chichester, West Sussex, PO19 8SQ, UK
111 River Street, Hoboken, NJ 07030-5774, USA

For details of our global editorial offices, for customer services and for information about how to apply for permission to reuse the copyright material in this book please see our website at www.wiley.com/wiley-blackwell

The right of the author to be identified as the author of this work has been asserted in accordance with the Copyright, Designs and Patents Act 1988.

Library of Congress Cataloging-in-Publication Data

Nichols, Gary.
   Sedimentology and stratigraphy / Gary Nichols. – 2nd ed.
      p. cm.
   Includes bibliographical references and index.
   ISBN 978-1-4051-3592-4 (pbk. : alk. paper) – ISBN 978-1-4051-9379-5 (hardcover : alk. paper) 1. Sedimentation and deposition. 2. Geology, Stratigraphic. I. Title.
QE571.N53 2009
551.3'03–dc22

                                              2008042948

A catalogue record for this book is available from the British Library.

Set in 9/11pt Photina by SPi Publisher Services, Pondicherry, India
Printed and bound in the United Kingdom

1   2009

Contents

Preface

There is pleasing symmetry about the fact that the backbone of the first edition of this book was written within the Antarctic Circle in gaps between fieldwork with the British Antarctic Survey, while the bulk of this second edition has been written from within the Arctic Circle during my tenure of a 2-year position as Professor of Geology at the University Centre on Svalbard. It is not that I have any great affinity for the polar regions, it just seems that I have almost literally gone to the ends of the Earth to find the peace and quiet that I need to write a book. Between my sojourns in these polar regions 10 years have passed, and both sedimentology and stratigraphy have moved on enough for a thorough update of the material to be required. Just as importantly, technology has moved on, and I can provide a much more satisfying range of illustrative material in digital form on a CD included with the text. Geology is a wonderfully visual science, and it is best appreciated at first hand in the field, but photographs of examples can also aid understanding. I am an unashamed geo-tourist, always looking for yet another example of a geological phenomenon, whether on fieldwork or on holiday. The photographs used in this book and accompanying CD-ROM were taken over a period of 20 years and include examples from many 'corners' of the globe.

AN UNDERGRADUATE TEXT

This book has been written for students who are studying geology at university and it is intended to provide them with an introduction to sedimentology and stratigraphy. It is hoped that the text is accessible to those who are completely new to the subject and that it will also provide a background in concepts and terminology used in more advanced work. The approach is largely descriptive and is intended to complement the more numerical treatment of the topics provided by books such as Leeder (1999). Sedimentary processes are covered in more detail in texts such as Allen (1997) and a much more detailed analysis of sedimentary environments and facies is provided by Reading (1996). For a more comprehensive treatment of some aspects of stratigraphy books such as Coe (2003) are recommended.

DEFINITIONS OF TERMS

This book does not include a glossary, but instead it is intended that terminology is explained and, where necessary, defined in context within the text. The first occurrence of a technical term is usually cast in **_bold italics_**, and it is at this point that an explanation is provided. To find the meaning of a term, the reader should consult the index and go to its first listed occurrence. There are differences of opinion about some terminology, but it is beyond the scope of this text to provide discussion of the issues: in most cases the most broadly accepted view has been adopted; in others simplicity and consistency within the book have taken precedence.

REFERENCES

The references chosen are not intended to be comprehensive for a topic, but merely a selection of a few relatively recent publications that can be used as a starting point for further information. Older sources are cited where these provide important primary accounts of a topic. At the end of each chapter there is a list of suggested further reading materials: these are mainly recent textbooks, compilations of papers in special publications and key review papers and are intended as a starting point for further general information about the topics covered in the chapter.

CROSS-REFERENCING AND THE CD-ROM

To reduce duplication of material, there is quite extensive cross-referencing within the text, indicated by the section number italicised in parentheses, for example (*2.3.4*). Relevant figures are indicated by, for example, 'Fig. 2.34'. The accompanying CD-ROM contains more illustrative material, principally photographs, than is provided within the book: specific reference to this material has not been made in the text, as the book is intended to be 'stand-alone'. In contrast, the CD-ROM is intended for use in conjunction with the book, and so the diagrams and photographs on it are not fully captioned or explained. An index on the CD-ROM contains information about each slide. All photographs used in the book and CD-ROM were taken by the author and all diagrams drafted by the author. A list of the locations of each of the photographs in the text is provided in an appendix on the CD-ROM.

Acknowledgements

Thanks to Phil Chapman for casually suggesting to the teenage younger brother of his friend Roger Nichols that he might like to study geology at 'A' level: this turned out to be the best piece of advice I ever received. The late Doug Shearman was an inspirational lecturer in sedimentology when I was a student at Imperial College, London, and he unwittingly made me committed to the idea of being an academic sedimentologist. (The greatest professional compliment ever paid to me was by Rick Sibson, a former colleague of Doug, who, after I had given a presentation at a conference nearly 20 years later, said 'there were shades of Doug Shearman in the talk you gave today'.) Peter Friend provided understated guidance to me as PhD project supervisor: I could not have a better academic pedigree than as a former research student of Peter. It has been a great pleasure to work with many different people in many different countries, all of whom have in some way provided me with some inspiration. Most importantly, they have made my whole experience of 25 years in geology a lot of fun. Thanks also to Davina for just about everything else.

Introduction: Sedimentology and Stratigraphy

Sedimentology is the study of the processes of formation, transport and deposition of material that accumulates as sediment in continental and marine environments and eventually forms sedimentary rocks. Stratigraphy is the study of rocks to determine the order and timing of events in Earth history: it provides the time frame that allows us to interpret sedimentary rocks in terms of dynamic evolving environments. The stratigraphic record of sedimentary rocks is the fundamental database for understanding the evolution of life, plate tectonics through time and global climate change.

1.1 SEDIMENTARY PROCESSES

The concept of interpreting rocks in terms of modern processes dates back to the 18th and 19th centuries ('the present is the key to the past'). 'Sedimentology' has existed as a distinct branch of the geological sciences for only a few decades. It developed as the observational elements of physical stratigraphy became more quantitative and the layers of strata were considered in terms of the physical, chemical and biological processes that formed them.

The nature of sedimentary material is very varied in origin, size, shape and composition. Particles such as grains and pebbles may be derived from the erosion of older rocks or directly ejected from volcanoes. Organisms form a very important source of material, ranging from microbial filaments encrusted with calcium carbonate to whole or broken shells, coral reefs, bones and plant debris. Direct precipitation of minerals from solution in water also contributes to sediments in some situations.

Formation of a body of sediment involves either the transport of particles to the site of deposition by gravity, water, air, ice or mass flows or the chemical or biological growth of the material in place. Accumulation of sediments in place is largely influenced by the chemistry, temperature and biological character of the setting. The processes of transport and deposition can be determined by looking at individual layers of sediment. The size, shape and distribution of particles all provide clues to the way in which the material was carried and deposited. Sedimentary structures such as ripples can be seen in sedimentary rocks and can be compared to ripples forming today, either in natural environments or in a laboratory tank.

Assuming that the laws that govern physical and chemical processes have not changed through time, detailed measurements of sedimentary rocks can be used to make estimates (to varying degrees of accuracy) of the physical, chemical and biological conditions that existed at the time of sedimentation. These conditions may include the salinity, depth and flow velocity in lake or seawater, the strength and direction of the wind in a desert and the tidal range in a shallow marine setting.

1.2 SEDIMENTARY ENVIRONMENTS AND FACIES

The environment at any point on the land or under the sea can be characterised by the physical and chemical processes that are active there and the organisms that live under those conditions at that time. As an example, a fluvial (river) environment includes a channel confining the flow of fresh water that carries and deposits gravelly or sandy material on bars in the channel (Fig. 1.1). When the river floods, water spreads relatively fine sediment over the floodplain where it is deposited in thin layers. Soils form and vegetation grows on the floodplain area. In a succession of sedimentary rocks (Fig. 1.2) the channel may be represented by a lens of sandstone or conglomerate that shows internal structures formed by deposition on the channel bars. The floodplain setting will be represented by thinly bedded mudrock and sandstone with roots and other evidence of soil formation.

In the description of sedimentary rocks in terms of depositional environments, the term 'facies' is often used. A rock ***facies*** is a body of rock with specified characteristics that reflect the conditions under which it was formed (Reading & Levell 1996). Describing the facies of a body of sediment involves documenting all the characteristics of its lithology, texture, sedimentary structures and fossil content that can aid in determining the processes of formation. By recognising associations of facies it is possible to establish the combinations of processes that were dominant; the characteristics of a depositional environment are determined by the processes that are present, and hence there is a link between facies associations and environments of deposition. The lens of sandstone in Fig. 1.2 may be shown to be a river channel if the floodplain deposits are found associated with it. However, recognition of a channel form on its own is not a sufficient basis to determine the depositional environment because channels filled with sand exist in other settings, including deltas,

Fig. 1.1 A modern depositional environment: a sandy river channel and vegetated floodplain.

Fig. 1.2 Sedimentary rocks interpreted as the deposits of a river channel (the lens of sandstones in the centre right of the view) scoured into mudstone deposited on a floodplain (the darker, thinly bedded strata below and to the side of the sandstone lens).

tidal environments and the deep sea floor: it is the association of different processes that provides the full picture of a depositional environment.

1.3 THE SPECTRUM OF ENVIRONMENTS AND FACIES

Every depositional environment has a unique combination of processes, and the products of these processes, the sedimentary rocks, will be a similarly unique assemblage. For convenience of description and interpretation, depositional environments are classified as, for example, a delta, an estuary or a shoreline, and subcategories of each are established, such as wave-dominated, tide-dominated and river-dominated deltas. This approach is in general use by sedimentary geologists and is followed in this book. It is, however, important to recognise that these environments of deposition are convenient categories or 'pigeonholes', and that the description of them tends to be of 'typical' examples. The reality is that every delta, for example, is different from its neighbour in space or time, that every deltaic deposit will also be unique, and although we categorise deltas into a number of types, our deposit is likely to fall somewhere in between these 'pigeonholes'. Sometimes it may not even be possible to conclusively distinguish between the deposits of a delta and an estuary, especially if the data set is incomplete, which it inevitably is when dealing with events of the past. However, by objectively considering each bed in terms of physical, chemical and biological processes, it is always possible to provide some indication of where and how a sedimentary rock was formed.

1.4 STRATIGRAPHY

Use of the term 'stratigraphy' dates back to d'Orbingy in 1852, but the concept of layers of rocks, or strata, representing a sequence of events in the past is much older. In 1667 Steno developed the principle of superposition: 'in a sequence of layered rocks, any layer is older than the layer next above it'. Stratigraphy can be considered as the relationship between rocks and time and the stratigrapher is concerned with the observation, description and interpretation of direct and tangible evidence in rocks to determine the history of the Earth. We all recognise that our planet is a dynamic place, where plate tectonics creates mountains and

oceans and where changes in the atmosphere affect the climate, perhaps even on a human time scale. To understand how these global systems work, we need a record of their past behaviour to analyse, and this is provided by the study of stratigraphy.

Stratigraphy provides the temporal framework for geological sciences. The relative ages of rocks, and hence the events that are recorded in those rocks, can be determined by simple stratigraphic relationships (younger rocks generally lie on top of older, as Steno recognised), the fossils that are preserved in strata and by measurements of processes such as the radioactive decay of elements that allow us to date some rock units. At one level, stratigraphy is about establishing a nomenclature for rock units of all ages and correlating them all over the world, but at another level it is about finding the evidence for climate change in the past or the movements of tectonic plates. One of the powerful tools we have for predicting future climate change is the record in the rock strata of local and global changes over periods of thousands to millions of years. Furthermore our understanding of evolutionary processes is in part derived from the study of fossils found in rocks of different ages that tell us about how forms of life have changed through time. Other aspects of stratigraphy provide the tools for finding new resources: for example, 'sequence stratigraphy' is a predictive technique, widely used in the hydrocarbon industry, that can be used to help to find new reserves of oil and gas.

The combination of sedimentology and stratigraphy allows us to build up pictures of the Earth's surface at different times in different places and relate them to each other. The character of the sedimentary rocks deposited might, for example, indicate that at one time a certain area was an arid landscape, with desert dunes and with washes of gravel coming from a nearby mountain range. In that same place, but at a later time, conditions allowed the formation of coral reefs in a shallow sea far away from any landmass, and we can find the record of this change by interpreting the rocks in terms of their processes and environments of deposition. Furthermore, we might establish that at the same time as there were shallow tropical seas in one place, there lay a deep ocean a few tens of kilometres away where fine sediment was deposited by ocean currents. We can thus build up pictures of the **_palaeogeography_**, the appearance of an area during some time in the past, and establish changes in palaeogeography through Earth history. To complete the picture, the distribution of different environments and their changes through time can be related to plate tectonics, because mountain building provides the source for much of the sediment, and plate movements also create the sedimentary basins where sediment accumulates.

1.5 THE STRUCTURE OF THIS BOOK

Sedimentology and stratigraphy can be considered together as a continuum of processes and products, both in space and time. Sedimentology is concerned primarily with the formation of sedimentary rocks but as soon as these beds of rock are looked at in terms of their temporal and spatial relationships the study has become stratigraphic. Similarly if the stratigrapher wishes to interpret layers of rock in terms of environments of the past the research is sedimentological. It is therefore appropriate to consider sedimentology and stratigraphy together at an introductory level.

The starting point taken in this book is the smallest elements, the particles of sand, pebbles, clay minerals, pieces of shell, algal filaments, chemical precipitates and other constituents that make up sediments (Chapters 2 and 3). An introduction to the petrographic analysis of sedimentary materials in hand specimen and under the microscope is included in these chapters. In Chapter 4 the processes of sediment transport and deposition are considered, followed by a section on the methodology of recording and analysing sedimentary data in the field in Chapter 5. Weathering and erosion is considered in Chapter 6 as an introduction to the processes which generate the clastic material that is deposited in many sedimentary environments. The following chapters (7 to 17) deal largely with different depositional environments, outlining the physical, chemical and biological processes that are active, the characteristics of the products of these processes and how they may be recognised in sedimentary rocks. Continental environments are covered in Chapters 8

to 10, followed by marine environments in Chapters 12 to 16 – the general theme being to start at the top, with the mountains, and end up in the deep oceans. Exceptions to this pattern are Chapter 7 on glacial environments and Chapter 17 on volcanic processes and products. Post-depositional processes, including lithification and the formation of hydrocarbons, are considered in Chapter 18. Chapters 19 to 23 are on different aspects of stratigraphy and are intended to provide an introduction to the principles of stratigraphic analysis using techniques such as lithostratigraphy, biostratigraphy and sequence stratigraphic correlation. The final chapters in the book provide a brief introduction to sedimentary basins and the large-scale tectonic and climatic controls on the sedimentary record.

Sedimentology and stratigraphy cannot be considered in isolation from other aspects of geology, and in particular, plate tectonics, petrology, palaeontology and geomorphology are complementary topics. Reference is made to these subjects in the text, but only a basic knowledge of these topics is assumed.

FURTHER READING

The following texts provide a general background to geology.

Chernicoff, S. & Whitney, D. (2007) *Geology: an Introduction to Physical Geology* (4th edition). Pearson/Prentice Hall, New Jersey.

Grotzinger, J., Jordan, T.H., Press, F. & Siever, R. (2007) *Understanding Earth* (5th edition). Freeman and Co., New York.

Lutgens, F.K. & Tarbuck, E.J. (2006) *Essentials of Geology* (9th edition). Pearson/Prentice Hall, New Jersey.

Smith, G.A. & Pun, A. (2006) *How Does the Earth Work? Physical Geology and the Process of Science.* Pearson/Prentice Hall, New Jersey.

Summerfield, M.A. (1991) *Global Geomorphology: an Introduction to the Study of Landforms.* Longman/Wiley, London/New York.

2

Terrigenous Clastic Sediments:
Gravel, Sand and Mud

Terrigenous clastic sediments and sedimentary rocks are composed of fragments that result from the weathering and erosion of older rocks. They are classified according to the sizes of clasts present and the composition of the material. Analysis of gravels and conglomerates can be carried out in the field and can reveal where the material came from and how it was transported. Sands and sandstones can also be described in the field, but for a complete analysis examination under a petrographic microscope is required to reveal the composition of individual grains and their relationships to each other. The finest sediments, silt and clay, can only be fully analysed using scanning electron microscopes and X-ray diffractometers. The proportions of different clast sizes and the textures of terrigenous clastic sediments and sedimentary rocks can provide information about the history of transport of the material and the environment of deposition.

2.1 CLASSIFICATION OF SEDIMENTS AND SEDIMENTARY ROCKS

A convenient division of all sedimentary rocks is shown in Fig. 2.1. Like most classification schemes of natural processes and products it includes anomalies (a deposit of chemically precipitated calcium carbonate would be classified as a limestone, not an evaporite) and arbitrary divisions (the definition of a limestone as a rock having more than 50% calcium carbonate), but it serves as a general framework.

Terrigenous clastic material This is material that is made up of particles or **clasts** derived from

pre-existing rocks. The clasts are principally detritus eroded from bedrock and are commonly made up largely of silicate minerals: the terms **detrital sediments** and **siliciclastic sediments** are also used for this material. Clasts range in size from clay particles measured in microns, to boulders metres across. Sandstones and conglomerates make up 20–25% of the sedimentary rocks in the stratigraphic record and mudrocks are 60% of the total.

Carbonates By definition, a limestone is any sedimentary rock containing over 50% calcium carbonate ($CaCO_3$). In the natural environment a principal source of calcium carbonate is from the hard parts of organisms, mainly invertebrates such

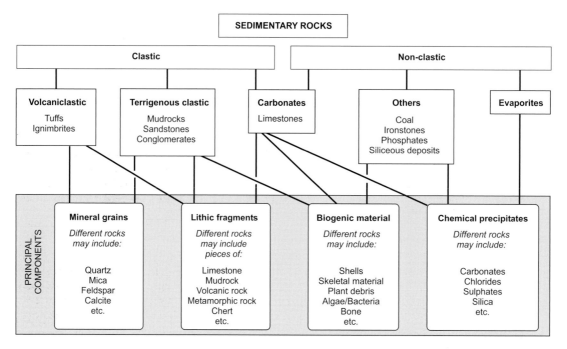

Fig. 2.1 A classification scheme for sediments and sedimentary rocks.

as molluscs. Limestones constitute 10–15% of the sedimentary rocks in the stratigraphic record.

Evaporites These are deposits formed by the precipitation of salts out of water due to evaporation.

Volcaniclastic sediments These are the products of volcanic eruptions or the result of the breakdown of volcanic rocks.

Others Other sediments and sedimentary rocks are sedimentary ironstone, phosphate sediments, organic deposits (coals and oil shales) and cherts (siliceous sedimentary rocks). These are volumetrically less common than the above, making up about 5% of the stratigraphic record, but some are of considerable economic importance.

In this chapter terrigenous clastic deposits are considered: the other types of sediment and sedimentary rock are covered in Chapter 3.

2.1.1 Terrigenous clastic sediments and sedimentary rocks

A distinction can be drawn between sediments (generally loose material) and sedimentary rocks which are lithified sediment: **lithification** is the process of 'turning into rock' (*18.2*). Mud, silt and sand are all loose **aggregates**; the addition of the suffix '-stone' (mudstone, siltstone, sandstone) indicates that the material has been lithified and is now a solid rock. Coarser, loose gravel material is named according to its size as granule, pebble, cobble and boulder aggregates, which become lithified into conglomerate (sometimes with the size range added as a prefix, e.g. 'pebble conglomerate').

A threefold division on the basis of grain size is used as the starting point to classify and name terrigenous clastic sediments and sedimentary rocks: gravel and conglomerate consist of clasts greater than 2 mm in diameter; sand-sized grains are between 2 mm and 1/16 mm (63 microns) across; mud (including clay and silt) is made up of particles less than 63 μm in diameter. There are variants on this scheme and there are a number of ways of providing subdivisions within these categories, but sedimentologists generally use the Wentworth Scale (Fig. 2.2) to define and name terrigenous clastic deposits.

mm	phi	Name		
256	−8	Boulders		
128	−7			
64	−6	Cobbles		*Gravel Conglomerate*
32	−5			
16	−4			
8	−3	Pebbles		
4	−2			
2	−1	Granules		
1	0	Very coarse sand		
0.5	1	Coarse sand		
0.25	2	Medium sand		*Sand Sandstone*
0.125	3	Fine sand		
0.063	4	Very fine sand		
0.031	5	Coarse silt		
0.0156	6	Medium silt		
0.0078	7	Fine silt		*Mud Mudrock*
0.0039	8	Very fine silt		
		Clay		

Fig. 2.2 The Udden–Wentworth grain-size scale for clastic sediments: the clast diameter in millimetres is used to define the different sizes on the scale, and the phi values are −log₂ of the grain diameter.

2.1.2 The Udden–Wentworth grain-size scale

Known generally as the **Wentworth Scale**, this is the scheme in most widespread use for the classification of aggregates particulate matter (Udden 1914; Wentworth 1922). The divisions on the scale are made on the basis of factors of two: for example, medium sand grains are 0.25 to 0.5 mm in diameter, coarse sand grains are 0.5 to 1.0 mm, very coarse sand 1.0 to 2.0 mm, etc. It is therefore a logarithmic progression, but a logarithm to the 'base two', as opposed to the 'base ten' of the more common 'log' scales. This scale has been chosen because these divisions appear to reflect the natural distribution of sedimentary particles and in a simple way it can be related to starting with a large block and repeatedly breaking it into two pieces.

Four basic divisions are recognised:

clay (<4 μm)
silt (4 μm to 63 μm)
sand (63 μm or 0.063 mm to 2.0 mm)
gravel/aggregates (>2.0 mm)

The **phi scale** is a numerical representation of the Wentworth Scale. The Greek letter 'φ' (phi) is often used as the unit for this scale. Using the logarithm base two, the grain size can be denoted on the phi scale as

$$\phi = -\log_2 (\text{grain diameter in mm})$$

The negative is used because it is conventional to represent grain sizes on a graph as decreasing from left to right (*2.5.1*). Using this formula, a grain diameter of 1 mm is 0φ: increasing the grain size, 2 mm is −1φ, 4 mm is −2φ, and so on; decreasing the grain size, 0.5 mm is +1φ, 0.25 mm is 2φ, etc.

2.2 GRAVEL AND CONGLOMERATE

Clasts over 2 mm in diameter are divided into granules, pebbles, cobbles and boulders (Fig. 2.2). Consolidated gravel is called **conglomerate** (Fig. 2.3) and when described will normally be named according to the dominant clast size: if most of the clasts are between 64 mm and 256 mm in diameter the rock would be called a cobble conglomerate. The term **breccia** is commonly used for conglomerate made up of clasts that are angular in shape (Fig. 2.4). In

Fig. 2.3 A conglomerate composed of well-rounded pebbles.

Fig. 2.4 A conglomerate (or breccia) made up of angular clasts.

some circumstances it is prudent to specify that a deposit is a 'sedimentary breccia' to distinguish it from a 'tectonic breccia' formed by the fragmentation of rock in fault zones. Mixtures of rounded and angular clasts are sometimes termed ***breccio-conglomerate***. Occasionally the noun ***rudite*** and the adjective ***rudaceous*** are used: these terms are synonymous with conglomerate and conglomeratic.

2.2.1 Composition of gravel and conglomerate

A more complete description of the nature of a gravel or conglomerate can be provided by considering the types of clast present. If all the clasts are of the same material (all of granite, for example), the conglomerate is considered to be ***monomict***. A ***polymict*** conglomerate is one that contains clasts of many different lithologies, and sometimes the term ***oligomict*** is used where there are just two or three clast types present.

Almost any lithology may be found as a clast in gravel and conglomerate. ***Resistant lithologies***, those which are less susceptible to physical and chemical breakdown, have a higher chance of being preserved as a clast in a conglomerate. Factors controlling the resistance of a rock type include the minerals present and the ease with which they are chemically or physically broken down in the environment. Some sandstones break up into sand-sized fragments when eroded because the grains are weakly cemented together. The most important factor controlling the varieties of clast found is the bedrock being eroded in the area. Gravel will be composed entirely of limestone clasts if the source area is made up only of

limestone bedrock. Recognition of the variety of clasts can therefore be a means of determining the source of a conglomeratic sedimentary rock (5.4.1).

2.2.2 Texture of conglomerate

Conglomerate beds are rarely composed entirely of gravel-sized material. Between the granules, pebbles, cobbles and boulders, finer sand and/or mud will often be present: this finer material between the large clasts is referred to as the ***matrix*** of the deposit. If there is a high proportion (over 20%) of matrix, the rock may be referred to as a ***sandy conglomerate*** or ***muddy conglomerate***, depending on the grain size of the matrix present (Fig. 2.5). An ***intraformational conglomerate*** is composed of clasts of the same material as the matrix and is formed as a result of reworking of lithified sediment soon after deposition.

The proportion of matrix present is an important factor in the texture of conglomeratic sedimentary rock, that is, the arrangement of different grain sizes within it. A distinction is commonly made between conglomerates that are ***clast-supported*** (Fig. 2.6), that is, with clasts touching each other throughout the rock, and those which are ***matrix-supported*** (Fig. 2.7), in which most of the clasts are completely surrounded by matrix. The term ***orthoconglomerate*** is sometimes used to indicate that the rock is clast-supported, and ***paraconglomerate*** for a matrix-supported texture. These textures are significant when determining the mode of transport and deposition of a conglomerate (e.g. on alluvial fans: 9.5).

The arrangement of the sizes of clasts in a conglomerate can also be important in interpretation of depositional processes. In a flow of water, pebbles are moved more easily than cobbles that in turn require less energy to move them than boulders. A deposit that is made up of boulders overlain by cobbles and then pebbles may be interpreted in some cases as having been formed from a flow that was decreasing in velocity. This sort of interpretation is one of the techniques used in determining the processes of transport and deposition of sedimentary rocks (4.2).

2.2.3 Shapes of clasts

The shapes of clasts in gravel and conglomerate are determined by the fracture properties of the bedrock they are derived from and the history of transport.

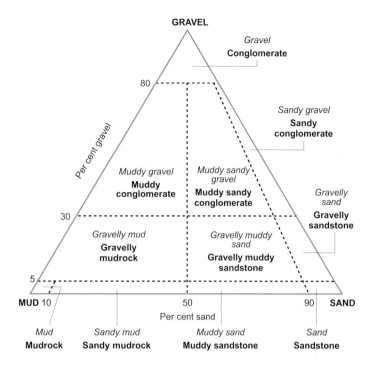

Fig. 2.5 Nomenclature used for mixtures of gravel, sand and mud in sediments and sedimentary rock.

Rocks with equally spaced fracture planes in all directions form cubic or **equant** blocks that form spherical clasts when the edges are rounded off (Fig. 2.8). Bedrock lithologies that break up into slabs, such as a well-bedded limestone or sandstone, form clasts with one axis shorter than the other two (Krumbein & Sloss 1951). This is termed an **oblate** or **discoid** form. Rod-shaped or **prolate** clasts are less common, forming mainly from metamorphic rocks with a strong linear fabric.

When discoid clasts are moved in a flow of water they are preferentially oriented and may stack up in a form known as **imbrication** (Figs 2.9 & 2.10). These stacks are arranged in positions that offer the least resistance to flow, which is with the discoid clasts dipping upstream. In this orientation, the water can flow most easily up the upstream side of the clast, whereas when clasts are oriented dipping down stream, flow at the edge of the clast causes it to be reoriented. The direction of imbrication of discoid

Fig. 2.6 A clast-supported conglomerate: the pebbles are all in contact with each other.

Fig. 2.7 A matrix-supported conglomerate: each pebble is surrounded by matrix.

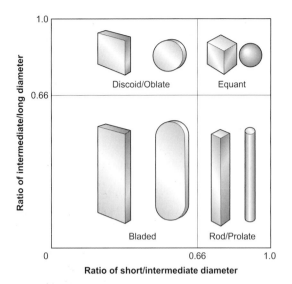

Fig. 2.8 The shape of clasts can be considered in terms of four end members, equant, rod, disc and blade. Equant and disc-shaped clasts are most common.

Fig. 2.9 A conglomerate bed showing imbrication of clasts due to deposition in a current flowing from left to right.

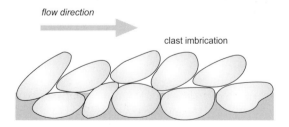

Fig. 2.10 The relationship between imbrication and flow direction as clasts settle in a stable orientation.

pebbles in a conglomerate can be used to indicate the direction of the flow that deposited the gravel. If a discoid clast is also elongate, the orientation of the longest axis can help to determine the mode of deposition: clasts deposited by a flow of water will tend to have their long axis oriented perpendicular to the flow, whereas glacially deposited clasts (*7.3.3*) will have the long axis oriented parallel to the ice flow.

2.3 SAND AND SANDSTONE

Sand grains are formed by the breakdown of pre-existing rocks by weathering and erosion (*6.4 & 6.5*), and from material that forms within the depositional environment. The breakdown products fall into two categories: ***detrital mineral grains***, eroded from pre-existing rocks, and sand-sized pieces of rock, or ***lithic fragments***. Grains that form within the depositional environment are principally biogenic in origin, that is, they are pieces of plant or animal, but there are some which are formed by chemical reactions.

Sand may be defined as a sediment consisting primarily of grains in the size range 63 μm to 2 mm and a ***sandstone*** is defined as a sedimentary rock with grains of these sizes. This size range is divided into five intervals: very fine, fine, medium, coarse and very coarse (Fig. 2.2). It should be noted that this nomenclature refers only to the size of the particles. Although many sandstones contain mainly quartz grains, the term sandstone carries no implication about the amount of quartz present in the rock and some sandstones contain no quartz at all. Similarly, the term ***arenite***, which is a sandstone with less than 15% matrix, does not imply any particular clast composition. Along with the adjective ***arenaceous*** to describe a rock as sandy, arenite has its etymological roots in the Latin word for sand, 'arena', also used to describe a stadium with a sandy floor.

2.3.1 Detrital mineral grains in sands and sandstones

A very large number of different minerals may occur in sands and in sandstones, and only the most common are described here.

Quartz

Quartz is the commonest mineral species found as grains in sandstone and siltstone. As a primary mineral it is a major constituent of granitic rocks, occurs in some igneous rocks of intermediate composition and is absent from basic igneous rock types. Metamorphic rocks such as gneisses formed from granitic material and many coarse-grained metasedimentary rocks contain a high proportion of quartz. Quartz also occurs in veins, precipitated by hot fluids associated with igneous and metamorphic processes. Quartz is a very stable mineral that is resistant to chemical breakdown at the Earth's surface. Grains of quartz may be broken or abraded during transport but with a hardness of 7 on **Mohs' scale** of hardness, quartz grains remain intact over long distances and long periods of transport. In hand specimen quartz grains show little variation: coloured varieties such as smoky or milky quartz and amethyst occur but mostly quartz is seen as clear grains.

Feldspar

Most igneous rocks contain feldspar as a major component. Feldspar is hence very common and is released in large quantities when granites, andesites, gabbros as well as some schists and gneisses break down. However, feldspar is susceptible to chemical alteration during weathering and, being softer than quartz, tends to be abraded and broken up during transport. Feldspars are only commonly found in circumstances where the chemical weathering of the bedrock has not been too intense and the transport pathway to the site of deposition is relatively short. Potassium feldspars are more common as detrital grains than sodium- and calcium-rich varieties, as they are chemically more stable when subjected to weathering (6.4).

Mica

The two commonest mica minerals, **biotite** and **muscovite**, are relatively abundant as detrital grains in sandstone, although muscovite is more resistant to weathering. They are derived from granitic to intermediate composition igneous rocks and from schists and gneisses where they have formed as metamorphic minerals. The platy shape of mica grains makes them distinctive in hand specimen and under the microscope. Micas tend to be concentrated in bands on bedding planes and often have a larger surface area than the other detrital grains in the sediment; this is because a platy grain has a lower settling velocity than an equant mineral grain of the same mass and volume so micas stay in temporary suspension longer than quartz or feldspar grains of the same mass.

Heavy minerals

The common minerals found in sands have densities of around 2.6 or $2.7 \, \text{g cm}^{-3}$: quartz has a density of $2.65 \, \text{g cm}^{-3}$, for example. Most sandstones contain a small proportion, commonly less than 1%, of minerals that have a greater density. These **heavy minerals** have densities greater than $2.85 \, \text{g cm}^{-3}$ and are traditionally separated from the bulk of the lighter minerals by using a liquid of that density which the common minerals will float in but the small proportion of dense minerals will sink. These minerals are uncommon and study of them is only possible after concentrating them by dense liquid separation. They are valuable in provenance studies (5.4.1) because they can be characteristic of a particular source area and are therefore valuable for studies of the sources of detritus. Common heavy minerals include zircon, tourmaline, rutile, apatite, garnet and a range of other metamorphic and igneous accessory minerals.

Miscellaneous minerals

Other minerals rarely occur in large quantities in sandstone. Most of the common minerals in igneous silicate rocks (e.g. olivine, pyroxenes and amphiboles) are all too readily broken down by chemical weathering. Oxides of iron are relatively abundant. Local concentrations of a particular mineral may occur when there is a nearby source.

2.3.2 Other components of sands and sandstones

Lithic fragments

The breakdown of pre-existing, fine- to medium-grained igneous, metamorphic and sedimentary rocks results in sand-sized fragments. Sand-sized lithic fragments are only found of fine to medium-grained rocks because by definition the mineral crystal and grains of a coarser-grained rock type are the size of sand grains or larger. Determination of the lithology of these fragments of rock usually requires petrographic analysis by thin-section examination (2.3.5) to identify the mineralogy and fabric.

Grains of igneous rocks such as basalt and rhyolite are susceptible to chemical alteration at the Earth's surface and are only commonly found in sands formed close to the source of the volcanic material. Beaches around volcanic islands may be black because they are made up almost entirely of lithic grains of basalt. Sandstone of this sort of composition is rare in the stratigraphic record, but grains of volcanic rock types may be common in sediments deposited in basins related to volcanic arcs or rift volcanism (Chapter 17).

Fragments of schists and pelitic (fine-grained) metamorphic rocks can be recognised under the microscope by the strong aligned fabric that these lithologies possess: pressure during metamorphism results in mineral grains becoming reoriented or growing into an alignment perpendicular to the stress field. Micas most clearly show this fabric, but quartz crystals in a metamorphic rock may also display a strong alignment. Rocks formed by the metamorphism of quartz-rich lithologies break down to relatively resistant grains that can be incorporated into a sandstone.

Lithic fragments of sedimentary rocks are generated when pre-existing strata are uplifted, weathered and eroded. Sand grains can be reworked by this process and individual grains may go through a number of cycles of erosion and redeposition (2.5.4). Finer-grained mudrock lithologies may break up to form sand-sized grains although their resistance to further breakdown during transport is largely dependent on the degree of lithification of the mudrock (18.2). Pieces of limestone are commonly found as lithic fragments in sandstone although a rock made up largely of calcareous grains would be classified as a limestone (3.1). One of the most common lithologies seen as a sand grain is chert (3.3), which being silica is a resistant material.

Biogenic particles

Small pieces of calcium carbonate found in sandstone are commonly broken shells of molluscs and other organisms that have calcareous hard parts. These **biogenic fragments** are common in sandstone deposited in shallow marine environments where these organisms are most abundant. If these calcareous fragments make up over 50% of the bulk of the rock it would be considered to be a limestone (the nature and occurrence of calcareous biogenic fragments is described in the next chapter: 3.1.3). Fragments of bone and teeth may be found in sandstones from a wide variety of environments but are rarely common. Wood, seeds and other parts of land plants may be preserved in sandstone deposited in continental and marine environments.

Authigenic minerals

Minerals that grow as crystals in a depositional environment are called **authigenic** minerals. They are distinct from all the detrital minerals that formed by igneous or metamorphic processes and were subsequently reworked into the sedimentary realm. Many carbonate minerals form authigenically and another important mineral formed in this way is glauconite/glaucony (11.5.1), a green iron silicate that forms in shallow marine environments.

Matrix

Fine-grained material occurring between the sand grains is referred to as matrix (2.2.2). In sands and sandstone the matrix is typically silt and clay-sized material, and it may wholly or partly fill the spaces between the grains. A distinction should be drawn between the matrix, which is material deposited along with the grains, and cement (18.2.2), which is chemically precipitated after deposition.

2.3.3 Sandstone nomenclature and classification

Full description of a sandstone usually includes some information concerning the types of grain present. Informal names such as **micaceous sandstone** are used when the rock clearly contains a significant amount of a distinctive mineral such as mica. Terms such as **calcareous sandstone** and **ferruginous sandstone** may also be used to indicate a particular chemical composition, in these cases a noticeable proportion of calcium carbonate and iron respectively. These names for a sandstone are useful and appropriate for field and hand-specimen descriptions, but when a full petrographic analysis is possible with a thin-section of the rock under a microscope, a more

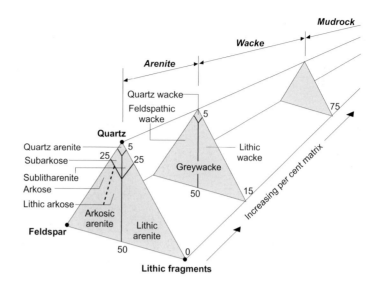

Fig. 2.11 The Pettijohn classification of sandstones, often referred to as a 'Toblerone plot' (Pettijohn 1975).

formal nomenclature is used. This is usually the Pettijohn et al. (1987) classification scheme (Fig. 2.11).

The Pettijohn sandstone classification combines textural criteria, the proportion of muddy matrix, with compositional criteria, the percentages of the three commonest components of sandstone: quartz, feldspar and lithic fragments. The triangular plot has these three components as the end members to form a '**Q, F, L**' triangle, which is commonly used in clastic sedimentology. To use this scheme for sandstone classification, the relative proportions of quartz, feldspar and lithic fragments must first be determined by visual estimation or by counting grains under a microscope: other components, such as mica or biogenic fragments, are disregarded. The third dimension of the classification diagram is used to display the texture of the rock, the relative proportions of clasts and matrix. In a sandstone the matrix is the silt and clay material that was deposited with the sand grains. The second stage is therefore to measure or estimate the amount of muddy matrix: if the amount of matrix present is less than 15% the rock is called an arenite, between 15% and 75% it is a **wacke** and if most of the volume of the rock is fine-grained matrix it is classified as a mudstone (2.4.1).

Quartz is the most common grain type present in most sandstones so this classification emphasises the presence of other grains. Only 25% feldspar need be present for the rock to be called a **feldspathic arenite**, **arkosic arenite** or **arkose** (these three terms are interchangeable when referring to sandstone rich in feldspar grains). By the same token, 25% of lithic fragments in a sandstone make it a **lithic arenite** by this scheme. Over 95% of quartz must be present for a rock to be classified as a **quartz arenite**; sandstone with intermediate percentages of feldspar or lithic grains is called subarkosic arenite and sublithic arenite. Wackes are similarly divided into **quartz wacke**, **feldspathic (arkosic) wacke** and **lithic wacke**, but without the subdivisions. If a grain type other than the three main components is present in significant quantities (at least 5% or 10%), a prefix may be used such as 'micaceous quartz arenite': note that such a rock would not necessarily contain 95% quartz as a proportion of all the grains present, but 95% of the quartz, feldspar and lithic fragments when they are added together.

The term **greywacke** has been used in the past for a sandstone that might also be called a feldspathic or lithic wacke. They are typically mixtures of rock fragments, quartz and feldspar grains with a matrix of clay and silt-sized particles.

2.3.4 Petrographic analysis of sands and sandstones

In sand-grade rocks, the nature of the individual grains and the relationship between these grains and the material between them is best seen in a **thin-section**

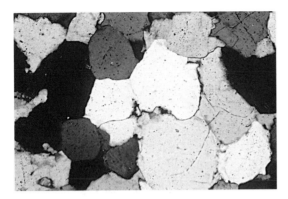

Fig. 2.12 A photomicrograph of a sandstone: the grains are all quartz but appear different shades of grey under crossed polars due to different orientations of the grains.

of the rock, a very thin (normally 30 microns) slice of the rock, which can be examined under a ***petrological/petrographic microscope*** (Fig. 2.12). Thin-section examination is a standard technique for the analysis of almost all types of rock, igneous and metamorphic as well as sedimentary, and the procedures form part of the training of most geologists.

The petrographic microscope

A thin-section of a rock is cemented onto a glass microscope slide and it is normal practice to cement a thin glass cover slip over the top of the rock slice to form a sandwich, but there are circumstances where the thin-section is left uncovered (*3.1.2*). The slide is placed on the microscope stage where a beam of white light is projected through the slide and up through the lenses to the eyepiece: this transmitted light microscopy is the normal technique for the examination of rocks, the main exceptions being ore minerals, which are examined using reflected light (this is because of the optical properties of the minerals concerned – see below). The majority of minerals are translucent when they are sliced to 30 microns thick, whatever their colour or appearance in hand specimen: this is particularly true of silicate and carbonate minerals, which are the groups of prime interest to the sedimentary geologist. It is therefore possible to view the ***optical properties*** of the minerals, the way they appear and interact with the light going through them, using a petrographic microscope.

Underneath the microscope stage the light beam passes through a ***polarising filter***, which only allows light waves vibrating in one plane to pass through it and hence through the thin-section. Toward the bottom of the eyepiece tube there is a second polarising filter that is retractable. This polarising filter is mounted perpendicular to the one below the stage, such that it only allows through light waves that are vibrating at ninety degrees to the lower one. If this second filter, known as the ***analysing filter***, is inserted across the lenses when there is no thin-section, or just plain glass, on the stage, then all the light from the beam will be cut out and it appears black. The same effect can be achieved with 'Polaroid' sunglasses: putting two Polaroid lenses at ninety degrees to each other should result in the blocking out of all light.

Other standard features on a petrographic microscope are a set of lenses at the end of the eyepiece tube that allow different magnifications of viewing to be achieved. The total magnification will be a multiple of one of these lenses and the eyepiece magnification. The eyepiece itself has a very fine cross-wire mounted in it: this acts as a frame of reference to be used when the orientation of the thin-section is changed by rotating the stage. The stage itself is graduated in degrees around the edge so that the amount of rotation can be measured. An optional feature within the eyepiece is a ***graticule***, a scale that allows measurements of features of the thin-section to be made if the magnification is known.

There are usually further tools for optical analyses on the microscope, such as additional lenses that can be inserted above and below the stage, and plates that can be introduced into the eyepiece tube. These are used when advanced petrographic techniques are employed to make more detailed analyses of minerals. However, at an introductory level of sedimentary petrography, such techniques are rarely used, and analysis can be carried out using only a limited range of the optical properties of minerals, which are described in the following sections.

2.3.5 Thin-section analysis of sandstones

Use of the following techniques will allow identification of the most frequently encountered minerals in sedimentary rocks. Only a very basic introduction to the principles and application of thin-section analysis is provided here. For more detailed and advanced petrographic analysis, reference should be made to an appropriate book on optical mineralogy (e.g. Gribble & Hall 1999; Nesse 2004), which should be used in conjunction with suitable reference books on sedimentary petrography, particularly colour guides such as Adams et al. (1984).

Grain shape

A distinctive shape can be a characterising feature of a mineral, for example members of the mica family, which usually appear long and thin if they have been cut perpendicular to their platy form. Minerals may also be elongate, needle-like or equant, but in all cases it must be remembered that the shape depends on the angle of the cut through the grain. Grain shape also provides information about the history of the sediment (*2.5.4*) so it is important to distinguish between grains that show crystal faces and those that show evidence of abrasion of the edges.

Relief

Relief is a measure of how strong the lines that mark the edges of the mineral, or minerals that comprise a grain, are and how clearly the grain stands out against the glass or the other grains around it. It is a visual appraisal of the refractive index of the mineral, which is in turn related to its density. A mineral such as quartz has a refractive index that is essentially the same as glass, so a grain of quartz 30 microns thick mounted on a microscope slide will only just be visible (the mounting medium – glue – normally has the same optical properties as the glass slide): it is therefore considered to have 'low relief'. In contrast, a grain of calcite against glass will appear to have very distinct, dark edges, because it is a denser mineral with a higher refractive index and therefore has a 'high relief'. Because a sedimentary grain will often be surrounded by a cement (*18.2.2*) the contrast with the cement is important, and a quartz grain will stand out very clearly if surrounded by a calcite cement. Certain 'heavy minerals', such as zircon, can readily be distinguished by their extremely high relief.

Cleavage

Not all minerals have a regular *cleavage*, a preferred fracture orientation determined by the crystal lattice structure, so the presence or absence of a cleavage when the mineral is viewed in thin-section can be a useful distinguishing feature. Quartz, for example, lacks a cleavage, but feldspars, which otherwise have many optical properties that are similar to quartz, commonly show clear, parallel lines of cleavage planes. However, the orientation of the mineral in the thin-section will have an important effect because if the cut is parallel to the cleavage planes it will appear as if the mineral does not have a cleavage. The angle between pairs of cleavage planes can be important distinguishing features (e.g. between minerals of the pyroxene family and the amphibole group of minerals). The cleavage is usually best seen under plane-polarised light and often becomes clearer if the intensity of the light shining through is reduced.

Colour and opacity

This property is assessed using plane-polarised light (i.e. without the analysing filter inserted). Some minerals are completely clear while others appear slightly cloudy, but are essentially still colourless: minerals that display distinct colours in hand specimen do not necessarily show any colours in thin-section (e.g. purple quartz or pink feldspar). Colours may be faint tints or much stronger hues, the most common being shades of green and brown (some amphiboles and micas), with rarer yellows and blues. (A note of caution: if a rock is rather poorly lithified, part of the process of manufacture of the thin-section is to inject a resin into the pore spaces between the grains to consolidate it; this resin is commonly dyed bright blue so that it can easily be distinguished from the original components of the rock – it is not a blue mineral!)

Some grains may appear black or very dark brown. The black grains are opaque minerals that do not allow any light through them even when cut to a thin slice. Oxides and sulphides are the commonest opaque minerals in sedimentary rocks, particularly iron oxides (such as haematite) and iron sulphide (pyrite), although others may occur. Black grains that have a brown edge, or grains that are dark brown throughout, are likely to be fragments of organic material.

Pleochroism

A grain of hornblende, a relatively common member of the amphibole group, may appear green or brown when viewed under plane-polarised light, but what is distinctive is that it changes from one colour to the other when the grain is moved by rotating the microscope stage. This phenomenon is known as **pleochroism** and is also seen in biotite mica and a number of other minerals. It is caused by variations in the degree of absorption of different wavelengths of light when the crystal lattice is at different orientations.

Birefringence colours

When the analysing lens is inserted across the objective/eyepiece tube, the appearance of the minerals in the thin-section changes dramatically. Grains that had appeared colourless under plane-polarised light take on a range of colours, black, white or shades of grey, and this is a consequence of the way the polarised light has interacted with the minerals. Non-opaque minerals can be divided into two groups: **isotropic minerals** have crystal lattices that do not have any effect on the pathway of light passing through them, whatever orientation they are in (halite is an example of an isotropic mineral); when light passes through a crystal of an **anisotropic mineral**, the pathway of the light is modified, and the degree to which it is affected depends on the orientation of the crystal. When a crystal of an isotropic mineral is viewed with both the polarising and analysing filters inserted (under **cross-polars**), it appears black. However, an anisotropic mineral will distort the light passing through it, and some of the light passes through the analyser. The mineral will then appear to have a colour, a **birefringence colour**, which will vary in hue and intensity depending on the mineral type and the orientation of the particular grain (and, in fact, the thickness of the slice, but thin-sections are normally cut to 30 microns, so this is not usually a consideration).

For any given mineral type there will be a 'maximum' birefringence colour on a spectrum of colours and hues that can be illustrated on a birefringence chart. In a general sense, minerals can be described as having one of the following: 'low' birefringence colours, which are greys (quartz and feldspars are examples), 'first order' colours (seen in micas), which are quite intense colours of the rainbow, and 'high order' colours, which are pale pinks and greens (common in carbonate minerals). Petrology reference books (e.g. Gribble & Hall 1999; Nesse 2004) include charts that show the birefringence colours for common minerals.

Angle of extinction

When the stage is rotated, the birefringence colour of a grain of an anisotropic mineral will vary as the crystal orientation is rotated with respect to the plane-polarised light. The grain will pass through a 'maximum' colour (although this may not be the maximum colour for this mineral, as this will depend on the three-dimensional orientation of the grain) and will pass through a point in the rotation when the grain is dark: this occurs when the crystal lattice is in an orientation when it does not influence the path of the polarised light. With some minerals the grain goes black – goes into **extinction** – when the grain is oriented with the plane of the polarised light parallel to a crystal face: this is referred to as parallel extinction. When viewed through the eyepiece of the microscope the grain will go into extinction when the crystal face is parallel to the vertical cross-wire. Many mineral types go into extinction at an angle to the plane of the polarised light: this can be measured by rotating a grain that has a crystal face parallel to the vertical cross-wire until it goes into extinction and measuring the angle against a reference point on the edge of the circular stage. Different types of feldspar can be distinguished on the basis of their extinction angle.

Twinning of crystals

Certain minerals commonly display a phenomenon known as **twinning**, when two crystals have formed adjacent to each other but with opposite orientations of the crystal lattice (i.e. mirror images). Twinned crystals may be difficult to recognise under plane-polarised light, but when viewed under crossed polars the two crystals will go into extinction at 180° to each other. Multiple twins may also occur, and in fact are a characteristic of plagioclase feldspars, and these are seen as having a distinctive striped appearance under crossed polars.

2.3.6 The commonest minerals in sedimentary rocks

Almost any mineral which is stable under surface conditions could occur as a detrital grain in a sedimentary rock. In practice, however, a relatively small number of minerals constitute the vast majority of

Mineral	Colour	Cleavage	Relief	Birefringence	Extinction	Other features
quartz	colourless	none	very low	weak	straight	unaltered
orthoclase	colourless	two at 90°	low	weak	slightly oblique	simple twins, often altered
microcline	colourless	two at 90°	low	weak	slightly oblique	cross-hatch twins, often altered
plagioclase	colourless	two at 93°	low	weak	oblique	lamellar twins, often altered
muscovite	colourless	perfect basal	moderate	strong	straight	flakes
biotite	brown or green	perfect basal	moderate	strong	straight	strongly pleochroic green-brown
hornblende	brown or green	two at 124°	moderate to high	moderate to strong	oblique up to 25°	pleochroic green-brown
glauconite	green	none	moderate	masked by mineral colour	straight	pleochroic green-yellow
calcite	colourless	two at 75°	high	very strong	symmetrical	often biogenic origin
dolomite	colourless	two at 75°	high	very strong	symmetrical	forms euhedral crystals

Fig. 2.13 The optical properties of the minerals most commonly found in sedimentary rocks.

grains in sandstones. The common ones are briefly described here, and their optical properties summarised in Fig. 2.13.

Quartz

Most sandstones and siltstones contain grains of quartz, which is chemically the simplest of the silicate minerals, an oxide of silicon. In thin-section grains are typically clear, low relief and do not show any cleavage; birefringence colours are grey. Quartz grains from a metamorphic source (and occasionally some igneous sources) may show a characteristic **undulose extinction**, that is, as the grain is rotated, the different parts go into extinction at different angles, but there is no sharp boundary between these areas. This phenomenon, known as **strained quartz**, is attributed to deformation of the crystal lattice, which gives the grain irregular optical properties and its presence can be used as an indicator of provenance (5.4.1).

Feldspars

Feldspars are silicate minerals that are principal components of most igneous and many metamorphic rocks: they are also relatively common in sandstones, especially those made up of detritus eroded directly from a bedrock such as a granite. Feldspar crystals are moderately elongate, clear or sometimes slightly cloudy and may show a well-developed cleavage. Relief is variable according to chemical composition, but is generally low, and birefringence colours are weak, shades of grey. Feldspars fall into two main groups, **potash feldspars** and the **plagioclase feldspars**.

Potash feldspars such as **orthoclase** are the most common as grains in sedimentary rocks. It can be difficult to distinguish orthoclase from quartz at first glance because the two minerals have a similar relief and low birefringence colours, but the feldspar will show a cleavage in some orientations, twinning may be seen under cross-polars, and it is often slightly cloudy under plane-polarised light. The cloudiness is due to chemical alteration of the feldspar, something that is not seen in quartz. Another mineral in this group is **microcline**, which is noteworthy because, under plane-polarised light, it shows a very distinctive cross-hatch pattern of fine, black and white stripes perpendicular to each other: although less common than orthoclase, it is very easy to recognise in thin-section.

Plagioclase feldspars are a family of minerals that have varying proportions of sodium and calcium in their composition: **_albite_** is the sodium-rich form, and **_anorthite_** the calcium-rich, with several others in between. The most characteristic distinguishing feature is the occurrence of multiple twins, which give the grains a very pronounced black and white striped appearance under crossed polars. The extinction angle varies with the composition, and is used as a way of distinguishing different minerals in the plagioclase group (Gribble & Hall 1999; Nesse 2004).

Micas

There are many varieties of mica, but two of the most frequently encountered forms are the white mica, muscovite, and the brown mica, biotite. Micas are **_phyllosilicates_**, that is, they have a crystal structure of thin sheets, and have a very well developed platy cleavage that causes the crystals to break up into very thin grains. If the platy grains lie parallel to the plane of the thin-section, they will appear hexagonal, but it is much more common to encounter grains that have been cut oblique to this and therefore show the cleavage very clearly in thin-section. The grains also appear elongate and may be bent: mica flakes are quite delicate and can get squeezed between harder grains when a sandstone is compacted (*18.3.1*). Biotite is usually very distinctive because of its shape, cleavage, brown colour and pleochroism (which may not always be present). It has bright, first-order birefringence colours, but these are often masked by the brown mineral colour: the extinction angle is 0° to 3°. The strong, bright birefringence colours of muscovite flakes are very striking under cross-polars, which along with the elongate shape and cleavage make this a distinctive mineral.

Other silicate minerals

In comparison to igneous rocks, sedimentary rocks contain a much smaller range of silicate minerals as common components. Whereas minerals belonging to the amphibole, pyroxene and olivine groups are essential minerals in igneous rocks of intermediate to mafic composition (i.e. containing moderate to relatively low proportions of SiO_2), these minerals are rare in sediments. Hornblende, an amphibole, is the most frequently encountered, but would normally be considered a 'heavy mineral' (see below), as would any minerals of the pyroxene group. Olivine, so common in gabbros and basalts, is very rare as a detrital

grain in a sandstone. This is because of the susceptibility of these silicate minerals to chemical breakdown at the Earth's surface, and they do not generally survive for long enough to be incorporated into a sediment.

Glauconite

This distinctive green mineral is unusual because, unlike other silicates, it does not originate from igneous or metamorphic sources. It forms in sediment on the sea floor and can accumulate to form significant proportions of some shallow marine deposits (*11.5.1*). Under plane-polarised light glauconite grains have a distinctive, strong green colour that is patchy and uneven over the area of the grain: this colour mottling is because the mineral normally occurs in an amorphous form, and other crystal properties are rarely seen.

Carbonate minerals

The most common minerals in this group are the calcium carbonates, calcite and aragonite, while dolomite (a magnesium–calcium carbonate) and siderite (iron carbonate) are also frequently encountered in sedimentary rocks. Calcium carbonate minerals are extremely common in sedimentary rocks, being the main constituents of limestone. Calcite and aragonite are indistinguishable in thin-section: like all sedimentary carbonates, these minerals have a high relief and crystals show two clear cleavage planes present at 75° to each other. Birefringence colours are pale, high-order greens and pinks. The form of calcite in a sedimentary rock varies considerably because much of it has a biogenic origin: the recognition of carbonate components in thin-section is considered in section 3.1.2.

Most dolomite is a diagenetic product (*18.4.2*), the result of alteration of a limestone that was originally composed of calcium carbonate minerals. When individual crystals can be seen they have a distinctive euhedral rhombic shape, and cleavage planes parallel to the crystal faces may be evident. The euhedral morphology can be a good clue, but identification of dolomite cannot be confirmed without chemical tests on the material (*3.1.2*). Siderite is very difficult to distinguish from calcite because most of its optical properties are identical. The best clue is often a slight yellow or brownish tinge to the grain, which is a result of alteration of some of the iron to oxides and hydroxides.

Oxides and sulphides

The vast majority of natural oxide and sulphide minerals are opaque, and simply appear as black grains under plane-polarised light. The iron oxide haematite is particularly common, occurring as particles that range down to a fine dust around the edges of grains and scattered in the matrix. The edges of haematite grains will often look brownish-red. Magnetite, also an iron oxide, occurs as a minor component of many igneous rocks and is quite distinctive because it occurs as euhedral, bipyramidal crystals, which appear as four or eight-sided, equant black grains in thin-section. Iron hydroxides, limonite and goethite, which are yellowish brown in hand specimen, appear to have brown edges in thin-section.

Pyrite is an iron sulphide that may crystallise within sediments. Although a metallic gold colour as a fully-formed crystal, fine particles of pyrite appear black, and in thin-section this mineral often appears as black specks, with the larger crystals showing the cubic crystal shape of the mineral. Locally, other sulphides and oxides can be present, for example the tin ore, cassiterite, which occurs as a *placer mineral* (minerals that concentrate at the bottom of a flow due to their higher density).

Heavy minerals

A thin-section of a sandstone is unlikely to contain many heavy mineral grains. Zircon is the most frequently encountered member of this group: it is an extremely resistant mineral that can survive weathering and long distances of transport. Grains are equant to elongate, colourless and easily recognised by their very high relief: the edges of a zircon grain will appear as thick, black lines. Other relatively common heavy minerals are rutile, apatite, tourmaline and sphene.

2.3.7 Lithic grains

Not every grain in a sandstone is an individual mineral: the breakdown of bedrock by weathering leads to the formation of sand-sized fragments of the original rock that can be incorporated into a sedi-

ment. The bedrock must itself be composed of crystals or particles that are smaller than sand-size: granite consists of crystals that are sand-sized or larger, and so cannot occur as lithic clasts in sands, but its fine-grained equivalent, rhyolite, can occur as grains. Lithic fragments of fine-grained metamorphic and sedimentary rocks can also be common.

Chert and chalcedony

Under plane-polarised light, *chert* (3.3) looks very much like quartz, because it is also composed of silica. The difference is that the silica in chert is in an amorphous or microcrystalline form: under cross-polars it therefore often appears to be highly speckled black, white and grey, with individual 'crystals' too small to be resolved under a normal petrographic microscope. *Chalcedony* is also a form of silica that can readily be identified in thin-section because it has a radial structure when viewed under cross-polars; fine black and white lines radiate from the centre, becoming lighter and darker as the grain is rotated.

Organic material

Carbonaceous material, the remains of plants, is brown in colour, varying from black and opaque to translucent reddish brown in thin-section. The paler grains can resemble a mineral, but are always black under cross-polars. The shape and size is extremely variable and some material may appear fibrous. Coal is a sedimentary rock made up largely of organic material: the thin-section study of coal is a specialised subject that can yield information about the vegetation that it formed from and its burial history.

Sedimentary rock fragments

Clasts of claystone, siltstone or limestone may be present in a sandstone, and a first stage of recognition of them is that they commonly appear rather 'dirty' under plane-polarised light. Very fine particles of clay and iron oxide in a lithic fragment will make it appear brownish in thin-section, and if the grain is made entirely of clay it may be dark brown. Siltstone is most commonly composed of quartz grains, which will be evident as black and white spots under crossed polars: individual silt grains may be identified if a high-power magnification is used to reveal the edges of the silt-sized clasts.

Igneous rock fragments

Fragments of fine-grained igneous rocks can occur as grains in a sandstone, especially in areas of deposition close to volcanic activity. Dark grains in hand specimens can be revealed by the microscope to contain tiny laths of pale feldspar crystals in a finer groundmass that appears dark under cross-polars and can be recognised as pieces of basalt. Basalt weathers readily, breaking down to clays and iron oxides, and these particles will give a brown, rusty rim to any grains that have been exposed for any length of time. With more extensive weathering, fine-grained igneous rocks will break down to clays (*2.4*) and the clast will appear brownish, turning dark and speckled under crossed polars.

Metamorphic rock fragments

Slates and fine-grained schists may be incorporated into sandstones if a metamorphic terrain is eroded. These rocks have a strong fabric, and break up into platy fragments that can be recognised by their shape as grains. This fabric also gives a pronounced alignment to the fine crystals that make up the grain, and this can be seen both in plane-polarised light and under crossed polars. Micas are common metamorphic minerals (e.g. in schists), so elongate, bright birefringence colour specks within the clast may be seen.

2.3.8 Matrix and cement

The material between the clasts will be one of, or a mixture of, matrix and cement. A matrix to a sandstone will be silt and/or clay-sized sediment. It can be difficult to determine the mineralogy of individual silt particles because of their small size, but they are commonly grains of quartz that will appear as black or white specks under crossed polars. Tiny flakes of mica or other phyllosilicate minerals may also be present in this size fraction, and their bright birefringence colours may be recognisable despite the small size of the laths. Clay-sized grains are too small to be identified individually with an optical microscope. Under plane-polarised light patches of clay minerals forming a matrix usually appear as amorphous masses of brownish colour. Under crossed polars the clays turn dark, but often the area of clay material appears very finely speckled as light passes through

individual grains. Analysis of the clay content of a matrix requires other techniques such as X-ray diffraction analysis (*2.4.4*).

A cement is precipitated out of fluids as part of the post-depositional history of the sediment. It will normally be crystalline material that fills, or partly fills, the gaps between the grains. The formation of cements and their varieties are considered in section 18.3.1.

2.3.9 Practical thin-section microscopy

Before putting a thin-section slide on a microscope stage, hold it up to the light and look for features such as evidence of lamination, usually seen as bands of lighter or darker, or larger and smaller grains. The rock might not be uniform in other ways, with a patchy distribution of grain sizes and types. Such features should be noted and compared to the hand specimen the thin-section has been cut from.

It is always best to start by looking at the slide using low magnification and under plane-polarised light. Lithic fragments and mineral grains can often be best distinguished from each other at this point, and certain distinctive, coloured minerals such as biotite and glauconite recognised. Individual grains can then be selected for investigation, and their mineral or lithological composition determined using the techniques described above. Once a few different grain types have been identified it is usually possible to scan the rest of the slide to see whether other clasts are more of the same or are different. For each clast type the following are then recorded:
- optical properties (shape, relief, cleavage, colour, pleochroism, birefringence colours, extinction angle, twinning)
- mineral name
- size range and mean size
- distribution (even, concentrated, associated with another clast type)
- estimate of percentage in the thin-section (either as a proportion of the clast types present, or a percentage of the whole rock, including cement and matrix).

The nature and proportion of the matrix must also be determined, and also the character and proportion of any cement that is present. The proportions of different clast types and of the cement/matrix then need to be estimated which add up to 100% and with this information the rock can then be named using an appropriate classification scheme (e.g. the Pettijohn classification, Fig. 2.11).

Point counting

To make a quantitative analysis of the components of a sedimentary rock some form of systematic determination of the proportions of the different clast types, matrix and cement is required. The commonest technique is to attach a *point counting* mechanism on to the stage of the microscope: this is a device that holds the thin-section slide and shifts the position of the slide to the side in a series of small increments. It is attached to a mechanical counter or to a computer such that each time a button or key is pressed, the slide moves sideways. The operator determines the clast type under the cross-wires at each step by pressing different buttons or keys. A series of transects across the slide is made until a sufficient number of points have been counted – typically not less than 300. The number of counts of each grain type, matrix and/or cement is then converted into a percentage. The size of the step, the magnification used and the number of categories of clast will be determined by the operator at the outset, depending on the grain-size range and clast types recognised in a preliminary examination of the thin-section.

2.4 CLAY, SILT AND MUDROCK

Fine-grained terrigenous clastic sedimentary rocks tend to receive less attention than any other group of deposits despite the fact that they are volumetrically the most common of all sedimentary rocks types (*2.1*). The grain size is generally too small for optical techniques of mineral determination and until scanning electron microscopes and X-ray diffraction analysis techniques (*2.4.4*) were developed little was known about the constituents of these sediments. In the field mudrocks do not often show the clear sedimentary and biogenic structures seen in coarser clastic rocks and limestone. Exposure is commonly poor because they do not generally form steep cliffs and soils support vegetation that covers the outcrop. This group of sediments therefore tends to be overlooked but, as will be seen in later sections concerning depositional environments and stratigraphy, they can provide as much information as any other sedimentary rock type.

2.4.1 Definitions of terms in mudrocks

Silt is defined as the grain size of material between 4 and 62 microns in diameter (Fig. 2.2). This size range is subdivided into coarse, medium, fine and very fine. The coarser grains of silt are just visible to the naked eye or with a hand lens. Finer silt is most readily distinguished from clay by touch, as it will feel 'gritty' if a small amount is ground between teeth, whereas clay feels smooth. *Clay* is a textural term to define the finest grade of clastic sedimentary particles, those less than 4 microns in diameter. Individual particles are not discernible to the naked eye and can only just be resolved with a high power optical microscope. *Clay minerals* are a group of phyllosilicate minerals that are the main constituents of clay-sized particles.

When clay- and silt-sized particles are mixed in unknown proportions as the main constituents in unconsolidated sediment we would call this material **mud**. The general term **mudrock** can be applied to any indurated sediment made up of silt and/or clay. If it can be determined that most of the particles (over two-thirds) are clay-sized the rock may then be called a *claystone* and if silt is the dominant size a *siltstone*; mixtures of more than one-third of each component are referred to as **mudstone** (Folk 1974; Blatt et al. 1980). The term **shale** is sometimes applied to any mudrock (e.g. by drilling engineers) but it is best to use this term only for mudrocks that show a *fissility*, which is a strong tendency to break in one direction, parallel to the bedding. (Note the distinction between shale and slate: the latter is a term used for fine-grained metamorphic rocks that break along one or more cleavage planes.)

2.4.2 Silt and siltstone

The mineralogy and textural parameters of silt are more difficult to determine than for sandstone because of the small particle size. Only coarser silt grains can be easily analysed using optical microscope techniques. Resistant minerals are most common at this size because other minerals will often have been broken down chemically before they are physically broken down to this size. Quartz is the most common mineral seen in silt deposits. Other minerals occurring in this grade of sediment include feldspars, muscovite, calcite and iron oxides amongst many other minor components. Silt-sized lithic fragments are only abundant in the 'rock flour' formed by glacial erosion (*7.3.4*).

In aqueous currents silt remains in suspension until the flow is very slow and deposition is therefore characteristic of low velocity flows or standing water with little wave action (*4.4*). Silt-sized particles can remain in suspension in air as dust for long periods and may be carried high into the atmosphere. Strong, persistent winds can carry silt-sized dust thousands of kilometres and deposit it as laterally extensive sheets (Pye 1987); wind-blown silt forming *loess* deposits appears to have been important during glacial periods (*7.6 & 7.7*).

2.4.3 Clay minerals

Clay minerals commonly form as breakdown products of feldspars and other silicate minerals. They are phyllosilicates with a layered crystal structure similar to that of micas and compositionally they are aluminosilicates. The crystal layers are made up of silica with aluminium and magnesium ions, with oxygen atoms linking the sheets (Fig. 2.14). Two patterns of layering occur, one with two layers, the **kandite group**, and the other with three layers, the **smectite group**. Of the many different clay minerals that occur in sedimentary rocks the four most common (Tucker 1991) are considered here (Fig. 2.14).

Kaolinite is the commonest member of the kandite group and is generally formed in soil profiles in warm, humid environments where acidic waters intensely leach bedrock lithologies such as granite. Clay minerals of the smectite group include the expandable or **swelling clays** such as **montmorillonite**, which can absorb water within their structure. Montmorillonite is a product of more moderate temperature conditions in soils with neutral to alkaline pH. It also forms under alkaline conditions in arid climates. Another three-layer clay mineral is **illite**, which is related to the mica group and is the most common clay mineral in sediments, forming in soils in temperate areas where leaching is limited. **Chlorite** is a three-layer clay mineral that forms most commonly in soils with moderate leaching under fairly acidic groundwater conditions and in soils in arid climates. Montmorillonite, illite and chlorite all form as a weathering product of volcanic rocks, particularly volcanic glass.

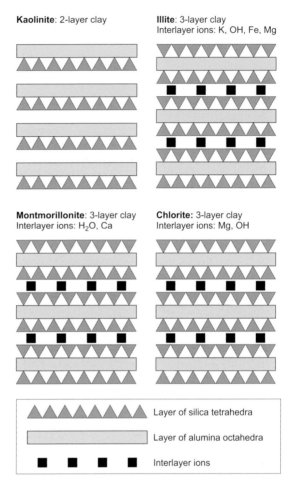

Fig. 2.14 The crystal lattice structure of some of the more common clay minerals.

2.4.4 Petrographic analysis of clay minerals

Identification and interpretation of clay minerals requires a higher technology approach than is needed for coarser sediment. There are two principal techniques: scanning electron microscopy and X-ray diffraction pattern analysis (Tucker 1988). An image from a sample under a **scanning electron microscope** (**SEM**) is generated from secondary electrons produced by a fine electron beam that scans the surface of the sample. Features only microns across can be imaged by this technique, providing much higher resolution than is possible under an optical microscope. It is therefore used for investigating the form of clay minerals and their relationship to other grains in a rock. The distinction between clay minerals deposited as detrital grains and those formed diagenetically (*18.3.1*) within the sediment can be most readily made using an SEM.

An *X-ray diffractometer* (*XRD*) operates by firing a beam of X-rays at a powder of a mineral or disaggregated clay and determining the angles at which the radiation is diffracted by the crystal lattice. The pattern of intensity of diffracted X-rays at different angles is characteristic of particular minerals and can be used to identify the mineral(s) present. X-ray diffractometer analysis is a relatively quick and easy method of semi-quantitatively determining the mineral composition of fine-grained sediment. It is also used to distinguish certain carbonate minerals (*3.1.1*) that have very similar optical properties.

2.4.5 Clay particle properties

The small size and platy shape of clay minerals means that they remain in suspension in quite weak fluid flows and only settle out when the flow is very sluggish or stationary. Clay particles are therefore present as suspended load in most currents of water and air but are only deposited when the flow ceases.

Once they come into contact with each other clay particles tend to stick together, they are *cohesive*. This cohesion can be considered to be partly due to a thin film of water between two small platy particles having a strong surface tension effect (in much the same way as two plates of glass can be held together by a thin film of water between them), but it is also a consequence of an electrostatic effect between clay minerals charged due to incomplete bonds in the mineral structure. As a result of these cohesive properties clay minerals in suspension tend to *flocculate* and form small aggregates of individual particles (Pejrup 2003). These flocculated groups have a greater settling velocity than individual clay particles and will be deposited out of suspension more rapidly. Flocculation is enhanced by saline water conditions and a change from fresh to saline water (e.g. at the mouth of a delta or in an estuary: *12.3 & 13.6*) results in clay deposition due to flocculation. Once clay particles are deposited the cohesion makes them resistant to remobilisation in a flow (*4.2.4*). This allows deposition and preservation of fine sediment

in areas that experience intermittent flows, such as tidal environments (*11.2*).

2.5 TEXTURES AND ANALYSIS OF TERRIGENOUS CLASTIC SEDIMENTARY ROCKS

The shapes of clasts, their degree of sorting and the proportions of clasts and matrix are all aspects of the texture of the material. A number of terms are used in the petrographic description of the texture of terrigenous clastic sediments and sedimentary rocks.

Clasts and matrix The fragments that make up a sedimentary rock are called clasts. They may range in size from silt through sand to gravel (granules, pebbles, cobbles and boulders). A distinction is usually made between the clasts and the matrix, the latter being finer-grained material that lies between the clasts. There is no absolute size range for the matrix: the matrix of a sandstone may be silt and clay-sized material, whereas the matrix of a conglomerate may be sand, silt or clay.

Sorting *Sorting* is a description of the distribution of clast sizes present: a well-sorted sediment is composed of clasts that mainly fall in one class on the Wentworth scale (e.g. medium sand); a poorly sorted deposit contains a wide range of clast sizes. Sorting is a function of the origin and transport history of the detritus. With increased transport distance or repeated agitation of a sediment, the different sizes tend to become separated. A visual estimate of the sorting may be made by comparison with a chart (Fig. 2.15) or calculated from grain-size distribution data (*2.5.1*).

Clast roundness During sediment transport the individual clasts will repeatedly come into contact with each other and stationary objects: sharp edges tend to be chipped off first, the abrasion smoothing the surface of the clast. A progressive rounding of the edges occurs with prolonged agitation of the sediment and hence the roundness is a function of the transport history of the material. Roundness is normally visually estimated (Fig. 2.16), but may also be calculated from the cross-sectional shape of a clast.

Clast sphericity In describing individual clasts, the dimensions can be considered in terms of closeness to a sphere (Fig. 2.16). Discoid or needle-like clasts have

Sorting description	'Standard deviation'
Very well sorted	< 0.35
Well sorted	= 0.35–0.5
Moderately well sorted	= 0.5–0.71
Moderately sorted	= 0.71–1.0
Poorly sorted	= 1.0–2.0
Very poorly sorted	> 2.0

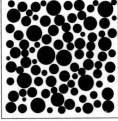

'Standard deviation' = 0.35   'Standard deviation' = 0.5

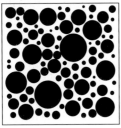 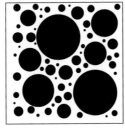

'Standard deviation' = 1.0   'Standard deviation' = 2.0

Fig. 2.15 Graphic illustration of sorting in clastic sediments. The sorting of a sediment can be determined precisely by granulometric analysis, but a visual estimate is more commonly carried out.

a low sphericity. **Sphericity** is an inherited feature, that is, it depends on the shapes of the fragments which formed during weathering. A slab-shaped clast will become more rounded during transport and become disc-shaped, but will generally retain its form with one axis much shorter than the other two.

Fabric If a rock has a tendency to break in a certain direction, or shows a strong alignment of elongate clasts, this is described as the **fabric** of the rock. Mudrock that breaks in a platy fashion is considered to have a shaly fabric (and may be called a shale), and sandstone that similarly breaks into thin slabs is sometimes referred to as being flaggy. Fabrics of this type are due to anisotropy in the arrangement of particles: a rock with an isotropic fabric would not show any preferred direction of fracture because it consists of evenly and randomly oriented particles.

2.5.1 Granulometric and clast-shape analysis

Quantitative assessment of the percentages of different grain sizes in clastic sediments and sedimentary rocks is called **granulometric analysis**. These data and measurements of the shape of clasts can be used in the description and interpretation of clastic sedimentary material (see Lewis & McConchie 1994). The techniques used will depend on the grain size of the material examined. Gravels are normally assessed by direct measurement in the field. A quadrant is laid over the loose material or on a surface of the conglomerate and each clast measured within the area of the quadrant. The size of quadrant required will depend on the approximate size of the clasts: a metre square is appropriate for pebble and cobble size material.

A sample of unconsolidated sand is collected or a piece of sandstone disaggregated by mechanical or chemical breakdown of the cement. The sand is then passed through a stack of sieves that have meshes at intervals of half or one unit on the 'phi' scale (*2.1.2*). All the sand that passes through the 500 micron (phi = 1) mesh sieve but is retained by the 250

	Well rounded	Rounded	Subrounded	Subangular	Angular	Very angular
Low sphericity						
High sphericity						

Fig. 2.16 Roundness and sphericity estimate comparison chart (from Pettijohn et al. 1987).

micron (phi = 2) mesh sieve will have the size range of medium sand. By weighing the contents of each sieve the distribution by weight of different size fractions can be determined.

It is not practical to sieve material finer than coarse silt, so the proportions of clay- and silt-sized material are determined by other means. Most laboratory techniques employed in the granulometric analysis of silt- and clay-size particles are based on settling velocity relationships predicted by Stokes' Law (4.2.5). A variety of methods using settling tubes and pipettes have been devised (Krumbein & Pettijohn 1938; Lewis & McConchie 1994), all based on the principle that particles of a given grain size will take a predictable period of time to settle a certain distance in a water-filled tube. Samples are siphoned off at time intervals, dried and weighed to determine the proportions of different clay and silt size ranges. These settling techniques do not fully take into account the effects of grain shape or density on settling velocity and care must be used in comparing the results of these analyses with grain-size distribution data obtained from more sophisticated techniques such as the **Coulter Counter**, which determines grain size on the basis of the electrical properties of grains suspended in a fluid, or a **laser granulometer**, which analyses the diffraction pattern of a laser beam created by small particles.

The results from all these grain-size analyses are plotted in one of three forms: a histogram of the weight percentages of each of the size fractions, a frequency curve, or a cumulative frequency curve (Fig. 2.17). Note in each case that the coarse sizes plot on the left and the finer material on the right of the graph. Each provides a graphic representation of the grain-size distribution and from them a value for the mean grain size and sorting (standard deviation from a normal distribution) can be calculated. Other values that can be calculated are the **skewness** of the distribution, an indicator of whether the grain-size histogram is symmetrical or is skewed to a higher percentage of coarser or finer material, and the **kurtosis**, a value that indicates whether the histogram has a sharp peak or a flat top (Pettijohn 1975; Lewis & McConchie 1994).

The grain-size distribution is determined to some extent by the processes of transport and distribution. Glacial sediments are normally very poorly sorted, river sediments moderately sorted and both beach and aeolian deposits are typically well sorted. The reasons for these differences are discussed in later chapters. In most circumstances the general sorting characteristics can be assessed in a qualitative way and there are many other features such as sedimentary structures that would allow the deposits of different environments to be distinguished. A quantitative granulometric analysis is therefore often unnecessary and may not provide much more information than is evident from other, quicker observations.

Moreover, determination of environment of deposition from granulometric data can be misleading under circumstances where material has been

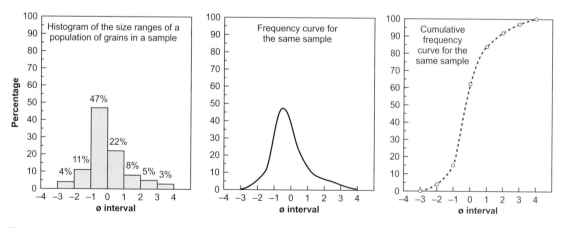

Fig. 2.17 Histogram, frequency distribution and cumulative frequency curves of grain size distribution data. Note that the grain size decreases from left to right.

reworked from older sediment. For examples, a river transporting material eroded from an outcrop of older sandstone formed in an aeolian environment will deposit very well-sorted material. The grain-size distribution characteristics would indicate deposition by aeolian processes, but the more reliable field evidence would better reflect the true environment of deposition from sedimentary structures and facies associations (5.6.3).

Granulometric analysis provides quantitative information when a comparison of the character is required from sediments deposited within a known environment, such as a beach or along a river. It is therefore most commonly used in the analysis and quantification of present-day processes of transport and deposition.

2.5.2 Clast-shape analysis

Attempts have been made to relate the shape of pebbles to the processes of transport and deposition. Analysis is carried out by measuring the longest, shortest and intermediate axes of a clast and calculating an index for its shape (approaching a sphere, a disc or a rod: Fig. 2.8). Although there may be some circumstances where clasts are sorted according to their shape, the main control on the shape of a pebble is the shape of the material eroded from the bedrock in the source area. If a rock breaks up into cubes after transport the rounded clasts will be spherical, and if the bedrock is thinly bedded and breaks up into slabs the resulting clasts will be discoid. No amount of rounding of the edge of a clast will change its fundamental dimensions. Clast-shape analysis is therefore most informative about the character of the rocks in the source area and provides little information about the depositional environment.

2.5.3 Maturity of terrigenous clastic material

A terrigenous clastic sediment or sedimentary rock can be described as having a certain degree of **maturity**. This refers to the extent to which the material has changed when compared with the starting material of the bedrock it was derived from. Maturity can be measured in terms of texture and composition. Normally a compositionally mature sediment is also texturally mature but there are exceptions, for example on a beach around a volcanic island where only mineralogically unstable components (basaltic rock and minerals) are available but the texture reflects an environment where there has been prolonged movement and grain abrasion by the action of waves and currents.

Textural maturity

The texture of sediment or sedimentary rock can be used to indicate something about the erosion, transport and depositional history. The determination of the **textural maturity** of a sediment or sedimentary rock can best be represented by a flow diagram (Fig. 2.18). Using this scheme for assessing maturity, any sandstone that is classified as a wacke is considered to be texturally immature. Arenites can be subdivided on the basis of the sorting and shape of the grains. If sorting is moderate to poor the sediment is considered to be submature, whereas well-sorted or very well-sorted sands are considered mature if the individual grains are angular to subrounded and supermature if rounded to well-rounded. The textural classification of the maturity is independent of composition of the sands. An assessment of the textural maturity of a sediment is most useful when comparing material derived from the same source as it may be expected that the maturity will increase as the amount of energy

TEXTURAL MATURITY OF SANDSTONES

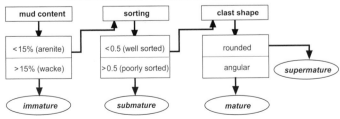

Fig. 2.18 Flow diagram of the determination of the textural maturity of a terrigenous clastic sediment or sedimentary rock.

input increases. For example, maturity often increases downstream in a river and once the same sediment reaches a beach the high wave energy will further increase the maturity. Care must be taken when comparing sediment from different sources as they are likely to have started with different grain size and shape distributions and are therefore not directly comparable. Sediments may also be recycled from older deposits, resulting in greater degrees of maturity (*2.5.4*).

Mineralogical maturity

Compositional maturity is a measure of the proportion of resistant or stable minerals present in the sediment. The proportion of highly resistant clasts such as quartz and siliceous lithic fragments in a sandstone, compared with the amount of less resistant, ***labile***, clast types present, such as feldspars, most other mineral types and lithic clasts, is considered when assessing compositional maturity. A sandstone is compositionally mature if the proportion of quartz grains is very high and it is a quartz arenite according to the Pettijohn classification scheme (Fig. 2.11): if the ratio of quartz, feldspar and lithic fragments meant that the composition falls in the lower part of the triangle it is a mineralogically immature sediment.

2.5.4 Cycles of sedimentation

Mineral grains and lithic clasts eroded from an igneous rock, such as a granite, are transported by a variety of processes (Chapter 4) to a point where they are deposited to form an accumulation of clastic sediment. Material formed in this way is referred to as a ***first cycle deposit*** because there has been one cycle of erosion transport and deposition. Once this sediment has been lithified into sedimentary rock, it may subsequently be uplifted by tectonic processes and be subject to erosion, transport and redeposition. The redeposited material is considered to be a ***second cycle deposit*** as the individual grains have gone through two ***cycles of sedimentation***. Clastic sediment may go through many cycles of sedimentation and each time the mineralogical and textural maturity of the clastic detritus increases. The only clast types that survive repeated weathering, erosion, transport and

redeposition are resistant minerals such as quartz and lithic fragments of chert. Certain heavy minerals (e.g. zircon) are also extremely resistant and the degree to which zircon grains are rounded may be used as an index of the number of cycles of sedimentation material has been subjected to.

2.6 TERRIGENOUS CLASTIC SEDIMENTS: SUMMARY

Terrigenous clastic gavels, sands and muds are widespread modern sediments and are found abundantly as conglomerate, sandstone and mudrock in successions of sedimentary rocks. They are composed mainly of the products of the breakdown of bedrock and may be transported by a variety of processes to depositional environments. The main textural and compositional features of sand and gravel can be readily determined in the field and in hand specimen. For detailed analysis of the composition and texture of sandstones, thin-sections are examined using a petrographic microscope. Investigation of mudrocks depends on submicroscopic and chemical analysis of the material. Sedimentary structures formed in clastic sediments provide further information about the conditions under which the material was deposited and provide the key to the palaeoenvironmental analysis discussed in later chapters of this book.

FURTHER READING

Adams, A., Mackenzie, W. & Guilford, C. (1984) *Atlas of Sedimentary Rocks under the Microscope*. Wiley, Chichester.

Blatt, H. (1982) *Sedimentary Petrology*. W.H. Freeman and Co, New York.

Blatt, H., Middleton, G.V. & Murray, R.C. (1980) *Origin of Sedimentary Rocks* (2nd edition). Prentice-Hall, Englewood Cliffs, New Jersey.

Chamley, H. (1989) *Clay Sedimentology*. Springer-Verlag, Berlin.

Leeder, M.R. (1999) *Sedimentology and Sedimentary Basins: from Turbulence to Tectonics*. Blackwell Science, Oxford.

Lewis, D.G. & McConchie, D. (1994) *Analytical Sedimentology*. Chapman and Hall, New York, London.

Pettijohn, F.J., Potter, P.E. & Siever, R. (1987) *Sand and Sandstone*. Springer-Verlag, New York.

Tucker, M.E. (2001) *Sedimentary Petrology* (3rd edition). Blackwell Science, Oxford.

3

Biogenic, Chemical and Volcanogenic Sediments

In areas where there is not a large supply of clastic detritus other processes are important in the accumulation of sediments. The hard parts of plants and animals ranging from microscopic algae to vertebrate bones make up deposits in many different environments. Of greatest significance are the many organisms that build shells and structures of calcium carbonate in life, and leave behind these hard parts when they die as calcareous sediments that form limestone. Chemical processes also play a part in the formation of limestone but are most important in the generation of evaporites, which are precipitated out of waters concentrated in salts. Volcaniclastic sediments are largely the products of primary volcanic processes of generation of ashes and deposition of them subaerially or under water. In areas of active volcanism these deposits can swamp all other sediment types. Of the miscellaneous deposits also considered in this chapter, most are primarily of biogenic origin (siliceous sediments, phosphates and carbonaceous deposits) while ironstones are chemical deposits.

3.1 LIMESTONE

Limestones are familiar and widespread rocks that form the peaks of mountains in the Himalayas, form characteristic karst landscapes and many spectacular gorges throughout the world. Limestone is also important in the built environment, being the construction material for structures ranging from the Pyramids of Egypt to many palaces and churches. As well as being a good building stone in many places, limestone is also important as a source of lime to make cement, and is hence a component of all concrete, brick and stone buildings and other structures, such as bridges and dams. Limestone strata are common through much of the stratigraphic record and include some very characteristic rock units, such as the Late Cretaceous Chalk, a relatively soft limestone that is found in many parts of the world. The origins of these rocks lie in a range of sedimentary environments: some form in continental settings, but the vast majority are the products of processes in shallow marine environments, where organisms play an important role in creating the sediment that ultimately forms limestone rocks.

Calcium carbonate ($CaCO_3$) is the principal compound in **limestones**, which are, by definition, rocks composed mainly of calcium carbonate. Limestones, and sediments that eventually solidify to form them, are referred to as *calcareous* (note that, although they are carbonate, they are not 'carbonaceous': this latter term is used for material that is rich in carbon, such as coal). Sedimentary rocks may also be made of carbonates of elements such as magnesium or iron, and there are also carbonates of dozens of elements occurring in nature (e.g. malachite and azurite are copper carbonates). This group of sediments and rocks are collectively known as *carbonates* to sedimentary geologists, and most carbonate rocks are sedimentary in origin. Exceptions to this are *marble*, which is a carbonate rock recrystallised under metamorphic conditions, and *carbonatite*, an uncommon carbonate-rich lava.

3.1.1 Carbonate mineralogy

Calcite

The most familiar and commonest carbonate mineral is *calcite* ($CaCO_3$). As a pure mineral it is colourless or white, and in the field it could be mistaken for quartz, although there are two very simple tests that can be used to distinguish calcite from quartz. First, there is a difference in hardness: calcite has a hardness of 3 on Mohs' scale, and hence it can easily be scratched with a pen-knife; quartz (hardness 7) is harder than a knife blade and will scratch the metal. Second, calcite reacts with dilute (10%) hydrochloric acid (HCl), whereas silicate minerals do not. A small dropper-bottle of dilute HCl is hence useful as a means of determining if a rock is calcareous, as most common carbonate minerals (except dolomite) will react with the acid to produce bubbles of carbon dioxide gas, especially if the surface has been powdered first by scratching with a knife. Although calcite sometimes occurs in its simple mineral form, it most commonly has a *biogenic* origin, that is, it has formed as a part of a plant or animal. A wide variety of organisms use calcium carbonate to form skeletal structures and shells and a lot of calcareous sediments and rocks are formed of material made in this way.

Magnesium ions can substitute for calcium in the crystal lattice of calcite, and two forms of calcite are recognised in nature: low-magnesium calcite (low-Mg calcite), which contains less than 4% Mg, and high-magnesium calcite (high-Mg calcite), which typically contains 11% to 19% Mg. The hard parts of many marine organisms are made of high-Mg calcite, for example echinoderms, barnacles and foraminifers, amongst others (see *3.13*). Strontium may substitute for calcium in the lattice and although it is in small quantities (less than 1%) it is important because strontium isotopes can be used in dating rocks (*21.3.1*).

Aragonite

There is no chemical difference between calcite and *aragonite*, but the two minerals differ in their mineral form: whereas calcite has a trigonal crystal form, aragonite has an orthorhombic crystal form. Aragonite has a more densely packed lattice structure and is slightly denser than calcite (a specific gravity of 2.95, as opposed to a range of 2.72–2.94 for calcite), and is slightly harder (3.5–4 on Mohs' scale). In practice, it is rarely possible to distinguish between the two, but the differences between them have some important consequences (18.2.2). Many invertebrates use aragonite to build their hard parts, including bivalves and corals.

Dolomite

Calcium magnesium carbonate ($CaMg(CO_3)_2$) is a common rock-forming mineral which is known as *dolomite*. Confusingly, a rock made up of this mineral is also called dolomite, and the term *dolostone* is now sometimes used for the lithology to distinguish it from dolomite, the mineral. The mineral is similar in appearance to calcite and aragonite, with a similar hardness to the latter. The only way that dolomite can be distinguished in hand specimen is by the use of the dilute HCl acid test: there is usually little or no reaction between cold HCl and dolomite. Although dolomite rock is quite widespread, it does not seem to be forming in large quantities today, so large bodies of dolomite rock are considered to be diagenetic (*18.4.2*).

Siderite

Siderite is iron carbonate ($FeCO_3$) with the same structure as calcite, and is very difficult to distinguish between iron and calcium carbonates on mineralogical grounds. It is rarely pure, often containing some magnesium or manganese substituted for iron in the lattice. Siderite forms within sediments as an early diagenetic mineral (18.2).

3.1.2 Carbonate petrography

All of these carbonate minerals have similar optical properties and it can be difficult to distinguish between them in thin-section using the usual optical tests. Their relief is high, and the birefringence colours are high-order pale green and pink. The cleavage is usually very distinct, and where two cleavage planes are visible they can be seen to intersect to form a rhombohedral pattern. Dolomite can be identified by adding a dye to the cut surface before a glass cover slip is put on the thin-section: **Alizarin Red-S** does not stain dolomite, but colours the other carbonates pink. A second chemical dye is also commonly used: **potassium ferricyanide** reacts with traces of iron in a carbonate to stain it blue, and on this basis it is possible to distinguish between **ferroan** calcite/aragonite/dolomite and **non-ferroan** forms of these minerals. The two stains may be used in combination, such that ferroan calcite/aragonite ends up purple, ferroan dolomite blue, non-ferroan calcite/aragonite pink and non-ferroan dolomite clear.

There is an alternative to making thin-sections of rocks made up primarily of carbonate minerals. It is possible to transfer the detail of a cut, flat surface of a block of limestone onto a sheet of acetate by etching the surface with dilute hydrochloric acid, then flooding the surface with acetone and finally applying the acetate film. Once the acetone has evaporated, the acetate is peeled off and the imprint of the rock surface can then be examined under the microscope. These **acetate peels** are a quick, easy way to look at the texture of the rock, and distinguish different clast types: the rock can also be stained in the same way as a thin-section.

3.1.3 Biomineralised carbonate sediments

Carbonate-forming organisms include both plants and animals. They may create hard parts out of calcite, in either its low-Mg or high-Mg forms, or aragonite, or sometimes a combination of both minerals. The **skeletal fragments** in carbonate sediments are whole or broken pieces of the hard body parts of organisms that use calcium carbonate minerals as part of their structure (Figs 3.1 & 3.2). Some of them have characteristic microstructures, which can be used to identify the organisms in thin-sections (Adams & Mackenzie 1998).

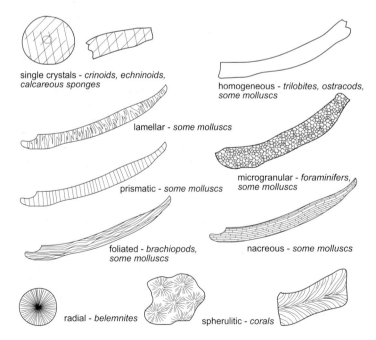

single crystals - *crinoids, echninoids, calcareous sponges*

homogeneous - *trilobites, ostracods, some molluscs*

lamellar - *some molluscs*

prismatic - *some molluscs*

microgranular - *foraminifers, some molluscs*

foliated - *brachiopods, some molluscs*

nacreous - *some molluscs*

radial - *belemnites*

spherulitic - *corals*

Fig. 3.1 Types of bioclast commonly found in limestones and other sedimentary rocks.

Fig. 3.2 Bioclastic debris on a beach consisting of the hard calcareous parts of a variety of organisms.

Fig. 3.3 Fossil gastropod shells in a limestone.

Carbonate-forming animals

The ***molluscs*** are a large group of organisms that have a fossil record back to the Cambrian and commonly have calcareous hard parts. ***Bivalve molluscs***, such as mussels, have a distinctive layered shell structure consisting of two or three layers of calcite, or aragonite, or both. Of the modern forms, some such as oysters and scallops are calcitic, but most of the rest are aragonitic: aragonite shells may have been the norm throughout their history, but no pre-Jurassic bivalve shells are preserved because of the instability of the mineral compared with the more stable form of calcium carbonate, calcite. ***Gastropods*** are molluscs with a similar long history: they also have a calcite or aragonite layered structure, and are distinctive for their coiled form (Fig. 3.3). The ***cephalopod*** molluscs include the modern *Nautilus* and the coiled, chambered ammonites, which were very common in Mesozoic times. Most cephalopods have a layered shell structure, and, in common with most other molluscs, this is a feature that may be recognisable in fragments of shells under the microscope. There is an important exception in the ***belemnites***, a cephalopod that had a cigar-shaped 'guard' of radial, fibrous calcite: these can be preserved in large numbers in Mesozoic sedimentary rocks.

The ***brachiopods*** are also shelly organisms with two shells and are hence superficially similar to bivalves. They are not common today but were very abundant in the Palaeozoic and Mesozoic. The shells are made up of low-magnesium calcite, and a two-layer structure of fibrous crystals may be completely preserved in brachiopod shells. The exoskeletons of arthropods, such as the *trilobites*, are made up of microscopic prisms of calcite that are elongate perpendicular to the edges of the plates. Although they may appear to be quite different, ***barnacles*** are also arthropods and have a similar internal structure to their skeletal material.

Another group of shelly organisms, the ***echinoids*** (sea urchins), can be easily recognised because they construct their hard body parts out of whole low-magnesium calcite crystals. Individual plates of echinoids are preserved in carbonate sediments. ***Crinoids*** (sea lilies) belong to the same phylum as echinoids (the Echinodermata) and are similar in the sense that they too construct their body parts out of whole calcite crystals, with the discs that make up the stem of a crinoid forming sizeable accumulations in Carboniferous sediments. In life the individual crystals in echinoid and crinoid body parts are perforated, but the pores are filled with growths of calcite that may also extend beyond the original limits of the skeletal element as an overgrowth (*18.2.2*). These large single crystals that make up echinoderm fragments make them easily recognisable in thin-section.

Foraminifera are small, single-celled marine organisms that range from a few tens of microns in diameter to tens of millimetres across. They are either floating in life (***planktonic***) or live on the sea floor (***benthic***) and most modern and ancient forms have hard outer parts (***tests***) made up of high- or low-magnesium calcite. Both modern sediments and ancient limestone beds have been found with huge concentrations of foraminifers such that they may form the bulk of the sediment.

Some of the largest calcium carbonate biogenic structures are built by **corals** (**Cnidaria**) which may be in the form of colonies many metres across or as solitary organisms. Calcite seems to have been the main crystal form in Palaeozoic corals, with aragonite crystals making the skeleton in younger corals. **Hermatypic corals** have a symbiotic relationship with algae that require clear, warm, shallow marine waters. These corals form more significant build-ups than the less common, **ahermatypic corals** that do not have algae and can exist in colder, deeper water. Another group of colonial organisms that may contribute to carbonate deposits are the **bryozoa**. These single-celled protozoans are seen mainly as encrusting organisms today but in the past they formed large colonies. The structure is made up of aragonite, high-magnesium calcite or a mixture of the two. The **sponges** (**Porifera**) are a further group of sedentary organisms that may form hard parts of calcite, although structures of silica or protein are also common. **Stromatoporoids** are calcareous sponges that were common in the Palaeozoic. Other calcareous structures associated with animals are the tubes of carbonate secreted by **serpulid** worms. These are a type of annelid worm that encrusts pebbles or the hard parts of other organisms with sinuous tubes of calcite or aragonite.

Carbonate-forming plants

Algae and microbial organisms are an important source of biogenic carbonate and are important contributors of fine-grained sediment in carbonate environments through much of the geological record (Riding 2000). Three types of alga are carbonate producers. **Red algae** (**rhodophyta**) are otherwise known as the coralline algae: some forms are found encrusting surfaces such as shell fragments and pebbles. They have a layered structure and are effective at binding soft substrate. The **green algae** (**chlorophyta**) have calcified stems and branches, often segmented, that contribute fine rods and grains of calcium carbonate to the sediment when the organism dies. **Nanoplankton** are planktonic **yellow-green algae** that are extremely important contributors to marine sediments in parts of the stratigraphic record. This group, the **chrysophyta**, include **coccoliths**, which are spherical bodies a few tens of microns across made up of plates. Coccoliths are an important constituent of pelagic limestone, including the Cretaceous Chalk.

Cyanobacteria are now classified separately to algae. The 'algal' mats formed by these organisms are more correctly called bacterial or **microbial mats**. In addition to sheet-like mats, columnar and domal forms are also known. The filaments and sticky surfaces of the cyanobacteria act as traps for fine-grained carbonate and as the structure grows it forms layered, flat or domed structures called **stromatolites** (Fig. 3.4), which are some of the earliest lifeforms on Earth. In contrast to stromatolites, **thrombolites** are cyanobacterial communities that have an irregular rather than layered form. **Oncoids** are irregular concentric structures millimetres to centimetres across formed of layers bound by cyanobacteria found as clasts within carbonate sediments. Other cyanobacteria bore into the surface of skeletal

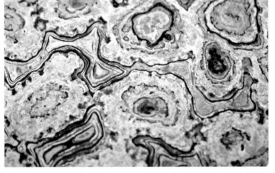

Fig. 3.4 Mounds of cyanobacteria form stromatolites, which are bulbous masses of calcium carbonate material at various scales: (top) modern stromatolites; (bottom) a cross-section through ancient stromatolites.

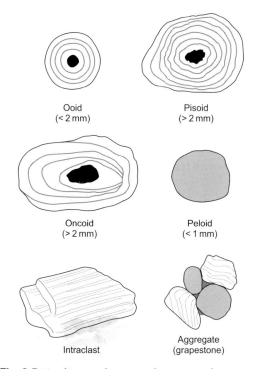

Ooid
(< 2 mm)

Pisoid
(> 2 mm)

Oncoid
(> 2 mm)

Peloid
(< 1 mm)

Intraclast

Aggregate
(grapestone)

Fig. 3.5 Non-biogenic fragments that occur in limestones.

debris and alter the original structure of a shell into a fine-grained micrite (*micritisation*).

3.1.4 Non-biogenic constituents of limestone

A variety of other types of grain also occur commonly in carbonate sediments and sedimentary rocks (Fig. 3.5). *Ooids* are spherical bodies of calcium carbonate less than 2 mm in diameter. They have an internal structure of concentric layers which suggests that they form by the precipitation of calcium carbonate around the surface of the sphere. At the centre of an ooid lies a nucleus that may be a fragment of other carbonate material or a clastic sand grain. Accumulations of ooids form shoals in shallow marine environments today and are components of limestone throughout the Phanerozoic. A rock made up of carbonate ooids is commonly referred to as an **oolitic limestone**, although this name does not form part of the Dunham classification of carbonate rocks (*3.1.6*). The origin of ooids has been the subject of much debate and the

present consensus is that they form by chemical precipitation out of agitated water saturated in calcium carbonate in warm waters (Tucker & Wright 1990). It is likely that bacteria also play a role in the process, especially in less agitated environments (Folk & Lynch 2001). Concentrically layered carbonate particles over 2 mm across are called **pisoids**: these are often more irregular in shape but are otherwise similar in form and origin to ooids.

Some round particles made up of fine-grained calcium carbonate found in sediments do not show any concentric structure and have apparently not grown in water in the same way as an ooid or pisoid. These **peloids** are commonly the **faecal pellets** of marine organisms such as gastropods and may be very abundant in some carbonate deposits, mostly as particles less than a millimetre across. In thin-section these pellets are internally homogeneous, and, if the rock underwent some early compaction, they may have become deformed, squashed between harder grains, making them difficult to distinguish from loose mud deposited as a matrix.

Intraclasts are fragments of calcium carbonate material that has been partly lithified and then broken up and reworked to form a clast which is incorporated into the sediment. This commonly occurs where lime mud dries out by subaerial exposure in a mud flat and is then reworked by a current. A conglomerate of flakes of carbonate mud can be formed in this way. Other settings where clasts of lithified calcium carbonate occur are associated with reefs where the framework of the reef is broken up by wave or storm action and redeposited (*15.3.2*). Carbonate grains consisting of several fragments cemented together are **aggregate grains**, which when they comprise a collection of rounded grains are known as **grapestones**.

3.1.5 Carbonate muds

Fine-grained calcium carbonate particles less than 4 microns across (cf. clay: *2.4*) are referred to as **lime mud, carbonate mud** or **micrite**. The source of this fine material may be purely chemical precipitation from water saturated in calcium carbonate, from the breakdown of skeletal fragments, or have an algal or bacterial origin. The small size of the particles usually makes it very difficult to determine the source. Lime mud is found in many carbonate-forming environments and can be the main constituent of limestone.

3.1.6 Classification of limestones

The **Dunham Classification** is the most widely used scheme for the description of limestone in the field, in hand specimen and in thin-section. The primary criterion used in this classification scheme is the texture, which is described in terms of the proportion of carbonate mud present and the framework of the rock (Fig. 3.6). The first stage in using the Dunham classification is to determine whether the fabric is matrix- or clast-supported. Matrix-supported limestone is divided into **carbonate mudstone** (less than 10% clasts) and **wackestone** (with more than 10% clasts). If the limestone is clast-supported it is termed a **packstone** if there is mud present or a **grainstone** if there is little or no matrix. A **boundstone** has an organic framework such as a coral colony. The original scheme (Dunham 1962) did not include the subdivision of boundstone into **bafflestone**, **bindstone** and **framestone**, which describes the type of organisms that build up the framework. These categories, along with the addition of *rudstone* (which are clast-supported limestone conglomerate) and **floatstone** (matrix-supported limestone conglomerate) were added by Embry & Klovan (1971) and James & Bourque (1992). Note that the terms rudstone and floatstone are used for carbonate intraformational conglomerate made up of material deposited in an adjacent part of the same environment and then redeposited (e.g. at the front of a reef: *15.3.2*). These should be distinguished from conglomerate made up of clasts of limestone eroded from older bedrock and deposited in a quite different setting, for example on an alluvial fan (*7.5*).

The nature of the grains or framework material forms the secondary part of the classification. A rock consisting entirely of ooids with no matrix would be an oolitic grainstone, one composed of about 75% broken shelly fragments in a matrix of carbonate mud is a bioclastic packstone, and rock composed mainly of large oyster shells termed a bioclastic rudstone. Naming a limestone using the combination of textural and compositional criteria in the Dunham scheme provides information about the likely conditions under which the sediment formed: for example, a coral boundstone forms under quite different conditions to a foraminiferal wackestone.

3.1.7 Petrographic analysis of carbonate rocks

Thin-section analysis of limestones and dolostones can reveal a great deal of information about the environment in which the sediment was deposited. Assessment of the proportions of carbonate mud and larger fragments provides an indication of the environment of deposition: a high proportion of fine-grained carbonate material suggests a relatively low-energy setting, whereas an absence of mud characterises higher-energy environments. The mud to fragmental component ratio is also the basis for classification using the Dunham scheme of carbonate mudstones, wackestones, packstones and grainstones. If it is not already evident from hand specimen, thin-sections will also reveal the presence of framework organisms such as corals and algae that form a boundstone fabric.

The nature of the fragmental material provides further evidence of the conditions under which the sediment was deposited: for example, high concentrations of ooids indicate shallow, wave-dominated coastal settings (*15.3.1*) whereas a rock composed of biogenic material that is all from the same group of organisms, such as bivalves or gastropods, is an indicator of a lagoonal setting (*15.2.2*). The degree to which the shelly material is broken up also reflects the energy of the setting or the amount of transport and reworking of the sediment. It is usually possible to determine the fossil group to which larger bioclasts belong from their overall shape and the internal structure (Fig. 3.1). Additional clues may also come from the mineral that the original bioclast was made of (Fig. 3.7): shells originally composed of aragonite tend to recrystallise and the primary fabric is lost; similarly, high-magnesium calcite commonly recrystallises and also results in bioclasts with a recrystallised fabric. Organisms such as many brachiopods and bivalves that were formed of low-magnesium calcite tend to retain their primary structure.

It should be noted, however, that all carbonate rocks are susceptible to diagenetic alteration (*18.4*) that can change both the mineralogy and the structure of the fragments and the carbonate mud. Diagenetic alteration can vary from a simple cementation of

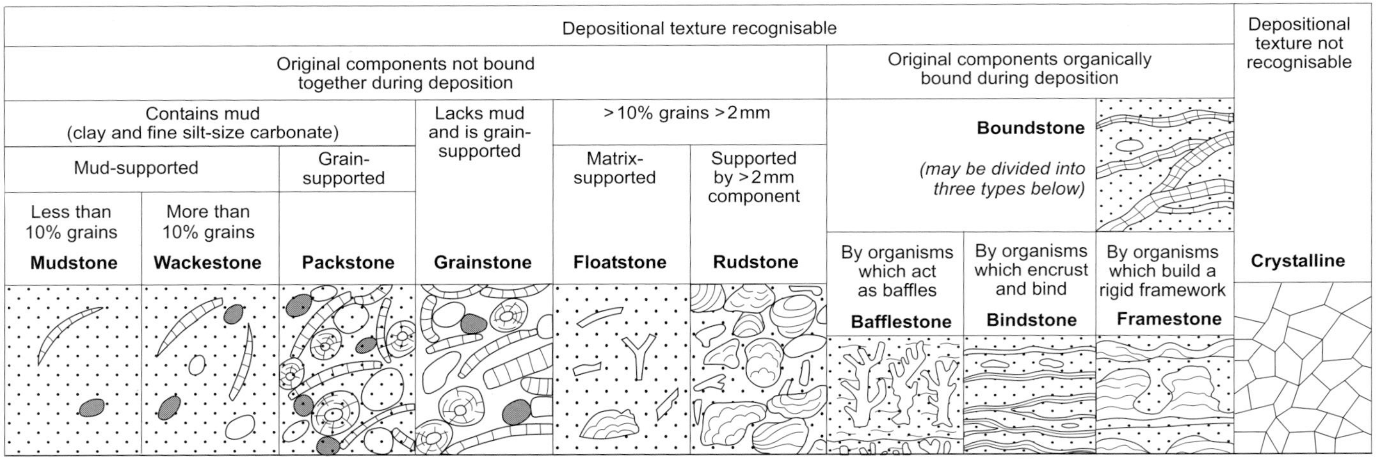

Fig. 3.6 The Dunham classification of carbonate sedimentary rocks (Dunham 1962) with modifications by Embry & Klovan (1971). This scheme is the most commonly used for description of limestones in the field and in hand specimen.

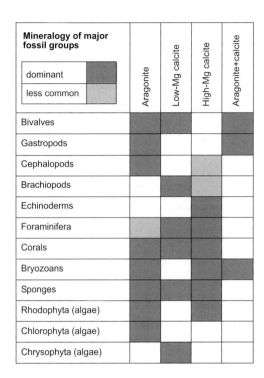

Mineralogy of major fossil groups	Aragonite	Low-Mg calcite	High-Mg calcite	Aragonite+calcite
dominant / less common				
Bivalves	■			■
Gastropods	■			■
Cephalopods	■		■	
Brachiopods		■	■	
Echinoderms			■	
Foraminifera	■	■	■	
Corals	■		■	
Bryozoans	■		■	■
Sponges	■		■	
Rhodophyta (algae)	■		■	
Chlorophyta (algae)	■			
Chrysophyta (algae)		■		

Fig. 3.7 The calcareous hard parts of organisms may be made up of aragonite, calcite in either its low- or high-magnesium forms, or mixtures of minerals.

the sediment with little alteration of the material to complete recrystallisation that obliterates all of the depositional fabric (*18.4.3*).

3.2 EVAPORITE MINERALS

These are minerals formed by precipitation out of solution as ions become more concentrated when water evaporates. On average, seawater contains $35\,g\,L^{-1}$ (35 parts per thousand) of dissolved ions, mainly chloride, sodium, sulphate, magnesium, calcium and potassium (Fig. 3.8). The chemistry of lake waters is variable, often with the same principal ions in different proportions. The combination of anions and cations into minerals occurs as they become concentrated and the water saturated with respect to a particular compound. The least soluble compounds are precipitated first, so calcium carbonate is first precipitated out of seawater, followed by calcium sulphate and sodium chloride as the waters become more

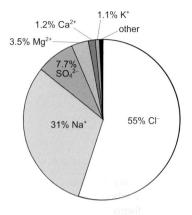

Fig. 3.8 The proportions of the principal ions in seawater of normal salinity and 'average' river water. (Data from Krauskopf 1979).

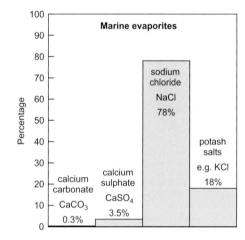

Fig. 3.9 The proportions of minerals precipitated by the evaporation of seawater of average composition.

concentrated. Potassium and magnesium chlorides will only precipitate once seawater has become very concentrated. The order of precipitation of evaporite minerals from seawater and the loss of water required for them to form are listed in Fig. 3.9, along with the mass formed per unit volume of seawater and the chemistry of the mineral.

3.2.1 Gypsum and anhydrite

The most commonly encountered evaporite minerals in sedimentary rocks are forms of calcium sulphate, either as **gypsum** or **anhydrite**. Calcium sulphate is precipitated from seawater once evaporation has concentrated the water to 19% of its original volume. Gypsum is the hydrous form of the mineral ($CaSO_4.2H_2O$). It precipitates at the surface under all but the most arid conditions but may become dehydrated to anhydrite on burial (*18.5*). Anhydrite has no water in the crystal structure ($CaSO_4$) and forms either by direct precipitation in arid shorelines (*15.2.3*) or as a result of alteration of gypsum by burial. It may become hydrated to gypsum if water is introduced. Primary gypsum occurs as elongate crystals of **selenite** when it forms from precipitation out of water. If it forms as a result of the rehydration of anhydrite it has a fine crystalline form in nodules of **alabaster**. Gypsum also occurs as a fibrous form in secondary veins.

Gypsum is readily distinguished from calcium carbonate minerals in the field because it is softer (hardness 2, easily scratched with a fingernail) and does not react with dilute HCl: it can be distinguished from halite by the fact that it does not taste salty. Crystals of gypsum have a low relief when they are viewed under the microscope, cleavage is usually well developed and the birefringence colours are low-order greys. Anhydrite is a harder (hardness 3.5), denser mineral than gypsum: it is commonly white in hand specimen, and is not easily scratched by a fingernail. In thin-section the high density means crystals have a relatively high relief; birefringence colours are moderate, higher-order colours than gypsum.

3.2.2 Halite

Halite (NaCl) precipitates out of seawater once it has been concentrated to 9.5% of its original volume (Fig. 3.10). It may occur as thick crystalline beds or as individual crystals that have a distinctive cubic symmetry, sometimes with a stepped crystal face (a **hopper crystal**). The high solubility of sodium chloride means that it is only preserved in rocks in the absence of dilute groundwater, which would dissolve it. Surface exposures of halite can be found in some arid regions where it is not removed by rainwater.

Naturally occurring halite is **rock salt**, so the simplest test to confirm the presence of the mineral is taste: the only mineral it might be confused with on this basis is sylvite (below), but this potassium

Fig. 3.10 White halite precipitated on the shores of the Dead Sea, Jordan, which has a higher concentration of ions than normal seawater.

chloride mineral has a more bitter taste than 'normal salt' and is much less common. Halite is soft (hardness 2.5, slightly more than gypsum but still scratched by a fingernail), white or colourless. In thin-section halite crystals may show a strong cleavage with planes at right angles and, being a cubic mineral, it is isotropic.

3.2.3 Other evaporite minerals

Evaporation of seawater can yield other minerals, which are rarely found in large amounts but can be economically important. In particular, potassium chloride, **sylvite** (KCl), is an important source of industrial potash that occurs associated with halite and is interpreted as the product of extreme evaporation of marine waters. However, evaporation of modern waters results in a number of different magnesium sulphate ($MgSO_4$) minerals rather than sylvite, and this has led to suggestions that the chemical composition of seawater has not been constant over hundreds of millions of years (Hardie 1996). Variations in the relative importance of meteoric waters (run-off from land) and hydrothermal waters (from mid-ocean ridge vents) are thought to be the reason for these variations in water chemistry, which either favour KCl or $MgSO_4$ precipitation at different times.

Saline lakes (10.3) generally contain the same dissolved ions as seawater, but the proportions are usually different, and this results in suites of evaporite minerals characteristic of different lake chemistries. Most of these minerals are sulphates, carbonates and bicarbonates of sodium and magnesium such as **trona** ($Na_2CO_3.NaHCO_3.2H_2O$), **mirabilite** ($Na_2SO_4.10H_2O$) and **epsomite** ($MgSO_4.7H_2O$). All are relatively soft minerals and, of course, all are soluble in water.

3.3 CHERTS

Cherts are fine-grained siliceous sedimentary rocks made up of silt-sized interlocking quartz crystals (**microquartz**) and chalcedony, a form of silica which is made up of radiating fibres a few tens to hundreds of microns long. Beds of chert form either as primary sediments or by diagenetic processes.

On the floors of seas and lakes the siliceous skeletons of microscopic organisms may accumulate to form a **siliceous ooze**. These organisms are **diatoms** in lakes and these may also accumulate in marine conditions, although **Radiolaria** are more commonly the main components of marine siliceous oozes. Radiolarians are **zooplankton** (microscopic animals with a planktonic lifestyle) and diatoms are **phytoplankton** (free-floating algae). Upon consolidation these oozes form beds of chert. The opaline silica (opal is cryptocrystalline silica with water in the mineral structure) of the diatoms and radiolaria is metastable and recrystallises to chalcedony or microquartz. Cherts formed from oozes are often thin bedded with a layering caused by variations in the proportions of clay-sized material present. They are most common in deep-ocean environments (*16.5.1*).

Diagenetic cherts are formed by the replacement of other material such as calcium carbonate by waters rich in silica flowing through the rock. The source of the silica is mainly biogenic with the opaline silica of diatoms, radiolarian and siliceous sponges being redistributed. Chert formed in this way occurs as nodules within a rock, such as the dark **flint** nodules that are common within the Cretaceous Chalk, and as nodules and irregular layers within other limestones and mudstones.

The dense internal structure of interlocking microquartz grains and fibres makes chert the hardest sedimentary rock. It breaks with a conchoidal fracture and can form fine shards when broken, a feature which made this rock very significant in the development of tools by early humans. The colour is variable, depending on the proportions of impurities: the presence of haematite causes the strong red colour of **jasper**, and traces of organic material result in grey or black chert. Thin-sections through chert reveal characteristic patterns of either radiating fibres in chalcedony or completely interlocking microquartz grains.

3.4 SEDIMENTARY PHOSPHATES

Calcium phosphate occurs in igneous rocks as the mineral **apatite**, which is a common accessory mineral in many granitic rocks. Some apatite is preserved in sediments as mineral grains, but generally phosphates occur in solution and are absorbed in the soil by plants or washed into the marine realm where it is taken up by plants and animals. Phosphorus is essential to lifeforms and is present in all living matter. Phosphatic material in the form of bone, teeth and fish scales occurs dispersed in many clastic and biogenic sedimentary rocks, but higher concentrations are uncommon, being found most frequently associated with shallow marine continental shelf deposits. Most occurrences occur where there is high organic productivity and low oxygen, but not fully anoxic conditions. Rocks with concentrations of phosphate (5% to 35% P_2O_5) are called **phosphorites** (*11.5.2*). Mineralogically, phosphorites are composed of **francolite**, which is a calcium phosphate (carbonate hydroxyl fluorapatite). In some cases the phosphate is in the form of **coprolites**, which are the fossilised faeces of fish or animals.

Apatite is clear, with a high relief and is found quite commonly as a heavy mineral in sandstones and may be identified in thin-section. Biogenic phosphorites occur as nodules or laminated beds made up of clay to fine pebble-size material that is usually brown or occasionally black in colour. They can be difficult to identify with certainty in the field, and in thin-section the amorphous brown form of the phosphate may be difficult to distinguish from carbonaceous material. Chemical analysis is the most reliable test.

3.5 SEDIMENTARY IRONSTONE

Iron is one of the most common elements on the planet, and is found in small to moderate amounts in almost all deposits. Sedimentary rocks that contain at least 15% iron are referred to as **ironstones** or **iron formations** in which the iron is in the form of oxides, hydroxides, carbonate, sulphides or silicates (Simonson 2003). Iron-rich deposits can occur in all types of depositional environment, and are known from some of the oldest rocks in the world: most of the iron ore mined today is from Precambrian rocks.

3.5.1 Iron minerals in sediments

Magnetite (Fe_3O_4) is a black mineral which occurs as an accessory mineral in igneous rocks and as detrital grains in sediments, but **haematite** (Fe_2O_3) is the most common oxide, bright red to black in colour, occurring as a weathering or alteration product in a wide variety of sediments and sedimentary rocks. **Goethite** is an iron hydroxide (FeO.OH) that is widespread in sediments as yellow-brown mineral, which may be a primary deposit in sediments, or is a weathering product of other iron-rich minerals, representing less oxidising conditions than haematite. Goethite forms as a precursor to haematite in desert environments giving desert sands their yellowish colour. The oxidation to haematite to give these sands the red colour seen in some ancient desert deposits may be a post-depositional process. **Limonite** (FeO.OH.nH$_2$O) is similar, a hydrated iron oxide that is amorphous. In thin-section iron oxides are opaque: magnetite is black and often euhedral whereas haematite occurs in a variety of forms and is red in reflected light. Goethite and limonite are yellow-brown in thin-section and are anisotropic.

Pyrite (FeS) is a common iron sulphide mineral that is found in igneous and metamorphic rocks as brassy cubic crystals ('fool's gold'). It is also common in sediments, but often occurs as finely disseminated particles that appear black, and may give a dark coloration to sediments. In thin-section it is opaque and if the crystals are large enough the cubic form may be seen. There are several silicate minerals that are iron-rich: **greenalite** and **chamosite** are phyllosilicate minerals (minerals with sheet-like layers in their crystal lattices) that are found in ironstones and formed either as authigenic (2.3.2) or diagenetic (18.2) products. **Glauconite** (**glaucony**) is also a phyllosilicate formed authigenically in shallow marine environments (2.3). The most common iron carbonate, siderite, is considered in section 3.1.1.

3.5.2 Formation of ironstones

Iron-rich sedimentary rocks are varied in character, ranging from mudstones rich in pyrite formed under reducing, low-energy conditions to oolitic ironstones deposited in more energetic settings. Most are thought to have originated in shallow marine or marginal marine environments, but it is not always clear whether the iron minerals found in the rocks are the original minerals formed at the time of deposition, or whether they are later diagenetic products. For example, the presence of ooids suggests agitated, and therefore probably oxygenated shallow water, conditions under which all iron minerals formed should be oxides or hydroxides. It is therefore likely that the iron silicates found in some shallow-marine ironstones (e.g. ooids of chamosite) may be altered goethite. It is generally thought that sedimentary ironstones form under conditions of lowered sedimentation rate of carbonate or terrigenous clastic material. Siderite-rich mudstones are most commonly associated with deposition in freshwater reducing conditions, such as non-saline marshes: where sulphate ions are available from seawater then iron sulphide forms in preference to iron carbonate.

3.5.3 Banded Iron Formations

Banded Iron Formations (BIFs) are an example of a type of sedimentary rock for which there is no equivalent forming today. All examples are from the Precambrian, and most are from the period 2.5 to 1.9 Ga, although there are some older examples as well (Trendall 2002). As their name suggests, BIFs consist of laminated or thin-bedded alternations of haematite-rich sediment and other material (Fig. 3.11), which is typically siltstone or chert (3.3) (Fralick & Barrett 1995). Individual layers may be traced for kilometres where exposure allows and units of BIF may be hundreds of metres thick and extend for hundreds of kilometres. The origin of BIFs is not fully understood, but they probably formed on widespread shelves or shallow basins, with the iron originating in muddy deposits on the sea floor, possibly in association with microbial activity. The source of the iron is thought to be either hydrothermal or a weathering product, and could only have been transported as dissolved iron if the ocean waters were not oxygenated. This is one of a number

Fig. 3.11 Thinly bedded banded iron formation (BIF) composed of alternating layers of iron-rich and silica-rich rock.

of lines of evidence that the atmosphere contained little or no oxygen through much of Precambrian times.

3.5.4 Ferromanganese deposits

Nodules or layers of ferromanganese oxyhydroxide form authigenically on the sea floor: they are black to dark brown in colour and range from a few millimetres to many centimetres across as nodules or as extensive laminated crusts on hard substrates. Although these **manganese nodules** form at any depth, they form very slowly and are only found concentrated in deep oceans (*16.5.4*) where the rate of deposition of any other sediment is even slower (Calvert 2003).

3.6 CARBONACEOUS (ORGANIC) DEPOSITS

Sediments and sedimentary rocks with a high proportion of organic matter are termed **carbonaceous** because they are rich in carbon (cf. calcareous – *3.1*). A deposit is considered to be carbonaceous if it contains a proportion of organic material that is significantly higher than average (>2% for mudrock, >0.2% for limestone, >0.05% for sandstone). Organic matter normally decomposes on the death of the plant or animal and is only preserved under conditions of limited oxygen availability, **anaerobic** conditions. Environments where this may happen are waterlogged swamps and bogs (*18.7.1*), stratified lakes (*10.2.1*) and marine waters with restricted circulation such as lagoons (*13.3.2*). Strata containing high

concentrations of organic material are of considerable economic importance: coal, oil and gas are all products of the diagenetic alteration of organic material deposited and preserved in sedimentary rocks, and the processes of formation of these naturally occurring hydrocarbons are considered further in 18.7.

3.6.1 Modern organic-rich deposits

Most of the dead remains of land plants decompose at the surface or within the soil as a result of oxidation, microbial or animal activity. Long-term preservation of dead vegetation is favoured by the wet, anaerobic conditions of mires, bogs and swamps and thick accumulations of **peat** may form. Peats are forming at the present day in a wide range of climatic zones from subarctic boggy regions to mangrove swamps in the tropics (McCabe 1984; Hazeldine 1989) and contain a range of plant types, from mosses in cool upland areas to trees in lowland fens and swamps. Thick peat deposits are most commonly associated with river floodplains (*9.3*), the upper parts of deltas (*12.3.1*) and with coastal plains (*13.2.2*). Pure peat will form only in areas that receive little clastic input. Regular flooding from rivers or the sea will introduce mud into the peat-forming environment and the resulting deposit will be a carbonaceous mudrock.

The accumulation of organic material in subaqueous environments is just as important as land deposits. **Sapropel** is the remains of planktonic algae, spores and very fine detritus from larger plants that accumulates underwater in anaerobic conditions: these deposits may form a sapropelic coal (*18.7.1*). Anaerobic conditions are also required to accumulate the organic material that ultimately forms liquid and gaseous hydrocarbons: these deposits are composed of the remains of zooplankton (microscopic animals), phytoplankton (floating microscopic algae) and bacteria. The formation of oil and gas from deposits of this type is considered in section 18.7.3.

3.6.2 Coal

If over two-thirds of a rock is solid organic matter it may be called a **coal**. Most economic coals have less than 10% non-organic, non-combustible material that is often referred to as **ash**. Coal can be readily recognised because it is black and has a low density.

Peat is heterogeneous because it is made up of different types of vegetation, and of the various different components (wood, leaves, seeds, etc.) of the plants. Moreover, the vegetation forming the peat may vary with time, depending on the predominance of either tree communities or herbaceous plants, and this will be reflected as layers in the beds of coal. A nomenclature for the description of different lithotypes of coal has therefore been developed as follows:

Vitrain: bright, shiny black coal that usually breaks cubically and mostly consists of woody tissue.

Durain: black or grey in colour, dull and rough coal that usually contains a lot of spore and detrital plant material.

Fusain: black, fibrous with a silky lustre, friable and soft coal that represents fossil charcoal.

Clarain: banded, layered coal that consists of alternations of the other three types.

Sapropelic coal has a conchoidal fracture and may have a dull black lustre (called **cannel coal**) or is black/brown in colour (known as **boghead coal**).

Microscopic examination of these lithotypes reveals that a number of different particle types can be recognised: these are called **macerals**, and are the organic equivalent of minerals in rocks. Macerals are examined by looking at the coal as polished surfaces in reflected light under a thin layer of oil. Three main groups of maceral are recognised: **vitrinite**, the origin of which is mainly cell walls of woody tissue and leaves, **liptinite**, which mainly comes from spores, cuticles and resins, and **inertinite**, which is burnt, oxidised or degraded plant material.

A further analysis that can be made is the **reflectance** of the different particles, which can be assessed by measuring the amount of light reflected from the polished surface. Liptinites generally have low reflectance, and inertinites have high reflectance, but vitrinite, which is by far the most common maceral in most coals, shows different reflectance depending on the **coal rank**. **Vitrinite reflectance** therefore can be used as a measure of the rank of the coal, and because coal rank increases with the temperature to which the material has been heated, vitrinite reflectance is a measure of the burial temperature of the bed. This is an analytical technique in basin analysis (*24.8*) that provides a measure of how deep a bed has been buried.

The coalification of carbonaceous matter into macerals and coal lithotypes takes place as a series of post-depositional bacteriological, chemical and physical processes that are considered further in section 18.7.2.

3.6.3 Oil shales and tar sands

Mudrocks that contain a high proportion of organic material that can be driven off as a liquid or gas by heating are called **oil shales**. The organic material is usually the remains of algae that have broken down during diagenesis to form **kerogen**, long-chain hydrocarbons that form **petroleum** (natural oil and gas) when they are heated. Oil shales are therefore important **source rocks** of the hydrocarbons that ultimately form concentrations of oil and gas. The environments in which they are formed must be anaerobic to prevent oxidisation of the organic material; suitable conditions are found in lakes and certain restricted shallow-marine environments (Eugster 1985). Oil shales are black and the presence of hydrocarbons may be detected by the smell of the rock and the fact that it will make a brown, oily stain on other materials.

Tar sands or **oil sands** are clastic sediments that are saturated with hydrocarbons and they are the exposed equivalents of subsurface oil reservoirs (*18.7.4*). The oil in tar sands is usually very viscous (bitumen), and may be almost solid, because the lighter components of the hydrocarbons that are present at depth are lost by biodegradation near the surface. The presence of the oil in the pores of the sediment prevents the formation of any cement, so tar sands remain unlithified, held together only by the bitumen that gives them a black or very dark brown colour.

3.7 VOLCANICLASTIC SEDIMENTARY ROCKS

Volcanic eruptions are the most obvious and spectacular examples of the formation of both igneous and sedimentary rocks on the Earth's surface. During eruption volcanoes produce a range of materials that include molten lava flowing from fissures in the volcano and particulate material that is ejected from the vent to form **volcaniclastic deposits** (Cas & Wright 1987). The location of volcanoes is related to the plate tectonic setting, mainly in the vicinity of plate margins and other areas of high heat flow in the crust. The presence of beds formed by volcanic processes can be an important indicator of the tectonic setting in which the sedimentary succession formed. Lavas are found close to the site of the eruption, but ash may be spread tens, hundreds or even thousands of kilometres away. Volcaniclastic material may therefore occur in any depositional environment and hence may be found associated with a wide variety of other sedimentary rocks (*Chapter 17*). Volcanic rocks are also of considerable value in stratigraphy as they may often be dated radiometrically (*21.1*), providing an absolute time constraint on the sedimentary succession.

3.7.1 Types of volcaniclastic rocks

The composition of the magma affects the style of eruption. Basaltic magmas tend to form volcanoes that produce large volumes of lava, but small amounts of volcanic ash. Volcanoes with more silicic magma are much more explosive, with large amounts of the molten rock being ejected from the volcano as particulate matter. The particles ejected are known as **pyroclastic** material, also collectively referred to as **tephra**. Note that the term pyroclastic is used for material ejected from the volcano as particles and **volcaniclastic** refers to any deposit that is mainly composed of volcanic detritus. Pyroclastic material may be individual crystals, pieces of volcanic rock (lithic fragments), or **pumice**, the highly vesicular, chilled, 'froth' of the molten rock. The size of the pyroclastic debris ranges from fine dust a few microns across to pieces that may be several metres across.

3.7.2 Nomenclature of volcaniclastic rocks

The textural classification of volcaniclastic deposits (Fig. 3.12) is a modification of the Wentworth scheme. Coarse material (over 64 mm) is divided into **volcanic blocks**, which were solid when erupted, and **volcanic bombs**, which were partly molten and have cooled in the air; consolidated into a rock these are referred to as **volcanic breccia** and **agglomerate** respectively. Granule to pebble-sized particles (2–64 mm) are called **lapilli** and form a **lapillistone**. **Accretionary lapilli** are spherical aggregates of fine ash formed during air fall. Sand-, silt- and clay-grade tephra is **ash** when unconsolidated and **tuff** upon lithification. Coarse ash/tuff is sand-sized and fine ash/tuff is silt- and clay-grade material. Compositional descriptions hinge on the relative proportions of crystals, lithic fragments and **vitric** material, which is fragments of volcanic glass formed when the molten rock cools very rapidly, sometimes forming pumice.

(a)

Clast size	Unconsolidated	Consolidated
>64 mm	Bombs	Agglomerate
	Blocks	Volcanic breccia
2-64 mm	Lapilli	Lapillistone
0.063-2 mm	Coarse ash	Coarse tuff (volcanic sandstone)
<0.063 mm	Fine ash	Fine tuff (volcanic mudstone)

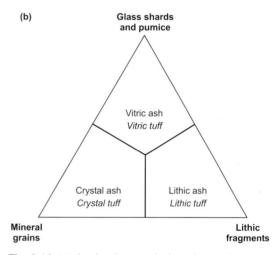

Fig. 3.12 (a) The classification of volcaniclastic sediments and sedimentary rocks based on the grain size of the material. (b) Nomenclature used for loose ash and consolidated tuff with different proportions of lithic, vitric and crystal components.

3.7.3 Recognition of volcaniclastic material

The origin of coarse-grained volcaniclastic sediments is usually easy to determine if the lithology of the larger clasts can be recognised as an igneous rock such as basalt. The tephra particles are usually angular, with the exception of rounded volcanic bombs, well-rounded accretionary lapilli found in some air fall ashes, and the distinctive shape of **fiamme**, glassy pumice fragments that may resemble a tuning fork when compacted. Another useful indicator is the uniform nature of the material, as mixing of tephra with other types of sediment occurs only by subsequent reworking. In general, volcaniclastic sediments with a basaltic composition are dark in colour, whereas more rhyolitic deposits are paler. Fine ash and tuff can be more difficult to identify with certainty in the field, especially if the material has been weathered. Brightly coloured green and orange strata sometimes form as a result of the alteration of ash beds. Characteristic sedimentary structures resulting from the processes of transport are considered further in Chapter 17 along with the environments of deposition of volcaniclastic sediments.

Petrographic analysis of volcaniclastic sediments is usually required to confirm the composition. In thin-section the composition of lithic fragments can be determined if a high magnification is used to identify the minerals that make up the rock fragments. Crystals of feldspar are usually common, especially if the deposit is a crystal tuff, and other silicate minerals may also be present as euhedral to subhedral crystal grains. Fiamme can be seen as clear, isotropic grains with characteristic shapes: volcanic glass is not stable, and in older tuffs the glass may have a very finely crystalline structure or will be altered to clay minerals.

FURTHER READING

Adams, A.E. & Mackenzie, W.S. (1998) *A Colour Atlas of Carbonate Sediments and Rocks under the Microscope*. Manson Publishing, London.

Braithwaite, C. (2005) *Carbonate Sediments and Rocks*. Whittles Publishing, Dunbeath.

Cas, R.A.F. & Wright, J.V. (1987) *Volcanic Successions: Modern and Ancient*. Unwin-Hyman, London.

Northolt, A.J.G. & Jarvis, I. (1990) *Phosphorite Research and Development*. Special Publication 52, Geological Society Publishing House, Bath.

Scholle, P.A. (1978) *A Color Illustrated Guide to Carbonate Rock Consituents, Textures, Cements and Porosities*. Memoir 27, American Association of Petroleum Geologists, Tulsa.

Scoffin, T.P. (1987) *Carbonate Sediments and Rocks*. Blackie, Glasgow, 274 pp.

Stow, D.A. (2005) *Sedimentary Rocks in the Field: a Colour Guide*. Manson, London.

Tucker, M.E. (2001) *Sedimentary Petrology* (3rd edition). Blackwell Science, Oxford.

Tucker, M.E. & Wright, V.P. (1990) *Carbonate Sedimentology*, Blackwell Scientific Publications, Oxford, 482 pp.

4

Processes of Transport and Sedimentary Structures

Most sedimentary deposits are the result of transport of material as particles. Movement of detritus may be purely due to gravity but more commonly it is the result of flow in water, air, ice or dense mixtures of sediment and water. The interaction of the sedimentary material with the transporting media results in the formation of bedforms, which may be preserved as sedimentary structures in rocks and hence provide a record of the processes occurring at the time of deposition. If the physical processes occurring in different modern environments are known and if the sedimentary rocks are interpreted in terms of those same processes it is possible to infer the probable environment of deposition. Understanding these processes and their products is therefore fundamental to sedimentology. In this chapter the main physical processes occurring in depositional environments are discussed. The nature of the deposits resulting from these processes and the main sedimentary structures formed by the interaction of the flow medium and the detritus are introduced. Many of these features occur in a number of different sedimentary environments and should be considered in the context of the environments in which they occur.

4.1 TRANSPORT MEDIA

Gravity The simplest mechanism of sediment transport is the movement of particles under gravity down a slope. **Rock falls** generate piles of sediment at the base of slopes, typically consisting mainly of coarse debris that is not subsequently reworked by other processes. These accumulations are seen as **scree** along the sides of valleys in mountainous areas. They build up as **talus cones** with a surface at the **angle of rest** of the gravel, the maximum angle at which the material is stable without clasts falling

further down slope. The slope angle for loose debris varies with the shape of the clasts and distribution of clast sizes, ranging from just over 30° for well-sorted sand to around 36° for angular gravel (Carson 1977; Bovis 2003). Scree deposits are localised in mountainous areas (6.5.1) and occasionally along coasts: they are rarely preserved in the stratigraphic record.

Water Transport of material in water is by far the most significant of all transport mechanisms. Water flows on the land surface in channels and as overland flow. Currents in seas are driven by wind, tides and oceanic circulation. These flows may be strong enough to carry coarse material along the base of the flow and finer material in suspension. Material may be carried in water hundreds or thousands of kilometres before being deposited. The mechanisms by which water moves this material are considered below.

Air Wind blowing over the land can pick up dust and sand and carry it large distances. The capacity of the wind to transport material is limited by the low density of air. As will be seen in section 4.2.2 the density contrast between the fluid medium and the clasts is critical to the effectiveness of the medium in moving sediment.

Ice Water and air are clearly fluid media but we can also consider ice as a fluid because over long time periods it moves across the land surface, albeit very slowly. Ice is therefore a rather high viscosity fluid that is capable of transporting large amounts of clastic debris. Movement of detritus by ice is significant in and around polar ice caps and in mountainous areas with glaciers (*7.3.2*). The volume of material moved by ice has been very great at times of extensive glaciation.

Dense sediment and water mixtures When there is a very high concentration of sediment in water the mixture forms a *debris flow*, which can be thought of as a slurry with a consistency similar to that of wet concrete. These dense mixtures behave in a different way to sediment dispersed in water and move under gravity over land or under water as debris flows (*4.5.1*). More dilute mixtures may also move under gravity in water as turbidity currents (*4.5.2*). These gravity-driven flow mechanisms are important as a means of transporting coarse material into the deep oceans.

4.2 THE BEHAVIOUR OF FLUIDS AND PARTICLES IN FLUIDS

A brief introduction to some aspects of *fluid dynamics*, the behaviour of moving fluids, is provided in this section to give some physical basis to the discussion of sediment transport and the formation of sedimentary structures in later sections. More comprehensive treatments of sedimentary fluid dynamics are provided in Allen (1994), Allen (1997) and Leeder (1999).

4.2.1 Laminar and turbulent flow

There are two types of fluid flow. In *laminar flows*, all molecules within the fluid move parallel to each other in the direction of transport: in a hetero-

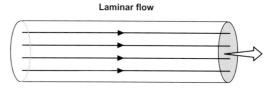

Laminar flow

At all points in the flow all molecules are moving downstream

Turbulent flow

At any point in the flow a molecule may be moving in any direction, but the net flow is downstream

Fig. 4.1 Laminar and turbulent flow of fluids through a tube.

geneous fluid almost no mixing occurs during laminar flow. In *turbulent flows*, molecules in the fluid move in all directions but with a net movement in the transport direction: heterogeneous fluids are thoroughly mixed in turbulent flows. Experiments using threads of dye in tubes show that the lines of flow are parallel at low flow rates, but at higher flow velocities the dye thread breaks up as the flow becomes turbulent (Fig. 4.1).

Flows can be assigned a parameter called a *Reynolds number* (Re), named after Osborne Reynolds who documented the distinction between laminar and turbulent motion in the late 19th century. This is a dimensionless quantity that indicates the extent to which a flow is laminar or turbulent. The Reynolds number is obtained by relating the following factors: the velocity of flow (v), the ratio between the density of the fluid and viscosity of the fluid (v – the fluid kinematic viscosity) and a 'characteristic length' (l – the diameter of a pipe or depth of flow in an open channel). The equation to define the Reynolds number is:

$$Re = v \cdot l / v$$

Fluid flow in pipes and channels is found to be laminar when the Reynolds value is low (<500) and turbulent at higher values (>2000). With increased velocity the flow is more likely to be turbulent and a transition from laminar to turbulent flow in the fluid occurs. Laminar flow occurs in debris flows, in

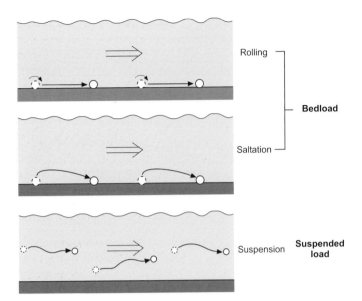

Fig. 4.2 Particles move in a flow by rolling and saltating (bedload) and in suspension (suspended load).

moving ice and in lava flows, all of which have high kinematic viscosities. Fluids with low kinematic viscosity, such as air, are turbulent at low velocities so all natural flows in air that can transport particles are turbulent. Water flows are only laminar at very low velocities or very shallow water depths, so turbulent flows are much more common in aqueous sediment transport and deposition processes. Most flows in water and air that are likely to carry significant volumes of sediment are turbulent.

4.2.2 Transport of particles in a fluid

Particles of any size may be moved in a fluid by one of three mechanisms (Fig. 4.2). **Rolling**: the clasts move by rolling along at the bottom of the air or water flow without losing contact with the bed surface. **Saltation**: the particles move in a series of jumps, periodically leaving the bed surface, and carried short distances within the body of the fluid before returning to the bed again. **Suspension**: turbulence within the flow produces sufficient upward motion to keep particles in the moving fluid more-or-less continually. Particles being carried by rolling and saltation are referred to as **bedload**, and the material in suspension is called the **suspended load**. At low current velocities in water only fine particles (fine silt and

clay) and low density particles are kept in suspension while sand-size particles move by rolling and some saltation. At higher flow rates all silt and some sand may be kept in suspension with granules and fine pebbles saltating and coarser material rolling. These processes are essentially the same in air and water but in air higher velocities are required to move particles of a given size because of the lower density and viscosity of air compared with water.

4.2.3 Entraining particles in a flow

Rolling grains are moved as a result of **frictional drag** between the flow and the clasts. However, to make grains saltate and therefore temporarily move upwards from the base of the flow a further force is required. This force is provided by the **Bernoulli effect**, which is the phenomenon that allows birds and aircraft to fly and yachts to sail 'close to the wind'. The Bernoulli effect can best be explained by considering flow of a fluid (air, water or any fluid medium) in a tube that is narrower at one end than the other (Fig. 4.3). The cross-sectional area of the tube is less at one end than the other, but in order to maintain a constant transport of the fluid along the tube the same amount must go in one end and out the other in a given time period. In order to get the same

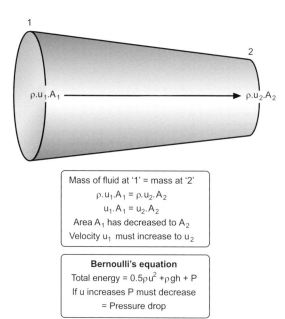

Mass of fluid at '1' = mass at '2'

$\rho.u_1.A_1 = \rho.u_2.A_2$

$u_1.A_1 = u_2.A_2$

Area A_1 has decreased to A_2

Velocity u_1 must increase to u_2

Bernoulli's equation

Total energy = $0.5\rho u^2 + \rho gh + P$

If u increases P must decrease

= Pressure drop

Fig. 4.3 Flow of a fluid through a tapered tube results in an increase in velocity at the narrow end where a pressure drop results.

amount of fluid through a smaller gap it must move at a greater velocity through the narrow end. This effect is familiar to anyone who has squeezed and constricted the end of a garden hose: the water comes out as a faster jet when the end of the hose is partly closed off.

The next thing to consider is the conservation of mass and energy along the length of the tube. The variables involved can be presented in the form of the **Bernoulli equation**:

$$\text{total energy} = \rho gh + \rho v^2/2 + p$$

where ρ is the density of the fluid, v the velocity, g the acceleration due to gravity, h is the height difference and p the pressure. The three terms in this equation are potential energy (ρgh), kinetic energy ($\rho v^2/2$) and pressure energy (p). This equation assumes no loss of energy due to frictional effects, so in reality the relationship is

$$\rho gh + \rho v^2/2 + p + E_{\text{loss}} = \text{constant}$$

The potential energy (ρgh) is constant because the difference in level between where the fluid is starting

from and where it is ending up are the same. Kinetic energy ($\rho v^2/2$) is changed as the velocity of the flow is increased or decreased. If the total energy in the system is to be conserved, there must be some change in the final term, the pressure energy (p). Pressure energy can be thought of as the energy that is stored when a fluid is compressed: a compressed fluid (such as a canister of a compressed gas) has a higher energy than an uncompressed one. Returning to the flow in the tapered tube, in order to balance the Bernoulli equation, the pressure energy (p) must be reduced to compensate for an increase in kinetic energy ($\rho v^2/2$) caused by the constriction of the flow at the end of the tube. This means that there is a reduction in pressure at the narrower end of the tube.

If these principles are now transferred to a flow along a channel (Fig. 4.4) a clast in the bottom of the channel will reduce the cross-section of the flow over it. The velocity over the clast will be greater than upstream and downstream of it and in order to balance the Bernoulli equation there must be a reduction in pressure over the clast. This reduction in pressure provides a temporary **lift force** that moves the clast off the bottom of the flow. The clast is then temporarily entrained in the moving fluid before falling under gravity back down onto the channel base in a single saltation event.

4.2.4 Grain size and flow velocity

The fluid velocity at which a particle becomes entrained in the flow can be referred to as the **critical velocity**. If the forces acting on a particle in a flow are considered then a simple relationship between the critical velocity and the mass of the particle would be expected. The drag force required to move a particle along in a flow will increase with mass, as will the lift force required to bring it up into the flow. A simple linear relationship between the flow velocity and the drag and lift forces can be applied to sand and gravel, but when fine grain sizes are involved things are more complicated.

The **Hjülstrom diagram** (Fig. 4.5) shows the relationship between water flow velocity and grain size and although this diagram has largely been superseded by the Shields diagram (Miller et al. 1977) it nevertheless demonstrates some important features of sediment movement in currents. The lower line on the graph displays the relationship between flow

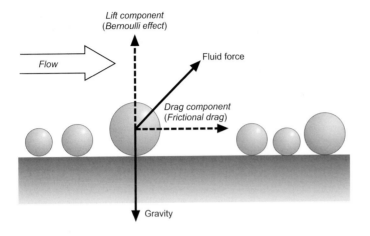

Fig. 4.4 The lift force resulting from the Bernoulli effect causes grains to be moved up from the base of the flow.

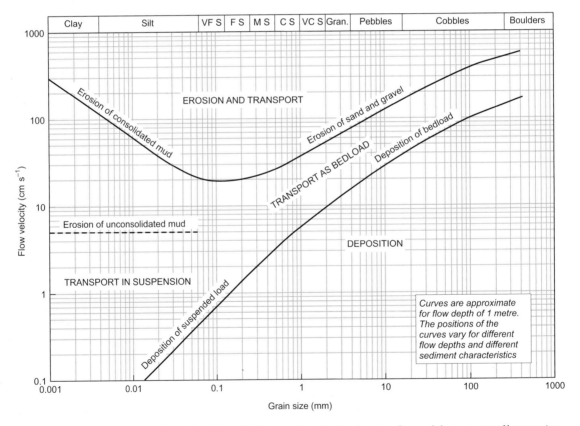

Fig. 4.5 The Hjülstrom diagram shows the relationship between the velocity of a water flow and the transport of loose grains. Once a grain has settled it requires more energy to start it moving than a grain that is already in motion. The cohesive properties of clay particles mean that fine-grained sediments require relatively high velocities to re-erode them once they are deposited, especially once they are compacted. (From Press & Siever 1986.)

velocity and particles that are already in motion. This shows that a pebble will come to rest at around 20 to 30 cm s^{-1}, a medium sand grain at 2 to 3 cm s^{-1}, and a clay particle when the flow velocity is effectively zero. The grain size of the particles in a flow therefore can be used as an indicator of the velocity at the time of deposition of the sediment if deposited as isolated particles. The upper, curved line shows the flow velocity required to move a particle from rest. On the right half of the graph this line parallels the first but at any given grain size the velocity required to initiate motion is higher than that to keep a particle moving. On the left side of the diagram, there is a sharp divergence of the lines: counter-intuitively, the smaller particles require a higher velocity to move them below coarse silt size. This is due to the properties of clay minerals that will dominate the fine fraction in a sediment. Clay minerals are cohesive (2.4.5) and once they are deposited they tend to stick together making it difficult to entrain them in a flow. Note that there are two lines for cohesive material. 'Unconsolidated' mud has settled but remains a sticky, plastic material. 'Consolidated' mud has had much more water expelled from it and is rigid.

The behaviour of fine particles in a flow as indicated by the Hjülstrom diagram has important consequences for deposition in natural depositional environments. Were it not for this behaviour, clay would be eroded in all conditions except standing water, but mud can accumulate in any setting where the flow stops for long enough for the clay particles to be deposited: resumption of flow does not re-entrain the deposited clay unless the velocity is relatively high. Alternations of mud and sand deposition are seen in environments where flow is intermittent, such as tidal settings (11.2).

4.2.5 Clast-size variations: graded bedding

The grain size in a bed is usually variable (2.5) and may show a pattern of an overall decrease in grain size from base to top, known as **normal grading**, or a pattern of increase in average size from base to top, called **reverse grading** (Fig. 4.6). Normal grading is the more commonly observed pattern and can result from the settling of particles out of suspension or as a consequence of a decrease in flow strength through time.

The settling velocity of particles in a fluid is determined by the size of the particle, the difference in the density between the particle and the fluid, and the fluid viscosity. The relationship, known as **Stokes Law**, can be expressed in an equation:

$$V = g \cdot D^2 \cdot (\rho_s - \rho_f)/18\mu$$

where V is the terminal settling velocity, D is the grain diameter, $(\rho_s - \rho_f)$ is the difference between the density of the particle (ρ_s) and the density of the fluid (ρ_f) and μ is the fluid viscosity; g is the acceleration due to gravity. One of the implications of this for sedimentary processes is that larger diameter clasts reach higher velocities and therefore grading of particles results from sediment falling out of suspension in standing water. Stokes Law only accurately predicts the settling velocity of small grains (fine sand or less) because turbulence created by the drag of larger grains falling through the fluid reduces the velocity. The shape of the particle is also a factor because the drag effect is greater for plate-like clasts and they therefore fall more slowly. It is for this reason that mica grains are commonly found concentrated at the tops of bed because they settle more slowly than quartz and other grains of equivalent mass.

A flow decreasing in velocity from 20 cm s^{-1} to 1 cm s^{-1} will initially deposit coarse sand but will progressively deposit medium and fine sand as the velocity drops. The sand bed formed from this decelerating flow will be normally graded, showing a reduction in grain size from coarse at the bottom to fine at the top. Conversely, an increase in flow velocity through time may result in an increase in grain size up through a bed, reverse grading, but flows that gradually increase in strength through time to produce reverse grading are less frequent. Grading can occur in a wide variety of depositional settings: normal grading is an important characteristic of many turbidity current deposits (4.5.2), but may also result from storms on continental shelves (14.2.1), overbank flooding in fluvial environments (9.3) and in delta-top settings (12.3.1).

It is useful to draw a distinction between **grading** that is a trend in grain size within a single bed and trends in grain size that occur through a number of beds. A pattern of several beds that start with a coarse clast size in the lowest bed and finer material in the

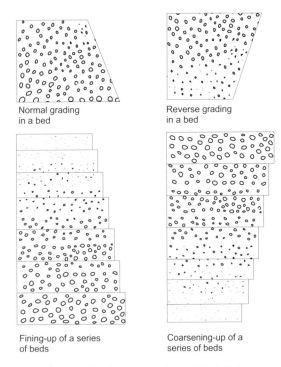

Normal grading
in a bed

Reverse grading
in a bed

Fining-up of a series
of beds

Coarsening-up of a
series of beds

Fig. 4.6 Normal and reverse grading within individual beds and fining-up and coarsening-up patterns in a
series of beds.

highest is considered to be **fining-upward**. The
reverse pattern with the coarsest bed at the top is a
coarsening-upward succession (Fig. 4.6). Note that
there can be circumstances where individual beds are
normally graded but are in a coarsening-up succession of beds.

4.2.6 Fluid density and particle size

A second important implication of Stokes Law is that
the forces acting on a grain are a function of the
viscosity and density of the fluid medium as well as
the mass of the particle. A clast falling through air will
travel faster than if it was falling through water
because the density contrast between particle and
fluid is greater and the fluid viscosity is lower.
Furthermore, higher viscosity fluids exert greater
drag and lift forces for a given flow velocity. Water
flows are able to transport clasts as large as boulders

at the velocities recorded in rivers, but even at the
very high wind strengths of storms the largest rock
and mineral particles carried are likely to be around a
millimetre. This limitation to the particle size carried
by air is one of the criteria that may be used to
distinguish material deposited by water from that
transported and deposited by wind. Higher viscosity
fluids such as ice and debris flows (dense slurries of
sediment and water) can transport boulders metres or
tens of metres across.

4.3 FLOWS, SEDIMENT AND BEDFORMS

A **bedform** is a morphological feature formed by the
interaction between a flow and cohesionless sediment
on a bed. Ripples in sand in a flowing stream and sand
dunes in deserts are both examples of bedforms, the
former resulting from flow in water, the latter by air-
flow. The patterns of ripples and dunes are products of
the action of the flow and the formation of bedforms
creates distinctive layering and structures within the
sediment that can be preserved in strata. Recognition
of sedimentary structures generated by bedforms provides information about the strength of the current,
the flow depth and the direction of sediment transport.

To explain how bedforms are generated some
further consideration of fluid dynamics is required (a
comprehensive account can be found in Leeder
1999). A fluid flowing over a surface can be divided
into a **free stream**, which is the portion of the flow
unaffected by boundary effects, a **boundary layer**,
within which the velocity starts to decrease due to
friction with the bed, and a **viscous sublayer**, a
region of reduced turbulence that is typically less
than a millimetre thick (Fig. 4.7). The thickness of
the viscous sublayer decreases with increasing flow
velocity but is independent of the flow depth. The
relationship between the thickness of the viscous sublayer and the size of grains on the bed of flow defines
an important property of the flow. If all the particles
are contained within the viscous sublayer the surface
is considered to be **hydraulically smooth**, and if
there are particles that project up through this layer
then the flow surface is **hydraulically rough**. As
will be seen in the following sections, processes within
the viscous sublayer and the effects of rough and
smooth surfaces are fundamental to the formation of
different bedforms.

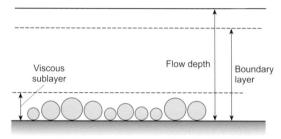

Smooth boundary:
Thick viscous sublayer (low velocities)
and/or small grain diameters

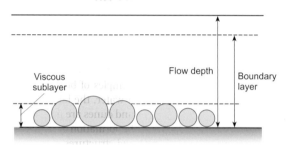

Rough boundary:
Thin viscous sublayer (high velocities)
and/or large grain diameters

Fig. 4.7 Layers within a flow and flow surface roughness: the viscous sublayer, the boundary layer within the flow and the flow depth.

The following sections are concerned mainly with the formation of bedforms in flowing water in rivers and seas, but many of the fluid dynamic principles also apply to aeolian (wind-blown) deposits: these are considered in more detail in Chapter 8.

4.3.1 Current ripples

Flow within the viscous sublayer is subject to irregularities known as **turbulent sweeps**, which move grains by rolling or saltation and create local clusters of grains. These clusters are only a few grains high but once they have formed they create steps or defects that influence the flow close to the bed surface. Flow can be visualised in terms of **streamlines** in the fluid, imaginary lines that indicate the direction of flow (Fig. 4.8). Streamlines lie parallel to a flat bed or the sides of a cylindrical pipe, but where there is an

irregularity such as a step in the bed caused by an accumulation of grains, the streamlines converge and there is an increased transport rate. At the top of the step, a streamline separates from the bed surface and a region of **boundary layer separation** forms between the **flow separation point** and the **flow attachment point** downstream (Fig. 4.8). Beneath this streamline lies a region called the **separation bubble** or **separation zone**. Expansion of flow over the step results in an increase in pressure (the Bernoulli effect, *4.2.3*) and the sediment transport rate is reduced, resulting in deposition on the lee side of the step.

Current ripples (Figs 4.9 & 4.10) are small bedforms formed by the effects of boundary layer separation on a bed of sand (Baas 1999). The small cluster of grains grows to form the **crest** of a ripple and separation occurs near this point. Sand grains roll or saltate up to the crest on the upstream **stoss side** of the ripple. Avalanching of grains occurs down the downstream or **lee side** of the ripple as accumulated grains become unstable at the crest. Grains that avalanche on the lee slope tend to come to rest at an angle close to the maximum critical slope angle for sand at around 30°. At the flow attachment point there are increased stresses on the bed, which result in erosion and the formation of a small scour, the **trough** of the ripple.

Current ripples and cross-lamination

A ripple migrates downstream as sand is added to the crest and accretes on the lee slope. This moves the crest and hence the separation point downstream, which in turn moves the attachment point and trough downstream as well. Scour in the trough and on the base of the stoss side supplies the sand, which moves up the gentle slope of the stoss side of the next ripple and so a whole train of ripple troughs and crests advance downstream. The sand that avalanches on the lee slope during this migration forms a series of layers at the angle of the slope. These thin, inclined layers of sand are called **cross-laminae**, which build up to form the sedimentary structure referred to as **cross-lamination** (Fig. 4.9).

When viewed from above current ripples show a variety of forms (Fig. 4.11). They may have relatively continuous straight to sinuous crests (**straight ripples** or **sinuous ripples**) or form a pattern of unconnected arcuate forms called **linguoid ripples**. The

1. Erosion in the trough of a bedform

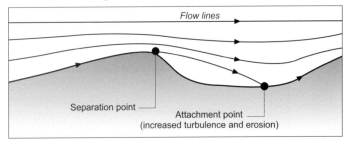

2. Development of counter-currents in lee of bedform

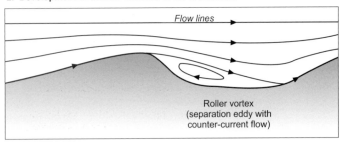

Fig. 4.8 Flow over a bedform: imaginary streamlines within the flow illustrate the separation of the flow at the brink of the bedform and the attachment point where the streamline meets the bed surface, where there is increased turbulence and erosion. A separation eddy may form in the lee of the bedform and produce a minor counter-current (reverse) flow.

Fig. 4.9 Current ripple cross-lamination in fine sandstone: the ripples migrated from right to left. The coin is 20 mm in diameter.

that all dipped in the same direction and lay in the same plane: this is ***planar cross-lamination***. Sinuous and linguoid ripples have lee slope surfaces that are curved, generating laminae that dip at an angle to the flow as well as downstream. As linguoid ripples migrate, curved cross-laminae are formed mainly in the trough-shaped low areas between adjacent ripple forms resulting in a pattern of ***trough cross-lamination*** (Fig. 4.9).

Creating and preserving cross-lamination

Current ripples migrate by the removal of sand from the stoss (upstream) side of the ripple and deposition on the lee side (downstream). If there is a fixed amount of sand available the ripple will migrate over the surface as a simple ripple form, with erosion in the troughs matching addition to the crests. These ***starved ripple*** forms are preserved if blanketed by mud. If the current is adding more sand particles than it is carrying away, the amount of sand deposited on the lee slope will be greater than that removed from the stoss side. There will be a net addition of sand to the ripple and it will grow as it migrates, but most importantly, the depth of scour in the trough is

relationship between the two forms appears to be related to both the duration of the flow and its velocity, with straight ripples tending to evolve into linguoid forms through time and at higher velocities (Baas 1994). Straight and linguoid ripple crests create different patterns of cross-lamination in three dimensions. A perfectly straight ripple would generate cross-laminae

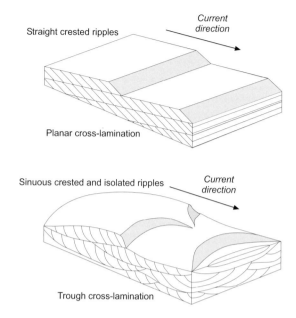

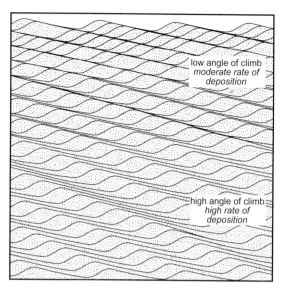

Fig. 4.12 Climbing ripples: in the lower part of the figure, more of the stoss side of the ripple is preserved, resulting in a steeper 'angle of climb'.

Fig. 4.10 Migrating straight crested ripples form planar cross-lamination. Sinuous or isolated (linguoid or lunate) ripples produce trough cross-lamination. (From Tucker 1991.)

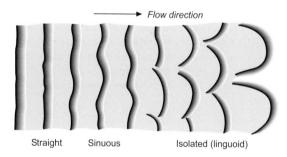

Fig. 4.11 In plan view current ripples may have straight, sinuous or isolated crests.

reduced leaving cross-laminae created by earlier migrating ripples preserved. In this way a layer of cross-laminated sand is generated.

When the rate of addition of sand is high there will be no net removal of sand from the stoss side and each ripple will migrate up the stoss side of the ripple form in front. These are ***climbing ripples*** (Allen 1972) (Fig. 4.12). When the addition of sediment from the current exceeds the forward movement of the ripple, deposition will occur on the stoss side as well as on the lee side. Climbing ripples are therefore indicators of rapid sedimentation as their formation depends upon the addition of sand to the flow at a rate equal to or greater than the rate of downstream migration of the ripples.

Constraints on current ripple formation

The formation of current ripples requires moderate flow velocities over a hydrodynamically smooth bed (see above). They only form in sands in which the dominant grain size is less than 0.6 mm (coarse sand grade) because bed roughness created by coarser sand creates turbulent mixing, which inhibits the small-scale flow separation required for ripple formation. Because ripple formation is controlled by processes within the viscous sublayer their formation is independent of water depth and current ripples may form in waters ranging from a few centimetres to kilometres deep. This is in contrast to most other subaqueous bedforms (subaqueous dunes, wave ripples), which are water-depth dependent.

Current ripples can be up to 40 mm high and the wavelengths (crest to crest or trough to trough distances) range up to 500 mm (Leeder 1999). The ratio of the wavelength to the height is typically between 10 and 40. There is some evidence of a relationship between the ripple wavelength and the grain size, approximately 1000 to 1 (Leeder 1999). It is important to note the upper limit to the dimensions of current ripples and to emphasise that ripples do not 'grow' into larger bedforms.

4.3.2 Dunes

Beds of sand in rivers, estuaries, beaches and marine environments also have bedforms that are distinctly larger than ripples. These large bedforms are called **dunes** (Fig. 4.13): the term 'megaripples' is also sometimes used, although this term fails to emphasise the fundamental hydrodynamic distinctions between ripple and dune bedforms. Evidence that these larger bedforms are not simply large ripples comes from measurement of the heights and wavelengths of all bedforms (Fig. 4.14). The data fall into clusters which do not overlap, indicating that they form by distinct processes which are not part of a continuum. The formation of dunes can be related to large-scale turbulence within the whole flow; once again flow separation is important, occurring at the dune crest, and scouring occurs at the reattachment point in the trough. The water depth controls the scale of the turbulent eddies in the flow and this in turn controls the height and wavelength of the dunes: there is a considerable amount of scatter in the data, but generally dunes are tens of centimetres high in water depths of a few metres, but are typically metres high in the water depths measured in tens of metres (Allen 1982; Leeder 1999).

Dunes and cross-bedding

The morphology of a subaqueous dune is similar to a ripple: there is a stoss side leading up to a crest and sand avalanches down the lee slope towards a trough (Figs 4.15 & 4.16). Migration of a subaqueous dune results in the construction of a succession of sloping layers formed by the avalanching on the lee slope and these are referred to as **cross-beds**. Flow separation creates a zone in front of the lee slope in which a **roller vortex** with reverse flow can form (Fig. 4.17). At low flow velocities these roller vortices are weakly developed and they do not rework the sand on the lee slope. The cross-beds formed simply lie at the angle of rest of the sand and as they build out into the trough the basal contact is angular (Fig. 4.17). Bedforms that develop at these velocities usually have low sinuosity crests, so the three-dimensional form of the structure is similar to planar cross-lamination. This is **planar cross-bedding** and the surface at the bottom of the cross-beds is flat and close to horizontal because of the absence of scouring in the trough. Cross-beds bound by horizontal surfaces are sometimes referred to as **tabular cross-bedding** (Fig. 4.18). Cross-beds may form a sharp angle at the base of the avalanche slope or may be asymptotic (tangential) to the horizontal (Fig. 4.17). At high flow velocities the roller

Fig. 4.13 Dune bedforms in an estuary: the most recent flow was from left to right and the upstream side of the dunes is covered with current ripples.

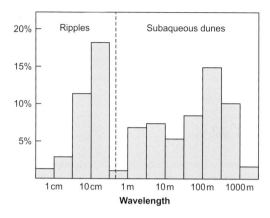

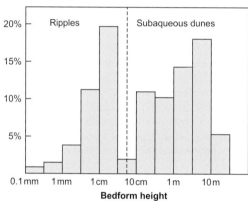

Fig. 4.14 Graphs of subaqueous ripple and subaqueous dune bedform wavelengths and heights showing the absence of overlap between ripple and dune-scale bedforms. (From Collinson et al. 2006.)

vortex is well developed creating a counter-current at the base of the slip face that may be strong enough to generate ripples (**counter-flow ripples**), which migrate a short distance up the toe of the lee slope (Fig. 4.17).

A further effect of the stronger flow is the creation of a marked scour pit at the reattachment point. The avalanche lee slope advances into this scoured trough so the bases of the cross-beds are marked by an undulating erosion surface. The crest of a subaqueous dune formed under these conditions will be highly sinuous or will have broken up into a series of linguoid dune forms. **Trough cross-bedding** (Fig. 4.15) formed by the migration of sinuous subaqueous dunes typically has asymptotic bottom contacts and an undulating lower boundary.

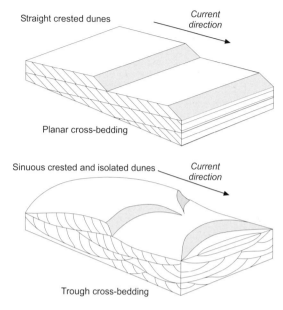

Fig. 4.15 Migrating straight crested dune bedforms form planar cross-bedding. Sinuous or isolated (linguoid or lunate) dune bedforms produce trough cross-bedding. (From Tucker 1991.)

Fig. 4.16 Subaqueous dune bedforms in a braided river.

Constraints on the formation of dunes

Dunes range in size from having wavelengths of about 600 mm and heights of a few tens of millimetres to wavelengths of hundreds of metres and heights of over ten metres. The smallest are larger than the biggest ripples. Dunes can form in a range of grain sizes from fine gravels to fine sands, but they are less well developed in finer deposits and do not occur in

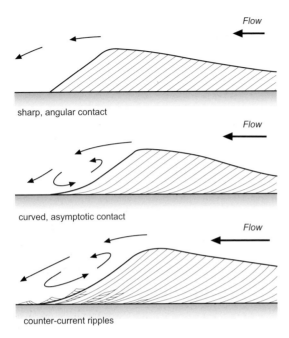

Fig. 4.17 The patterns of cross-beds are determined by the shape of the bedforms resulting from different flow conditions.

Fig. 4.18 Planar tabular cross-stratification with tangential bases to the cross-beds (the scale bar is in inches and is 100 mm long).

very fine sands or silts. This grain size limitation is thought to be related to the increased suspended load in the flow if the finer grain sizes are dominant: the suspended load suppresses turbulence in the flow and flow separation does not occur (Leeder 1999). The formation of dunes also requires flow to be sustained for long enough for the structure to build up, and to

form cross-bedding the dune must migrate. Dune-scale cross-bedding therefore cannot be generated by short-lived flow events. Dunes are most commonly encountered in river channels, deltas, estuaries and shallow marine environments where there are relatively strong, sustained flows.

4.3.3 Bar forms

Bars are bedforms occurring within channels that are of a larger scale than dunes: they have width and height dimensions of the same order of magnitude as the channel within which they are formed (Bridge 2003). Bars can be made up of sandy sediment, gravelly material or mixtures of coarse grain sizes. In a sandy channel the surfaces of bar forms are covered with subaqueous dune bedforms, which migrate over the bar surface and result in the formation of units of cross-bedded sands. A bar form deposit is therefore typically a cross-bedded sandstone as a lens-shaped body. The downstream edge of a bar can be steep and develop its own slip-face, resulting in large-scale cross-stratification in both sandstones and conglomerates. Bars in channels are classified in terms of their position within the channel (side and alternate bars at the margins, mid-channel bars in the centre and point bars on bends: Collinson et al. 2006) and their shape (9.2).

4.3.4 Plane bedding and planar lamination

Horizontal layering in sands deposited from a flow is referred to as **plane bedding** in sediments and produces a sedimentary structure called **planar lamination** in sedimentary rocks. As noted above, current ripples only form if the grains are smaller than the thickness of the viscous sublayer: if the bed is rough, the small-scale flow separation required for ripple formation does not occur and the grains simply roll and saltate along the surface. Plane beds form in coarser sands at relatively low flow velocities (close to the threshold for movement – 4.2.4), but as the flow speed increases dune bedforms start to be generated. The horizontal planar lamination produced under these circumstances tends to be rather poorly defined.

Plane bedding is also observed at higher flow velocities in very fine- to coarse-grained sands: ripple and dune bedforms become washed out with an increase

Fig. 4.19 Horizontal lamination in sandstone beds.

in flow speed as the formation of flow separation is suppressed at higher velocities. These **plane beds** produce well-defined planar lamination with laminae that are typically 5–20 grains thick (Bridge 1978) (Fig. 4.19). The bed surface is also marked by elongate ridges a few grain diameters high separated by furrows oriented parallel to the flow direction. This feature is referred to as **primary current lineation** (often abbreviated to '**pcl**') and it is formed by sweeps within the viscous sublayer (Fig. 4.7) that push grains aside to form ridges a few grains high which lie parallel to the flow direction. The formation of sweeps is subdued when the bed surface is rough and primary current lineation is therefore less well defined in coarser sands. Primary current lineation is seen on the surfaces of planar beds as parallel lines of main grains which form very slight ridges, and may often be rather indistinct.

4.3.5 Supercritical flow

Flow may be considered to be **subcritical**, often with a smooth water surface, or **supercritical**, with an uneven surface of wave crests and troughs. These flow states relate to a parameter, the **Froude number** (*Fr*), which is a relationship between the flow velocity (*v*) and the flow depth (*h*), with '*g*' the acceleration due to gravity:

$$Fr = v/\sqrt{g \cdot h}$$

The Froude number can be considered to be a ratio of the flow velocity to the velocity of a wave in the flow (Leeder 1999). When the value is less than one, the flow is subcritical and a wave can propagate upstream because it is travelling faster than the flow. If the Froude number is greater than one this indicates that the flow is too fast for a wave to propagate upstream and the flow is supercritical. In natural flows a sudden change in the height of the surface of the flow, a **hydraulic jump**, is seen at the transition from thin, supercritical flow to thicker, subcritical flow.

Where the Froude number of a flow is close to one, standing waves may temporarily form on the surface of the water before steepening and breaking in an upstream direction. Sand on the bed develops a bedform surface parallel to the standing wave, and as the flow steepens sediment accumulates on the upstream side of the bedform. These bedforms are called **antidunes**, and, if preserved, **antidune cross-bedding** would be stratification dipping upstream. However, such preservation is rarely seen because as the wave breaks, the antidune bedform is often reworked, and as the flow velocity subsequently drops the sediment is reworked into upper stage plane beds by subcritical flow. Well-documented occurrences of antidune cross-stratification are known from pyroclastic surge deposits (*17.2.3*), where high velocity flow is accompanied by very high rates of sedimentation (Schminke et al. 1973).

4.3.6 Bedform stability diagram

The relationship between the grain size of the sediment and the flow velocity is summarised on Fig. 4.20. This **bedform stability diagram** indicates the bedform that will occur for a given grain size and velocity and has been constructed from experimental data (modified from Southard 1991, and Allen 1997). It should be noted that the upper boundary of the ripple field is sharp, but the other boundaries between the fields are gradational and there is an overlap where either of two bedforms may be stable. Note also that the scales are logarithmic on both axes. Two general flow regimes are recognised: a **lower flow regime** in which ripples, dunes and lower plane beds are stable and an **upper flow regime** where plane beds and antidunes form. Flow in the lower flow regime is always subcritical and the change to supercritical flow lies within the antidune field.

The fields in the bedform stability diagram in Fig. 4.20 are for a certain water depth (25 to 40 cm)

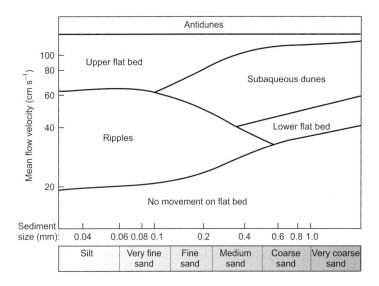

Fig. 4.20 A bedform stability diagram which shows how the type of bedform that is stable varies with both the grain size of the sediment and the velocity of the flow.

and for clear water at a particular temperature (10°C), and the boundaries will change if the flow depth is varied, or if the density of the water is varied by changing the temperature, salinity or by addition of suspended load. Bedform stability diagrams can be used in conjunction with sedimentary structures in sandstone beds to provide an estimate of the velocity, or recognise changes in the velocity, of the flow that deposited the sand. For example, a bed of medium sand that was plane-bedded at the base, cross-bedded in the middle and ripple cross-laminated at the top could be interpreted in terms of a decrease in flow velocity during the deposition of the bed.

4.4 WAVES

A **wave** is a disturbance travelling through a gas, liquid or solid which involves the transfer of energy between particles. In their simplest form, waves do not involve transport of mass, and a wave form involves an **oscillatory motion** of the surface of the water without any net horizontal water movement. The waveform moves across the water surface in the manner seen when a pebble is dropped into still water. When a wave enters very shallow water the amplitude increases and then the wave breaks creating the horizontal movement of waves seen on the beaches of lakes and seas.

A single wave can be generated in a water body such as a lake or ocean as a result of an input of energy by an earthquake, landslide or similar phenomenon. Tsunamis are waves produced by single events, and these are considered further in section 11.3.2. Continuous trains of waves are formed by wind acting on the surface of a water body, which may range in size from a pond to an ocean. The height and energy of waves is determined by the strength of the wind and the **fetch**, the expanse of water across which the wave-generating wind blows. Waves generated in open oceans can travel well beyond the areas they were generated.

4.4.1 Formation of wave ripples

The oscillatory motion of the top surface of a water body produced by waves generates a circular pathway for water molecules in the top layer (Fig. 4.21). This motion sets up a series of circular cells in the water below. With increasing depth internal friction reduces the motion and the effect of the surface waves dies out. The depth to which surface waves affect a water body is referred to as the **wave base** (11.3). In shallow water, the base of the water body interacts with the waves. Friction causes the circular motion at the surface to become transformed into an elliptical pathway, which is flattened at the base into a horizontal oscillation. This horizontal oscillation may generate wave ripples in sediment. If the water motion is purely oscillatory the ripples formed are symmetrical, but a superimposed current can result in asymmetric wave ripples.

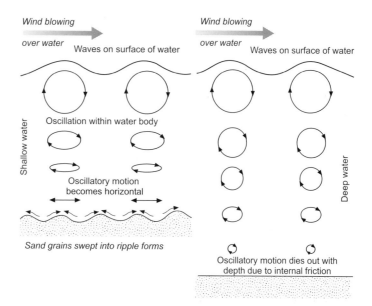

Fig. 4.21 The formation of wave ripples in sediment is produced by oscillatory motion in the water column due to wave ripples on the surface of the water. Note that there is no overall lateral movement of the water, or of the sediment. In deep water the internal friction reduces the oscillation and wave ripples do not form in the sediment.

At low energies **rolling grain ripples** form (Fig. 4.22). The peak velocity of grain motion is at the mid-point of each oscillation, reducing to zero at the edges. This sweeps grains away from the middle, where a trough forms, to the edges where ripple crests build up. Rolling grain ripples are characterised by broad troughs and sharp crests. At higher energies grains can be kept temporarily in suspension during each oscillation. Small clouds of grains are swept from the troughs onto the crests where they fall out of suspension. These **vortex ripples** (Fig. 4.22) have more rounded crests but are otherwise symmetrical.

4.4.2 Characteristics of wave ripples

In plan view wave ripples have long, straight to gently sinuous crests which may bifurcate (split) (Fig. 4.23); these characteristics may be seen on the bedding planes of sedimentary rocks. In cross-section wave ripples are generally symmetrical in profile, laminae within each ripple dip in both directions and are overlapping (Fig. 4.24). These characteristics may be preserved in cross-lamination generated by the accumulation of sediment influenced by waves (Fig. 4.25). Wave ripples can form in any non-cohesive sediment and are principally seen in coarse silts and sand of all grades. If the wave energy is high enough wave ripples can form in granules and pebbles, forming gravel ripples with wavelengths of several metres and heights of tens of centimetres.

4.4.3 Distinguishing wave and current ripples

Distinguishing between wave and current ripples can be critical to the interpretation of palaeoenvironments. Wave ripples are formed only in relatively shallow water in the absence of strong currents, whereas current ripples may form as a result of water flow in any depth in any subaqueous environment. These distinctions allow deposits from a shallow lake (*10.7.2*) or lagoon (*13.3.2*) to be distinguished from offshore (*14.2.1*) or deep marine environments (*14.2.1*), for example. The two different ripple types can be distinguished in the field on the basis of their shapes and geometries. In plan view wave ripples have long, straight to sinuous crests which may bifurcate (divide) whereas current ripples are commonly very sinuous and broken up into short, curved crests. When viewed from the side wave ripples are symmetrical with cross-laminae dipping in both directions either side of the crests. In contrast, current ripples are asymmetrical with cross-laminae dipping only in one direction, the only exception

Rolling grain ripples: Low energy

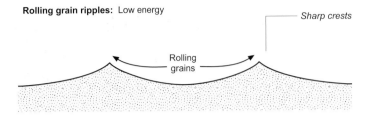

Vortex ripples: High energy

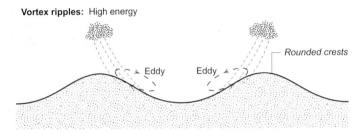

Fig. 4.22 Forms of wave ripple: rolling grain ripples produced when the oscillatory motion is capable only of moving the grains on the bed surface and vortex ripples are formed by higher energy waves relative to the grain size of the sediment.

Fig. 4.23 Wave ripples in sand seen in plan view: note the symmetrical form, straight crests and bifurcating crest lines.

Fig. 4.25 Wave ripple cross-lamination in sandstone (pen is 18 cm long).

being climbing ripples which have distinctly asymmetric dipping laminae.

In addition to the wave and current bedforms and sedimentary structures described in this chapter there are also features called 'hummocky and swaley cross-stratification'. These features are thought to be characteristic of storm activity on continental shelves and are considered separately in the chapter on this depositional setting (*14.2.1*).

wave ripple cross-lamination

laminae dip in both directions in the same layer

Fig. 4.24 Internal stratification in wave ripples showing cross-lamination in opposite directions within the same layer. The wavelength may vary from a few centimetres to tens of centimetres.

4.5 MASS FLOWS

Mixtures of detritus and fluid that move under gravity are known collectively as ***mass flows***, *gravity flows* or *density currents* (Middleton & Hampton 1973). A number of different mechanisms are involved and all require a slope to provide the potential energy to drive the flow. This slope may be the surface over which the flow occurs, but a gravity flow will also move on a horizontal surface if it thins downflow, in which case the potential energy is provided by the difference in height between the tops of the upstream and the downstream parts of the flow.

4.5.1 Debris flows

Debris flows are dense, viscous mixtures of sediment and water in which the volume and mass of sediment exceeds that of water (Major 2003). A dense, viscous mixture of this sort will typically have a low Reynolds number so the flow is likely to be laminar (*4.2.1*). In the absence of turbulence no dynamic sorting of material into different sizes occurs during flow and the resulting deposit is very poorly sorted. Some sorting may develop by slow settling and locally there may be reverse grading produced by shear at the bed boundary. Material of any size from clay to large boulders may be present.

Debris flows occur on land, principally in arid environments where water supply is sparse (such as some alluvial fans, *9.5*) and in submarine environments where they transport material down continental slopes (*16.1.2*) and locally on some coarse-grained delta slopes (*12.4.4*). Deposition occurs when internal friction becomes too great and the flow 'freezes' (Fig. 4.26). There may be little change in the thickness of the deposit in a proximal to distal direction and the clast size distribution may be the same throughout the deposit. The deposits of debris flows on land are typically matrix-supported conglomerates although clast-supported deposits also occur if the relative proportion of large clasts is high in the sediment mixture. They are poorly sorted and show a chaotic fabric, i.e. there is usually no preferred orientation to the clasts (Fig. 4.27), except within zones of shearing that may form at the base of the flow. When a debris flow travels through water it may partly mix with it and the top part of the flow may become dilute. The tops of subaqueous debris flows are therefore characterised by a gradation up into better sorted, graded sediment, which may have the characteristics of a turbidite (see below).

Fig. 4.26 A muddy debris flow in a desert wadi.

Fig. 4.27 A debris-flow deposit is characteristically poorly sorted, matrix-supported conglomerate.

4.5.2 Turbidity currents

Turbidity currents are gravity-driven turbid mixtures of sediment temporarily suspended in water. They are less dense mixtures than debris flows and with a relatively high Reynolds number are usually turbulent flows (*4.2.1*). The name is derived from their characteristics of being opaque mixtures of sediment and water (turbid) and not the turbulent flow. They flow down slopes or over a horizontal surface provided that the thickness of the flow is greater upflow than it is downflow. The deposit of a turbidity current is a ***turbidite***. The sediment mixture may contain gravel, sand and mud in concentrations as little as a few parts per thousand or up to 10% by weight: at the high concentrations the flows may not be turbulent and are not always referred to as turbidity currents. The volumes of material involved in a single flow event can be anything up to tens of cubic kilometres, which is spread out by the flow and deposited as a layer a few millimetres to tens of metres thick. Turbidity currents, and hence turbidites, can occur in water anywhere that there is a supply of sediment and a slope. They are common in deep lakes (*10.2.3*), and may occur on continental shelves (*14.1*), but are most abundant in deep marine environments, where turbidites are the dominant clastic deposit (*16.1.2*). The association with deep marine environments may lead to the assumption that all turbidites are deep marine deposits, but they are not an indicator of depth as turbidity currents are a process that can occur in shallow water as well.

Sediment that is initially in suspension in the turbidity current (Fig. 4.28) starts to come into contact with the underlying surface where it may come to a halt or move by rolling and suspension. In doing so it comes out of suspension and the density of the flow is reduced. Flow in a turbidity current is maintained by the density contrast between the sediment–water mix and the water, and if this contrast is reduced, the flow slows down. At the head of the flow (Fig. 4.28) tur-

bulent mixing of the current with the water dilutes the turbidity current and also reduces the density contrast. As more sediment is deposited from the decelerating flow a deposit accumulates and the flow eventually comes to a halt when the flow has spread out as a thin, even sheet.

Low- and medium-density turbidity currents

The first material to be deposited from a turbidity current will be the coarsest as this will fall out of suspension first. Therefore a turbidite is characteristically normally graded (*4.2.9*). Other sedimentary structures within the graded bed reflect the changing processes that occur during the flow and these vary according to the density of the initial mixture. Low- to medium-density turbidity currents will ideally form a succession known as a ***Bouma sequence*** (Fig. 4.29), named after the geologist who first described them (Bouma 1962). Five divisions are recognised within the Bouma sequence, referred to as 'a' to 'e' divisions and annotated T_a, T_b, and so on.

T_a This lowest part consists of poorly sorted, structureless sand: on the scoured base deposition occurs rapidly from suspension with reduced turbulence inhibiting the formation of bedforms.

T_b Laminated sand characterises this layer, the grain size is normally finer than in 'a' and the material is better sorted: the parallel laminae are generated by the separation of grains in upper flow regime transport (*4.3.4*).

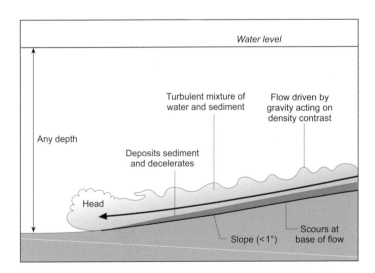

Fig. 4.28 A turbidity current is a turbulent mixture of sediment and water that deposits a graded bed – a turbidite.

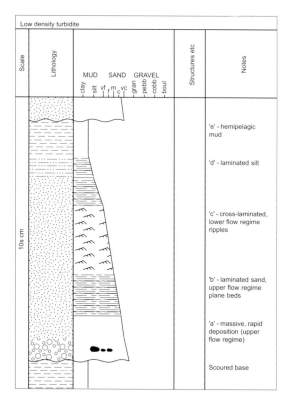

Fig. 4.29 The 'Bouma sequence' in a turbidite deposit.

T$_c$ Cross-laminated medium to fine sand, sometimes with climbing ripple lamination, form the middle division of the Bouma sequence: these characteristics indicate moderate flow velocities within the ripple bedform stability field (*4.3.6*) and high sedimentation rates. Convolute lamination (*18.1.2*) can also occur in this division.

T$_d$ Fine sand and silt in this layer are the products of waning flow in the turbidity current: horizontal laminae may occur but the lamination is commonly less well defined than in the 'b' layer.

T$_e$ The top part of the turbidite consists of fine-grained sediment of silt and clay grade: it is deposited from suspension after the turbidity current has come to rest and is therefore a hemipelagic deposit (*16.5.3*).

Turbidity currents are waning flows, that is, they decrease velocity through time as they deposit material, but this means that they also decrease velocity with distance from the source. There is therefore a decrease in the grain size deposited with distance (Stow 1994). The lower parts of the Bouma sequence are only present in the more proximal parts of the flow. With distance the lower divisions are progressively lost as the flow carries only finer sediment (Fig. 4.30) and only the 'c' to 'e' or perhaps just 'd' and 'e' parts of the Bouma sequence are deposited. In the more proximal regions the flow turbulence may be strong enough to cause scouring and completely remove the upper parts of a previously deposited bed. The 'd' and 'e' divisions may therefore be absent due

Fig. 4.30 Proximal to distal changes in the deposits formed by turbidity currents. The lower, coarser parts of the Bouma sequence are only deposited in the more proximal regions where the flow also has a greater tendency to scour into the underlying beds.

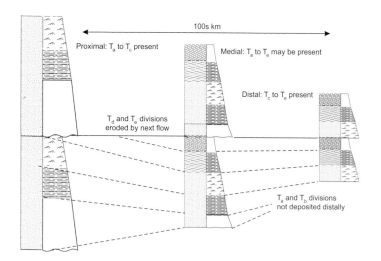

to this erosion and the eroded sediment may be incorporated into the overlying deposit as mud clasts. The complete T_a to T_e sequence is therefore only likely to occur in certain parts of the deposit, and even there intermediate divisions may be absent due, for example, to rapid deposition preventing ripple formation in T_c. Complete T_{a-e} Bouma sequences are in fact rather rare.

High-density turbidity currents

Under conditions where there is a higher density of material in the mixture the processes in the flow and hence of the characteristics of the deposit are different from those described above. High-density turbidity currents have a bulk density of at least $1.1\,\text{g cm}^{-3}$ (Pickering et al. 1989). The turbidites deposited by these flows have a thicker coarse unit at their base, which can be divided into three divisions (Fig. 4.31). Divisions S_1 and S_2 are traction deposits of coarse material, with the upper part, S_2, representing the 'freezing' of the traction flow. Overlying this is a unit, S_3, that is characterised by fluid-escape structures indicating rapid deposition of sediment. The upper part of the succession is more similar to the Bouma Sequence, with T_t equivalent to T_b and T_c and overlain by T_d and T_e: this upper part therefore reflects deposition from a lower density flow once most of the sediment had already been deposited in the 'S' division. The characteristics of high-density turbidites were described by Lowe (1982), after whom the succession is sometimes named.

4.5.3 Grain flows

Avalanches are mechanisms of mass transport down a steep slope, which are also known as *grain flows*. Particles in a grain flow are kept apart in the fluid medium by repeated grain to grain collisions and grain flows rapidly 'freeze' as soon as the kinetic energy of the particles falls below a critical value. This mechanism is most effective in well-sorted material falling under gravity down a steep slope such as the slip face of an aeolian dune. When the particles in the flow are in temporary suspension there is a tendency for the finer grains to fall between the coarser ones, a process known as *kinetic sieving*, which results in a slight reverse grading in the layer once it is deposited. Although most common on a small scale in sands, grain flows may also occur in coarser, grav-

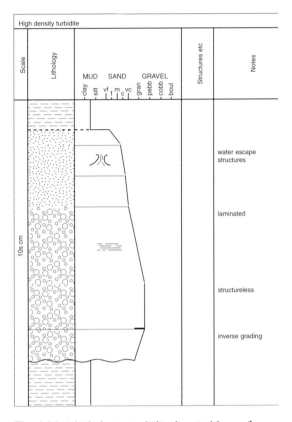

Fig. 4.31 A high-density turbidite deposited from a flow with a high proportion of entrained sediment.

elly material in a steep subaqueous setting such as the foreset of a Gilbert-type delta (*12.4.4*).

4.6 MUDCRACKS

Clay-rich sediment is cohesive and the individual particles tend to stick to each other as the sediment dries out. As water is lost the volume reduces and clusters of clay minerals pull apart developing cracks in the surface. Under subaerial conditions a polygonal pattern of cracks develops when muddy sediment dries out completely: these are *desiccation cracks* (Fig. 4.32). The spacing of desiccation cracks depends upon the thickness of the layer of wet mud, with a broader spacing occurring in thicker deposits. In cross-section desiccation cracks taper downwards and the upper edges may roll up if all of the moisture in the mud is driven off. The edges of desiccation

Fig. 4.32 Mudcracks caused by subaerial desiccation of mud.

Fig. 4.33 Syneresis cracks in mudrock, believed to be formed by subaqueous shrinkage.

cracks are easily removed by later currents and may be preserved as **mud-chips** or **mud-flakes** in the overlying sediment. Desiccation cracks are most clearly preserved in sedimentary rocks when the cracks are filled with silt or sand washed in by water or blown in by the wind. The presence of desiccation cracks is a very reliable indicator of the exposure of the sediment to subaerial conditions.

Syneresis cracks are shrinkage cracks that form under water in clayey sediments (Tanner 2003). As the clay layer settles and compacts it shrinks to form single cracks in the surface of the mud. In contrast to desiccation cracks, syneresis cracks are not polygonal but are simple, straight or slightly curved tapering cracks (Fig. 4.33). These subaqueous shrinkage cracks have been formed experimentally and have been reported in sedimentary rocks, although some of these occurrences have been re-interpreted as desiccation cracks (Astin 1991). Neither desiccation cracks nor syneresis cracks form in silt or sand because these coarser materials are not cohesive.

4.7 EROSIONAL SEDIMENTARY STRUCTURES

A turbulent flow over the surface of sediment that has recently been deposited can result in the partial and localised removal of sediment. Scouring may form a **channel** which confines the flow, most commonly seen on land as rivers, but similar confined flows can occur in many other depositional settings, right down to the deep sea floor. One of the criteria for recognising the deposits of channelised flow within strata is the presence of an erosional scour surface

that marks the base of the channel. The size of channels can range from features less than a metre deep and only metres across to large-scale structures many tens of metres deep and kilometres to tens of kilometres in width. The size usually distinguishes channels from other scour features (see below), although the key criterion is that a channel confines the flow, whereas other scours do not.

Small-scale erosional features on a bed surface are referred to as **sole marks** (Fig. 4.34). They are preserved in the rock record when another layer of sediment is deposited on top leaving the feature on the bedding plane. Sole marks may be divided into those that form as a result of turbulence in the water causing erosion (**scour marks**) and impressions formed by objects carried in the water flow (**tool marks**) (Allen 1982). They may be found in a very wide range of depositional environments, but are particularly common in successions of turbidites where the sole mark is preserved as a cast at the base of the overlying turbidite.

Scour marks Turbulent eddies in a flow erode into the underlying bed and create a distinctive erosional scour called a *flute cast*. Flute casts are asymmetric in cross-section with one steep edge opposite a tapered edge. In plan view they are narrower at one end, widening out onto the tapered edge. The steep, narrow end of the flute marks the point where the eddy initially eroded into the bed and the tapered, wider edge marks the passage of the eddy as it is swept away by the current. The size can vary from a few centimetres to tens of centimetres across. As with many sole marks it is as common to find the cast of the feature formed by the infilling of the depression as

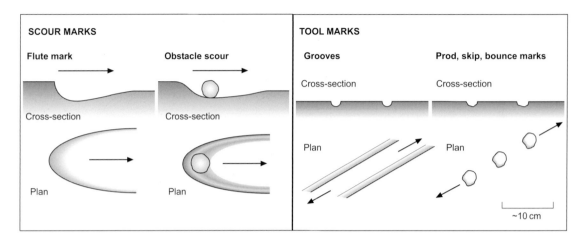

Fig. 4.34 Sole marks found on the bottoms of beds: flute marks and obstacle scours are formed by flow turbulence; groove and bounce marks are formed by objects transported at the base of the flow.

it is to find the depression itself (Fig. 4.34). The asymmetry of flute marks means that they can be used as palaeocurrent indicators where they are preserved as casts on the base of the bed (5.3.1). An obstacle on the bed surface such as a pebble or shell can produce eddies that scour into the bed (***obstacle scours***). Linear features on the bed surface caused by turbulence are elongate ***ridges and furrows*** if on the scale of millimetres or ***gutter casts*** if the troughs are a matter of centimetres wide and deep, extending for several metres along the bed surface.

Tool marks An object being carried in a flow over a bed can create marks on the bed surface. ***Grooves*** are sharply defined elongate marks created by an object (tool) being dragged along the bed. Grooves are sharply defined features in contrast to ***chevrons***, which form when the sediment is still very soft. An object saltating (4.2.2) in the flow may produce marks known variously as ***prod***, ***skip*** or ***bounce marks*** at the points where it lands. These marks are often seen in lines along the bedding plane. The shape and size of all tool marks is determined by the form of the object which created them, and irregular shaped fragments, such as fossils, may produce distinctive marks.

4.8 TERMINOLOGY FOR SEDIMENTARY STRUCTURES AND BEDS

When describing layers of sedimentary rock it is useful to indicate how thick the beds are, and this can be done by simply stating the measurements in millimetres, centimetres or metres. This, however, can be cumbersome sometimes, and it may be easier to describe the beds as 'thick' or 'thin'. In an attempt to standardize this terminology, there is a generally agreed set of 'definitions' for bed thickness (Fig. 4.35). A *bed* is a unit of sediment which is generally uniform in character and contains no distinctive breaks: it may be graded (4.2.5), or contain different sedimentary structures. The base may be erosional if there is scouring, for example at the base of a channel, sharp, or sometimes gradational. Alternations of thin layers of different lithologies are described as ***interbedded*** and are usually considered as a single unit, rather than as separate beds.

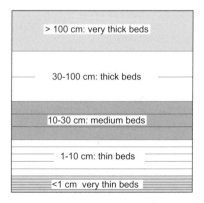

Fig. 4.35 Bed thickness terminology.

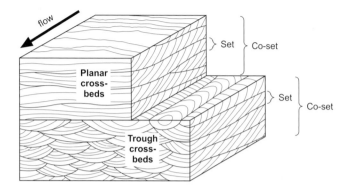

Fig. 4.36 Terminology used for sets and co-sets of cross-stratification.

In common with many other fields of geology, there is some variation in the use of the terminology to describe bedforms and sedimentary structures. The approach used here follows that of Collinson et al. (2006). **Cross-stratification** is any layering in a sediment or sedimentary rock that is oriented at an angle to the depositional horizontal. These inclined strata most commonly form in sand and gravel by the migration of bedforms and may be preserved if there is net accumulation. If the bedform is a ripple the resulting structure is referred to as ***cross-lamination***. Ripples are limited in crest height to about 30 mm so cross-laminated beds do not exceed this thickness. Migration of dune bedforms produces ***cross-bedding***, which may be tens of centimetres to tens of metres in thickness. Cross-stratification is the more general term and is used for inclined stratification generated by processes other than the migration of bedforms, for example the inclined surfaces formed on the inner bank of a river by point-bar migration (9.2.2). A single unit of cross-laminated, cross-bedded or cross-stratified sediment is referred to as a ***bed-set***. Where a bed contains more than one set of the same type of structure, the stack of sets is called a ***co-set*** (Fig. 4.36).

Mixtures of sand and mud occur in environments that experience variations in current or wave activity or sediment supply due to changing current strength or wave power. For example, tidal settings (11.2) display regular changes in energy in different parts of the tidal cycle, allowing sand to be transported and deposited at some stages and mud to be deposited from suspension at others. This may lead to simple alternations of layers of sand and mud but if ripples form in

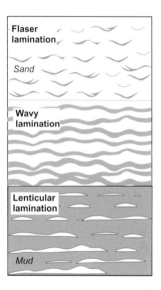

Fig. 4.37 Lenticular, wavy and flaser bedding in deposits that are mixtures of sand and mud.

the sands due to either current or wave activity then an array of sedimentary structures (Fig. 4.37) may result depending on the proportions of mud and sand. **Flaser bedding** is characterised by isolated thin drapes of mud amongst the cross-laminae of a sand. **Lenticular bedding** is composed of isolated ripples of sand completely surrounded by mud, and intermediate forms made up of approximately equal proportions of sand and mud are called ***wavy bedding*** (Reineck & Singh 1980).

4.9 SEDIMENTARY STRUCTURES AND SEDIMENTARY ENVIRONMENTS

Bernoulli's equation, Stokes Law, Reynolds and Froude numbers may seem far removed from sedimentary rocks exposed in a cliff but if we are to interpret those rocks in terms of the processes that formed them a little knowledge of fluid dynamics is useful. Understanding what sedimentary structures mean in terms of physical processes is one of the starting points for the analysis of sedimentary rocks in terms of environment of deposition. Most of the sedimentary structures described are familiar from terrigenous clastic rocks but it is important to remember that any particulate matter interacts with the fluid medium it is transported in and many of these features also occur commonly in calcareous sediments made up of bioclastic debris and in volcaniclastic rocks. The next chapter introduces the concepts used in palaeoenvironmental analysis and is followed by chapters that consider the processes and products of different environments in more detail.

FURTHER READING

Allen, J.R.L. (1982) *Sedimentary Structures: their Character and Physical Basis*, Vol. 1. *Developments in Sedimentology*. Elsevier, Amsterdam.

Allen, J.R.L. (1985) *Principles of Physical Sedimentology*. Unwin-Hyman, London.

Allen, P.A. (1997) *Earth Surface Processes*. Blackwell Science, Oxford, 404 pp.

Collinson, J., Mountney, N. & Thompson, D. (2006) *Sedimentary Structures*. Terra Publishing, London.

Leeder, M.R. (1999) *Sedimentology and Sedimentary Basins: from Turbulence to Tectonics*. Blackwell Science, Oxford.

Pye, K. (Ed.) (1994) *Sediment Transport and Depositional Processes*. Blackwell Scientific Publications, Oxford.

5

Field Sedimentology, Facies and Environments

The methodology of analysing sedimentary rocks, recording data and interpreting them in terms of processes and environments are considered in a general sense in this chapter. Geology is like any other science: the value of the interpretations that come from the results is determined by the quality of the data collected. The description of rocks in hand specimen and thin-section has been considered in previous chapters, but a very high proportion of the data used in sedimentological analysis comes from fieldwork, during which the characteristics of strata are analysed at a larger scale. The character of sediment in any depositional environment will be determined by the physical, chemical and biological processes that have occurred during the formation, transport and deposition of the sediment. In the subsequent chapters the range of depositional environments is considered in terms of the processes that occur in each and the character of the sediment deposited. By way of introduction to these chapters the concepts of depositional environments and sedimentary facies are considered here. The examples used relate to processes and products in environments considered in more detail in subsequent chapters.

5.1 FIELD SEDIMENTOLOGY

A large part of modern sedimentology is the interpretation of sediments and sedimentary rocks in terms of processes of transport and deposition and how they are distributed in space and time in sedimentary environments. To carry out this sort of sedimentological analysis some data are required, and this is mainly collected from exposures of rocks, although data from the subsurface are becoming increasingly important (2.2). A satisfactory analysis of sedimen-

tary environments and their stratigraphic context requires a sound basis of field data, so the first part of this chapter is concerned with practical aspects of sedimentological fieldwork, followed by an introduction to the methods used in interpretation.

5.1.1 Field equipment

Only a few tools are needed for field studies in sedimentology and stratigraphy. A notebook to record data is essential and a strong hard-backed book made with weather-resistant paper is strongly recommended. Also essential is a hand lens (10 × magnification), a compass–clinometer and a geological hammer. If a sedimentary log is going to be recorded (*5.2*), a measuring tape or metre stick is also essential and if proforma log sheets are to be used, a clipboard is needed. For the collection of samples, small, strong, plastic bags and a marker pen are necessary. A small bottle containing dilute hydrochloric acid is very useful to test for the presence of calcium carbonate in the field (*3.1.1*). It is good to have some form of grain-size comparator. Small cards with a printed visual chart of grain sizes can be bought, but some sedimentologists prefer to make a comparator by gluing sand of different grain sizes on to areas of a small piece of card or Perspex. The advantage of these comparators made with real grains is that they make it possible to compare by touch as well as visually.

The most 'hi-tech' items taken in the field are likely to be a camera and a GPS (Global Positioning Satellite) receiver. Photographs are very useful for providing a record of the features seen in the field, but only if a note is kept of where every photograph was taken, and it is also important that supplementary sketches are made. Global Positioning Satellite receivers have become standard equipment for field geologists, and can be a quick and effective way of determining locations. They are used alongside a compass–clinometer, and are not a replacement for it: a GPS unit will not normally have a clinometer on it, and a compass will work without batteries.

5.1.2 Field studies: mapping and logging

The organisation of a field programme of sedimentary studies will depend on the objectives of the project. When an area with sedimentary rock units is mapped the character of the beds exposed in different places is described in terms used in this book. To describe the lithology the Dunham classification (*3.1.6*) can be used for limestones, and the Pettijohn classification for sandstones (*2.3.3*). Other features to be noted are bed thicknesses, sedimentary structures, fossils (both body and trace fossils – *11.7*), rock colour and any other characteristics such as weathering, degree of consolidation and so on. Field guides such as

Tucker (2003) and Stow (2005) provide a check-list of features to be noted. Once different formations have been recognised (*19.3.3*) it is normal for a graphic sedimentary log (*5.2*) to be measured and recorded from a suitable location within each formation. Although it is sufficient to regard a rock unit as simply 'red sandstone' for the purposes of drawing a geological map, any report accompanying the map should attempt to reconstruct the geological history of the area. At this stage some knowledge of the detailed character of the sandstone will be required, and sufficient information will have to be gathered to be able to interpret the sandstone in terms of environment of deposition (*5.7*).

An in-depth study may involve recording a lot of data from sedimentary rocks, either to see how a particular unit may vary geographically, or to see how the sedimentary character of a unit varies vertically (i.e. through time) – or both. The data for these palaeoenvironmental (*5.7*) or stratigraphic (Chapter 19) studies need to be collected in a systematic and efficient way, and for this purpose the graphic sedimentary log is the main method of recording data. A sedimentologist may spend a lot of time recording and drawing these logs, in conditions which vary from sunny beaches to wind-swept mountainsides (or even a warehouse in an industrial city – *22.3*), but the methodology is essentially the same in every instance. In conjunction with the data recorded on logs, other information such as palaeocurrent data will be collected, along with samples for petrographic and palaeontological analyses.

5.2 GRAPHIC SEDIMENTARY LOGS

A *sedimentary log* is a graphical method for representing a series of beds of sediments or sedimentary rocks. There are many different schemes in use, but they are all variants on a theme. The format presented here (Fig. 5.1) closely follows that of Tucker (1996); other commonly used formats are illustrated in Collinson et al. (2006). The objective of any graphic sedimentary log should be to present the data in a way which is easy to recognise and interpret using simple symbols and abbreviations that should be understandable without reference to a key (although a key should always be included to avoid ambiguity).

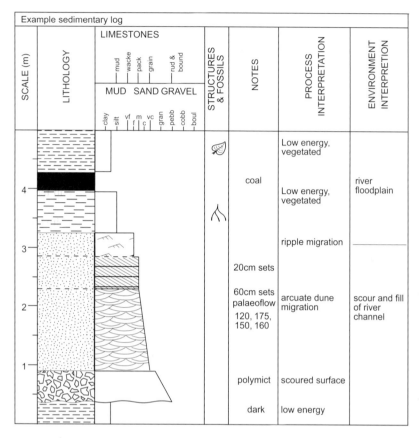

Fig. 5.1 An example of a graphic sedimentary log: this form of presentation is widely used to summarise features in successions of sediments and sedimentary rocks.

5.2.1 Drawing a graphic sedimentary log

The vertical scale used is determined by the amount of detail required. If information on beds a centimetre thick is needed then a scale of 1:10 is appropriate. A log drawn through tens or hundreds of metres may be drawn at 1:100 if beds less than 10 cm thick need not be recorded individually. Intermediate scales are also used, with 1:20 and 1:50 usually preferred in order to make scale conversion easy. Summary logs that provide only an outline of a succession of strata may be drawn at a scale of 1:500 or 1:1000.

Most of the symbols for lithologies in common use are more-or-less standardised: dots are used for sands and sandstone, bricks for limestone, and so on (Fig. 5.2). The scheme can be modified to suit the succession under description, for example, by the superimposition of the letter 'G' to indicate a glauconitic sandstone, by adding dots to the brickwork to represent a sandy limestone, and so on. In many schemes the lithology is shown in a single column. Alongside the lithology column (to the right) there is space for additional information about the sediment type and for the recording of sedimentary structures (see below). A horizontal scale is used to indicate the grain size in clastic sediments. The Dunham classification for limestones can also be represented using this type of scale. This scheme gives a quick visual impression of any trends in grain size in normal or reverse graded beds, and in fining-upwards or coarsening-upwards successions of beds.

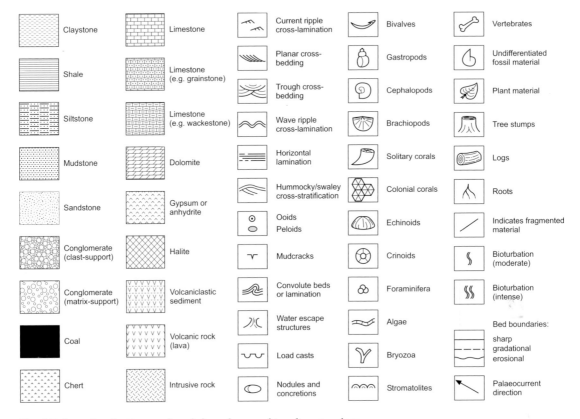

Fig. 5.2 Examples of patterns and symbols used on graphic sedimentary logs.

By convention the symbols used to represent sedimentary structures bear a close resemblance to the appearance of the feature in the field or in core (Fig. 5.2). This representation is somewhat stylised for the sake of simplicity and, again, symbols can be adapted to suit individual circumstances. Where space allows symbols can be placed within the bed but may also be drawn alongside on the right. Bed boundaries may be sharp/erosional, where the upper bed cuts down into the lower one, or transitional/gradational, in which there is a gradual change from one lithology to another. Any other details about the succession of beds can also be recorded on the graphic log (Fig. 5.1). Palaeocurrent data may be presented as a series of arrows oriented in the direction of palaeoflow measured or summarised for a unit as a rose diagram (5.3.3) alongside the log. Colour is normally recorded in words or abbreviations and any further remarks or observations may be

simply written alongside the graphic log in an appropriate place.

5.2.2 Presentation of graphic sedimentary logs

It is common practise to draw a log in the field and then redraft it at a later stage. The field log can be drawn straight on to a proforma log sheet (such as Fig. 5.3), but it can be more convenient to draw a sketch log in a field notebook. The field log does not have to be drawn to scale, but the thickness of every bed must be recorded so that a properly scaled version of the log can be drawn later. Sketch logs in the field notebook are also a quick and convenient way of recording sedimentological data, even when there are no plans to present graphic logs in a report. As is usually the case in geology, sketches and graphical

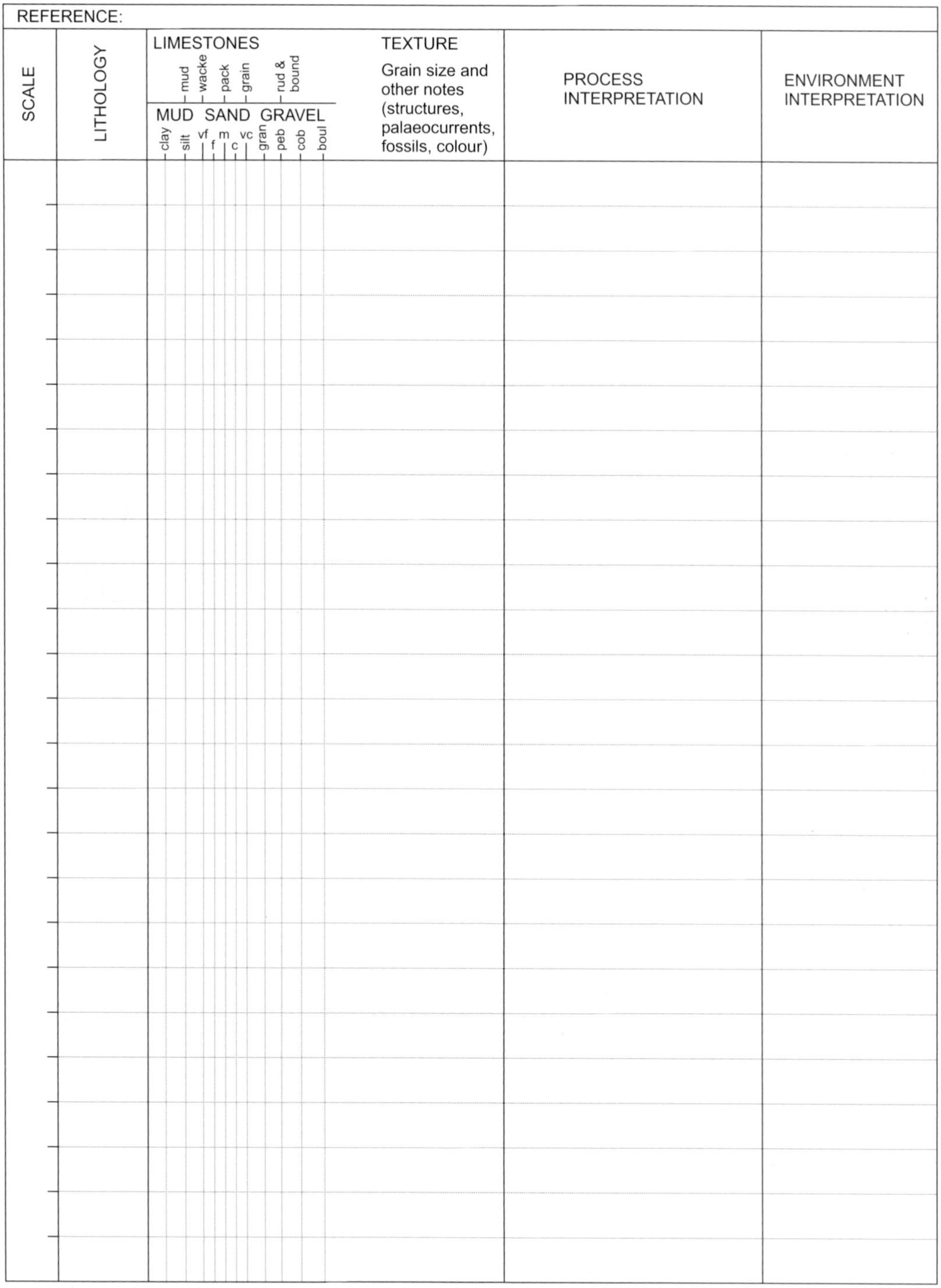

		LIMESTONES										TEXTURE	PROCESS INTERPRETATION	ENVIRONMENT INTERPRETATION

REFERENCE:

SCALE | LITHOLOGY | LIMESTONES | TEXTURE — Grain size and other notes (structures, palaeocurrents, fossils, colour) | PROCESS INTERPRETATION | ENVIRONMENT INTERPRETATION

mud / wacke / pack / grain / rud & bound

MUD SAND GRAVEL

clay silt | vf f m c vc | gran peb cob boul

Fig. 5.3 A proforma sheet for constructing graphic sedimentary logs.

representations of data, especially field data, are a quicker and more effective way of recording information than words. Interpretation of the information in terms of processes and environment (facies analysis – 5.6.1) is normally carried out back in the laboratory.

Computer-aided graphic log presentation has become widespread in recent years, including both dedicated log drawing packages and standard drawing packages (www.sedlog.com). These can provide clear images for presentation purposes, and are used in publications, but must be used with some care to ensure that logs do not become over-simplified with the loss of detailed information. For fieldwork, there is no substitute for the graphic log drawn by pen and pencil and the log drawn in the field must still be considered to be the fundamental raw data.

A number of sedimentary logs can be presented on a single sheet and linked together along surfaces of correlation, using either lithostratigraphic or sequence stratigraphic principles (see Chapters 19 and 23). These *fence diagrams* can be simple correlation panels if all the log locations fall along a line, but can also be used to show relationships and correlations in three-dimensions.

5.2.3 Other graphical presentations: sketches and photographs

A graphic log is a one-dimensional representation of beds of sedimentary rock that is the only presentation possible with drill-core (22.3.2) and is perfectly adequate for the most simple 'layer-cake' strata (beds that do not vary in thickness or characteristics laterally). Where an exposure of beds reveals that there is significant lateral variation, for example, river channel and overbank deposits in a fluvial environment, a single, vertical log does not adequately represent the nature of the deposits. A two-dimensional representation is required in the form of a section drawn of a natural or artificial exposure in a cliff or cutting (Fig. 5.4).

A carefully drawn, annotated sketch section showing all the main sedimentological features (bedding, cross-stratification, and so on) is normally satisfactory and may be supplemented by a photograph. Photographs (Fig. 5.5) can be used as a template for a field sketch, and now that digital cameras, laptops and portable printers are all available, the image can be produced in the field. However, a photograph should never be considered as a substitute for a field sketch: sedimentological features are rarely as clear on a photograph as they are in the field and a lot of information can be lost if important features and relationships are not drawn at the time. A good geological sketch need not be a work of art. Geological features should be clearly and prominently represented while incidental objects like trees and bushes can often be ignored. All sketches and photographs must include a scale of some form and the orientation of the view must be recorded and annotated to highlight key geological features.

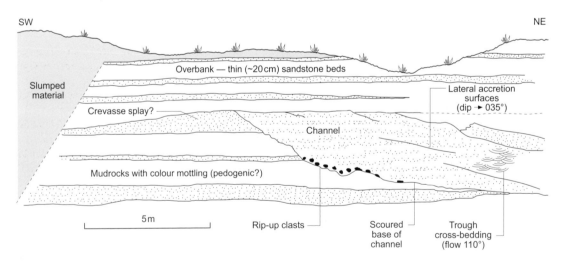

Fig. 5.4 An example of an annotated sketch illustrating sedimentary features observed in the field.

Fig. 5.5 A field photograph of sedimentary rocks: an irregular lower surface of the thick sandstone unit in the upper part of the cliff marks the base of a river channel.

Further information on the field description of sedimentary rocks is provided in Tucker (2003) and Stow (2005).

5.3 PALAEOCURRENTS

A *palaeocurrent indicator* is evidence for the direction of flow at the time the sediment was deposited, and may also be referred to as the *palaeoflow*. Palaeoflow data are used in conjunction with facies analysis (5.6.2) and provenance studies (5.4.1) to make palaeogeographic reconstructions (5.7). The data are routinely collected when making a sedimentary log, but additional palaeocurrent data may also be collected from localities that have not been logged in order to increase the size of the data set.

5.3.1 Palaeocurrent indicators

Two groups of palaeocurrent indicators in sedimentary structures can be distinguished (Miall 1999). *Unidirectional indicators* are features that give the direction of flow.

1 Cross-lamination (4.3.1) is produced by ripples migrating in the direction of the flow of the current. The dip direction of the cross-laminae is measured.

2 Cross-bedding (4.3.2) is formed by the migration of aeolian and subaqueous dunes and the direction of dip of the lee slope is approximately the direction of flow.

The direction of dip of the cross-strata in cross-bedding is measured.

3 Large-scale cross-bedding and cross-stratification formed by large bars in river channels (9.2.1) and shallow marine settings (14.3.1), or the progradation of foresets of Gilbert-type deltas (12.4.2) is an indicator of flow direction. The direction of dip of the cross-strata is measured. An exception is epsilon cross-stratification produced by point-bar accumulation, which lies perpendicular to flow direction (9.2.2).

4 Clast imbrication is formed when discoid gravel clasts become oriented in strong flows into a stable position with one of the two longer axes dipping upstream when viewed side-on (Fig. 2.9). Note that this is opposite to the measured direction in cross-stratification.

5 Flute casts (4.7) are local scours in the substrata generated by vortices within a flow. As the turbulent vortex forms it is carried along by the flow and lifted up, away from the basal surface to leave an asymmetric mark on the floor of the flow, with the steep edge on the upstream side. The direction along the axis of the scour away from the steep edge is measured.

Flow axis indicators are structures that provide information about the axis of the current but do not differentiate between upstream and downstream directions. They are nevertheless useful in combination with unidirectional indicators, for example, grooves and flutes may be associated with turbidites (4.5.2).

1 Primary current lineations (4.3.4) on bedding planes are measured by determining the orientation of the lines of grains.

2 Groove casts (4.7) are elongate scours caused by the indentation of a particle carried within a flow that give the flow axis.

3 Elongate clast orientation may provide information if needle-like minerals, elongate fossils such as belemnites, or pieces of wood show a parallel alignment in the flow.

4 Channel and scour margins can be used as indicators because the cut bank of a channel lies parallel to the direction of flow.

5.3.2 Measuring palaeocurrents

The most commonly used features for determining palaeoflow are cross-stratification, at various scales.

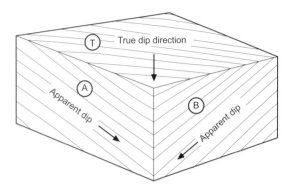

Fig. 5.6 The true direction of dip of planes (e.g. planar cross-beds) cannot be determined from a single vertical face (faces A or B): a true dip can be calculated from two different apparent dip measurements or measured directly from the horizontal surface (T).

Fig. 5.7 Trough cross-bedding seen in plan view: flow is interpreted as being away from the camera.

The measurement of the direction of dip of an inclined surface is not always straightforward, especially if the surface is curved in three dimensions as is the case with trough cross-stratification. Normally an exposure of cross-bedding that has two vertical faces at right angles is needed (Fig. 5.6), or a horizontal surface cuts through the cross-bedding (Fig. 5.7). In all cases a single vertical cut through the cross-stratification is unsatisfactory because this only gives an apparent dip, which is not necessarily the direction of flow.

Imbrication of discoid pebbles is a useful palaeoflow indicator in conglomerates, and if clasts protrude from the rock face, it is usually possible to directly measure the direction of dip of clasts. It must be remembered that imbricated clasts dip upstream, so the direction of dip of the clasts will be 180 degrees from the direction of palaeoflow (Figs 2.9 & 2.10). Linear features such as grooves and primary current lineations are the easiest things to measure by recording their direction on the bedding surface, but they do not provide a unidirectional flow indicator. The positions of the edges of scours and channels provide an indication of the orientation of a confined flow: three-dimensional exposures are needed to make a satisfactory estimate of a channel orientation, and other features such as cross-bedding will be needed to obtain a flow direction.

The procedure for the collection and interpretation of palaeocurrent data becomes more complex if the strata have been deformed. The direction has to be recorded as a plunge with respect to the orientation of the bedding, and this direction must then be rotated back to the depositional horizontal using stereonet techniques (Collinson et al. 2006).

In answer to the question of how many data points are required to carry out palaeocurrent analysis, it is tempting to say 'as many as possible'. The statistical validity of the mean will be improved with more data, but if only a general trend of flow is required for the project in hand, then fewer will be required. A detailed palaeoenvironmental analysis (5.7) is likely to require many tens or hundreds of readings. In general, a mean based on less than 10 readings would be considered to be unreliable, but sometimes only a few data points are available, and any data are better than none. Although every effort should be made to obtain reliable readings, the quality of exposure does not always make this possible, and sometimes the palaeocurrent reading will be known to be rather approximate. Once again, anything may be better than nothing, but the degree of confidence in the data should be noted. (One technique is to use numbers for good quality flow indicators, e.g. 275°, 290°, etc., but use points of the compass for the less reliable readings, e.g. WNW.)

There are several important considerations when collecting palaeocurrent data. Firstly it is absolutely essential to record the nature of the palaeocurrent indicator that has been recorded (trough cross-bedding, flute marks, primary current lineation, and so on). Secondly, the facies (5.6.1) of the beds that contain the palaeoflow indicators is also critical: the deposits of a river channel will have current indicators that reflect the river flow, but in overbank deposits the flow may have been perpendicular to the river channel (9.3). Lastly, not all palaeoflow indicators have the same 'rank': due to the irregularities of flow in a channel, a ripple on a bar may be oriented in almost any direction, but the downstream face of a large sandy or gravelly bar will produce cross-bedding that is close to the direction of flow of the river. It is therefore good practice to separate palaeoflow indicators into their different ranks when carrying out analyses of the data.

5.3.3 Presentation and analysis of directional data

Directional data are commonly collected and used in geology. Palaeocurrents are most frequently encountered in sedimentology, but similar data are collected in structural analyses. Once a set of data has been collected it is useful to be able to determine parameters such as the mean direction and the spread about the mean (or standard deviation). The procedure used for calculating the mean of a set of directional data is described below. Palaeocurrent data are normally plotted on a ***rose diagram*** (Fig. 5.8). This is a circular histogram on which directional data are

plotted. The calculated mean can also be added. The base used is a circle divided up with radii at 10° or 20° intervals and containing a series of concentric circles. The data are firstly grouped into blocks of 10° or 20° (001°–020°, 021–040°, etc.) and the number that fall within each range is marked by gradations out from the centre of the circular histogram. In this example (Fig. 5.8) three readings are between 261° and 270°, five between 251° and 260°, and so on. The scale from the centre to the perimeter of the circle should be marked, and the total number 'N' in the data set indicated.

5.3.4 Calculating the mean of palaeocurrent data

Calculating the mean of a set of directional data is not as straightforward as, for example, determining the average of a set of bed thickness measurements. Palaeocurrents are measured as a bearing on a circle and determining the average of a set of bearings by adding them together and dividing by the number of readings does not give a meaningful result: to illustrate why, two bearings of 010° and 350° obviously have a mean of 000°/360°, but simple addition and division by two gives an answer of 180°, the opposite

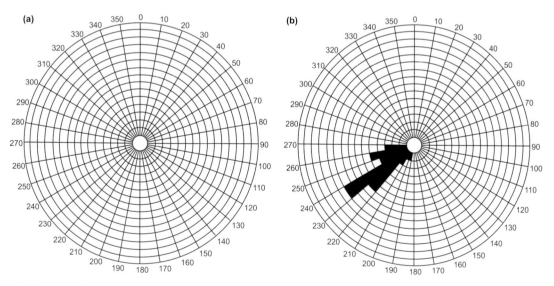

Fig. 5.8 A rose diagram is used to graphically summarise directional data such as palaeocurrent information: the example on the right shows data indicating a flow to the south west.

direction. Calculation of the circular mean and circular variance of sets of palaeocurrent data can be carried out with a calculator or by using a computer program. The mathematical basis for the calculation (Swan & Sandilands 1995) is as follows.

In order to mathematically handle directional data it is first necessary to translate the bearings into rectangular co-ordinates and express all the values in terms of x and y axes (Fig. 5.9).

1 For each bearing θ, determine the x and y values, where $x = \sin\theta$ and $y = \cos\theta$.

2 Add all the x values together and determine the mean.

3 Add all the y values together and determine the mean.

The result will be a mean value for the average direction expressed in rectangular co-ordinates, with the values of x and y each between -1 and $+1$. To determine the bearing that this represents use $\theta = \tan^{-1}(y/x)$. This value of θ will be between $+90$ and -90. To correct this to a true bearing, it is necessary to determine which quadrant the mean will lie in.

The spread of the data around the calculated mean is proportional to the length of the line r (Fig. 5.9). If the end lies very close to the perimeter of the circle, as happens when all the data are very close together, r will have a value close to 1. If the line r is very short it is because the data have a wide spread: as an extreme example, the mean of 000°, 090°, 180° and 270° would result in a line of length 0 as the mean values of x and y for this group would lie at the centre of the circle. The length of the line r is calculated using Pythagoras' theorem

$$r = \nu(x^2) + (y^2)$$

5.4 COLLECTION OF ROCK SAMPLES

Field studies only provide a portion of the information that may be gleaned from sedimentary rocks, so it is routine to collect samples for further analysis. Material may be required for palaeontological studies, to determine the biostratigraphic age of the strata (20.4), or for mineralogical and geochemical analyses. Thin-sections are used to investigate the texture and composition of the rock in detail, or the sample may be disaggregated to assess the heavy mineral content or dissolved to undertake chemical analyses. A number of these procedures are used in the determination of provenance.

The size and condition of the sample collected will depend on the intended use of the material, but for most purposes pieces that are about 50 mm across will be adequate. It is good practice to collect samples that are 'fresh', i.e. with the weathered surface removed. The orientation of the sample with respect to the bedding should usually be recorded by marking an arrow on the sample that is perpendicular to the bedding planes and points in the direction of younging (19.3.1). Every sample should be given a unique identification number at the time that it is collected in the field, and its location recorded in the field notebook. If collected as part of the process of recording a sedimentary log, the position of the sample in the logged succession should be recorded.

Samples should always be placed individually in appropriate bags – usually strong, sealable plastic bags. If you want to be really organised, write out the sample numbers on small pieces of heavy-duty adhesive tape before setting off for the field and attach the pieces of tape to a sheet of acetate. Each number is written on two pieces of tape, one to be attached to the sample, the other on to the plastic bag that the

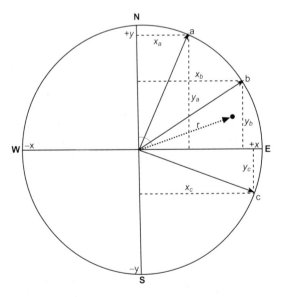

Fig. 5.9 Directions measured from palaeoflow can be considered in terms of 'x' and 'y' co-ordinates: see text for discussion.

sample is placed in. The advantage of this procedure is that sample numbers will not be missed or duplicated by mistake, and they will always be legible, even in the most unfavourable field conditions.

5.4.1 Provenance studies

Information about the source of sediment, or ***provenance*** of the material, may be obtained from an examination of the clast types present (Pettijohn 1975; Basu 2003). If a clast present in a sediment can be recognised as being characteristic of a particular source area by its petrology or chemistry then its provenance can be established. In some circumstances this makes it possible to establish the palaeogeographical location of a source area and provides information about the timing and processes of erosion in uplifted areas (6.7) (Dickinson & Suczek 1979).

Provenance studies are generally relatively easy to carry out in coarser clastic sediments because a pebble or cobble may be readily recognised as having been eroded from a particular bedrock lithology. Many rock types may have characteristic textures and compositions that allow them to be identified with confidence. It is more difficult to determine the provenance where all the clasts are sand-sized because many of the grains may be individual minerals that could have come from a variety of sources. Quartz grains in sandstones may have been derived from granite bedrock, a range of different metamorphic rocks or reworked from older sandstone lithologies, so although very common, quartz is often of little value in determining provenance. It has been found that certain heavy minerals (2.3.1) are very good indicators of the origin of the sand (Fig. 5.10). Provenance studies in sandstones are therefore often carried out by separating the heavy minerals from the bulk of the grains and identifying them individually (Mange & Maurer 1992). This procedure is called ***heavy mineral analysis*** and it can be an effective way of determining the source of the sediment (Morton et al. 1991; Morton & Hallsworth 1994; Morton 2003).

Clay mineral analysis is also sometimes used in provenance studies because certain clay minerals are characteristically formed by the weathering of particular bedrock types (Blatt 1985): for example, weathering of basaltic rocks produces the clay minerals in the smectite group (2.4.3). Analysis of mud and mudrocks can also be used to determine the average chemical composition of large continental areas. Large rivers may drain a large proportion of a continental landmass, and hence transport and deposit material eroded from that same area. A sample of mud from a river mouth is therefore a proxy for sampling the continental landmass, and much simpler than trying to collect representative, and proportionate, rock samples from that same area. This is a useful tool for comparing different continents and can be used on ancient mudrocks to compare potential sources of detritus. In particular, geochemical fingerprinting using Rare Earth Elements and isotopic dating using the neodymium–samarium system (21.2.3) can be used for this purpose.

5.5 DESCRIPTION OF CORE

Most of the world's fossil fuels and mineral resources are extracted from below the ground within sedimentary rocks. There are techniques for 'remotely' determining the nature of subsurface strata (Chapter 22), but hard evidence of the nature of strata tens, hundreds or thousands of metres below the surface can come only from drilling boreholes. Drilling is undertaken by the oil and gas industry, by companies prospecting mineral resources and coal, for water

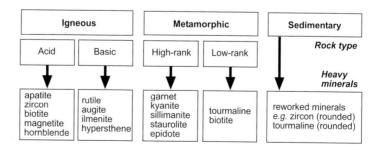

Fig. 5.10 Some of the heavy minerals that can be used as provenance indicators.

resources and for pure academic research purposes. When a hole is drilled it is not necessarily the case that core will be cut. Oil companies tend to rely on geophysical techniques to analyse the strata (22.4) and only cut core if details of particular horizons are required. In contrast, an exploration programme for coal will typically involve cutting core through the entire hole because they need to know precisely where coal beds are and sample them for quality.

After it has been cut, core is stored in boxes in lengths of about a metre. The core cut by the oil companies is typically between 100 and 200 mm diameter and is split vertically to provide a flat face (Fig. 5.11), but coal core is left whole, and is usually narrower, only 60 mm in diameter. When compared with outcrop, the obvious drawback of any core is that it is so narrow, and provides only a one-dimensional sample of the strata. Features that can be picked out by looking at two-dimensional exposure across a quarry or cliff face, such as river channels, reefs, or even some cross-bedding, can only be imagined when looking at core. This limitation tends to

Fig. 5.11 When drilling through strata it is possible to recover cylinders of rock that are cut vertically to reveal the details of the beds.

hamper interpretation of the strata. There is, however, a distinct advantage of core in that it usually provides a continuity of vertical section over tens or hundreds of metres. Even some of the best natural exposures of strata do not provide this 100% coverage, because beds are weathered away or covered by scree or vegetation.

Sedimentary data from core are recorded in the same way as strata in outcrop by using graphic sedimentary logs. Although recording field data is still important where it is possible to do so, it is also fair to say that geologists in industry working on sedimentary rocks will probably spend more time logging core than doing fieldwork.

5.6 INTERPRETING PAST DEPOSITIONAL ENVIRONMENTS

Sediments accumulate in a wide range of settings that can be defined in terms of their geomorphology, such as rivers, lakes, coasts, shallow seas, and so on. The physical, chemical and biological processes that shape and characterise those environments are well known through studies of physical geography and ecology. Those same processes determine the character of the sediment deposited in these settings. A fundamental part of sedimentology is the interpretation of sedimentary rocks in terms of the transport and depositional processes and then determining the environment in which they were deposited. In doing so a sedimentologist attempts to establish the conditions on the surface of the Earth at different times in different places and hence build up a picture of the history of the surface of the planet.

5.6.1 The concept of 'facies'

The term 'facies' is widely used in geology, particularly in the study of sedimentology in which **sedimentary facies** refers to the sum of the characteristics of a sedimentary unit (Middleton 1973). These characteristics include the dimensions, sedimentary structures, grain sizes and types, colour and biogenic content of the sedimentary rock. An example would be 'cross-bedded medium sandstone': this would be a rock consisting mainly of sand grains of medium grade, exhibiting cross-bedding as the primary sedimentary structure. Not all aspects of the rock are necessarily indicated in the facies name and in other instances it may be important to emphasise different characteristics. In other situations the facies name for a very similar rock might be 'red, micaceous sandstone' if the colour and grain types were considered to be more important than the grain size and sedimentary structures. The full range of the characteristics of a rock would be given in the facies description that would form part of any study of sedimentary rocks.

If the description is confined to the physical and chemical characteristics of a rock this is referred to as the **lithofacies**. In cases where the observations concentrate on the fauna and flora present, this is termed a **biofacies** description, and a study that focuses on the trace fossils (11.7) in the rock would be a description of the **ichnofacies**. As an example a single rock unit may be described in terms of its lithofacies as a grey bioclastic packstone, as having a biofacies of echinoid and crinoids and with a 'Cruziana' ichnofacies: the sum of these and other characteristics would constitute the sedimentary facies.

5.6.2 Facies analysis

The facies concept is not just a convenient means of describing rocks and grouping sedimentary rocks seen in the field, it also forms the basis for **facies analysis**, a rigorous, scientific approach to the interpretation of strata (Anderton 1985; Reading & Levell 1996; Walker 1992; 2006). The lithofacies characteristics are determined by the physical and chemical processes of transport and deposition of the sediments and the biofacies and ichnofacies provide information about the palaeoecology during and after deposition. By interpreting the sediment in terms of the physical, chemical and ecological conditions at the time of deposition it becomes possible to reconstruct **palaeoenvironments**, i.e. environments of the past.

The reconstruction of past sedimentary environments through facies analysis can sometimes be a very simple exercise, but on other occasions it may require a complex consideration of many factors before a tentative deduction can be made. It is a straightforward process where the rock has characteristics that are unique to a particular environment. As far as we know hermatypic corals have only ever grown in shallow, clear and fairly warm seawater: the presence of these fossil corals in life position in a sedimentary rock may therefore be used to indicate

that the sediments were deposited in shallow, clear, warm, seawater. The analysis is more complicated if the sediments are the products of processes that can occur in a range of settings. For example, cross-bedded sandstone can form during deposition in deserts, in rivers, deltas, lakes, beaches and shallow seas: a 'cross-bedded sandstone' lithofacies would therefore not provide us with an indicator of a specific environment.

Interpretation of facies should be objective and based only on the recognition of the processes that formed the beds. So, from the presence of symmetrical ripple structures in a fine sandstone it can be deduced that the bed was formed under shallow water with wind over the surface of the water creating waves that stirred the sand to form symmetrical wave ripples. The 'shallow water' interpretation is made because wave ripples do not form in deep water (11.3) but the presence of ripples alone does not indicate whether the water was in a lake, lagoon or shallow-marine shelf environment. The facies should therefore be referred to as 'symmetrically rippled sandstone' or perhaps 'wave rippled sandstone', but not 'lacustrine sandstone' because further information is required before that interpretation can be made.

5.6.3 Facies associations

The characteristics of an environment are determined by the combination of processes which occur there. A lagoon, for example, is an area of low energy, shallow water with periodic influxes of sand from the sea, and is a specific ecological niche where only certain organisms live due to enhanced or reduced salinity. The facies produced by these processes will be muds deposited from standing water, sands with wave ripples formed by wind over shallow water and a biofacies of restricted fauna. These different facies form a **facies association** that reflects the depositional environment (Collinson 1969; Reading & Levell 1996).

When a succession of beds are analysed in this way, it is usually evident that there are patterns in the distribution of facies. For example, on Fig. 5.12, do beds of the 'bioturbated mudstone' occur more commonly with (above or below) the 'laminated siltstone' or the 'wave rippled medium sandstone'? Which of these three occurs with the 'coal' facies? When

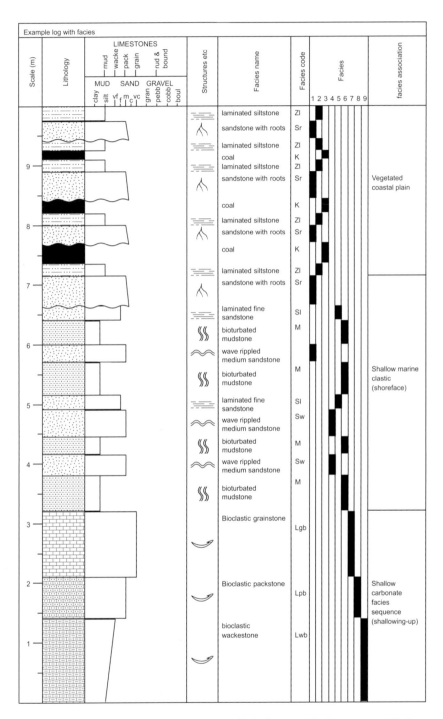

Fig. 5.12 A graphic sedimentary log with facies information added. The names for facies are usually descriptive. Facies codes are most useful where they are an abbreviation of the facies description. The use of columns for each facies allows for trends and patterns in facies and associations to be readily recognised.

attempting to establish associations of facies it is useful to bear in mind the processes of formation of each. Of the four examples of facies just mentioned the 'bioturbated mudstone' and the 'wave rippled medium sandstone' both probably represent deposition in a subaqueous, possibly marine, environment whereas 'medium sandstone with rootlets' and 'coal' would both have formed in a subaerial setting. Two facies associations may therefore be established if, as would be expected, the pair of subaqueously deposited facies tend to occur together, as do the pair of subaerially formed facies.

The procedure of facies analysis therefore can be thought of as a two-stage process. First, there is the recognition of facies that can be interpreted in terms of processes. Second, the facies are grouped into facies associations that reflect combinations of processes and therefore environments of deposition (Fig. 5.12). The temporal and spatial relationships between depositional facies as observed in the present day and recorded in sedimentary rocks were recognised by Walther (1894). Walther's Law can be simply summarised as stating that if one facies is found superimposed on another without a break in a stratigraphic succession those two facies would have been deposited adjacent to each other at any one time. This means that sandstone beds formed in a desert by aeolian dunes might be expected to be found over or under layers of evaporates deposited in an ephemeral desert lake because these deposits may be found adjacent to each other in a desert environment (Fig. 5.13). However, it would be surprising to find sandstones formed in a desert setting overlain by mudstones deposited in deep seas: if such is found, it would indicate that there was a break in the stratigraphic succession, i.e. an unconformity representing a period of time when erosion occurred and/or sea level changed (*2.3*).

5.6.4 Facies sequences/successions

A ***facies sequence*** or ***facies succession*** is a facies association in which the facies occur in a particular order (Reading & Levell 1996). They occur when there is a repetition of a series of processes as a response to regular changes in conditions. If, for example, a bioclastic wackestone facies is always overlain by a bioclastic packstone facies, which is in turn always overlain by a bioclastic grainstone

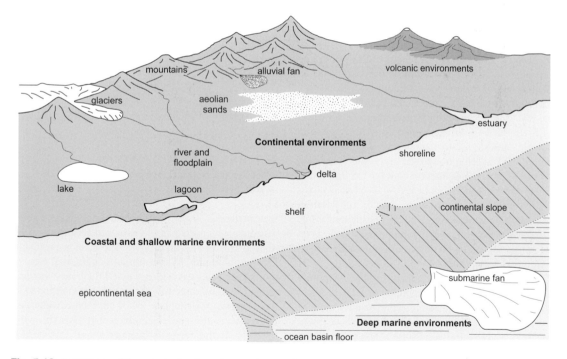

Fig. 5.13 A summary of the principal sedimentary environments.

(Fig. 5.12), these three facies may be considered to be a facies sequence. Such a pattern may result from repeated shallowing-up due to deposition on shoals of bioclastic sands and muds in a shallow marine environment (*Chapter 14*). Recognition of patterns of facies can be on the basis of visual inspection of graphic sedimentary logs or by using a statistical approach to determining the order in which facies occur in a succession, such as a **Markov analysis** (Swan & Sandilands 1995; Waltham 2000). This technique requires a transition grid to be set up with all the facies along both the horizontal and vertical axis of a table: each time a transition occurs from one facies to another (e.g. from bioclastic wackestone to bioclastic packstone facies) in a vertical succession this is entered on to the grid. Facies sequences/successions show up as higher than average transitions from one facies to another.

5.6.5 Facies names and facies codes

Once facies have been defined then they are given a name. There are no rules for naming facies, but it makes sense to use names that are more-or-less descriptive, such as 'bioturbated mudstone', 'trough cross-bedded sandstone' or 'foraminiferal wackestone'. This is preferable to 'Facies A', 'Facies B', 'Facies C', and so on, because these letters provide no clue as to the nature of the facies. A compromise has to be reached between having a name that adequately describes the facies but which is not too cumbersome. A general rule would be to provide sufficient adjectives to distinguish the facies from each other but no more. For example, 'mudstone facies' is perfectly adequate if only one mudrock facies is recognised in the succession. On the other hand, the distinction between 'trough cross-bedded coarse sandstone facies' and 'planar cross-bedded medium sandstone facies' may be important in the analysis of successions of shallow marine sandstone. Facies schemes are therefore variable, with definitions and names depending on the circumstances demanded by the rocks being examined.

The names for facies should normally be purely descriptive but it is quite acceptable to refer to facies associations in terms of the interpreted environment of deposition. An association of facies such as 'symmetrically rippled fine sandstone', 'black laminated mudstone' and 'grey graded siltstone' may have been interpreted as having been deposited in a lake

on the basis of the facies characteristics, and perhaps some biofacies information indicating that the fauna are freshwater. This association of facies may therefore be referred to as a 'lacustrine facies association' and be distinguished from other continental facies associations deposited in river channels ('fluvial channel facies association') and as overbank deposits ('floodplain facies association').

It can be convenient to have shortened versions of the facies names, for example for annotating sedimentary logs (Fig. 5.12). Miall (1978) suggested a scheme of letter codes for fluvial sediments that can be adapted for any type of deposit. In this scheme the first letter indicates the grain size ('S' for sand, 'G' for gravel, for example), and one or two suffix letters to reflect other features such as sedimentary structures: Sxl is 'cross-laminated sandstone', for example. There are no rules for the code letters used, and there are many variants on this theme (some workers use the letter 'Z' for silts, for example) including similar schemes for carbonate rocks based on the Dunham classification (*3.1.6*). As a general guideline it is best to develop a system that is consistent, with all sandstone facies starting with the letter 'S' for example, and which uses abbreviations that can be readily interpreted.

There is an additional graphical scheme for displaying facies on sedimentary logs (Fig. 5.12): columns alongside the log are used for each facies to indicate their vertical extent. An advantage of this form of presentation is that if the order of the columns is chosen carefully, for example with more shallow marine to the left and deeper marine on the right for shelf environments, trends through time can be identified on the logs.

5.7 RECONSTRUCTING PALAEOENVIRONMENTS IN SPACE AND TIME

One of the objectives of sedimentological studies is to try to create a reconstruction of what an area would have looked like at the time of deposition of a particular stratigraphic unit. Was it a tidally influenced estuary and, if so, from which direction did the rivers flow and where was the shoreline? If the beds are interpreted as lake deposits, was the lake fed by glacial meltwater and where were the glaciers? Which way was the wind blowing in the desert to produce those cross-bedded sandstones, and where were the evaporitic salt pans that we see in some modern desert basins? The process of reconstructing these palaeoenvironments depends on the integration of various pieces of sedimentological and palaeontological information.

5.7.1 Palaeoenvironments in space

The first prerequisite of any palaeoenvironmental analysis is a ***stratigraphic framework***, that is, a means of determining which strata are of approximately the same age in different areas, which are older and which are younger. For this we require some means of dating and correlating rocks, and this involves a range of techniques that will be considered in Chapters 19 to 23. However, once we have established that we do have rocks that we know to be of approximately the same age across an area, we can apply three of the techniques discussed in this chapter and consider them together.

First, there is the distribution of facies and facies associations. If we can recognise where there are the deposits of an ancient river, where the delta was and the location of the shoreline on the basis of the characteristics of the sedimentary rocks, then this will provide most of the information we need to draw a picture of how the landscape looked at that time. This information can be supplemented by a second technique, which is the analysis of palaeocurrent data, which can provide more detailed information about the direction of flow of the ancient rivers and the positions of the delta channels relative to the ancient shoreline. Third, provenance data can help us establish where the detritus came from, and help confirm that the rivers and deltas were indeed connected (if they contained sands of different provenance it would indicate that they were separate systems).

This sort of analysis is extremely useful in making predictions about the characteristics of rocks that cannot be seen because they are covered by younger strata. Palaeoenvironmental reconstructions are therefore more than just an academic exercise, they are a predictive tool that can be used to assess the distribution of the subsurface geology and help search for aquifers, hydrocarbon accumulations and mineral deposits.

5.7.2 Palaeoenvironments in time

Over thousands and millions of years of geological time, climate changes, plates move, mountains rise and the global sea level changes. The record of all these events is contained within sedimentary rocks, because the changes will affect environments that will in turn determine the character of the sedimentary rocks deposited. If we can establish that an area that had once been a coastal plain of peat swamps changed to being a region of shallow sandy seas, then we can infer that either the sea level rose or the land subsided. Similarly if a lake that had been a site of mud deposition became a place where coarse detritus from a mountainside formed an alluvial fan, we may conclude that there might have been a tectonic uplift in the area. Our palaeoenvironmental reconstructions therefore provide a series of pictures of the Earth's surface that we can then interpret in terms of large- and small-scale events. When palaeoenvironmental analysis is combined with stratigraphy in this way, the field of study is known as basin analysis and is concerned with the behaviour of the Earth's crust and its interaction with the atmosphere and hydrosphere. This topic is considered briefly in Chapter 24.

As stated above, one of the objectives of facies analysis is to determine the environment of deposition of successions of rocks in the sedimentary record. A general assumption is made that the range of sedimentary environments which exist today (Fig. 5.13) have existed in the past. In broad outline this is the case, but it should be noted that there is evidence from the stratigraphic record of conditions that existed during periods of Earth history that have no modern counterparts.

5.8 SUMMARY: FACIES AND ENVIRONMENTS

An objective, scientific approach is essential for successful facies analysis. A succession of sedimentary strata should be first described in terms of the lithofacies (and sometimes biofacies and ichnofacies) present, at which stage interpretations of the processes of deposition can be made. The facies can then be grouped into lithofacies associations which can be interpreted in terms of depositional environments on the basis of the combinations of physical, chemical and biological processes that have been identified from analysis of the facies. There are facies associations and sequences that commonly occur in particular environments and these are illustrated in the following chapters as 'typical' of these environments. However, there is a danger of making mistakes by 'pigeonholing', that is, trying to match a succession of rocks to a particular 'facies model'. Although general characteristics usually give a good clue to the depositional environment, small details can be vital and must not be overlooked.

FURTHER READING

Collinson, J., Mountney, N. & Thompson, D. (2006) *Sedimentary Structures*. Terra Publishing, London.

Reading, H.G. (Ed.) (1996) *Sedimentary Environments: Processes, Facies and Stratigraphy*. Blackwell Scientific Publications, Oxford.

Stow, D.A.V. (2005) *Sedimentary Rocks in the Field: a Colour Guide*. Manson, London.

Tucker, M.E. (2003) *Sedimentary Rocks in the Field* (3rd edition). Wiley, Chichester.

Walker, R.G. (2006) Facies models revisited: introduction. In: *Facies Models Revisited* (Eds Walker, R.G. & Posamentier, H.). Special Publication 84, Society of Economic Paleontologists and Mineralogists, Tulsa, OK; 1–17.

6

Continents: Sources of Sediment

The ultimate source of the clastic and chemical deposits on land and in the oceans is the continental realm, where weathering and erosion generate the sediment that is carried as bedload, in suspension or as dissolved salts to environments of deposition. Thermal and tectonic processes in the Earth's mantle and crust generate regions of uplift and subsidence, which respectively act as sources and sinks for sediment. Weathering and erosion processes acting on bedrock exposed in uplifted regions are strongly controlled by climate and topography. Rates of denudation and sediment flux into areas of sediment accumulation are therefore determined by a complex system that involves tectonic, thermal and isostatic uplift, chemical and mechanical weathering processes, and erosion by gravity, water, wind and ice. Climate, and climatic controls on vegetation, play a critical role in this Earth System, which can be considered as a set of linked tectonic, climatic and surface denudation processes. In this chapter some knowledge of plate tectonics is assumed.

6.1 FROM SOURCE OF SEDIMENT TO FORMATION OF STRATA

In the creation of sediments and sedimentary rocks the ultimate source of most sediment is bedrock exposed on the continents (Fig. 6.1). The starting point is the uplift of pre-existing bedrock of igneous, metamorphic or sedimentary origin. Once elevated this bedrock undergoes weathering at the land surface to create clastic detritus and release ions into solution in surface and near-surface waters. Erosion follows, the process of removal of the weathered material from the bedrock surface, allowing the transport of material as dissolved or par-

ticulate matter by a variety of mechanisms. Eventually the sediment will be deposited by physical, chemical and biogenic processes in a sedimentary environment on land or in the sea. The final stage is the lithification (18.2) of the sediment to form sedimentary rocks, which may then be exposed at the surface by tectonic processes. These processes are part of the sequence of events referred to as the ***rock cycle***.

In this chapter the first steps in the chain of events in Fig. 6.1 are discussed, starting with the uplift of continental crust, and then considering the processes of weathering and erosion, which result in the denudation of the landscape. The interactions

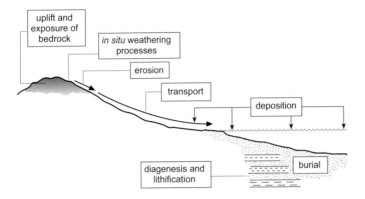

Fig. 6.1 The pathway of processes involved in the formation of a succession of clastic sedimentary rocks, part of the rock cycle.

between lithospheric behaviour, climate, weathering and erosion are then considered in terms of the Earth Systems that are the sources of sedimentary material.

6.2 MOUNTAIN-BUILDING PROCESSES

Plate tectonic theory provides a framework of understanding the processes that lead to the formation of mountains, as well as providing an explanation for how all the main morphological features of the crust have formed throughout most of Earth history (Kearey & Vine 1996; Fowler 2005). Plate movements and associated igneous activity create the topographic contours of the surface of the Earth that are then modified by erosion and deposition. Areas of high ground on the surface of the globe today can be related to plate boundaries (Fig. 6.2). For example, the Himalayas is an **orogenic belt**, a mountain chain formed as a result of the collision of the continental plates of India and Asia, and the Andes have a core of igneous rocks related to the subduction of oceanic crust of the east Pacific beneath South America. High ground also occurs on the flanks of major rifts, such as the East African Rift Valley, where the crust is pulling apart. Interpretation of the stratigraphic record indicates that the same mountain-building processes have occurred in the past: the Highlands of Scotland and the Appalachians of northeast USA are the relics of plate collisions resulting from the closure of past oceans. Similarly, past subduction zones and related magmatic belts can be recognised in the Western Cordillera of western North America. Plate tectonic processes are therefore the principal mechanisms for generating uplift of the crust and creating areas of high ground that provide the source of clastic sedimentary material.

However, not all vertical movements of the crust can be related to the horizontal movement of plates. The mantle has an uneven temperature distribution within it, and there are some areas of the crust that are underlain by relatively hot mantle, and other places where the mantle below is cooler. The hot regions are known as 'plumes', upwelling masses of buoyant mantle that in some instances can be on a large scale – 'superplumes' that probably originate from the core–mantle boundary. Above the hot buoyant mass of a superplume the continental crust is uplifted on a vast scale to generate high plateau areas, such as seen in southern Africa today. Plateaux like these are distant from any plate boundary, but are important areas of erosion and generation of detritus for supply to sedimentary basins.

6.3 GLOBAL CLIMATE

The climate belts around the world are principally controlled by latitude (Fig. 6.3). The amount of energy from the Sun per unit area is less in polar regions than in the equatorial zones so there is a temperature gradient from each pole to the Equator. These temperature variations determine the atmospheric pressure belts: high pressure regions occur at the poles where cold air sinks and low pressure at the Equator where the air is heated up, expands and rises. These differences in pressure give rise to winds, which move air masses between areas of high pressure in the subtropical and polar zones to regions of low pressure in between them. The **Coriolis force** imparted by the rotation of the globe influences these air movements to produce a

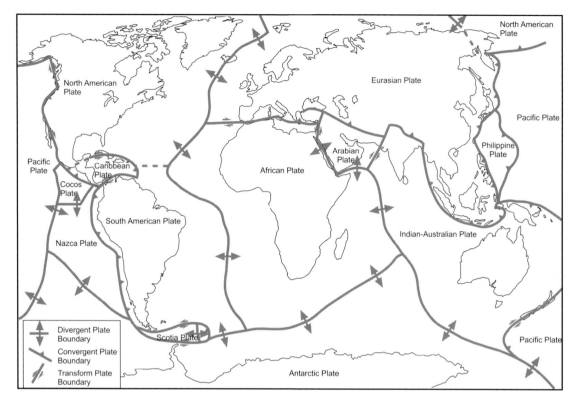

Fig. 6.2 The boundaries of the present-day principal tectonic plates.

basic pattern of winds around the Earth. The Coriolis force is a consequence of the movement of any body travelling towards or away from the poles over the surface of a rotating sphere, such that any moving object – an air mass, water in the ocean, or an airplane – will be deflected to the right in the northern hemisphere and the left in the southern hemisphere.

The combination of temperature distribution and wind belts gives rise to four main climate zones. *Polar regions* lie mainly north and south of the Arctic and Antarctic circles. They are regions of high pressure and low temperatures with conditions above freezing only part of the year, if at all. Between about 60° and 30° either side of the Equator lie the *temperate*, moist mid-latitude climate belts which have strongly seasonal climates and moderate levels of precipitation. The *dry subtropical* belts are variable in width depending on the configuration of land masses in the latitudes of the tropics of Cancer and Capricorn. Over large continental areas these dry areas are regions of high pressure, high temperatures and

low precipitation. In the middle lies the *wet equatorial* zone of high rainfall and high temperatures.

These climate zones are not uniform in width around the world and have different local climatic characteristics that are determined by the extent of continental land masses and the elevation of the land. As both the positions and height of continents vary through geological time due to plate movements, palaeoclimate belts can be related to the modern belts in only a relatively simplistic way unless complex climate modelling is carried out.

6.4 WEATHERING PROCESSES

Rock that is close to the land surface is subject to physical and chemical modification by a number of different *weathering processes* (Fig. 6.4). These processes generally start with water percolating down into joints formed by stress release as the rock comes close to the surface, and are most intense at the

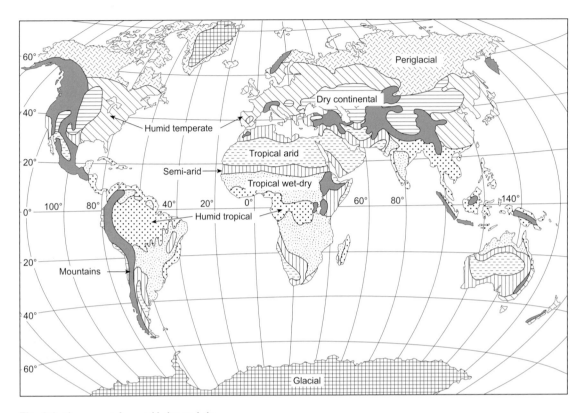

Fig. 6.3 The present-day world climate belts.

surface and in the soil profile. Weathering is the breakdown and alteration of bedrock by mechanical and chemical processes that create a **regolith** (layer of loose material), which is then available for transport away from the site (Fig. 6.1).

6.4.1 Physical weathering

These are processes that break the solid rock into pieces and may separate the different minerals without involving any chemical reactions. The most important agents in this process are as follows.

Freeze–thaw action

Water entering cracks in rock expands upon freezing, forcing the cracks to widen; this process is also known as **frost shattering** and it is extremely effective in

areas that regularly fluctuate around 0°C, such as high mountains in temperate climates and in polar regions (Fig. 6.5).

Salt growth

Seawater or other water containing dissolved salts may also penetrate into cracks, especially in coastal areas. Upon evaporation of the water, salt crystals form and their growth generates localised, but significant, forces that can further open cracks in the rock.

Temperature changes

Changes in temperature probably play a role in the physical breakdown of rock. Rapid changes in temperature occur in some desert areas where the temperature can fluctuate by several tens of degrees Celsius between

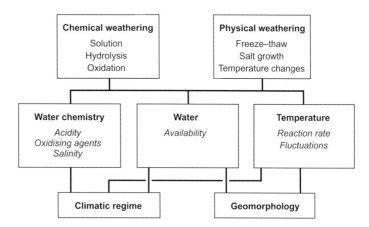

Fig. 6.4 The principal weathering processes and their controls.

Fig. 6.5 Frost shattering of a boulder (50 cm across) in a polar climate setting.

day and night; if different minerals expand and contract at different rates, the internal forces created could cause the rock to split. This process is referred to as *exfoliation*, as thin layers break off the surface of the rock.

6.4.2 Chemical weathering

These processes involve changes to the minerals that make up a rock. The reactions that can take place are as follows.

Solution

Most rock-forming silicate minerals have very low solubility in pure water at the temperatures at the Earth's surface and so most rock types are not susceptible to rapid solution. It is only under conditions of strongly alkaline waters that silica becomes moderately soluble. Carbonate minerals are moderately soluble, especially if the *groundwater* (water passing through bedrock close to the surface) is acidic. Most soluble are evaporite minerals such as halite (sodium chloride) and gypsum, which locally can form an important component of sedimentary bedrock.

Hydrolysis

Hydrolysis reactions depend upon the dissociation of H_2O into H^+ and OH^- ions that occurs when there is an acidifying agent present. Natural acids that are important in promoting hydrolysis include carbonic acid (formed by the solution of carbon dioxide in water) and humic acids, a range of acids formed by the bacterial breakdown of organic matter in soils. Many silicates undergo hydrolysis reactions, for example the formation of kaolinite (a clay mineral) from orthoclase (a feldspar) by reaction with water.

Oxidation

The most widespread evidence of oxidation is the formation of iron oxides and hydroxides from minerals containing iron. The distinctive red-orange rust colour of ferric iron oxides may be seen in many rocks exposed at the surface, even though the amount of iron present may be very small.

6.4.3 The products of weathering

Material produced by weathering and erosion of material exposed on continental land masses is referred to as **terrigenous** (meaning derived from land). Weathered material on the surface is an important component of the regolith that occurs on top of the bedrock in most places. Terrigenous clastic detritus comprises minerals weathered out of bedrock, lithic fragments and new minerals formed by weathering processes.

Rock-forming minerals can be categorised in terms of their stability in the surface environment (Fig. 6.6). Stable minerals such as quartz are relatively unaffected by chemical weathering processes and physical weathering simply separates the quartz crystals from each other and from other minerals in the rock. Micas and orthoclase feldspars are relatively resistant to these processes, whereas plagioclase feldspars, amphiboles, pyroxenes and olivines all react very readily under surface conditions and are only rarely carried away from the site of weathering in an unaltered state. The most important products of the chemical weathering of silicates are clay minerals (*2.4.3*). A wide range of clay minerals form as a result of the breakdown of different bedrock minerals under different chemical conditions; the most common are kaolinite, illite, chlorite and montmorillonite. Oxides of aluminium (bauxite) and iron (mainly haematite) also form under conditions of extreme chemical weathering.

In places where chemical weathering is subdued, lithic fragments may form an important component of the detritus generated by physical processes. The nature of these fragments will directly reflect the bedrock type and can include any lithology found at the Earth's surface. Some lithologies do not last very long as fragments: rocks made of evaporite minerals are readily dissolved and other lithologies are very fragile making them susceptible to break-up. Detritus composed of basaltic lithic fragments can form around volcanoes and broken up limestone can make up an important clastic component of some shallow marine environments.

6.4.4 Soil development

Soil formation is an important stage in the transformation of bedrock and regolith into detritus available for transport and deposition. *In situ* (in place) physical and chemical weathering of bedrock creates a soil that may be further modified by biogenic processes (Fig. 6.7). The roots of plants penetrating into bedrock can enhance break-up of the underlying rock and the accumulation of vegetation (**humus**) leads to a change in the chemistry of the surface waters as humic acids

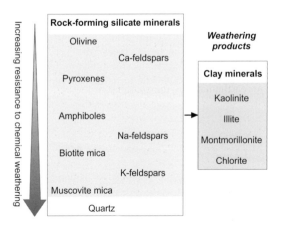

Fig. 6.6 The relative stability of common silicate minerals under chemical weathering.

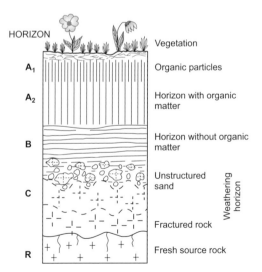

Fig. 6.7 An *in situ* soil profile with a division into different horizons according to presence of organic matter and degree of breakdown of the regolith.

form. Soil profiles become thicker through time as bedrock is broken up and organic matter accumulates, but a soil is also subject to erosion. Movement under gravity and by the action of flowing water may remove part or all of a soil profile. These erosion processes may be acute on slopes and important on flatter-lying ground where gullying may occur. The soil becomes disaggregated and contributes detritus to rivers. In temperate and humid tropical environments most of the sediment carried in rivers is likely to have been part of a soil profile at some stage.

Continental depositional environments are also sites of soil formation, especially the floodplains of rivers. These soils may become buried by overlying layers of sediment and are preserved in the stratigraphic record as fossil soils (palaeosols: 9.7).

6.5 EROSION AND TRANSPORT

Weathering is the *in situ* breakdown of bedrock and erosion is the removal of regolith material. Loose material on the land surface may be transported downslope under gravity, it may be washed by water, blown away by wind, scoured by ice or moved by a combination of these processes. Falls, slides and slumps are responsible for moving vast quantities of material downslope in mountain areas but they do not move detritus very far, only down to the floor of the valleys. The transport of detritus over greater distances normally involves water, although ice and wind also play an important role in some environments (Chapters 7 & 8).

6.5.1 Erosion and transport under gravity

On steep slopes in mountainous areas and along cliffs movements downslope under gravity are commonly the first stages in the erosion and transport of weathered material.

Downslope movement

There is a spectrum of processes of movement of material downslope (Fig. 6.8). A *landslide* is a coherent mass of bedrock that has moved downslope without significantly breaking up in the process. Many thousands of cubic metres of rock can be translated downhill retaining the internal structure and stratigra-

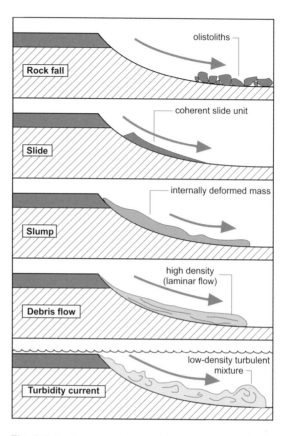

Fig. 6.8 Mechanisms of gravity-driven transport on slopes. Rock falls and slides do not necessarily include water, whereas slumps, debris flows and turbidity currents all include water to increasing degrees.

phy of the unit. If the rock breaks up during its movement it is a *rock fall*, which accumulates as a chaotic mass of material at the base of the slope. These movements of material under gravity alone may be triggered by an earthquake, by undercutting at the base of the slope, or by other mechanisms, such as waterlogging of a potentially unstable slope by a heavy rainfall.

Movement downslope may also occur when the regolith is lubricated by water and there is *soil creep*. This is a much slower process than falls and slides and may not be perceptible unless a hillside is monitored over a number of years. A process that may be considered to be intermediate between creep movement and slides is *slumping*. Slumps are instantaneous events like slides but the material is plastic due to saturation by water and it deforms during movement downslope.

With sufficient water a slump may break up into a debris flow (*4.5.1*).

Scree and talus cones

In mountain areas weathered detritus falls as grains, pebbles and boulders down mountainsides to accumulate near the bottom of the slope. These accumulations of **scree** are often reworked by water, ice and wind but sometimes remain preserved as **talus cones**, i.e. concentrations of debris at the base of gullies (Fig. 6.9) (Tanner & Hubert 1991). These deposits are characteristically made up of angular to very angular clasts because transport distances are very short, typically only a few hundred metres, so there is little opportunity for the edges of the clasts to become abraded. A small amount of sorting and stratification may result from percolating water flushing smaller particles down through the pile of sediment, but generally scree deposits are poorly sorted and crudely stratified. Bedding is therefore difficult to see in talus deposits but where it can be seen the layers are close to the angle of rest of loose aggregate material (around 30°). Talus deposits are distinct from alluvial fans (*9.5*) because water does not play a role in the transport and deposition.

6.5.2 Erosion and transport by water

Erosion by water on hillsides is initially as a **sheet wash**, i.e. unconfined surface run-off down a slope following rain. This overground flow may pick up loose debris from the surface and erode the regolith. The quantity of water involved and its carrying capacity depends not only on the amount of rainfall but also the characteristics of the surface: water runs faster down a steep slope, vegetation tends to reduce flow and trap debris and a porous substrate results in infiltration of the surface water. Surface run-off is therefore most effective at carrying detritus during flash-flood events on steep, impermeable slopes in sparsely vegetated arid regions. Vegetation cover and thicker, permeable soils in temperate and tropical climates tend to reduce the transport capacity of surface run-off.

Sheet wash becomes concentrated into rills and gullies that confine the flow and as these gullies coalesce into channels the headwaters of streams and rivers are established. Rivers erode into regolith and bedrock as the turbulent flow scours at the floor and margins of the channel, weakening them until pieces fall off into the stream. Flow over soluble bedrock such as limestone also gradually removes material in solution. Eroded material may be carried away in the stream flow as bedload, in suspension, or in solution; the confluence of streams forms larger rivers, which may feed alluvial fans, fluvial environments of deposition, lakes or seas.

6.5.3 Erosion and transport by wind

Winds are the result of atmospheric pressure differences that are partly due to global temperature distributions (*8.1.1*), and also local variations in pressure due to the temperature of water masses that

Fig. 6.9 A scree slope or talus cone in a mountain area with strong physical weathering.

move with ocean currents, heat absorbed by land masses and cold air over high glaciated mountain regions. A complex and shifting pattern of regions of high pressure (anticyclones) and low pressure (depressions) regions generates winds all over the surface of the Earth. Winds experienced at the present day range up to storm force winds of $100\,km\,h^{-1}$ to hurricanes that are twice that velocity.

Winds are capable of picking up loose clay, silt and sand-sized debris from the land surface. Wind erosion is most effective where the land surface is not bound by plants and hence it is prevalent where vegetation is sparse, in cold regions, such as near the poles and in high mountains, and dry deserts. Dry floodplains of rivers, sandy beaches and exposed sand banks in rivers in any climate setting may also be susceptible to wind erosion. Eroded fine material (up to sand grade) can be carried over distances of hundreds or thousands of kilometres by the wind (Schutz 1980; Pye 1987). The size of material carried is related to the strength (velocity) of the air current. The processes of transport and deposition by aeolian processes are considered in Chapter 8.

6.5.4 Erosion and transport by ice

Glaciers in temperate mountain regions make a very significant contribution to the erosion and transport of bedrock and regolith. The rate of erosion is between two and ten times greater in glaciated mountain areas than in comparable unglaciated regions (Einsele 2000). In contrast, glaciers and ice sheets in polar regions tend to inhibit the erosion of material because the ice is frozen to the bedrock: movement of the ice in these polar 'cold-based' glaciers is mainly by shearing within the ice body (7.2.1). In temperate (warm-based) glaciers, erosion of the bedrock by ice occurs by two processes, abrasion and plucking.

Glacial abrasion occurs by the frictional action of blocks of material embedded in the ice ('tools') on the bedrock. These tools cut grooves, *glacial striae*, in the bedrock a few millimetres deep and elongate parallel to the direction of ice movement: striae can hence be used to determine the pathways of ice flow long after the ice has melted. The scouring process creates *rock flour*, clay and silt-sized debris that is incorporated into the ice.

Glacial plucking is most common where a glacier flows over an obstacle. On the up-flow side of the obstacle abrasion occurs but on the down-flow side the ice dislodges blocks that range from centimetres to metres across. The blocks plucked by the ice and subsequently incorporated into the glacier are often loosened by subglacial freeze–thaw action (6.4.1). The landforms created by this combination of glacial abrasion and plucking are called *roche moutonée*, apparently because they resemble sheep from a (very) great distance.

6.6 DENUDATION AND LANDSCAPE EVOLUTION

The lowering of the land surface by the combination of weathering and erosion is termed *denudation*. Weathering and erosion processes are to some extent interdependent: it is the combination of these processes that are of most relevance to sedimentary geology, namely the rates and magnitudes at which denudation occurs and the implications that this has on the supply of material to sedimentary environments. Rates of denudation are determined by a combination of topographic and climatic factors, which in turn influence soil development and vegetation, both of which also affect weathering and erosion. In addition, different bedrock lithologies respond in different ways to these combinations of physical, chemical and biological processes.

6.6.1 Topography and relief

A distinction needs to be made between the altitude of a terrain and its *relief*, which is the change in the height of the ground over the area. A plateau region may be thousands of metres above sea level but if it is flat there may be little difference in the rates of denudation across the plateau and a lowland region with a comparable climate. With increasing relief the mechanical denudation rate increases as erosion processes are more efficient. Rock falls and landslides are clearly more frequent on steep slopes than in areas of subdued topography: stream flow and overland water flow are faster across steeper slopes and hence have more erosive power. A deeply incised topography consisting of steep sided valleys separated by narrow ridges provides the greatest area of steep slopes for bedrock and regolith to be eroded.

Relief tends to be greatest in areas that are undergoing uplift due to tectonic activity and thermal doming due to hot-spots in the mantle (Kearey & Vine 1996; Fowler 2005). Rejuvenation of the landscape by uplift occurs mainly around plate boundaries, particularly convergent margins such as orogenic belts. In tectonically stable areas the relief is subdued due to weathering and erosion resulting in a low, gentle topography. The cratonic centres of continental plates are typically regions of low relief and hence rates of denudation are low.

6.6.2 Climate controls on denudation processes

Chemical weathering processes are affected by factors that control the rate and the pathway of the reactions. First, water is essential to all chemical weathering processes and hence these reactions are suppressed where water is scarce (e.g. in deserts). Temperature is also important, because most chemical reactions are more vigorous at higher temperatures; hot climates therefore favour chemical weathering. Finally, water chemistry affects the reactions: the presence of acids enhances hydrolysis and dissolved oxidising agents facilitate oxidation reactions (Einsele 2000). The rates and efficiency of the reactions vary with different bedrock types.

Rates of erosion are climatically controlled because the availability of water is important to the removal of regolith by sheetwash and the extent to which rivers and streams erode soil and bedrock. Temperature is also significant: the presence of ice is important in mountains because wet-based, rapidly moving glaciers are more efficient at moving detritus than rivers. Denudation rates are therefore related to climatic regime, and general patterns can be recognised in each of the main global climate belts.

Wet tropical regions

In hot, wet, tropical areas, chemical weathering is enhanced because of the higher temperatures and abundance of water. Bedrock in these areas is typically deeply weathered and highly altered at the surface: seemingly resistant lithologies such as granite are reduced to quartz grains and clay as the feldspars and other silicate minerals are altered by surface weathering processes. In general, chemical weather-ing results in fine-grained detritus and partial solution of the bedrock. High rainfall gives rise to high discharge in streams, although the dense permanent vegetation in these settings reduces soil erosion by surface water, even on quite steep slopes.

Arid subtropical regions

The limited availability of water in arid regions means that chemical weathering processes are subdued. The bedrock is frequently barren of soil or vegetation cover, so when rainfall does occur it has little residence time on the land surface, and hence little time for chemical alteration to take place. Mechanical breakdown can be significant, especially in desert regions where cold nights and warm days promote freeze–thaw action, using whatever water is available. Exfoliation also occurs as a result of temperature changes. However, the absence of soil and vegetation means that infrequent but violent rainstorms can be very effective at removing surface detritus: **flash-floods** carry higher amounts of detritus than equivalent volumes of water occurring steadily over a longer time. Fine-grained debris is removed from the regolith by **wind ablation**, which is significant in barren desert areas.

Polar and cold mountain regions

Chemical weathering is less significant in cold, dry regions where chemical reactions are slower. In these areas physical weathering processes are more effective, although these too rely on the presence of water. The products of weathering in cold mountains are typically debris of the bedrock, broken up but with little or no change in the mineral composition. A granite breaks down into gravel clasts, plus grains of quartz, feldspar and other rock-forming minerals. Most of the products of physical weathering are hence coarse material with little clay generated or solution of the rock. Mountain glaciers are very powerful agents of erosion as they move downslope over rock, but in polar regions the ice is permanently frozen to bedrock and erosion due to glacial action is minimal (Chapter 7). **Periglacial** regions (areas that border glaciers) have a seasonal cover of snow that melts in the summer months. However, the ground may remain frozen at depths of a few metres all year round (**permafrost** – *7.4.4*) and water accumulating near the surface may eventually saturate the regolith and promote slumping on slopes. Repeated freezing and thawing of the regolith may also lead to creep downslope. Wind ablation is important because of the sparse vegetation cover in subarctic areas.

Temperate regions

In temperate climates both physical and chemical weathering processes tend to be subdued. Erosion is generally more vigorous under wetter climates, but on the other hand, vegetation, which is usually denser in humid climates, tends to stabilise the surface and can reduce erosion. The rate of denudation of limestone terrains is strongly climate-controlled, for in humid temperate or tropical regions the rate of denudation is ten times higher than in arid subtropical and subarctic regions (Einsele 2000).

6.6.3 Bedrock lithology and denudation

The type of bedrock is a fundamental control on the rates and patterns of denudation. The main factor is the rate at which weathering processes break down the rock to make material available for erosion. The greatest variability is seen in humid climates where chemical weathering processes are dominant because different lithologies are broken down, and hence eroded, at widely different rates. The proportions of the rock-forming silicate minerals (Fig. 6.6) are the main factor: quartz-rich rocks are least susceptible to

breakdown, whereas mafic rocks such as basalts are rapidly weathered and eroded. Large amounts of clay minerals are generated by the denudation of terrains such as volcanic arcs, which are composed mainly of basaltic to andesitic rocks. Under extreme chemical weathering of silicate rocks deep lateritic soils develop: **laterites** are red soils composed mainly of iron oxides and aluminium oxides.

Limestone bedrock is primarily weathered by dissolution, and the pattern of denudation is therefore dominated by development of **karst** scenery (Fig. 6.10). Solution related to joints and fractures in the rock leads to the formation of deep, steep-sided canyons on the surface and cave systems underground. Little clastic detritus is generated from the denudation of limestone terrains: conglomerates of limestone clasts may form near the site of erosion, but most of the material is in solution, with sand-sized detritus largely absent.

A characteristic scenery also forms where the bedrock is poorly lithified: **badland** terrains (Fig. 6.11) form by the deep erosion of weakly consolidated sandstones and mudstones as large amounts of detritus are carried away.

6.6.4 Soils and denudation

Soil development has an important role in weathering processes. First, water is retained in soils and hence the thickness of the soil profile influences how much water is available: if the soil profile is too thin it does

Fig. 6.10 Erosion by solution in beds of limestone results in a karst landscape.

Fig. 6.11 Badlands scenery formed by the erosion of weak mudrock beds.

not retain sufficient water for chemical weathering reactions in the bedrock to be effective, but if it is too thick it is able to store and lose water through plant evapotranspiration, hence reducing the availability of water for weathering reactions. Second, biochemical reactions in soils create acids, collectively known as humic acids, which increase rates of solution of carbonate bedrock. Third, soils are host to plants and animals, which also play a role in breaking down bedrock, especially roots that can penetrate deep into the rock and widen fractures. Although many soil processes may enhance weathering, soil development can inhibit erosion by hosting a vegetation cover that protects the bedrock.

6.6.5 Vegetation and denudation

The types of vegetation and the coverage they have over the land surface are determined by the climate regime, which is in turn influenced by the latitude and altitude. A dense vegetation cover is very effective at protecting the bedrock and its overlying regolith from erosion by rain impact and overland flow of water. Even steep mountain slopes can be effectively stabilised by plants. In tropical regions destruction of the vegetation cover by natural events such as fires or **anthropogenic activity** (man-made effects) such as logging can have a catastrophic effect on erosion: the bedrock beneath the plant roots may be very deeply weathered and the regolith susceptible to being

washed away by rainfall or floods. The effects of wind action on the regolith are also reduced where a vegetation cover binds fine detritus into soil. A sparse plant cover in cold or arid regions leaves the regolith exposed to erosion by water and wind. In deserts overland flow following storms may be very infrequent but in the absence of much plant life a lot of loose debris may be washed away in a single flash flood.

The nature of the vegetation colonising the land surface has changed considerably through geological time. Four stages in the development of land plants are significant in terms of sedimentological processes (Fig. 6.12) (Schumm 1968).

1 Pre-Silurian: there was no land vegetation at this time so it can be assumed that the denudation rates of continental areas were generally higher than they are today.

2 Silurian to mid-Cretaceous: the main plant groups were ferns, conifers and lycopods with relatively simple roots systems with a limited binding effect on the regolith.

3 Mid-Cretaceous to mid-Cenozoic: angiosperms (flowering plants) became important and had more complex root systems that were more effective at binding the soil.

4 Mid-Cenozoic to present: the evolution of grasses meant that there was now a widespread plant type which covered large areas of land surface with a dense fibrous root system that very effectively binds soil.

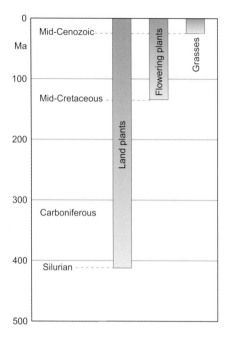

Fig. 6.12 The development of land plants through time: grasses, which are very effective at binding soil and stabilising the land surface, did not become widespread until the mid-Cenozoic.

6.7 TECTONICS AND DENUDATION

The creation of the topography of the continental land surface is fundamentally controlled by plate tectonic processes and mantle behaviour but surface processes, particularly erosion, play an important role in modifying the landscape. Climate plays an important role in weathering and erosion processes, and hence there is a climatic control on the interaction between erosion and tectonics (Burbank & Pinter 1999).

Denudation results in the removal of material from the uplifted bedrock and this reduces the mass of material in these areas. This removal of mass results in *isostatic uplift*. This process occurring in relatively rigid crust overlying mobile mantle is analogous to a block of ice floating on water: 10% of the ice will be above the water level, but if some of the exposed ice is removed from the top, the whole block will move up in the water so that there is still 10% of the mass above the water line. In mountain belts, there is an underlying mass of thickened crust that forms a bulge or root down into the mantle: erosion of

material from the top results in an isostatic readjustment and the whole crustal mass moving upwards (Fig. 6.13). Rates of denudation tend to be greater in areas of steep relief, and as a mountain belt grows, the steep topography created by tectonic uplift is subject to large amounts of erosion. Once the tectonic uplift ceases, surface processes start to reduce the topography. Through time, denudation followed by isostatic readjustment would remove mass from the top until the base of the root becomes level with the rest of the crust around. In this way, the mountains of an orogenic belt can be completely obliterated as denudation reduces the area to normal crustal thickness.

However, denudation does not occur evenly and it is possible to envisage an apparently paradoxical situation whereby denudation actually causes uplift. If the initial topography created is a large plateau, erosion will start by rivers incising into the plateau and removing mass from the valleys, but without significant erosion of the areas between them. The mass of the area will be reduced, and so isostatic uplift of the whole plateau occurs, including the areas between the rivers that have not been denuded (Fig. 6.13). This denudation-related uplift will continue until the valleys expand and the interfluve areas start to become eroded as fast as the valleys themselves.

Climate may control rates of denudation, but in turn the climate in an area can be determined by the presence of topography. A mountain belt may create a *rain shadow* effect (Fig. 6.14), as moisture-laden air is forced upwards and generates rainfall on the upwind side of the range: the winds that pass over the mountains are then dry, resulting in a more arid climate on the downwind side. This orographic effect results in a sharp climatic division across a mountain range, and hence a difference in the amount of erosion on either side.

From the foregoing it can be appreciated that the form of Earth's surface now (and at any time in the past) is as much a product of surface processes as tectonic forces, and that the two systems operate via a series of feedback mechanisms that influence each other. For example, it has been suggested that the creation of the Himalayas caused a change in weather patterns in Asia, strengthening the Asian monsoon, the pattern of intense seasonal rainfall across southern Asia (Raymo & Ruddiman 1992). This resulted in increased erosion of the Himalayas that triggered isostatic uplift. To the north, the Tibetan plateau lies in the rain-shadow of this weather system, and is a much drier area with less erosion: the

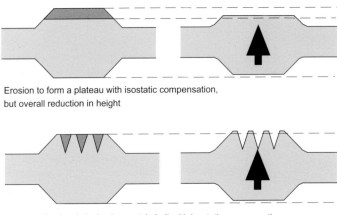

Erosion to form a plateau with isostatic compensation,
but overall reduction in height

Erosion of a deeply incised mountain belt with isostatic compensation
results in uplift of the peaks

Fig. 6.13 Uplift due to thickening of the crust followed by erosion results in isostatic compensation as the load of the rock mass eroded is removed. If the erosion is uneven then locally the removal of mass from valleys can result in uplift of the mountain peaks between.

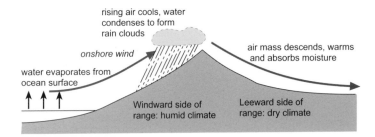

Fig. 6.14 The rain shadow effect in mountain belts: moisture in air blown from the sea falls as rain as the air mass cools over a mountain range.

area has been uplifted, but unlike the Himalayas has not been deeply dissected by rivers.

6.8 MEASURING RATES OF DENUDATION

The development of techniques that allow ***thermo-chronology***, the temperature history of rocks, to be carried out has made it possible to make estimates of the long-term rates of erosion in the past. The principal technique used is known as fission-track dating, and this is carried out on minerals such as apatite and zircon, both of which occur as accessory minerals in

many igneous and metamorphic rocks and as heavy minerals in sandstones. The basis of ***fission-track dating*** is that the radioactive decay of uranium isotopes in the mineral grains releases alpha particles that pass through the lattice of the crystal, leaving a trace of their path – these are the 'fission tracks'. If the crystal is heated, to over 110°C in the case of apatite, 300°C in zircon, the lines of the tracks become obscured as the heat anneals the lattice. As the crystal cools, new tracks start to form, and the longer the period of time since cooling below the annealing point, the more fission tracks there will be in the crystals. ***Apatite Fission Track Analysis*** (AFTA), and the less commonly used ***Zircon Fission Track***

Analysis are therefore thermochronological techniques which make it possible to determine at what date in the past a crystal was at a certain temperature.

Converting thermochronology data into rates of erosion requires knowledge of the **geothermal gradient**, that is, the change in temperature with depth in the crust. In many parts of the world, the temperature increases by about 25° for every thousand metres with depth, a geothermal gradient of 25° per kilometre, but is higher in places where there is volcanic activity. A rock that is at 4 km depth will be at about 120°C (assuming a surface temperature of 20°C and an increase of 25°C with every kilometre), and therefore all the fission tracks in apatite crystals will be annealed. Tectonic movements may cause the body of rock to be uplifted, and then, as the rock above the sample is removed by erosion, it will start to cool as it comes closer to the land surface. Fission tracks can then start to form in the apatite crystals in the sample, and continue to form until all the rock above has been eroded away and the sample is at the surface, available for collection and analysis. Measurement of the fission tracks can therefore tell us when the sample was at a certain depth, and hence how long it has taken to erode the rocks above: this provides us with an indication of the rate of erosion.

Thermochronological techniques also include the use of Ar–Ar dating (*21.2.2*) and there are relatively new techniques such as U/Th–He. Using combinations of approaches makes it possible to determine the dates when the rocks were at different temperatures and hence different depths. Statistical modelling of fission track data can be used to create a geothermal history of rock samples and hence a history of erosion in an area.

6.9 DENUDATION AND SEDIMENT SUPPLY: SUMMARY

The flux of material as bedload, suspended load and ions in solution to depositional environments is a primary control on the character of the sediments and facies that ultimately form. Thick successions of evaporite minerals cannot precipitate in lacustrine environments (Chapter 10) without an abundant supply of the relevant anions and cations from rivers draining nearby uplands. The characteristics of deltaic facies are fundamentally controlled by the grain size of the sediment supplied (*12.4*), and, in fact, a delta can only form if there is sufficient sediment supply in the first place. Carbonate-forming environments on shallow marine shelves can exist only in places where there is a reduced flux of terrigenous clastic material (Chapter 15). The starting point in any holistic view of depositional systems is therefore the source of the sediment and the linked tectonic and climatic processes that ultimately control the denudation of continental landmasses.

FURTHER READING

Einsele, G. (2000) *Sedimentary Basins, Evolution, Facies and Sediment Budget* (2nd edition). Springer-Verlag, Berlin.
Molnar, P. & England, P. (1990) Late Cenozoic uplift of mountains ranges and global climate change: chicken or egg? *Nature*, **346**, 29–34.
Ollier, C.D. (1984) *Weathering*. Longman, London.
Selby, M.J. (1994) Hillslope sediment transport and deposition. In: *Sediment Transport and Depositional Processes* (Ed. Pye, K.). Blackwell Scientific Publications, Oxford; 61–88.

7

Glacial Environments

Glaciers are important agents of erosion of bedrock and mechanisms of transport of detritus in mountain regions. Deposition of this material on land produces characteristic landforms and distinctive sediment character, but these continental glacial deposits generally have a low preservation potential in the long term and are rarely incorporated into the stratigraphic record. Glacial processes which bring sediment into the marine environment generate deposits that have a much higher chance of long-term preservation, and recognition of the characteristics of these sediments can provide important clues about past climates. The polar ice caps contain most of the world's ice and any climate variations that result in changes in the volumes of the continental ice caps have a profound effect on global sea level.

7.1 DISTRIBUTION OF GLACIAL ENVIRONMENTS

Ice accumulates in areas where the addition of snow each year exceeds the losses due to melting, evaporation or wind deflation. The climate is clearly a controlling factor, as these conditions can be maintained only in areas where there is either a large amount of winter snow that is not matched by summer thaw, or in places that are cold most of the time, irrespective of the amount of precipitation. There are areas of permanent ice at almost all latitudes, including within the tropics, and there are two main types of glacial terrains: temperate (or mountain) glaciers and polar ice caps.

Temperate or mountain glaciers form in areas of relatively high altitude where precipitation in the winter is mainly in the form of snow. Accumulating snow compacts and starts to form ice especially in the upper parts of valleys, and a glacier forms if the summer melt is insufficient to remove all of the mass added each winter. These conditions can exist at any latitude if the mountains are high enough. Once formed, the weight of snow accumulating in the upper part of the glacier (the **accumulation zone** of the glacier) causes it to move downslope, where it reaches lower altitudes and higher temperatures. The lower part of the glacier is the **ablation zone** where the glacier melts during the summer (Hambrey & Glasser 2003) (Fig. 7.1). Under stable climatic conditions an equilibrium develops between accumulation at the head and melting at the front, with the glacier moving downslope all the time, but the

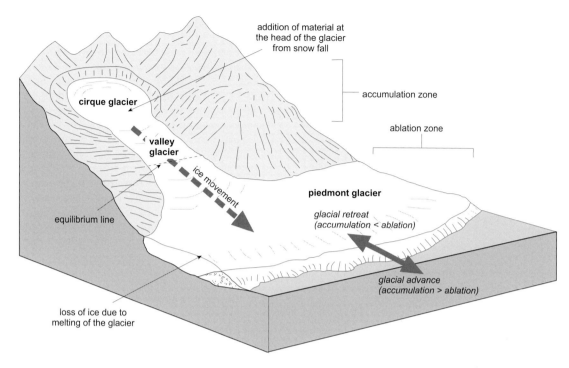

Fig. 7.1 Snowfall adds to the mass of a glacier in the accumulation zone and as the glacier advances downslope it enters the ablation zone where mass is lost due to ice melting. Glacial advance or retreat is governed by the balance between these two processes.

positions of the head and snout remain fixed. A cooling of the climate reduces the rate of melting and there will be *glacial advance* down the valley, whereas under a warmer climate the melting will exceed the rate of addition of snow and there will be *glacial retreat* (but note that the ice is still moving downslope within the glacier). Mountain glaciers do not usually reach sea level in temperate areas, except in places where there is high precipitation, which adds a lot of material at the head of the glacier: these glaciers move rapidly downslope and loss may be both by melting and calving of icebergs into the sea.

Polar glaciers occur at the north and south poles, which are regions of low precipitation (Antarctica is the driest continent): the addition to the glaciers from snow is quite small each year, but the year-round low temperatures mean that little melting occurs. Permanent ice in the polar continental areas forms large *ice sheets* and domed *ice caps* covering tens to hundreds of thousands of square kilometres. These may completely or partially bury the topography, and the hills

or mountains that protrude above the ice as areas of bare rock are called *nunataks* (Fig. 7.2). In polar regions the ice extends from the highlands of the land areas down to sea level, where glaciers feed *ice*

Fig. 7.2 Hills and ridges of bare rock (known as nunataks) surrounded by glaciers and ice sheets in a high-latitude polar glacial area.

Fig. 7.3 Floating ice, including icebergs, is formed by calving of ice from a glacier.

shelves, areas of floating ice extending out into the shallow marine realm. At the freezing point of pure water (0°C) ice has a density of $0.92\,\mathrm{g\ cm^{-3}}$ and therefore floats on both fresh water (density $1.0\,\mathrm{g\ cm^{-3}}$) and seawater (density $1.025\,\mathrm{g\ cm^{-3}}$). At the front of these ice shelves the ice breaks up to form floating masses, **icebergs** (Fig. 7.3), which drift in the ocean currents and wind for hundreds or thousands of kilometres before completely melting.

7.2 GLACIAL ICE

Ice is a solid, but under pressure it will behave in a ductile manner and flow by moving away from the point of higher pressure. Pressure is provided by the weight of ice above any particular point and the ice will flow slowly as an extremely viscous fluid (*4.2.1*). Glacier ice moves at rates which vary from as little as a few metres per year to hundreds of metres a year. Different parts of a body of ice move at different rates because of different pressure gradients, resulting in movement by **internal deformation** within the ice mass. Typically the flow rate is greatest at the surface of the ice decreasing downwards towards the base of the glacier, and valley-confined glaciers have greatest flow in the middle of the valley, decreasing towards the margins.

7.2.1 Thermal regimes of glaciers

In cold, polar regions glaciers and ice caps lie on ground that is permanently frozen (Fig. 7.4). The ice is therefore frozen to the ground and these **cold glaciers** move entirely by internal deformation, with the upper layers of the ice body shearing over

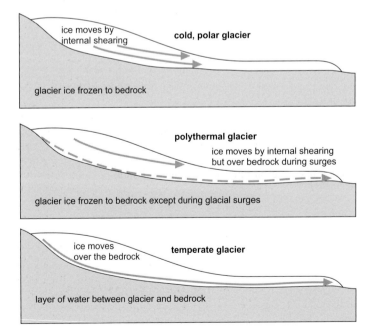

Fig. 7.4 The thermal regimes of glaciers are determined by the climatic setting: glaciers frozen to bedrock tend to occur in polar regions, while temperate glaciers occur in mountains in lower latitudes.

the lower parts. Because little or no movement takes place at the interface of the ice and the substrate, the glacier does not remove material from the valley floor or sides by glacial erosion. Cold glaciers are therefore less important than polythermal and temperate glaciers in terms of erosion and transport of sediment. Material carried by cold glaciers is largely detritus that has fallen under gravity down the upper part of the valley sides and comes to rest on the top of the glacier.

Polythermal glaciers are cold-based most of the time, but as snow and ice accumulate in the upper part of the glacier, the pressure near the base of it increases to the point where it melts (the **pressure melting point**, which decreases with increasing pressure). When this happens there is a **glacial surge** as the body of ice moves by **basal sliding** rapidly downslope and during this phase the glacier is capable of eroding bedrock (Hambrey & Glasser 2003). The glacier returns to equilibrium as it reaches a position downslope where the pressure is no longer sufficient to cause basal melting and the glacial snout breaks up and melts. The surge may take place over a matter of months and the retreat of the snout to its former position takes a few years. Detritus eroded during the surge is released during the subsequent retreat, so this process is capable of delivering sediment even though the glacier is frozen to the bedrock most of the time.

Temperate glaciers are typical of mountainous regions in lower latitudes. The ice is above the pressure melting point throughout the glacier and it is able to slide easily over the underlying bedrock (Fig. 7.4). Glacial action is an important erosional mechanism in mountainous areas with temperate glaciers, with glacial abrasion and glacial plucking generating detritus ranging from fine-grained rock flour to large blocks of bedrock. The action of temperate glaciers provides an important source of detritus that is carried downstream by rivers to supply other depositional environments.

7.3 GLACIERS

In high mountain areas small **cirque glaciers** (Fig. 7.1) form in protected hollows on mountain sides and are found at high altitudes all over the world, even within a few degrees of latitude from the

Fig. 7.5 A valley glacier in a temperate mountain region partially covered by a carapace of detritus.

Equator. Major mountain ranges in moderate and high latitudes also contain **valley glaciers**, bodies of ice that are confined within the valley sides (Fig. 7.5). In high latitudes valley glaciers may be fed by larger bodies of ice at higher altitudes, which are ice caps that wholly or partially blanket the higher parts of the mountains. The lower slopes of a mountain range may be the site of formation of a **piedmont glacier**, where valley glaciers may merge and spread out as a body of ice hundreds of metres thick.

7.3.1 Erosional glacial features

The geomorphological features associated with the glaciations of the past few hundred thousand years are largely found in upland areas and therefore will not be preserved in the geological record: cirques, U-shaped valleys and hanging valleys are evidence of past glaciation, which, in the framework of geological time, are ephemeral, lasting only until they are themselves eroded away. Smaller scale evidence such as glacial striae produced by ice movement over bedrock may be seen on exposed surfaces, including roche moutonée (6.5.4). Pieces of bedrock incorporated into a glacier by plucking may retain striae, and contact between clasts within the ice also results in scratch marks on the surfaces of sand and gravel transported and deposited by ice. These clast surface features are important criteria for the recognition of pre-Quaternary glacial deposits.

7.3.2 Transport by continental glaciers

Debris is incorporated into a moving ice mass by two main mechanisms: **supraglacial debris**, which accumulates on the surface of a glacier as a result of detritus falling down the sides of the glacial valley, and **basal debris**, which is entrained by processes of abrasion and plucking from bedrock by moving ice. Supraglacial debris is dominantly coarser-grained material with a low proportion of fine-grained sediment. Basal debris has a wider range of grain sizes, including fine-grained rock flour (6.5.4) produced by abrasion processes.

This basal debris of very fine to coarse material tends to be most abundant in polythermal glaciers because the alternation of pressure melting and freezing of the ice in contact with the bedrock exerts a strong freeze–thaw weathering effect (6.4.1). Melt water between the glacier and the bedrock forms a lubrication zone allowing the ice to move more freely and there is less erosion by the ice. Cold glaciers move only by internal deformation and hence do not erode bedrock. Cold and temperate glaciers therefore carry mainly coarser-grained supraglacial debris (Hambrey & Glasser 2003).

During movement of a glacier the ice mass undergoes deformation, internal folding and thrust faulting that can mix some of the basal and supraglacial debris into the main body of the glacier. In addition, the merging of two or more glaciers brings detritus from the margins of each into the centre of the combined glacier. Some modification of the debris occurs where it is carried along in the basal layer, with abrasion and fracturing of clasts occurring: water in channels within and at the base of the ice (**englacial channels** and **subglacial channels**) may also sort sediment carried in temperate glaciers. Supraglacial detritus is usually unmodified during transport and retains the poorly sorted, angular character of rockfall deposits.

7.3.3 Deposition by continental glaciers

The general term for all deposits directly deposited by ice is **till** if it is unconsolidated or **tillite** if it is lithified. These terms are genetic, that is, they imply a process of deposition and should therefore not be used as purely descriptive terms: for example, a bed may be described as a matrix-supported conglomerate (2.2.2), but because a deposit of this description could be formed by a number of different mechanisms in different environments (e.g. on alluvial fans, 9.5 and associated with submarine slumps, 16.1.2), the beds may or may not be interpreted as a tillite. To overcome this problem, the terms **diamicton** and **diamictite** are commonly used to describe unlithified and lithified deposits of poorly sorted material in an objective way, without necessarily implying that the deposits are glacial in origin. (It is noteworthy, however, that these terms, along with **diamict** for both unlithified and lithifed material, are rarely used by sedimentologists for deposits of pre-Quaternary age, and hence their use tends to be associated with glacial facies.)

Tills can be divided into a number of different types depending on their origin (Fig. 7.6). **Meltout tills** are deposited by melting ice as accumulations of material at a glacier front. **Lodgement tills** are formed by the plastering of debris at the base of a moving glacier, and the shearing process during the ice movement may result in a flow-parallel clast orientation fabric. Collectively meltout and lodgement tills are sometimes called **basal tills**. **Flow tills** are accumulations of glacial sediment reworked by gravity flows.

7.3.4 Characteristics of glacially transported material

Glacial erosion processes result in a wide range of sizes of detrital particles. As the ice movement is a laminar flow there is no opportunity for different parts of the

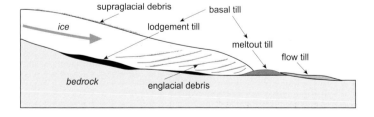

Fig. 7.6 Till deposits result from the accumulation of debris above, below and in front of a glacier.

ice body to mix and hence no sorting of material carried by the glacier will take place. Glacially transported debris is therefore typically very poorly sorted. Fragments plucked by the ice will be angular and debris carried within ice will not undergo any further abrasion, and only material on the top of an ice body will be subject to weathering processes. In addition to the poor sorting, debris carried by glaciers is very angular and the overall texture is therefore very immature. The constituents of tills and tillites are the products of weathering in cold environments, where physical weathering processes break up the rock but chemical weathering does not play an important role. For this reason, the mineral composition of the deposit may be very similar to that of the bedrock and unaltered lithic fragments are common. Clay minerals are often rather uncommon even in the fine-grained fraction of a till because clays form principally by the chemical weathering of minerals and in glacial environments this breakdown process is suppressed.

The fine-grained rock flour formed by glacial abrasion is different in composition to similar grade sediment produced by other mechanisms of weathering and erosion. Rock flour consists of very small fragments of many different minerals. In contrast the same sized material produced by chemical weathering typically consists of clay minerals and fine-grained quartz. Unlike clay minerals the fine particles in rock flour do not flocculate (*2.4.5*) and tend to remain in suspension for much longer periods of time. This high proportion of suspended sediment gives the characteristic green to white colour to lakes fed by glacial melt waters.

Material carried by a glacier is not necessarily all the result of glacial erosion. Valley sides in cold regions are subject to extensive freeze–thaw weathering (*6.4.1*), the products of which fall down the valley sides onto the top surface of the glacier. In more temperate regions detritus may also be washed down the valley sides by overland flow and by streams, which are active during the summer thaw. Streams may also form on the surface of a glacier or ice sheet during warmer periods and their action may contribute to the transport of debris.

7.4 CONTINENTAL GLACIAL DEPOSITION

Modern landscapes formerly covered by Quaternary ice sheets display a wide variety of depositional landforms (Fig. 7.7), which have been extensively studied and described by glacial geomorphologists (e.g. Hambrey 1994; Benn & Evans 1998). The depositional

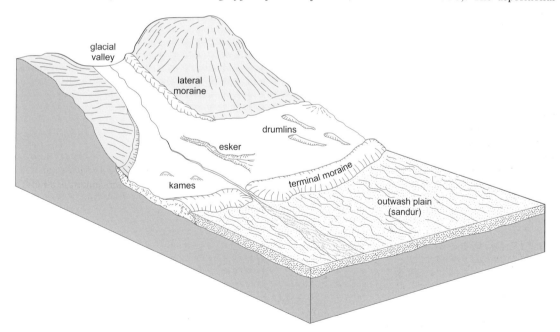

Fig. 7.7 Glacial landforms and glacial deposits in continental glaciated areas.

characteristics of these features provide information about glacial processes in the past few tens of thousands of years of Earth history and provide a basis for understanding the origins of the landscapes around us. However, most continental glacial deposits (Fig 7.8) are unlikely to be preserved in the strati-graphic record in the long term. This is because they mainly occur in areas that are only regions of deposition as a consequence of the glacial processes: many of the modern glacial landscapes are undergoing erosion and over time the continental glacial deposits will be reworked and removed. Glacial deposits recognized in pre-Quaternary strata are mostly marine in origin.

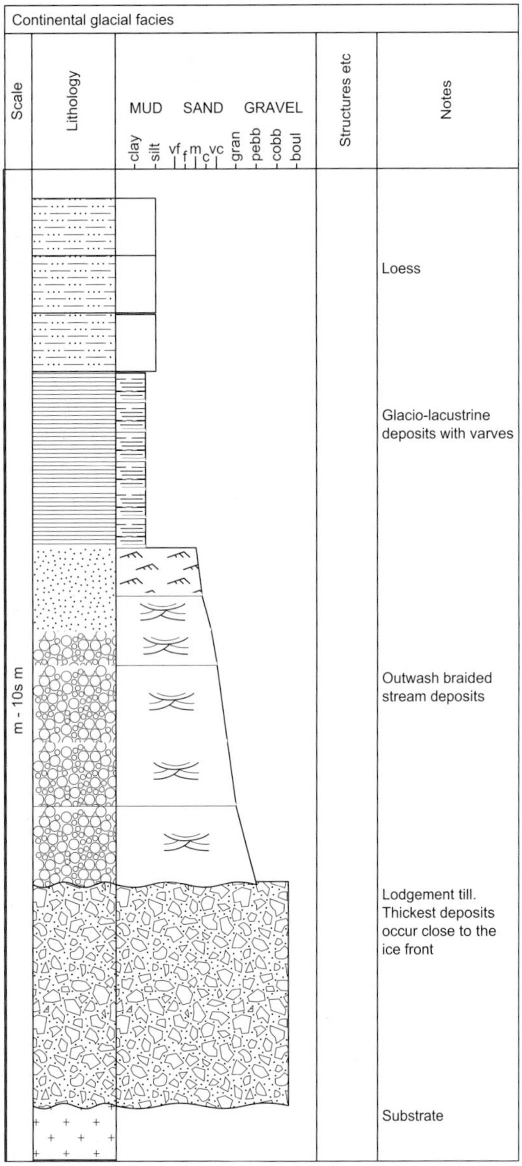

7.4.1 Moraines

Accumulations of till formed directly at the margins of a glacier are known as **moraine**. Several different types of moraine can be recognised (Benn & Evans 1998). **Terminal** or **end moraines** mark the limit of glacial advance and are typically ridges that lie across the valley. **Push moraines** are formed where a glacier front acts as a bulldozer scraping sediment from the valley floor and piling it up at the glacier front. **Dump moraines** form at the snout of the glacier where the melting of the ice keeps pace with glacial advance. If a glacier retreats the melting releases the detritus that has accumulated at the sides of the glacier where it is deposited as a **lateral moraine** (Figs 7.7 & 7.9). Lateral moraines form ridges along the sides of glaciated valleys, parallel to the valley walls. Where two glaciers in tributary valleys converge detritus from the sides of each is trapped in the centre of the amalgamated glacier (Fig. 7.10) and the resulting deposit upon ice retreat is a **medial moraine** along the centre of a glaciated valley. When a cold glacier retreats, the snout of the glacier is often left with a carapace of detritus left behind as the glacier front has been

Continental glacial facies					
Scale	Lithology	MUD SAND GRAVEL clay silt vf f m c vc gran pebb cobb boul		Structures etc	Notes

Loess

Glacio-lacustrine deposits with varves

Outwash braided stream deposits

Lodgement till. Thickest deposits occur close to the ice front

Substrate

(Scale note: m – 10s m)

Fig. 7.8 Graphic sedimentary log illustrating some of the deposits of continental glaciers.

Fig. 7.9 A lateral moraine left by the retreat of a valley glacier.

Fig. 7.10 A dark ridge of material within a valley glacier that will form a medial moraine when the ice retreats (viewed from the air).

melting. A few metres thickness of rock debris forms effective insulation and prevents the ice below it from melting. These *ice-cored moraines* (*ablation moraines*) give the impression of being much larger volumes of detritus than they really are because most of their bulk is made up of ice.

7.4.2 Other glacial landforms

Lodgement tills deposited beneath a glacier may form sheets that can be tens of metres thick, or show irregular ridges known as ribbed moraines. These tills also form smoothed mounds known as *drumlins*, which are oval-shaped hills tens of metres high and hundreds of metres to kilometres long, with the elongation in the direction of ice flow. In temperate glaciers partial melting of the ice results in rivers flowing in tunnels within or beneath the ice, carrying with them any detritus held by the ice that melted. The deposits of these rivers form sinuous ridges of material known as *eskers* (Fig. 7.7) and they are typically a few metres to tens of metres high, tens to hundreds of metres wide and stretch for kilometres across the area formerly covered by an ice sheet (Warren & Ashley 1994). The deposits are bars of gravel and sand that form cross-bedded and horizontally stratified lenses within the esker body. They may be distinguished from river deposits by the absence of associated overbank sediments (*9.3*) and by internal deformation (slump folds and faults) that forms when the sand and gravel layers collapse as the ice around the tunnel

melts. *Kames* and *kame terraces* are mounds or ridges of sediment formed by the collapse of crevasse fills, sediment formed in lakes lying on the top of the glacier or the products of the collapse of the edge of a glacier.

7.4.3 Outwash plains

As the front of a glacier or ice sheet melts it releases large volumes of water along with any detritus being carried by the ice (Fig. 7.10). Rivers flow away from the ice front over the broad area of the *outwash plain*, also known by their Icelandic name *sandur* (plural *sandar*). The rivers transport and deposit in the same manner as a braided river (*9.2.1*) (Boothroyd & Ashley 1975; Boothroyd & Nummedal 1978; Maizels 1993; Russell & Knudsen 2002). The large volumes of water and detritus associated with the melting of a glacier mean that the outwash plain is a very active region with river channels depositing sediment rapidly to form a thick, extensive braid plain deposit (*9.2.1*). Outwash plain deposits (Fig 7.8) can be distinguished from other braided river deposits by their association with other glacial features such as moraines.

The most spectacular events associated with glacial sedimentation are sudden glacial outburst events known by their Icelandic name as *jökulhlaups*. The outburst can be either the result of the failure of a natural dam holding back a lake at the front of the glacier or a consequence of melting associated with a subglacial volcanic eruption. Very large volumes of meltwater create a dramatic surge of water and sediment, which may include some very large blocks onto the outwash plain. Deposits of glacial outbursts are thick beds of sand and gravel that are massive and poorly sorted or cross-bedded and stratified (Maizels 1989; Russell & Knudsen 2002). Reworking of this material by 'normal' fluvial processes on the outwash plain may occur.

The absence of widespread vegetation under the cold climatic conditions means that fine-grained sediment on the outwash plain remains exposed and is subject to aeolian reworking. Sand may be blown into accumulations within and marginal to the outwash plain, forming deposits with the characteristics of wind-blown sediment (*8.3*). Silt- and clay-sized grains may be transported long distances and be widely distributed: accumulations of wind-blown silt of

Quaternary age (loess – 8.6.2) are thought to be sourced from periglacial environments. Clasts exposed on the outwash plain may be abraded by wind-blown sediment to form ventifacts (8.2).

7.4.4 Periglacial areas

In polar regions the areas that lie adjacent to ice masses are referred to as the *periglacial zone* (6.6.2). In this area the temperature is below zero for much of the year and the ground is largely frozen to create a region of *permafrost*. Only the soil and sediment near the surface thaws during the summer, and to a depth of only a few tens of centimetres, below which the ground remains perennially frozen. The thin layer of thawed material is often waterlogged because the water cannot drain away into the frozen subsurface. This upper mobile layer can be unstable on slopes and will slump or flow downslope. Other features of regions of permafrost are *patterned ground* (Fig. 7.11), which is composed of polygons of gravelly deposits formed by repeated freezing and thawing of the upper mobile layer, and *ice wedges*, which are cracks in the ground formed by ice that subsequently become filled with sediment.

7.5 MARINE GLACIAL ENVIRONMENTS

Where a continental ice sheet reaches the shoreline the ice may extend out to sea as an *ice shelf* (Figs 7.12 & 7.13). Modern ice shelves around the Antarctic con-

Fig. 7.11 In periglacial areas, freeze–thaw processes in the surface of the permafrost form polygonal patterns.

tinent extend hundreds of kilometres out to sea forming areas of floating ice which cover several hundred thousand square kilometres (Drewry 1986). These ice shelves partially act to buffer the seaward flow of continental ice: melting of the floating ice of an ice shelf does not add any volume to the oceans, but if they are removed then more continental ice will flow into the sea and this will cause sea level rise. Ice shelves such as those around the Antarctic contain relatively small amounts of sediment because there is little exposed rock to provide supraglacial detritus, so the main source is basal debris. Ice shelves break up at the edges to form icebergs and melt at the base in contact with seawater. Ice in a marine setting also occurs where temperate or poythermal valley glaciers flow down to sea level: these *tidewater glaciers* can contain large amounts of both supraglacial and basal debris. *Sea ice* is frozen seawater and does not contain any sedimentary material except for wind-blown dust.

7.5.1 Erosional features associated with marine glaciers

Where continental ice from an ice sheet or valley glacier reaches the shoreline of a shallow shelf the ice may be grounded on the sea floor. The movement of the ice mass and drifting icebergs may locally scour the sea floor, resulting in grooves in soft sediment that may be metres deep and hundreds of metres long. Meltwater flowing subglacially may be under considerable pressure and can erode channels into the sea-floor sediment beneath the ice, forming *tunnel valleys* that subsequently may be filled with deposits from the flowing water. The tunnel-valley deposits and the glacial scours features are preserved within shallow-marine strata in places such as continental shelves that have been covered with ice.

7.5.2 Marine glacial deposits

The terms till and tillite are also used to describe unconsolidated and lithifed marine glacial, *glaciomarine*, deposits (Fig. 7.14). The primary characteristics of the material are the same as the glacial sediment associated with continental glaciation. The detritus released from the bottom of an ice shelf forms *till sheets* (Fig. 7.12), which may be thick and extensive beneath a long-lived shelf (Miller 1996). These

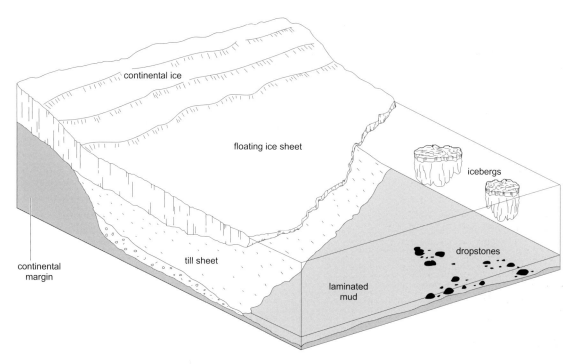

Fig. 7.12 At continental margins in polar areas, continental ice feeds floating ice sheets that eventually melt releasing detritus to form a till sheet and calve to form icebergs, which may carry and deposit dropstones.

Fig. 7.13 An ice shelf at the edge of a continental glaciated area.

deposits may be divided into those deposited close to the ice front (***ice-proximal glaciomarine sediments***), which are typically poorly sorted diamictons with little or no stratification or other sedimentary structures, and ***ice-distal glaciomarine sediments***, which are composed mainly of sediment released from icebergs. The more distal glaciomarine deposits are subject to reworking by shallow marine processes (Chapter 11): waves and currents produce a grain-size sorting of material, sand may be reworked to form wave and current bedforms and the finer-grained material may be transported in suspension to be deposited as laminated mud. Mixing of the glacially derived material with other sediment, such as biogenic material, can also occur.

The edges of ice shelves break up to form icebergs that can travel many hundreds of kilometres out into the open sea, driven by wind and ocean currents, but they often carry relatively little detritus. Icebergs formed at the front of tidewater glaciers are generally small, but may be laden with sediment. As an iceberg melts, this debris will gradually be released and deposited as ***dropstones*** in open marine sediments. Dropstones can be anything up to boulder size and their size is in marked contrast to surrounding fine-grained, pelagic deposits (*16.5.1*). Although rarely found in deep marine strata, dropstones are important indicators of the presence of ice shelves and hence provide evidence of past global climate conditions. However,

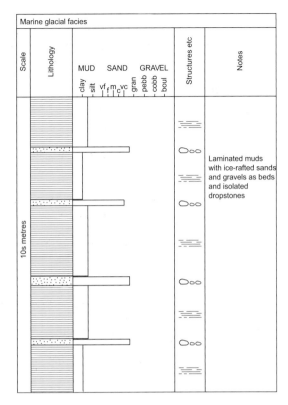

Fig. 7.14 Glaciomarine deposits are typically laminated mudrocks with sparse coarser debris derived from icebergs.

similar deposits can also result from sediment released from floating vegetation.

7.6 DISTRIBUTION OF GLACIAL DEPOSITS

Quaternary valley and piedmont glaciers form distinctive moraines but are largely confined to upland areas that are presently undergoing erosion. In these upland areas glacial and periglacial deposits such as moraines, eskers, kames, and so on have a very poor preservation potential in the long term. Of more interest from the point of view of the stratigraphic record are the tills formed in lowland continental areas and in marine environments as these are much more likely to lie in regions of net accumulation in a sedimentary basin. The volume of material deposited by ice sheets and ice shelves is also considerably greater than that associated with upland glaciation.

Extensive ice sheets are today confined to the polar regions within the Arctic and Antarctic circles. During the glacial episodes of the Quaternary the polar ice caps extended further into lower latitudes. The sea level was lower during glacial periods and many parts of the continental shelves were under ice. Upland glacial regions were also more extensive, with ice reaching beyond the immediate vicinity of the mountain glaciers. The growth of polar ice caps is known to be related to global changes in climate, with the ice at its most extensive when the globe was several degrees cooler. Other glacial episodes are known from the stratigraphic record to have occurred in the late Carboniferous and Permian (the Gondwana glaciation in the southern hemisphere), in the early Palaeozoic and in the Proterozoic.

7.7 ICE, CLIMATE AND TECTONICS

7.7.1 Glaciation and global climate

The continental ice caps at and near the poles contain the vast majority of the ice on the planet. The Antarctic ice cap covers almost all of the continent and has a fringe of floating ice shelves; much of the ice in the Arctic is sea ice in the Arctic Ocean, but Greenland also has a large ice cap. Compared with these polar regions, the ice in the mountain glaciers is of little global significance, although individual continental ice masses are important parts of local environments.

Evidence from the distribution of glacial sediments and sedimentary rocks indicates that there have been a number of periods during Earth history when the polar ice caps covered much larger areas than at present. The best documented glacial periods are from the Quaternary, a time of fluctuating global temperatures that has experienced advances and retreats of the polar ice caps a number of times over the past few hundred thousand years. The causes of the global changes in climate that lead to the ice caps growing and shrinking are complex and are considered further in Chapter 23. When the polar ice melts the water released adds to the volume of water in the oceans, and the sea level rises worldwide: it is estimated that complete melting of the Antarctic ice sheet would result in a global sea level rise of over 50 m, while the Greenland ice cap would add 7 m to world sea levels (Hambrey & Glasser 2003). The effects of sea level changes on sedimentation are covered in Chapter 23.

7.7.2 Glacial rebound – isostasy

During periods of glaciation the ice layer on the continents may be hundreds to thousands of metres thick. This mass of ice creates an extra load on the crust that forces the base of the crust down into the mantle. When the ice melts and the ice is removed, there is an isostatic uplift of the crust (6.7). The rate of melting is typically much faster than the isostatic uplift and consequently the crust continues to go up for thousands of years after the ice has melted. This effect is seen in many areas, such as Scandinavia, which were covered during the last ice age and are still undergoing uplift of a few millimetres a year. The effects of this so called **glacial rebound** are most clearly seen around coasts, where **raised beaches** provide evidence of the position of the land relative to the sea thousands of years ago, prior to uplift of the land.

7.8 SUMMARY OF GLACIAL ENVIRONMENTS

Glacial deposits are compositionally immature and tills are typically composed of detritus that simply represents broken up and powdered bedrock from beneath the glacier. Reworked glacial deposits on outwash plains may show a slightly higher compositional and textural maturity. There is a paucity of clay minerals in the fine-grained fraction because of the absence of chemical weathering processes in cold regions. Continental glacial deposits have a relatively low preservation potential in the stratigraphic record, but erosion by ice in mountainous areas is an important process in supplying detritus to other depositional environments. Glaciomarine deposits are more

commonly preserved, including dropstones which may provide a record of periods of glaciation in the past. The volume of continental ice in polar areas is closely linked to global sea level, so the history of past glaciations is an important key to understanding variations in the global climate.

Characteristics of glacial deposits
- lithologies – conglomerate, sandstone and mudstone
- mineralogy – variable, compositionally immature
- texture – extremely poorly sorted in till to poorly sorted in fluvio-glacial facies
- bed geometry – bedding absent to indistinct in many continental deposits, glaciomarine deposits may be laminated
- sedimentary structures – usually none in tills, cross-bedding in fluvio-glacial facies
- palaeocurrents – orientation of clasts can indicate ice flow direction
- fossils – normally absent in continental deposits, may be present in glaciomarine facies
- colour – variable, but deposits are not usually oxidised
- facies associations – may be associated with fluvial facies or with shallow-marine deposits

FURTHER READING

Benn, D.I. & Evans, D.J.A. (1998) *Glaciers and Glaciation.* Arnold, London.

Dowdeswell, J.A. & Scourse, J.D. (Eds) (1990) *Glaciomarine Environments: Processes and Sediments.* Special Publication 53, Geological Society Publishing House, Bath.

Hambrey, M.J. 1994. *Glacial Envionments.* UCL Press, London.

Knight, P. (Ed.) (2006) *Glacier Science and Environmental Change.* Blackwell Science, Oxford.

Miller, J.M.G. (1996) Glacial sediments. In: *Sedimentary Environments: Processes, Facies and Stratigraphy* (Ed. Reading, H.G.). Blackwell Scientific Publications, Oxford; 454–484.

8

Aeolian Environments

Aeolian sedimentary processes are those involving transport and deposition of material by the wind. The whole of the surface of the globe is affected by the wind to varying degrees, but aeolian deposits are only dominant in a relatively restricted range of settings. The most obvious aeolian environments are the large sandy deserts in hot, dry areas of continents, but there are significant accumulations of wind-borne material associated with sandy beaches and periglacial sand flats. Almost all depositional environments include a component of material that has been blown in as airborne dust, including the deep marine environments, and thick accumulations of wind-blown dust are known from Quaternary strata. Aeolian sands deposited in desert environments have distinctive characteristics that range from the microscopic grain morphology to the scale of cross-stratification. Recognition of these features provides important palaeoenvironmental information that can be used in subsurface exploration because aeolian sandstones are good hydrocarbon reservoirs and aquifers.

8.1 AEOLIAN TRANSPORT

The term *aeolian* (or *eolian* in North American usage) is used to describe the processes of transport of fine sediment up to sand size by the wind, and *aeolian environments* are those in which the deposits are made up mainly of wind-blown material.

8.1.1 Global wind patterns

The wind is a movement of air from one part of the Earth's surface to another and is driven by differences in air pressure between two places. Air masses move from areas of high pressure towards areas of low pressure, and the speed at which the air moves will be determined by the pressure difference. The circulation of air in the atmosphere is ultimately driven by temperature differences. The main contrast in temperature is between the Equator, which receives the most energy from the Sun, and the poles, which receive the least. Heat is transferred between these regions by air movement (as well as oceanic circulation). Hot air at the Equator rises, while cold air at the poles sinks, so the overall pattern is for a circulation cell to be set up with the warm air from the Equator travelling at high

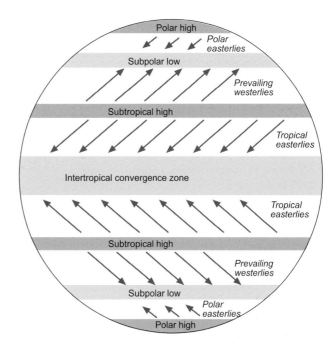

Fig. 8.1 The distribution of high- and low-pressure belts at different latitudes creates wind patterns that are deflected by the Coriolis force.

altitudes towards the poles and a complementary movement of cold air back to the Equator closer to ground level. This simple pattern is, however, complicated by two other factors. First, the circulation pattern breaks up into smaller cells, three in each hemisphere. Second, the Coriolis force (6.3) deflects the pathway of the air mass from simple north–south directions. The result is the pattern of winds shown in Fig. 8.1, although these patterns are modified and influenced by local topographic effects. Air masses blowing over mountain ranges are forced upwards and are cooled, and similarly the air is chilled when winds blow over ice caps: this results in *katabatic winds*, which are strong, cold air masses moving down mountain slopes or off the edges of ice masses.

8.1.2 Aeolian transport processes

A flow of air over a loose grain of sand exerts a lift force on the particle (4.2.3) and with increasing velocity the force may increase to the point where the grain rolls or saltates (4.2.2). The strength of the lift force is proportional to both the velocity of the flow and the density of the medium. Air has a density of $1.3\,\text{kg}\,\text{m}^{-3}$, which is three orders of magnitude less than that of water ($1000\,\text{kg}\,\text{m}^{-3}$) so, whereas water flows of only a few tens of centimetres a second can cause movement of sand grains, much higher velocities are required for the wind to move the same grains. Winds of $55\,\text{m}\,\text{s}^{-1}$ or more are recorded during hurricanes, but strong winds over land areas are typically around $30\,\text{m}\,\text{s}^{-1}$, and at these velocities the upper limit on the size of quartz grains moved by the wind is around a half a millimetre in diameter, that is, medium sand size storms (Pye 1987; Nickling 1994). This provides an important criterion for the recognition of aeolian deposits in the stratigraphic record: deposits consisting of grains coarser than coarse sand are unlikely to be aeolian deposits.

At high wind velocities silt- and clay-sized particles are carried as suspended load. This aeolian dust can become entrained in the wind in large quantities in dry areas to create dust storms that can carry airborne sediment large distances away from its origin. The dust will remain in suspension until the wind speed drops and the fine sediment starts to fall to the ground or onto a water surface. Significant accumulations of aeolian dust are relatively uncommon (*8.6.2*), but airborne material can be literally carried around the world by winds and be deposited in all depositional environments.

8.2 DESERTS AND ERGS

A ***desert*** is a continental area that receives little precipitation: they are ***arid*** areas that receive less than $250\,\text{mm yr}^{-1}$ precipitation. (Areas that receive average precipitation of between 250 and $500\,\text{mm yr}^{-1}$ are defined as ***semi-arid*** and are not usually considered to be true deserts.) This definition of a desert does consider temperature to be a factor, for, although the 'classic' deserts of the world today, such as the Sahara, are hot as well as dry places, there are also many dry areas that are also cold, including 'polar deserts' of high latitudes. The shortage of water limits the quantity and diversity of life in a desert: only a relatively limited range of plants and animals have adapted to live under these dry conditions and large parts of a desert surface are devoid of vegetation. The lack of vegetation is an important influence on surface processes because without a plant cover detritus lies loose on the surface where it is subject to aeolian activity.

An ***erg*** is an area where sand has accumulated as a result of aeolian processes (Brookfield 1992): these regions are also sometimes inappropriately referred to as a 'sand sea'. Ergs are prominent features of some deserts, but in fact most deserts are not sandy but are large barren areas known as ***rocky deserts***. Rocky deserts are areas of ***deflation***, that is, removal of material, and as such are not depositional environments. However, pebbles, cobbles and boulders that lie on the surface may subsequently be preserved if they are covered by other sediment, and these clasts may show evidence of their history in a rocky desert. Rocks in a desert are subject to a sand-blasting effect as sand and dust particles are blown against the surface by the wind: this erosive effect on the faces produces a characteristic clast shape, which is called a ***zweikanter*** if two faces are polished smooth, or ***dreikanter*** if there are three polished faces, with angled edges between each face. Long exposure of a rock surface in the oxidising conditions of a desert also results in the development of a dark, surface patina of iron and manganese oxides known as a ***desert varnish*** (Fig. 8.2).

Fig. 8.2 Pebbles in a stony desert with a shiny desert varnish of iron and manganese oxides.

8.3 CHARACTERISTICS OF WIND-BLOWN PARTICLES

8.3.1 Texture of aeolian particles

When two grains collide in the air they do so with greater impact than they would experience under water because air, being a much lower density medium than water, does not cushion the impact to the same extent. The collisions are hence relatively high energy and one or both of the grains may be damaged in the process. The most vulnerable parts of a grain are angular edges, which will tend to get chipped off, and with multiple impacts the grains gradually become more rounded as more of the edges are smoothed off. Sand grains that have undergone a sustained period of aeolian transport therefore become well-rounded, or even very well-rounded. Grain roundness is therefore a characteristic that can easily be seen in hand specimen using a hand lens, or will be evident under the microscope if a thin-section is cut of an aeolian sandstone. Inspection using a hand lens reveals another feature, which is more obvious if the grains are examined by scanning electron microscopy (SEM) (*2.4.4*): the grain surfaces will have a dull, matt appearance that under high magnification is a frosting of the rounded surface. This is a further consequence of the impacts suffered during transport and **grain surface frosting** is also a characteristic of aeolian processes. Aeolian dust shows similar grain characteristics but features on these sizes of grains can be recognised only if viewed under the high magnifications of an SEM.

A wind blowing at a relatively steady velocity can transport grains only up to a particular size threshold (Nickling 1994), and large, heavier grains are left behind. Grains close to the threshold for transport are carried as bedload and deposited as ripples and dunes (*4.3.1 & 4.3.2*), whereas finer grains remain in suspension and are carried away. This effective and selective separation of grains during transport means that aeolian deposits are typically well-sorted (*2.5*). Sands in dunes are normally fine to medium grained, with no coarser grains present and most of the finer grains winnowed away by the wind. This **winnowing effect**, the selective removal of finer grains from the sediment in a flow, also occurs in water flows, but is more effective in the lower density and viscosity medium of air.

A clastic deposit that consists of only sand-sized material, which is well sorted and with well-rounded grains, is considered to be texturally mature (*2.5.3*). Aeolian sandstones are, in fact, one of very few instances where granulometric analysis (*2.5.1*) provides useful information about the depositional environment of the deposit. There is, however, a need for caution when using petrographic characteristics alone as an indicator of environment of deposition. Consider an area of bedrock made up of sandstone deposited in a desert tens or hundreds of millions of years ago. After deposition it was buried and lithified, then uplifted and eroded. The sand that is being weathered off this bedrock will have the characteristics of the deposits of an aeolian environment, but is presently being transported and deposited by streams and rivers in a very different climatic and depositional setting. In these circumstances the sands have features that have been inherited from the earlier stage, or cycle, of deposition (*2.5.4*).

8.3.2 Composition of aeolian deposits

The abrasive effect of grain impacts during aeolian transport also has an effect on the grain types found in wind-blown deposits. When a relatively hard mineral, such as quartz, collides with a less robust mineral, for example mica, the latter will tend to suffer more damage. Abrasion during transport by wind therefore selectively breaks down the more labile grains, that is, the ones more susceptible to change. A mixture of different grain types becomes reduced to a grain assemblage that consists of very resistant minerals such as quartz and similarly robust lithic fragments such as chert. Other common minerals, for example feldspar, are likely to be less common in aeolian sandstones, and weak grains such as mica are very rare. A deposit with this grain assemblage is considered to be compositionally mature (*2.5.3*), and this is a common characteristic of aeolian sandstone. Most modern and ancient wind-deposited sands are quartz arenites.

In places where loose carbonate material is exposed on beaches, the sand-sized and finer sediment can be transported and redeposited by the wind. If the wind direction is onshore, wind-blown carbonate sands can accumulate and build up dune bedforms. Dunes built up of carbonate detritus have many of the same characteristics as a quartz-sand dune, and are several metres high with slip faces dipping at around 30° creating large-scale cross-bedding. The clasts may be ooids, bioclasts or pellets, depending upon what is available on the beach, and are well-rounded and well-sorted; if the clasts are bioclastic they will commonly have a relatively low density, so wind-blown grains may be very coarse sand or granule size. Wind-blown carbonates may accumulate in temperate as well as tropical settings: they are most commonly found near to coasts, but may also occur tens of kilometres inland. Loose carbonate grains on land are subject to wetting by the rain and subsequent drying in the sun; this leads to local dissolution and re-precipitation of carbonate, which results in rapid formation of cements and lithification of the sediment. Aeolian carbonate deposits are therefore more stable features than dunes made of quartz sand. Lithified wind-blown carbonate deposits are termed *aeolianites*, and these may be locally important components of coastal deposition (McKee & Ward 1983).

8.4 AEOLIAN BEDFORMS

The processes of transport and deposition by wind produce bedforms that are in some ways similar to subaqueous bedforms (4.3), but with some important differences that can be used to help distinguish aeolian from subaqueous sands. Three groups can be separated on the basis of their size: aeolian ripples, dunes and draas. Each appears to be a distinct class of bedform with no transitional forms and a plot of the range of sizes for each (Fig. 8.3) shows that they fall into three distinct fields (Wilson 1972).

8.4.1 Aeolian ripple bedforms

As wind blows across a bed of sand, grains will move by saltation forming a thin carpet of moving sand grains. The grains are only in temporary suspension, and as each grain lands, it has sufficient energy to knock impacted grains up into the free stream of air, continuing the process of saltation. Irregularities in the surface of the sand and the turbulence of the air flow will create patches where the grains are slightly more piled up. Grains in these piles will be more susceptible to being picked up by the flow and at a constant wind velocity all medium sand grains will move about the same distance each time they saltate. The result is a series of piles of grains aligned perpendicular to the wind and spaced equal distances apart. These are the crests of **aeolian ripples** (Figs 8.4 & 8.5). The troughs in between are shadow zones where

grains will not be picked up by the air flow and where few saltating grains land.

Aeolian ripples have extremely variable wavelengths (crest to crest distance) ranging from a few centimetres to several metres. Ripple heights (bottom of the trough to the top of a crest) range from less than a centimetre to more than ten centimetres. Coarser grains tend to be concentrated at the crests, where the finer grains are winnowed away by the wind, and as aeolian ripples migrate they may form a layer of inversely graded sand. Where a crest becomes well developed grains may avalanche down into the adjacent trough forming cross-lamination, but this is less common in aeolian ripples than in their subaqueous counterparts.

8.4.2 Aeolian dune bedforms

Aeolian dunes are bedforms that range from 3 m to 600 m in wavelength and are between 10 cm and 100 m high. They migrate by the saltation of sand up the stoss (upwind) side of the dune to the crest (Fig. 8.6). This saltation may result in the formation of aeolian ripples which are commonly seen on the stoss sides of dunes (Fig. 8.7). Sand accumulating at the crest of the dune is unstable and will cascade down the lee slope as an avalanche or grain flow (Fig. 8.8) (4.5.3) to form an inclined layer of sand (Fig. 8.6). Repeated avalanches build up a set of crossbeds that may be preserved if there is a net accumulation of sand. At high wind speeds some sand grains are in temporary suspension and are blown directly over the crest of the dune and fall out onto the lee slope. These **grain fall** deposits accumulate on

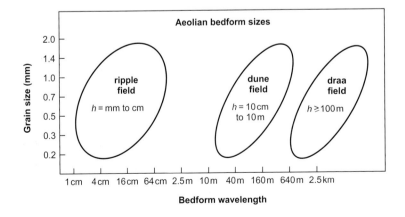

Fig. 8.3 Aeolian ripples, dunes and draas are three distinct types of aeolian bedform.

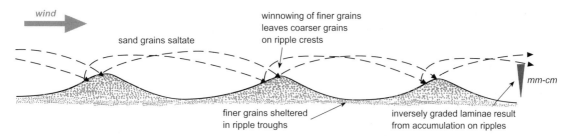

Fig. 8.4 Aeolian ripples form by sand grains saltating: finer grains are winnowed from the crests creating a slight inverse grading between the trough and the crest of the ripple that may be preserved in laminae.

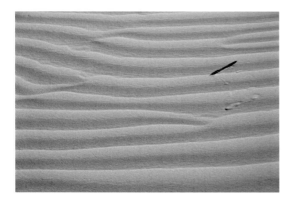

Fig. 8.5 Aeolian ripples in modern desert sands: the pen is 18 cm long.

the lee slope, but they will usually be reworked from the upper slope by avalanching: some may be preserved at the toe bedded with grain flow deposits.

The orientation and form (planar or trough) of the cross-bedding will depend on the type of dune (Figs 8.9 & 8.10) (McKee 1979; Wasson & Hyde 1983). Planar cross-beds will form by the migration of **transverse dunes**, straight-crested forms aligned perpendicular to the prevailing wind direction. Transverse dunes form where there is an abundant supply

of sand and as the sand supply decreases there is a transition to **barchan dunes**, which are lunate structures with arcuate slip faces forming trough cross-bedding. Under circumstances where there are two prominent wind directions at approximately 90° to each other, **linear** or **seif dunes** form. The deposits of these linear dunes are characterised by cross-bedding reflecting avalanching down both sides of the dune and hence oriented in different directions. In areas of multiple wind directions **star dunes** have slip faces in many orientations and hence the cross-bedding directions display a similar variability.

There are circumstances in which the whole dune bedform is preserved but more commonly the upper part of the dune is removed as more aeolian sand is deposited in an accumulating succession. The size of the set of cross-beds formed by the migration of aeolian dunes can vary from around a metre to ten or twenty metres (Fig. 8.11). Such large scale cross-bedding is common in aeolian deposits but is seen less frequently in subaqueous sands, which are typically cross-bedded in sets a few tens of centimetres to metres thick.

8.4.3 Draa bedforms

When an erg is viewed from high altitudes in aerial photographs or satellite images, it is possible to see a

Fig. 8.6 Aeolian dunes migrate as sand blown up the stoss (upwind) side is either blown off the crest to fall as grainfall on the lee side or moves by grain flow down the lee slope.

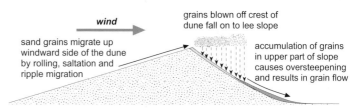

Fig. 8.7 Aeolian ripples superimposed on an aeolian dune.

Fig. 8.8 Grain flow on the lee slope of an aeolian dune.

pattern of structures that are an order of magnitude larger than dunes. The surface of the erg shows an undulation on a scale of hundreds of metres to kilometres in wavelength and tens to hundreds of metres

in amplitude. These structures are known as **_draas_** and there is again evidence that they are a distinct, larger bedform separate from the dunes that may be superimposed on them (Wilson 1972). Draas are usually made up of dunes on the stoss and lee sides, but a single slip face may develop on some lee slopes. They show a similar variability of shape to dunes with star, linear and transverse forms.

8.4.4 Palaeowind directions

The slip faces of aeolian dunes generally face downwind so by measuring the direction of dip of crossbeds formed by the migration of aeolian dunes it is possible to determine the direction of the prevailing wind at the time of deposition (Fig. 8.9). Results can be presented as a rose diagram (5.3.3). The variability of the readings obtained from cross-beds will depend upon the type of dune (McKee 1979). Transverse dunes generate cross-beds with little variability in orientation, whereas the curved faces of barchan dunes produce cross-beds that may vary by as much as 45° from the actual downwind direction. Multiple directions of cross-bedding result from the numerous slip faces of a star dune. In all cases the confidence with which the palaeowind direction can be inferred from cross-bedding orientations is improved with the more readings that are taken.

Wind directions are normally expressed in terms of the direction the wind blows from, that is, a southwesterly wind is one that is blowing from the southwest towards the northeast and will generate dune cross-bedding which dips towards the northeast. Note that this form of expression of direction is different from that of water currents that are normally presented in terms of the direction the flow is towards.

8.5 DESERT ENVIRONMENTS

Aeolian sands form one part of an arid zone facies association that also includes ephemeral lake deposits and alluvial fan and/or ephemeral river sediments (Figs 8.12 & 8.13). In these dry areas, sediment is brought into the basin by rivers that bring weathered detritus from the surrounding catchment areas and deposit poorly sorted mixtures of sediment on alluvial

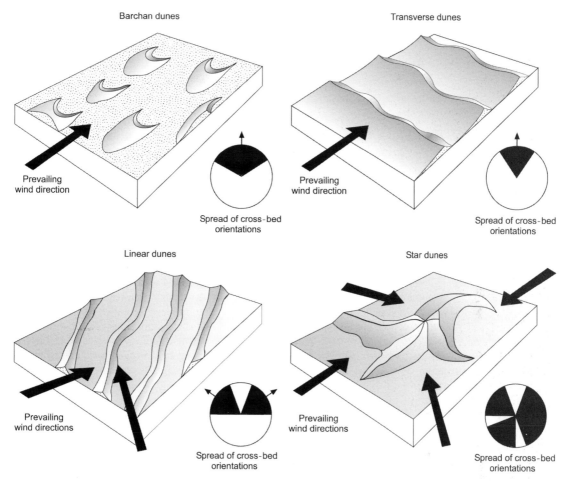

Fig. 8.9 Four of the main aeolian dune types, their forms determined by the direction of the prevailing wind(s) and the availability of sand. The small 'rose diagrams' indicate the likely distribution of palaeowind indicators if the dunes resulted in cross-bedded sandstone.

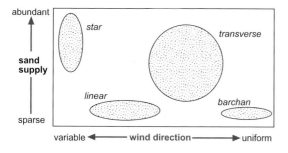

Fig. 8.10 Sand supply and the variability of prevailing wind directions control the types of dunes formed.

fans (9.5) or associated with the channels of ephemeral rivers (9.2.3). The sandy component of these deposits is subsequently reworked by the wind and redeposited in other parts of the basin in aeolian dune complexes. Water from these rivers and fans ponds in the basin to form ephemeral lakes and these temporary lakes dry up to leave deposits of mud and evaporite minerals (10.4). Through time the positions of the ephemeral lakes, sand dunes and the alluvial fans will change, and the deposits of these three subenvironments will be preserved as intercalated beds in the succession of strata (5.6.3).

Fig. 8.11 Aeolian dune cross-bedding in sands deposited in a desert: the view is approximately 5 m high.

The dominant factor in determining the distribution and extent of sandy deserts is climate. Arid conditions are necessary to inhibit the development of plants and a soil that would stabilise loose sediment, and an absence of abundant surface water prevents sediment from being reworked and removed by fluvial processes. Sand accumulates to form an erg where there are local or regional depressions: these may range from small build-ups of sand adjacent to topo-

graphy to broad areas of the continent covering many thousands of square kilometres.

8.5.1 Water table

The land surface in sandy deserts is mainly dry, but if the substrate is porous sediment or rock there will be **groundwater** below the surface. The level below the surface of this groundwater, the **water table**, is determined by the amount of water that is charging the water-bearing strata, the **aquifer**, and the relative level of the nearest lake or sea (Fig. 8.14). Charge to the aquifer is from areas around the desert that receive rainfall, and direct precipitation on the desert itself. The level of a lake in these settings will be largely determined by the climate and, if the erg borders the ocean, the sea level will be controlled by a number of local and global factors (23.8). A rise in the water table will affect aeolian processes in the erg if it comes up to the level of the interdune areas: wet interdune sediment will not be picked up by the wind, so it becomes stable and not available for aeolian reworking (Fig. 8.14). A rise in water table therefore tends to promote the accumulation of sediment within the erg. Conversely, a fall in water table from

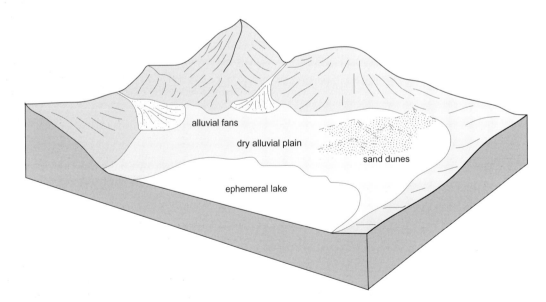

Fig. 8.12 Depositional environments in arid regions: coarse material is deposited on alluvial fans, sand accumulates to form aeolian dunes and occasional rainfall feeds ephemeral lakes where mud and evaporite minerals are deposited.

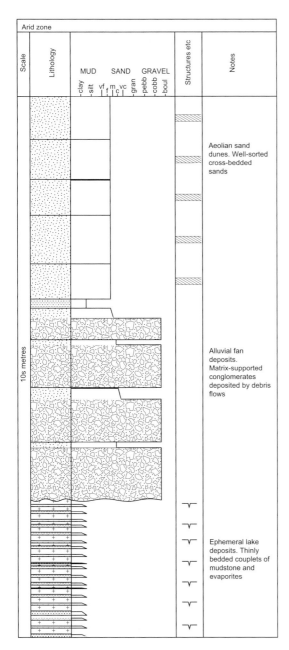

Fig. 8.13 Graphic sedimentary log of the arid-zone environments shown in Fig. 8.12.

the level of the interdunes will make more sediment available to be transported by the wind and this material may be removed from the area of the erg, that is, there will be net erosion. The relationship

between sea level and sediment accumulation in other environments is considered in Chapter 23.

8.5.2 Global climate variations

The formation of ergs requires an appropriate configuration of topography and wind patterns within a suitable climate belt. Modern sandy deserts are in the warm subtropical regions, which have predominantly dry, offshore wind patterns: most, in fact, lie to the western sides of continents in belts of westerly winds that have lost all of their moisture while crossing the eastern side of the continent (Fig. 8.15). Although similar conditions are likely to have existed at many times and places through Earth history, the number and extent of sandy desert areas are likely to have varied as plate movements rearranged the continents.

Global climate is also known to have changed through time. There have been periods of 'greenhouse' conditions when the temperature worldwide was warmer, and 'icehouse' periods when the whole world was cooler. During ice ages large ice sheets formed on one or both of the polar regions. The increased areas of ice created larger areas of high pressure, and there would have been steep pressure gradients between the expanded polar regions and the lower pressure equatorial belt. These conditions resulted in belts of strong winds in the subtropical regions, and hence increased potential for aeolian transport and deposition (Fig. 8.16). The large ergs of some modern deserts may be relics from the Pleistocene when they were very active, but have since become largely immobile. It is also notable that there are extensive aeolian deposits in the Permian of northern Europe, a time of Gondwana glaciation in the southern hemisphere.

8.5.3 Colour in desert sediments

The sands in modern deserts such as the Sahara are generally yellow. This colour is due to the presence of iron minerals, which occur as very fine coatings to the sand grains, particularly the iron hydroxide goethite (3.5.1) $(Fe(OH)_x)$, which is a dull yellow mineral. Oxidation of goethite forms the common iron oxide mineral haematite, Fe_2O_3, and this very common mineral has a strong red colour when it is very finely disseminated as a coating on sand grains.

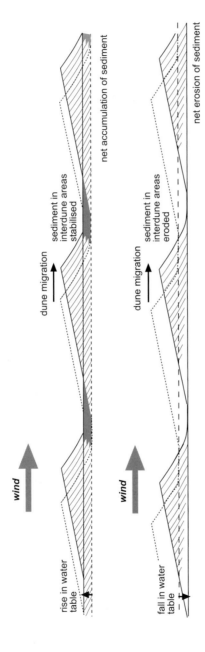

Fig. 8.14 The preservation of aeolian dune deposits is influenced by the level of the groundwater table: if the water table is high the interdune areas are wet and the sand is not reworked by the wind.

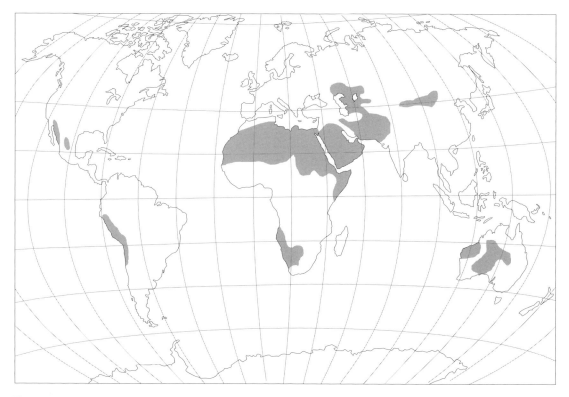

Fig. 8.15 The global distribution of modern deserts: most lie within 40° of the Equator.

Some modern desert sands are strongly oxidized, with haematite coatings on the grains giving them a vivid orange-red colour. Green, grey and black sediments are unlikely to be found in desert environments because these colours are due to reduced iron oxide (FeO, which is green) and the presence of dark organic material preserved in the sediment: preservation of these materials is unlikely in an environment that is mainly dry and exposed to the air, and therefore strongly oxidising.

Sedimentary successions composed of strongly reddened sandstone and mudstone are sometimes referred to as **red beds** (Turner 1980). It is tempting to use the presence of haematite as evidence that the sediments were deposited in an oxidising continental environment. However, some caution is needed when using the colour of the beds as an indicator of depositional environment. First, not all desert deposits are red in the first place, and second, the colour may be the result of oxidation after deposition (a diagenetic process – 18.2.4). There are also cases of sediments that

are deposited in other environments which are also shades of red: for example, mud deposited in deep oceans (16.5) may contain aeolian dust that includes haematite, giving the sediment a reddish colour.

8.5.4 Life in deserts and fossils in aeolian deposits

The absence of regular supplies of water in deserts makes them harsh environments for most plants and animals. A few specialised plants are able to survive long periods of drought and these form the bottom of the food chain for insects, reptiles, birds and mammals, but none occur in large quantities. The interdune areas are the most favourable places to support life because these can be places where water temporarily ponds and are the closest points to any groundwater if the water table is relatively close to the surface. In terms of fossil preservation, the paucity of organisms in deserts is compounded by the fact that

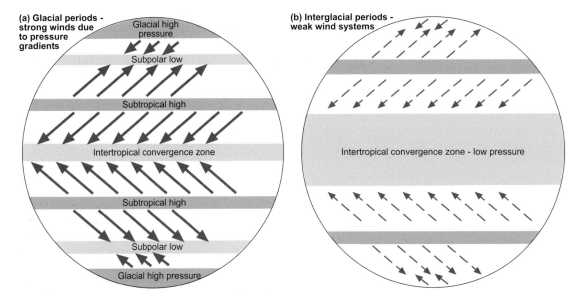

Fig. 8.16 During glacial periods the regions of polar high pressure are larger, creating stronger pressure gradients and hence stronger winds. In the absence of large high pressure areas at the poles in interglacial periods the pressure gradients are weaker and winds are consequently less strong.

they are also strongly oxidising environments, so plants and animals are likely to completely decompose and leave no fossil material. Only the most resistant animal remains, such as the bones of large animals such as dinosaurs, have much potential to be preserved in aeolian environments. Trace fossils (*11.7*) are also rare because few animals live on active sand dunes, but there is the possibility of walking traces being preserved in fine sediment in wet interdune areas. Strata deposited in desert environments are therefore likely to be barren of any fossils.

8.6 AEOLIAN DEPOSITS OUTSIDE DESERTS

8.6.1 Aeolian dust deposits

There are deposits of Quaternary age in eastern Europe, North America and China that are interpreted as accumulations of wind-blown dust (Pye 1987). These deposits, known as **loess**, locally occur in beds several metres thick made up predominantly of well-sorted silt-sized material, with little clay or sand-sized material present. The origin of loess is related to episodes of retreat of ice sheets, as large amounts of loose detritus

carried in the ice were released. In the cold periglacial environment in front of the receding ice colonisation by plants and stabilisation of the soil would have been slow, so the glacial debris was exposed on the outwash plains, where wind picked up and transported the silt-sized dust. This dust was probably transported over large parts of the globe but accumulated as loess deposits in some places. Similar processes probably occurred during other glacial episodes in Earth history, but pre-Quaternary loess deposits have not been recognised. The preservation potential of loess is likely to be quite low because it is soft, loose material that is easily reworked and mixed with other sediment.

Volcanism is an important source of dust in the atmosphere. Explosive eruptions can send plumes of volcanic ash high up into the atmosphere where it is distributed by wind. Coarser ash tends to be deposited close to the volcano (although in very large eruptions this can be hundreds of kilometres away – *17.6.2*), while the silt-sized ash particles can be transported around the world. Large amounts of atmospheric dust from eruptions can darken the sky, and it will gradually fall as fine sediment. A further source of atmospheric dust is from fires that propel soot (fine carbon) up into the air, where it can be redistributed by the wind. Despite the fine grain size, soot, volcanic and terrigenous dust can all be distinguished by geochemical analysis.

Aeolian dust is dispersed worldwide, but most of it ends up in other marine and continental depositional environments where it mixes with other sediment and its origin cannot easily be determined. In most places the proportion of aeolian dust is very low compared with other sediment being deposited, but there are some environments where terrigenous clastic deposition is very low, and the main source of silt and clay can be aeolian dust. Limestones formed in carbonate-forming environments can usually be shown to contain a residue of dust if the calcium carbonate is dissolved, and dust settled on ice sheets and glaciers may be seen as layers within the ice. The parts of the deep oceans that are distant from any continental margin receive very little sediment (*16.5*): airborne dust that settles through the water column can therefore be an important component of deep ocean deposits.

8.6.2 Aeolian sands in other environments

Beach dunes

Sand dunes built up by aeolian action can form adjacent to beaches in any climatic setting. In the intertidal zone of a foreshore loose sediment is subaerially exposed at low tide, and as it dries out it is available to be picked up and redeposited by the wind. Beach dune ridges form where the foreshore sediments are mainly sandy, exposed at low tide and subject to removal by onshore winds. The sand then accumulates at the head of the beach, either as a simple narrow ridge or sometimes extending for hundreds of metres inland. In humid climates the dunes become colonised by grasses, shrubs and trees that stabilise the sand and allow the ridges to build up metres to tens of metres thickness. The roots of these plants and burrowing animals disrupt any depositional stratification, so the cross-bedding characteristic of desert dunes may not be preserved in beach dune ridges. The association of beach dune ridges with other coastal facies is discussed in *13.2.1*.

Periglacial deposits

Glacial outwash areas (*7.4.3*) are places where loose detritus that has been released from melting ice remains exposed on the surface for long periods of time because plant growth and soil formation is slow in periglacial regions. Wind blowing over the outwash plain can pick up sand and redeposit it locally, usually against topographic features such as the side of a valley. These patches of aeolian sand may therefore occur intercalated with fluvio-glacial facies (*7.4.3*), but rarely form large deposits.

8.7 SUMMARY

Aeolian deposits occur mainly in arid environments where surface water is intermittent and there is little plant cover. Sands deposited in these desert areas are characteristically both compositionally and mineralogically mature with large-scale cross-bedding formed by the migration of dune bedforms. Oxidising conditions in deserts preclude the preservation of much fossil material, and sediments are typically red–yellow colours. Associated facies in arid regions are mud and evaporites deposited in ephemeral lakes and poorly sorted fluvial and alluvial fan deposits. Aeolian deposits are less common outside of desert environments, occurring as local sandy facies associated with beaches and glaciers, and as dust distributed over large distances into many different environments, but, apart from Quaternary loess, rarely in significant quantities.

Characteristics of aeolian deposits
- lithologies – sand and silt only
- mineralogy – mainly quartz, with rare examples of carbonate or other grains
- texture – well- to very well-sorted silt to medium sand
- fossils – rare in desert dune deposits, occasional vertebrate bones
- bed geometry – sheets or lenses of sand
- sedimentary structures – large-scale dune cross-bedding and parallel stratification in sands
- palaeocurrents – dune orientations reconstructed from cross-bedding indicate wind direction
- colour – yellow to red due to iron hydroxides and oxides
- facies associations – occur with alluvial fans, ephemeral river and lake facies in deserts, also with beach deposits or glacial outwash facies

FURTHER READING

Glennie. K.W. (1987) Desert sedimentary environments, past and present – a summary. *Sedimentary Geology*, **50**, 135–165.

Kocurek, G.A. (1996) Desert aeolian systems. In: *Sedimentary Environments: Processes, Facies and Stratigraphy* (Ed. Reading, H.G.). Blackwell Science, Oxford; 125–153.

Livingstone, I., Wiggs, G.F.S. & Weaver, C.M. (2007) Geomorphology of desert sand dunes: A review of recent progress. *Earth-Science Reviews*, **80**, 239–257.

Mountney, N.P. (2006) Eolian facies models. In: *Facies Models Revisited* (Eds Walker, R.G., & Posamentier, H.). Special Publication 84, Society of Economic Paleontologists and Mineralogists, Tulsa, OK; 19–83.

Pye, K. & Lancaster, N. (Eds) (1993) *Aeolian Sediments Ancient and Modern*. Special Publication 16, International Association of Sedimentologists. Blackwell Science, Oxford.

9

Rivers and Alluvial Fans

Rivers are an important feature of most landscapes, acting as the principal mechanism for the transport of weathered debris away from upland areas and carrying it to lakes and seas, where much of the clastic sediment is deposited. River systems can also be depositional, accumulating sediment within channels and on floodplains. The grain size and the sedimentary structures in the river channel deposits are determined by the supply of detritus, the gradient of the river, the total discharge and seasonal variations in flow. Overbank deposition consists mainly of finer-grained sediment, and organic activity on alluvial plains contributes to the formation of soils, which can be recognised in the stratigraphic record as palaeosols. Water flows over the land surface also occur as unconfined sheetfloods and debris flows that form alluvial fans at the edges of alluvial plains. Fluvial and alluvial deposits in the stratigraphic record provide evidence of tectonic activity and indications of the palaeoclimate at the time of deposition. Comparisons between modern and ancient river systems should be carried out with care because continental environments have changed dramatically through geological time as land plant and animal communities have evolved.

9.1 FLUVIAL AND ALLUVIAL SYSTEMS

Three geomorphological zones can be recognised within fluvial and alluvial systems (Fig. 9.1). In the **erosional zone** the streams are actively downcutting, removing bedrock from the valley floor and from the valley sides via downslope movement of material into the stream bed. In the **transfer zone**, the gradient is lower, streams and rivers are not actively eroding, but nor is this a site of deposition. The lower part of the system is the **depositional zone**, where sediment is deposited in the river channels and on the floodplains of a fluvial system or on the surface of an alluvial fan. These three components are not present in all systems: some may be wholly erosional as far as the sea or a lake, and others may not display a transfer zone. The erosional part of a fluvial system contributes a substantial proportion of the clastic sediment provided for deposition in other sedimentary environments, and is considered in Chapter 6: the depositional zone is the subject of this chapter.

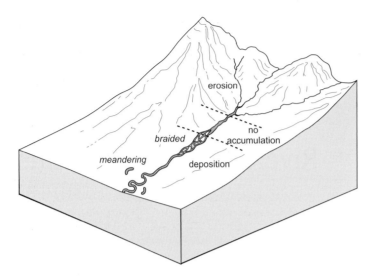

Fig. 9.1 The geomorphological zones in alluvial and fluvial systems: in general braided rivers tend to occur in more proximal areas and meandering rivers occur further downstream.

Water flow in rivers and streams is normally confined to **channels**, which are depressions or scours in the land surface that contain the flow. The **overbank** area or **floodplain** is the area of land between or beyond the channels that (apart from rain) receives water only when the river is in flood. Together the channel and overbank settings comprise the **fluvial** environment. **Alluvial** is a more general term for land surface processes that involve the flow of water. It includes features such as a water-lain fan of detritus (an alluvial fan – 9.5) that are not necessarily related to rivers. An **alluvial plain** is a general term for a low-relief continental area where sediment is accumulating, which may include the floodplains of individual rivers.

9.1.1 Catchment and discharge

The area of ground that supplies water to a river system is the **catchment area** (sometimes also referred to as the **drainage basin**). Rivers and streams are mainly fed by surface run-off and groundwater from subsurface aquifers in the catchment area following periods of rain. Soils act as a sponge soaking up moisture and gradually releasing it out into the streams. A continuous supply of water can be provided if rainfall is frequent enough to stop the soils drying out. Two factors are important in controlling the supply of water to a river system. First, the size of the catchment area: a small area has a more limited

capacity for storing water in the soil and as groundwater than a large catchment area. The second factor is the climate: catchment areas in temperate or tropical regions where there is regular rainfall remain wet throughout the year and keep the river supplied with water.

A large river system with a catchment area that experiences year-round rainfall is constantly supplied with water and the **discharge** (the volume of water flowing in a river in a time period) shows only a moderate variation through the year. These are called **perennial** fluvial systems. In contrast, rivers that have much smaller drainage areas and/or seasonal rainfall may have highly variable discharge. If the rivers are dry for long periods of time and only experience flow after there has been sufficient rain in the catchment area they are considered to be **ephemeral** rivers.

9.1.2 Flow in channels

The main characteristic of a fluvial system is that most of the time the flow is concentrated within channels. When the water level is well below the level of the channel banks it is at **low flow stage**. A river with water flowing close to or at the level of the bank is at **high flow stage** or **bank-full flow**. At times when the volume of water being supplied to a particular section of the river exceeds the volume that can be contained within the channel, the river **floods** and **overbank flow** occurs on the floodplain adjacent to the channel (Fig. 9.2).

As water flows in a channel it is slowed down by friction with the floor of the channel, the banks and the air above. These frictional effects decrease away from the edges of the flow to the deepest part of the channel where there is the highest velocity flow. The line of the deepest part of the channel is called the **thalweg**. The existence of the thalweg and its position in a channel is important to the scouring of the banks and the sites of deposition in all channels.

9.2 RIVER FORMS

Rivers in the depositional tract can have a variety of forms, with the principal variables being: (a) how straight or sinuous the channel is; (b) the presence or absence of depositonal bars of sand or gravel within the channel; (c) the number of separate channels that are present in a stretch of the river. A number of 'end-member' river types can be recognised (Miall 1978; Cant 1982), with all variations and intermediates between them possible (Fig. 9.3). A straight channel without bars is the simplest form but is relatively uncommon. A **braided river** contains mid-channel bars that are covered at bank-full flow, in contrast to an **anastomosing** (also known as **anabranching**) river, which consists of multiple, interconnected channels that are separated by areas of floodplain (Makaske 2001). Both braided and anastomosing river channels can be sinuous, and sinuous rivers

Fig. 9.2 A sandy river channel and adjacent overbank area: the river is at low-flow stage exposing areas of sand deposited in the channel.

that have depositional bars only on the insides of bends are called **meandering**.

When considering the deposits of ancient rivers, the processes of deposition on the mid-channel bars in braided streams and the deposition on the inner banks of meandering river bends are found to be important mechanisms for accumulating sediment. 'Braided' and 'meandering' are therefore useful ways of categorising ancient fluvial deposits, but considerable variations in and combinations of these main themes exist both in modern and ancient systems. Furthermore, not all rivers are filled by deposition out of flow in the channels themselves (9.2.4).

Anastomosing or anabranching rivers are seen today mostly in places where the banks are stabilised by vegetation, which inhibits the lateral migration of channels (Smith & Smith 1980; Smith 1983), but anastomosing rivers are also known from more arid regions with sparse vegetation. The positions of channels tend to remain fairly fixed but new channels may develop as a consequence of flooding as the water makes a new course across the floodplain, leaving an old channel abandoned. Recognition of anastomosing rivers in the stratigraphic record is problematic because the key feature is that there are several separate active channels. In ancient deposits it is not possible to unequivocally demonstrate that two or more channels were active at the same time and a similar pattern may form as a result of a single channel repeatedly changing position (9.2.4).

9.2.1 Bedload (braided) rivers

Rivers with a high proportion of sediment carried by rolling and saltation along the channel floor are referred to as **bedload rivers**. Where the bedload is deposited as bars (4.3.3) of sand or gravel in the channel the flow is divided to give the river a braided form (Figs 9.4 & 9.5). The bars in a braided river channel are exposed at low flow stages, but are covered when the flow is at bank-full level. Flow is generally strongest between the bars and the coarsest material will be transported and deposited on the channel floor to form an accumulation of larger clasts, or **coarse lag** (Fig. 9.6). The bars within the channel may vary in shape and size: **longitudinal bars** are elongate along the axis of the channel, and bars that are wider than they are long, spreading across the channel are called **transverse bars** and

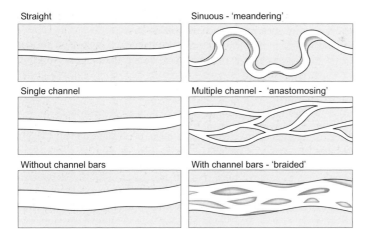

Straight

Sinuous - 'meandering'

Single channel

Multiple channel - 'anastomosing'

Without channel bars

With channel bars - 'braided'

Fig. 9.3 Several types of river can be distinguished, based on whether the river channel is straight or sinuous (meandering), has one or multiple channels (anastomosing), and has in-channel bars (braided). Combinations of these forms can often occur.

crescentic bars with their apex pointing downstream are **linguoid bars** (Smith 1978; Church & Jones 1982). Bars may consist of sand, gravel or a mixture of both ranges of clast size (**compound bars**).

Movement of the bedload occurs mainly at high flow stages when the bars are submerged in water. Sediment is brought downstream to a bar by the river flow and erosion of the upstream side of the bar may occur. In bars composed of gravelly material the clasts accumulate as inclined parallel layers on the downstream bar faces; some accretion may also occur on the lateral margins of the bar. Longitudinal bars have low relief and their migration forms deposits showing a poorly defined low-angle cross-stratification in a downstream direction. Transverse and linguoid bars have a higher relief and generate well-defined cross-stratification dipping downstream. The deposits of a migrating gravel

bar in a braided river therefore form beds of cross-stratified granules, pebbles or cobbles that lithify to form a conglomerate. In sandy braided rivers the bars are seen to comprise a complex of subaqueous dunes over the bar surface (Fig. 9.7). These subaqueous dunes migrate over the surface of the bar in the stream current to build up stacks of cross-bedded sands. Arcuate (linguoid) subaqueous dunes normally dominate, creating trough cross-bedding, but straight-crested subaqueous dunes producing planar cross-bedded sands also occur. Compound bars comprise cross-stratified gravel with lenses of cross-bedded sand or there may be lenses of gravel in sandy bar deposits.

Bars continue to migrate until the channel moves sideways leaving the bar out of the main flow of the water (Fig. 9.8). It will subsequently be covered by overbank deposits or the bars of another channel

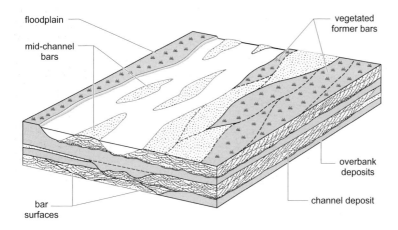

floodplain

mid-channel bars

vegetated former bars

overbank deposits

channel deposit

bar surfaces

Fig. 9.4 Main morphological features of a braided river. Deposition of sand and/or gravel occurs on mid-channel bars.

Fig. 9.5 Mid-channel gravel bars in a braided river.

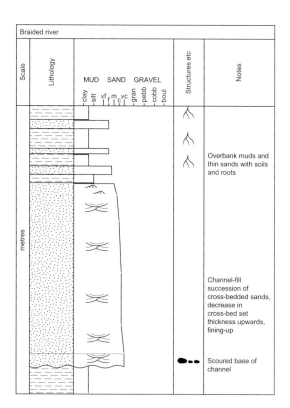

Fig. 9.6 A schematic graphic sedimentary log of braided river deposits.

(Fig. 9.9). A characteristic sedimentary succession (Fig. 9.6) formed by deposition in a braided river environment can be described. At the base there will

be an erosion surface representing the base of the channel and this will be overlain by a basal lag of coarse clasts deposited on the channel floor. In a gravelly braided river the bar deposits will commonly consist of cross-stratified granules, pebbles or rarely cobbles in a single set. A sandy bar composed of stacked sets of subaqueous dune deposits will form a succession of cross-bedded sands. As the flow is stronger in the lower part of the channel the subaqueous dunes, and hence the cross-beds, tend to be larger at the bottom of the bar, decreasing in set size upwards. Finer sands or silts on the top of a bar deposit represent the abandonment of the bar when it is no longer actively moving. There is therefore an overall fining-up of this channel-fill succession (Fig. 9.6). The thickness may represent the depth of the original channel if

Fig. 9.7 Sandy dune bedforms on a mid-channel bar in a braided river.

Fig. 9.8 This large braided river has moved laterally from right to left.

it is complete, but it is common for the top part to be eroded by the scour of a later channel.

In regions where braided rivers repeatedly change position on the alluvial plain, a broad, extensive region of gravelly bar deposits many times wider than the river channel will result. These **braidplains** are found in areas with very wet climates or where there is little vegetation to stabilise the river banks (e.g. glacial outwash areas: *7.4.3*). The succession built up in this setting will consist of stacks of cross-stratified conglomerate, and it can be difficult to identify the scour surfaces that mark the base of a

channel and hence recognise individual channel-fill successions.

9.2.2 Mixed load (meandering) rivers

In plan view the thalweg (*9.1.2*) in a river is not straight even if the channel banks are straight and parallel (Fig. 9.10): it will follow a sinuous path, moving from side to side along the length of the channel. In any part of the river the bank closest to the thalweg has relatively fast flowing water against it

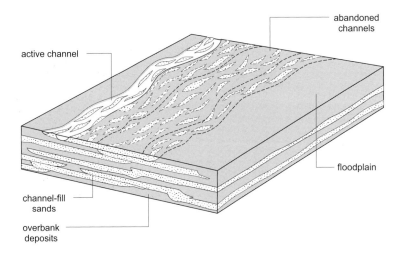

active channel

abandoned channels

floodplain

channel-fill sands

overbank deposits

Fig. 9.9 Depositional architecture of a braided river: lateral migration of the channel and the abandonment of bars leads to the build-up of channel-fill successions.

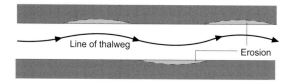

Fig. 9.10 Flow in a river follows the sinuous thalweg resulting in erosion of the bank in places.

while the opposite bank has slower flowing water alongside. Meanders develop by the erosion of the bank closest to the thalweg, accompanied by deposition on the opposite side of the channel where the flow is sluggish and the bedload can no longer be carried. With continued erosion of the outer bank and deposition of bedload on the inner bank the channel develops a bend and meander loops are formed (Figs 9.11 & 9.12). A distinction between river sinuosity and meandering form should be recognised: a river is considered to be sinuous if the distance measured along a stretch of channel divided by the direct distance between those points is greater than 1.5; a river is considered to be meandering if there is accumulation of sediment on the inside of bends, as described below.

Meandering rivers transport and deposit a mixture of suspended and bedload (***mixed load***) (Schumm 1981). The bedload is carried by the flow in the channel, with the coarsest material carried in the deepest parts of the channel. Finer bedload is also

carried in shallower parts of the flow and is deposited along the inner bend of a meander loop where friction reduces the flow velocity. The deposits of a meander bend have a characteristic profile of coarser material at the base, becoming progressively finer-grained up the inner bank (Fig. 9.11). The faster flow in the deeper parts of the channel forms subaqueous dunes in the sediment that develop trough or planar cross-bedding as the sand accumulates. Higher up on the inner bank where the flow is slower, ripples form in the finer sand, producing cross-lamination. A channel moving sideways by erosion on the outer bank and deposition on the inner bank is undergoing ***lateral migration***, and the deposit on the inner bank is referred to as a ***point bar***. A point bar deposit will show a fining-up from coarser material at the base to finer at the top (Fig. 9.13) and it may also show larger scale cross-bedding at the base and smaller sets of cross-lamination nearer to the top. As the channel migrates the top of the point bar becomes the edge of the floodplain and the fining-upward succession of the point bar will be capped by overbank deposits.

Stages in the lateral migration of the point bar of a meandering river can sometimes be recognised as inclined surfaces within the channel-fill succession (Fig. 9.11). These ***lateral accretion surfaces*** are most distinct when there has been an episode of low discharge allowing a layer of finer sediment to be deposited on the point-bar surface (Allen 1965; Bridge 2003; Collinson et al. 2006). These surfaces

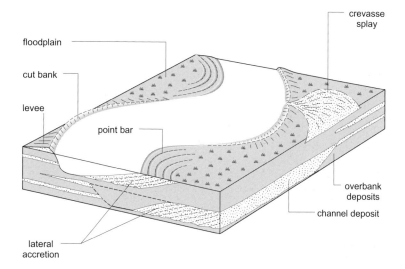

Fig. 9.11 Main morphological features of a meandering river. Deposition occurs on the point bar on the inner side of a bend while erosion occurs on the opposite cut bank. Levees form when flood waters rapidly deposit sediment close to the bank and crevasse splays are created when the levee is breached.

Fig. 9.12 The point bars on the inside bends of this meandering river have been exposed during a period of low flow in the channel.

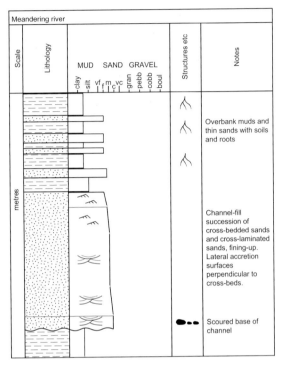

Fig. 9.13 A schematic graphic sedimentary log of meandering river deposits.

are low angle, less than 15°, and, because they represent the point-bar surface, are inclined from the river bank towards the deepest part of the channel – i.e. perpendicular to the flow direction. The scale of the cross-stratification will therefore be larger (as much as the channel depth) than other cross-bedding, and it will be perpendicular to any other palaeoflow indicators, such as cross-bedding produced by dune migration and ripple cross-lamination. The recognition of lateral accretion surfaces (also know as *epsilon cross-stratification*; Allen 1965) within the fining-up succession of a channel-fill deposit is therefore a reliable indication that the river channel was meandering. The outer bend of a meander loop will be a bank made up of floodplain deposits (9.3) that will be mainly muddy sediment. Dried mud is very cohesive (2.4.5) and pieces of the muddy bank material will not easily disintegrate when they form clasts carried by the river flow. These *mud clasts* will be deposited along with sand in the deeper parts of the channel, and will be preserved in the basal part of the channel-fill succession (Fig. 9.13).

During periods of high-stage flow, water may take a short-cut over the top of a point bar. This flow may become concentrated into a *chute channel* (Fig. 9.14) that cuts across the top of the inner bank of the meander. Chute channels may be semi-permanent features of a point bar, but they are only active during high-stage flow. They may be recognised in the deposits of a meandering river as a scour that cuts through lateral accretion surfaces.

Fig. 9.14 A pale band across the inside of this meander bend marks the path of a chute channel that cuts across the point bar.

The river flow may also take a short-cut between meander loops when the river floods: this may result in a new section of channel developing, and the longer loop of the meander built becoming abandoned

(Fig. 9.15). The abandoned meander loop becomes isolated as an ***oxbow lake*** (Fig. 9.15) and will remain as an area of standing water until it becomes filled up by deposition from floods and/or choked by vegetation. The deposits of an oxbow lake may be recognised in ancient fluvial sediments as channel fills made up of fine-grained, sometimes carbonaceous, sediment (Fig. 9.16).

9.2.3 Ephemeral rivers

In temperate or tropical climatic settings that have rainfall throughout the year, there is little variation in river flow, but in regions with strongly seasonal rainfall, due to a monsoonal climate, or with seasonal snow-melt in a high mountain or circumpolar area, discharge in a river system can be variable at different times of the year. During the dry season, smaller streams may dry up completely. In deserts (8.2) where the rainfall is irregular, whole river systems may be dry for years between rainstorm events that lead to temporary flow. Many alluvial fans (9.5) are also ephemeral.

In upland areas with dry climates, weathering results in detritus remaining on the hillslope or clasts may move by gravity down to the valley floor. Accumulation may continue for many years until there is a rainstorm of sufficient magnitude to create a flow of water that moves the detritus as bedload in the river or as a debris flow (4.5.1). The flow may carry the

Fig. 9.16 A channel is commonly not filled with sand: in this case the form of a channel is picked out by steep banks on either side, but the fill of the channel is mainly mud.

sediment many kilometres along normally dry channels cut into an alluvial plain. The deposits of these ephemeral flows are characteristically poorly sorted, consisting of angular or subangular gravel clasts in a matrix of sand and mud. Gravel clasts may develop imbrication, horizontal stratification may form and the deposits are often normally graded as the flow decreases strength through time. Longitudinal bars may develop and create some low-angle cross-stratification, but other bar and dune forms do not usually form. The deposits are restricted by the width of the channel but the channel may migrate laterally or there may be multiple channels on an alluvial plain

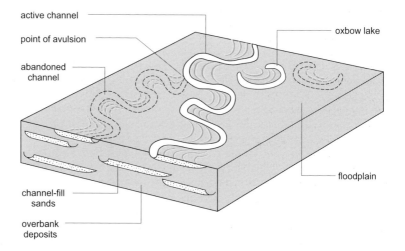

Fig. 9.15 Depositional architecture of a meandering river: sandstone bodies formed by the lateral migration of the river channel remain isolated when the channel avulses or is cut-off to form an oxbow lake.

that merge to form a more extensive deposit. The term **wadi** is commonly used for a river or stream in a desert with ephemeral flow and the resulting deposits are therefore sometimes referred to as **wadi gravels**.

9.2.4 Channel-filling processes

The channel-fill succession in both meandering and braided rivers described above is built up as a result of sideways movement or lateral migration of the active part of the channel. Accumulation and possible preservation of river channel deposits can occur only if the river changes its position in some way, either by shifting sideways, as above, or if the channel changes position on the floodplain, a process known as **avulsion**. When a river **avulses** part of the old river course is completely abandoned and a new channel is scoured into the land surface (Fig. 9.15). Oxbow lakes are an example of abandonment of a short stretch, but much longer tracts of any type of river channel may be involved. When avulsion occurs the flow in the old river course reduces in volume and slows down, and the bedload will be deposited. A decrease in the amount of water supplied limits the capacity of the channel to carry sediment and the water gradually becomes sluggish, depositing its suspended load. Abandonment of the old river channel will leave it with sluggish water containing only suspended load as all the bedload is diverted into the new course. Abandoned and empty stretches of river channel are unlikely to remain empty for very long because when the river floods from its new course it will carry sediment across the floodplain to the old channel where sediment will gradually accumulate. The final fill of any river channel is therefore most likely to be fine-grained overbank sedimentation related to a different river course. Channels entirely filled with mud may be very difficult to distinguish from overbank sediments in the stratigraphic record.

Recognition of channels is one of the key criteria for identifying the deposits of fluvial systems within a sedimentary succession. However, the cut banks of channel margins are not always easy to recognise. The lateral migration of the river channel may result in a succession of point bar or mid-channel bar deposits that is hundreds of metres across, even though the channel itself may be only a few tens of metres wide at any time. This deposit may be wider than the outcrop exposed and in some cases the rivers migrate laterally across the whole floodplain, leaving channel margins at the edges of the valley. It is therefore often necessary to use the characteristics of the vertical successions deposited within channels (Figs 9.6 & 9.13) as indicators of fluvial depositional environments.

9.2.5 Trends in fluvial systems

There is normally a general trend of reduction in gradient of a river downstream through the depositional tract. The slope of the river and the discharge affect the velocity of the flow, which in turn controls the ability of the river to scour and the size of the material that can be carried as bedload and suspended load. Gravelly braided rivers have the steepest depositional gradient (although the angle is typically less than half a degree) and bars of pebbles, cobbles and boulders form. Finer debris is mostly carried through to the lower reaches of the river. At lower gradients the sandy bedload is deposited on bars in braided rivers, the flow having decreased sufficiently to deposit most of the gravel upstream. A meandering pattern tends to develop at very gentle gradients (around a hundredth of a degree) in rivers carrying fine-grained sediment as mixed bedload and suspended material (Collinson 1986).

The erosional tracts of rivers exhibit a **tributary** drainage pattern as small streams merge into the trunk channel (a **dendritic** pattern; Fig. 9.1). This pattern may extend into the depositional tract. Most rivers flow as a single channel to a lake margin or the shoreline of a sea, where a delta or estuary may be formed. However, rivers in relatively arid regions may lose so much water through evaporation and soak-away into the dry floodplain that they dry up before reaching a standing water body. In some **enclosed** (or **endorheic**) **basins** (which do not have an outlet to the open ocean) with an arid climate there may not be a permanent lake (10.4). Due to the loss of water, the channels become smaller downstream and end in splays of water and sediment called **terminal fans** (Friend 1978). Rivers that show these characteristics may be referred to as **fluvial distributary systems** (Nichols & Fisher 2007), although it should be noted that it is mainly sediment that is being distributed. At any time most of the water flow will be in one principal channel, with other, minor channels splitting off from it (a **bifurcating pattern**): a minor channel may subsequently take over as the main flow route, or a new channel develops as a result of avulsion. Through time the channels occupy different radial positions and the deposits form a fan-shaped body of sediment (see also alluvial fans, 9.5).

9.3 FLOODPLAIN DEPOSITION

The areas between and beyond the river channels are as important as the channels themselves from the point of view of sediment accumulation. When the discharge exceeds the capacity of the channel, water flows over the banks and out onto the floodplain where overbank or floodplain deposition occurs. Most of the sediment carried out onto the floodplain is suspended load that will be mainly clay- and silt-sized debris but may include fine sand if the flow is rapid enough to carry sand in suspension. As water leaves the confines of the channel it spreads out and loses velocity very quickly. The drop in velocity prompts the deposition of the sandy and silty suspended load, leaving only clay in suspension (Hughes & Lewin 1982). The sand and silt is deposited as a thin sheet over the floodplain, which may show current ripple or horizontal lamination: rapid deposition may result in the formation of climbing ripple cross-lamination *(4.3.1)*. The remaining suspended load will be deposited as the floodwaters dry out and soak away after the flow has subsided.

Sheets of sand and silt deposited during floods are thickest near to the channel bank because coarser suspended load is dumped quickly by the floodwaters as soon as they start flowing away from the channel. Repeated deposition of sand close to the channel edge leads to the formation of a *levée*, a bank of sediment at the channel edge which is higher than the level of the floodplain (Fig. 9.11). Through time the level of the bottom of the channel can become raised by sedimentation in the channel and the level of water at bankfull flow becomes higher than the floodplain level. When the levée breaks, water laden with sediment is carried out onto the floodplain to form a *crevasse splay* (Fig. 9.11), a low cone of sediment formed by water flowing through the breach in the bank and out onto the floodplain (O'Brien & Wells 1986). The breach in the levée does not occur instantaneously but as a gradually deepening and widening conduit for water to pass out onto the floodplain. Initially only a small amount of water and sediment will pass through but the volume of water and the grain size

of the detritus carried increase until the breach reaches full size. Crevasse splay deposits are therefore characterised by an initial upward coarsening of the sediment particle size. They are typically lenticular in three dimensions. Channels within crevasse splays may develop into new river channels and carry progressively more water until avulsion occurs.

The primary depositional structures commonly observed in floodplain sediments are:
1 very thin and thin beds normally graded from sand to mud;
2 evidence of initial rapid flow (plane parallel lamination) quickly waning and accompanied by rapid deposition (climbing ripple lamination);
3 thin sheets of sediment, often only a few centimetres thick but extending for tens to hundreds of metres;
4 erosion at the base of the overbank sheet sandstone beds is normally localised to areas near the channel where the flow is most vigorous;
5 evidence of soil formation (9.7).

There is usually a general trend towards the deposition of more overbank sediments further downstream in a fluvial system. In the upper parts of the fluvial depositional tract, the river valley is likely to be narrow, and as braided rivers laterally migrate from side to side across the valley any floodplain deposits will be reworked by channel erosion. Floodplain deposits therefore sometimes have a lower chance of being preserved associated with braided river facies. In the wider alluvial plain normally associated with the lower parts of the fluvial depositional tract, meandering river deposits are commonly associated with a higher proportion of floodplain facies.

9.4 PATTERNS IN FLUVIAL DEPOSITS

9.4.1 Architecture of fluvial successions

The three-dimensional arrangement of channel and overbank deposits in a fluvial succession is commonly referred to as the *architecture* of the beds. The architecture is described in terms of the shape and size of the sand or gravel beds deposited in channels and the proportion of 'in-channel' deposits relative to the finer overbank facies. The thickness of the channel-fill deposit is determined by the depth of the rivers and their width is governed by the processes of avulsion and lateral migration of the channel. There is a tendency for nearly all rivers (meandering and braided) to shift sideways through time by erosion of one bank and deposition on the opposite side. Lateral migration continues until avulsion of the river causes the channel to be abandoned. If avulsion is frequent, there is less time for lateral migration to occur and the architecture will be characterised by narrow channel deposits (Fig. 9.17). Avulsion is frequent in rivers that are in regions of tectonic activity, where frequent faulting and related earthquakes affect the river course, and in settings where overbank flooding is frequent, resulting in weaker banks that make it easier for the river to change course.

Lateral migration is slowed down if the river banks are stable. Bank stability is governed by the nature of the floodplain: muddy floodplain deposits form stable banks because clay is cohesive and is not easily eroded. The type and abundance of vegetation are also important because dense vegetation, particularly grass with its fibrous roots, can very effectively bind the soils of a floodplain and stabilise the river banks. Vegetation also causes increased surface roughness, which slows overland flow. In arid or cold regions where vegetation is sparse, bank stability is decreased and flows on the floodplain are faster and therefore more likely to erode.

Rates of subsidence and the quantity of sediment supplied to the floodplain also affect the architecture of fluvial deposits (Fig. 9.17). With rapid subsidence and high sediment supply, aggradation on the floodplain will result in a high proportion of fine deposits. In regions of slow subsidence and reduced sediment supply to the overbank areas relatively more in-channel deposits will be preserved (Bridge & Leeder 1979).

9.4.2 Palaeocurrents in fluvial systems

Palaeocurrent data are a very valuable aid to the reconstruction of the palaeogeography of fluvial deposits. It may be used to determine the location of the source area from which the sediment was derived and it is possible to indicate the general position of the mouth of the river and hence the shoreline. Sedimentary structures that can be used as flow indicators in fluvial deposits include the orientation of channel margins, cross-bedding in sandstone and clast imbrication in conglomerate. An individual cross-bed is

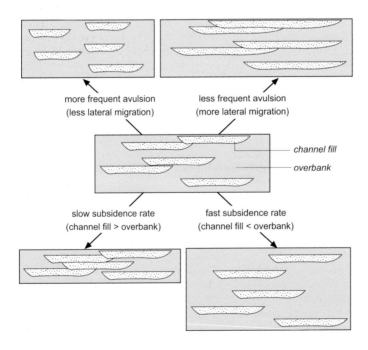

more frequent avulsion
(less lateral migration)

less frequent avulsion
(more lateral migration)

channel fill

overbank

slow subsidence rate
(channel fill > overbank)

fast subsidence rate
(channel fill < overbank)

Fig. 9.17 The architecture of fluvial deposits is determined by the rates of subsidence and frequency of avulsion.

formed by migration of a bar or dune bedform, but these features may be migrating obliquely to the main channel flow. Palaeoflow directions determined from cross-beds in braided river bar deposits can show a variance of around 60° either side of the mean channel flow. The sinuous character of a meandering river channel will also result in flow indications that will range at least 90° either side of the overall flow direction of the river. Large numbers of measurements from cross-bedding are therefore required to obtain a mean value that will approximate to the overall flow direction in the channel. It is also important to distinguish between channel and overbank facies, because flow directions in the latter will often be perpendicular to the channel.

9.4.3 Fluvial deposits and palaeogeography

Within ancient fluvial deposits, the recognition of different fluvial depositional styles (e.g. braided and meandering channel fill) along with changes in grain size of deposit can be used to reconstruct the palaeogeography and provide evidence of changes through time. It may be expected that a conglomerate deposited by a pebbly braided river will have, down palaeoflow, equivalent age sandstone beds deposited in a sandy braided river, and that this may in turn pass down palaeoflow to finer grained deposits with the characteristics of deposition by a meandering river (Fig. 9.1). In additional to these spatial variations in the fluvial deposits, a change from braided river deposits up through the succession (and therefore through time) to meandering river deposits may indicate a decrease in the gradient of the river and/or a reduction in the discharge in the river system.

Rivers vary in size from small streams only metres in width and tens of centimetres deep to rivers tens of kilometres wide and tens of metres deep. This range in channel size over several orders of magnitude is also seen in channel-fill deposits in fluvial successions and the dimensions of deposits can be used to infer the size of the river, and hence the size of the drainage basin from which it was supplied. Provenance studies on fluvial sediments provide more details of the drainage basin area, indicating the types of bedrock that were exposed at the time of deposition and helping to build up a palaeogeographical picture. Information about the palaeoclimate can also be determined if ephemeral and perennial flow conditions can be established from the character of the fluvial deposits, but the most sensitive indicators in continental facies of palaeoclimate are palaeosols (9.7).

9.5 ALLUVIAL FANS

Alluvial fans are cones of detritus that form at a break in slope at the edge of an alluvial plain. They are formed by deposition from a flow of water and sediment coming from an erosional realm adjacent to the basin. The term alluvial fan has been used in geological and geographical literature to describe a wide variety of deposits with an approximately conical shape, including deltas and large distributary river systems. Some authors (e.g. Blair & McPherson 1994) restrict usage of the term to deposits that are unchannelised (i.e. not river deposits) and occur on relatively steep slopes, greater than 1°. However, lower angle cones of detritus deposited by rivers at a basin margin are also generally considered to be alluvial fans (see Harvey et al. 2005).

The 'classic' modern alluvial fans described from places such as Death Valley in California, USA (Blair & McPherson 1994: Fig. 9.18) occur in arid and semi-arid environments. However, alluvial fans also form today in much wetter settings (see Harvey et al. 2005), and alluvial fan deposits occurring in the stratigraphic record may have been deposited in a wide range of climatic regimes. Larger scale deposits of sediment such as cones of glacial outwash deposited by braided rivers have also been considered to be alluvial fans (or 'humid' fans – Boothroyd & Nummedal 1978) and even larger deposits formed by the lateral migration of a river to produce a cone of detritus have also been considered to be types of alluvial fans (or *megafans* – Wells & Dorr 1987; Horton & DeCelles 2001).

Scree cones formed primarily of rock fall and rock avalanche are commonly associated with alluvial fan deposits at the basin margin. Sediment bodies that consist of a mixture of talus deposits (4.1) and debris-flow deposits (4.5.1) are sometimes called *colluvial fans*: these features are common in subpolar regions where gravity processes are augmented by wet mass flows of debris (Fig. 9.19).

9.5.1 Morphology of alluvial fans

Alluvial fans form where there is a distinct break in topography between the high ground of the drainage

Fig. 9.18 Alluvial fans in the Death Valley, USA, a region with a hot, arid climate.

Fig. 9.19 A colluvial fan, a mixture of scree and debris flows in a cold, relatively dry setting in the Arctic.

basin and the flatter sedimentary basin floor (Fig. 9.20). A ***feeder canyon*** funnels the drainage to the basin margin: at this point the valley opens out and there is a change in gradient allowing water and sediment to spread out. The flow quickly loses energy and deposits the sediment load. Repeated depositional events will build up a deposit that has the form of a segment of a cone radiating from the feeder canyon. On a typical alluvial fan, a number of morphological features can be recognised (Fig. 9.20). The ***fan apex*** is the highest, most proximal point adjacent to the feeder canyon from which the fan form radiates. A ***fan-head canyon*** may be incised into the fan surface near the apex. The depositional slope will usually be steepest in the proximal area: the slope over most of the fan may be only a degree or so, but this is a

relatively steep depositional surface and there is a distinct break in slope at the ***fan toe***, the limit of the deposition of coarse detritus at the edge of the alluvial fan. The fan deposits are thickest at the apex and taper as a conical wedge towards the toe.

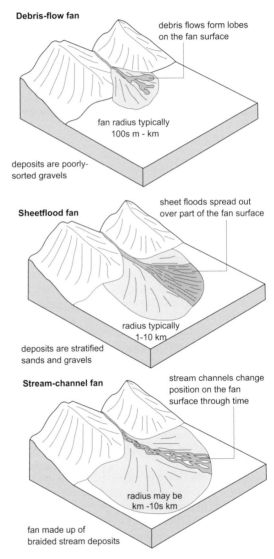

Fig. 9.20 Types of alluvial fan: debris-flow dominated, sheetflood and stream-channel types – mixtures of these processes can occur on a single fan.

9.5.2 Processes of deposition on alluvial fans

The processes of deposition on an alluvial fan will be determined by the availability of water, the amount and type of sediment being carried from the feeder canyon, and the gradient on the fan surface (Fig. 9.20). Where there is a dense mixture of water and sediment, transport and deposition are by debris flow (4.5.1), a viscous slurry of material that spreads out on the fan surface as a lobe. Debris flows do not travel far and a small, relatively steep, alluvial fan cone is built up if this is the dominant process. With more water available, the mixture of sediment and water is more dilute: deposition will be either by unconfined sheetfloods (see below), or flow will be constrained to channels on the surface. Dilute, water-lain fan deposits form fans with shallower slopes and greater radial extent (around 10 km).

Subaerial debris flows

A mixture consisting of a large amount of detritus and a small quantity of water flows as a dense slurry with a consistency similar to a wet concrete mix. Due to the high density and viscosity the flow will be laminar and it will continue to flow over the land surface as a viscous mass until it runs out of momentum, usually when the gradient decreases or the flow loses water content. Beds deposited by debris flows may be tens of centimetres to metres thick and will show very little thinning in a downflow direction. Clasts of all sizes, from clay particles to boulders, can be carried in a debris flow and because of the lack of turbulence there is no sorting of the grain sizes within the flow. The clasts are also commonly randomly oriented, with the exception of some elongate clasts that may be re-aligned parallel to the flow, and clasts in the basal part of the flow where friction with the underlying substrate results in a crude horizontal stratification and parallel alignment of clasts.

The characteristics of a bed deposited by a debris flow are (Fig. 9.21):

1 the conglomerate normally has a matrix-supported fabric – the clasts are mainly not in contact with each other and are almost entirely separated by the finer matrix;

2 sorting of the conglomerate into different clast sizes within or between beds is usually very poor;

3 the clasts may show a crude alignment parallel to flow in the basal sheared layer but otherwise the beds are structureless with clasts randomly oriented;

4 outsize clasts that may be metres across may occur within a debris flow unit (Fig. 9.22);

5 beds deposited by debris flows are tens of centimetres to metres thick.

Sheetflood deposition

When the catchment area of an alluvial fan is inundated with water by a heavy rainstorm, the loose detritus is moved as bedload and in suspension out onto the fan surface. The flow then spreads out over a portion of the fan as a **sheetflood**, a rapid, supercritical, turbulent flow that occurs on slopes of about 3° to 5° (Blair 2000b). Under these upper flow regime conditions (4.3.6) most of the pebbles, cobbles and boulders are carried as bedload, but finer pebbles and granules may be partially in suspension along with sand and finer sediment. These flows usually only last for an hour or so, and standing waves intermittently form in the flow, creating antidune bedforms (4.3.5) in the gravel bedload. Cross-stratification dipping up-flow generated by the antidune bedforms may be preserved, but more often the bedform is washed out as the standing wave breaks down. The most common style of bedding seen in sheetflood facies are **depositional couplets** of coarse gravel deposited as bedload when standing waves are forming, overlain by finer gravel and sand deposited from suspension as the wave is washed out. The formation and destruction of standing waves occurs repeatedly during a sheetflood event. These couplets are typically 5–20 cm thick and occur in packages tens of centimetres to a couple of metres thick formed by individual flow events. Individual sheetflood deposits may be hundreds of metres wide and stretch from the apex to the toe of the fan, but individual couplets within them are typically only a few metres across. There appears to be little difference between sheetflood deposits on proximal and distal parts of a fan, and most of the fine sediment is carried in suspension beyond the fan (Blair 2000b).

The characteristics of a sheetflood deposit on an alluvial fan are (Fig. 9.21):

1 sheet geometry of beds that are tens of centimetres to a couple of metres thick;

2 beds are very well stratified with distinct couplets of coarser gravel and sandy, finer gravel (Fig. 9.23);

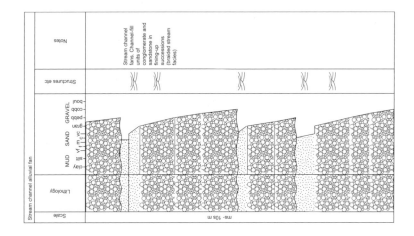

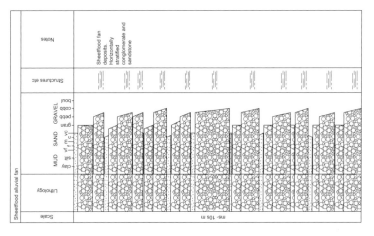

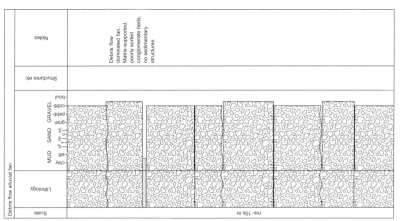

Fig. 9.21 Schematic sedimentary logs through debris-flow, sheetflood and stream-channel alluvial fan deposits.

Fig. 9.22 A debris flow on an alluvial fan: the conglomerate is poorly sorted, with the larger clasts completely surrounded by a matrix of finer sediment.

Fig. 9.23 Sheetflood deposits on an alluvial fan showing well-developed stratification.

3 imbrication of clasts is common, and up-stream cross-stratification formed by antidunes may also be preserved;
4 the sediment is poorly sorted, but silt and clay sized material is largely absent;
5 beds may show normal grading due to waning flow.

Fluvial deposits forming alluvial fans

The river emerging from the feeder canyon may continue to flow as a confined channel on the alluvial plain. The abrupt reduction in gradient as flow occurs on the low slope of the plain promotes deposition of gravel on bars within the channel to create a braided depositional form. Deposition in the channel by a number of high-discharge events will eventually cause the channel to become choked with sediment, and the active flow will move, either by a process of gradual lateral migration or by avulsion. Through time the position of the braided river channel migrates over the whole fan surface, depositing a more-or-less continuous sheet of gravel. The radius of the fan formed will be determined by the length over which the channel is depositing gravel: sand and finer suspended load will be carried further out onto the alluvial plain.

The overall shape of the sediment body formed will be similar to that of a fan formed by sheetflood deposits, but the radius is not limited by the extent that unconfined sheetflood processes can transport sediment, and these fans can therefore be over 10 km from apex to toe. Distinct channels may be preserved,

but individual beds often have a sheet geometry, the result of lateral amalgamation of channel deposits. Beds are sharp-based, with clast-supported conglomerate fining up to sandstone: sedimentary structures are those of a braided river, including imbrication and cross-stratification in gravels and cross-bedded sandstone (Fig. 9.21).

9.5.3 Modification of alluvial fan deposits

Deposition on alluvial fans in arid regions occurs very infrequently (on a human time scale). The sheetfloods and debris flows that deposit sediment normally last only a matter of hours and these events are separated by tens or hundreds of years. Between depositional episodes, less intense rainfall events in the catchment will result in water flowing on the fan as superficial, non-depositing streams. These flows can locally winnow out sand and mud from between the gravel clasts, removing the matrix of the deposit, and if the spaces are not filled in later an **open framework** or **matrix-free conglomerate** may be preserved. A more significant modification of the alluvial fan surface is by streams that become established on the fan surface between depositional episodes. These rework debris flow and sheetflood deposits, form a channel and remove some material from the fan surface and on many modern alluvial fans this is an important process (Blair & McPherson 1994). These steep channels have the form of a braided river with bars of gravel redeposited or left uneroded within the channel. Alluvial fan surfaces are also subject to modification by soil processes, and aeolian processes can winnow the surface, removing fine-grained material from it. A desert varnish (*8.2*) may be seen on gravels on fans formed in arid environments.

9.5.4 Controls on alluvial fan deposition

Although alluvial fan deposits are not the most significant in a sedimentary basin in terms of volume, they are important because fan deposition is sensitive to tectonic and climatic controls. Alluvial fans develop at the margins of sedimentary basins and these can be sites of tectonic activity, with faults along the basin margin creating uplift of the catchment area and subsidence in the basin (see Chapter 24). It is therefore possible to see evidence of tectonic activity within an alluvial fan succession, such as an influx of coarse detritus onto the fan resulting from renewed tectonic uplift (Heward 1978; Nichols 1987). Analysis of the bed thicknesses and clast sizes within beds can therefore be used as a means of identifying periods of tectonic uplift in the high ground adjacent to the basin. A change in climate can also result in changes in the processes of deposition on a fan (Harvey et al. 2005): for example, with an increase in rainfall more water is available and this may result in a predominance of sheetflood and stream-channel processes, with less debris-flow events occurring. The character of the conglomerates deposited on the fan will reflect this climatic change, with more clast-supported and fewer matrix-supported conglomerate beds. A further factor controlling fan deposition is the nature of the bedrock in the catchment area: lithologies that weather to form a lot of mud will tend to generate muddy debris flows, whereas more resistant rocks will break down to sand and gravel, which is transported and deposited by sheetflood and stream-channel processes (Blair 2000a, Nichols & Thompson 2005).

9.6 FOSSILS IN FLUVIAL AND ALLUVIAL ENVIRONMENTS

In comparison to marine settings the terrestrial environment has a poor potential for the preservation of fossil plants or animals. An organism that dies on the land surface is susceptible to scavenging by carrion or the tissue will be broken down by oxidation. Preservation occurs only if the organism has very resilient parts (e.g. teeth and bones of vertebrates) or if the plant or animal is covered by sediment soon after death. Faunal remains are therefore relatively rare, occurring as scattered bones or teeth of vertebrates, but plant fossils are more common and may be locally abundant. Fossilised tree; stumps may be preserved *in situ* (in place) in overbank deposits as a result of flood events that partially buried the tree; other plant parts such as pieces of branches and leaves occur within beds of both channel and overbank sediments. The most abundant plant fossil remains are those of pollen and seeds (*palynomorphs*) that are highly resistant to breakdown and can survive long periods of transport before being deposited and preserved. This makes them particularly useful for dating and correlation of terrestrial deposits (*20.5.3*).

The footprints of animals in soft mud have a good preservation potential if the mud dries hard and is later covered with sand. These are examples of trace fossils (*11.7*) that in continental environments are largely restricted to floodplains and alluvial plains. Trace fossils in these environments may range from the tracks of animals such as dinosaurs to the burrows and nests of insects such as beetles, bees and ants (Hasiotis 2002). These traces provide information about the palaeoenvironment, such as the level of the palaeowater table: an ant or termite nest will be constructed only in dry sediment, so the presence of these and other structures formed by insects is a reliable indicator of how wet or dry the land surface was and hence provides some information about the palaeoclimate. Trace fossils in continental environments have also provided valuable information about the morphology of extinct organisms. Footprints of dinosaurs, for example, can provide an indication of the way that the animal walked in a way that the skeletons (which are often incomplete) cannot.

9.7 SOILS AND PALAEOSOLS

9.7.1 Soils

A soil is formed by physical, chemical and biological processes that act on sediment, regolith or rock exposed at the land surface (Retallack 2001). Collectively these soil-forming processes are known as ***pedogenesis***. Within a layer of sediment the principal physical processes are the movement of water down from or up to the surface and the formation of vertical cracks by the shrinkage of clays. Chemical processes are closely associated with the vertical water movement as they involve the transfer of dissolved material from one layer to another, the formation of new minerals and the breakdown of some original mineral material. The activity of plants is evident in most soils by the presence of roots and the accumulation of decaying organic matter within the soil. The activity of animals can have a considerable impact, as vertebrates, worms and insects may all move through the soil mixing the layers and aerating it.

Soils can be classified according to (Mack et al. 1993):
- the degree of alteration (weathering) of the parent material;
- the precipitation of soluble minerals such as calcite and gypsum;

- oxidising/reducing conditions (***redox conditions***), particularly with respect to iron minerals;
- the development of layering (***horizonation***);
- the redistribution of clays, iron and organic material into these different layers (***illuviation***);
- the amount of organic matter that is preserved.

Twelve basic types of soil can be recognised using the US Soil Survey taxonomy (Retallack 2001) (Fig. 9.24). Some of these soil types can be related to the climatic conditions under which they form: gelisols indicate a cold climate whereas aridisols are characteristic of arid conditions, oxisols form most commonly under humid, tropical conditions and vertisols form in subhumid to semi-arid climates with pronounced seasonality. Particular hydrological conditions are required for some soils, such as the

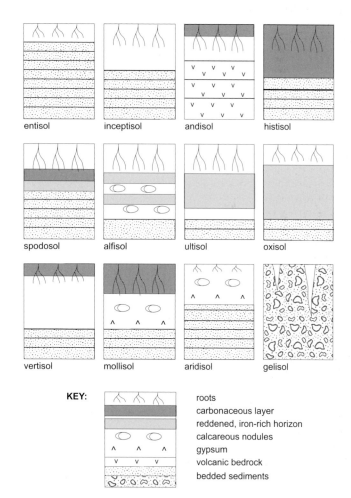

Fig. 9.24 Twelve major soil types recognised by the US Soil Survey.

waterlogged setting that histosols (peaty soils) form in. Other types are indicative of the degree of the maturity of the soil profile (and hence the time over which the soil has developed); entisols are very immature and inceptisols show more development, but are less mature than the other types lower in the list. The type of vegetation is an important factor in some cases: spodosols, alfisols and ultisols are soils formed in forests, whereas mollisols are grassland soils. Finally, the formation of andisols is restricted to volcanic substrates.

9.7.2 Palaeosols

A *palaeosol* is a fossil soil. Many of the characteristics of modern soils noted above can be recognised in soils that formed in the geological past (Mack et al. 1993; Retallack 2001). These features include the presence of fossilised roots, the burrows of soil-modifying organisms, vertical cracks in the sediment and layers enriched or depleted in certain minerals. The study of palaeosols provides important information about ancient landscapes and in particular they can indicate the palaeoclimate, the type of vegetation growing and the time period during which a land surface was exposed.

The precipitation of calcium carbonate within the soil is a conspicuous feature of some aridisols that form in semi-arid to arid climates. These *calcrete* soils form by the movement of water through the soil profile precipitating calcium carbonate as root encrustations (*rhizocretions*) and as small soil nodules (*glaebules*) (Wright & Tucker 1991). The nodules grow and coalesce as precipitation continues to form a fully developed calcrete, which consists of a dense layer of calcium carbonate near to the surface with *tepee structures*, i.e. domes in the layer formed by the expansion of the calcium carbonate as it is precipitated (Allen 1974) (Fig. 9.25). The stages in the development of a calcrete soil profile are easily recognised in palaeosols, so if the rate of development of a mature profile can be measured, the time over which an ancient profile formed can be estimated (Leeder 1975).

The passage of time can also be indicated by other palaeosol types: entisols and inceptisols indicate that the time available for soil formation on a particular surface was relatively short, whereas other, more mature categories of palaeosol require a longer period

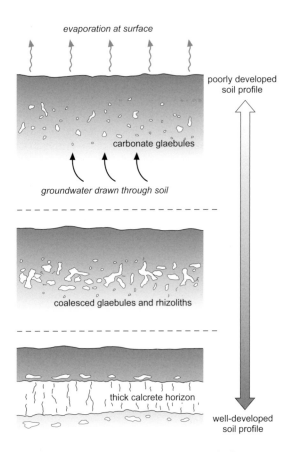

Fig. 9.25 A calcrete forms by precipitation of calcium carbonate within a soil in an arid or semi-arid environment.

of exposure of the surface. These distinctions become useful when attempting to assess rates of deposition on, for example, a floodplain surface: entisols would indicate relatively rapid deposition, with little time for soil development before flooding deposited more sediment on the surface, whereas a well-developed spodosol, alfisol or ultisol suggests a much longer period of time before the surface was covered with younger sediment. However, it should be noted that the time taken for any soil profile to develop varies considerably with temperature, rainfall and the availability of different minerals so time estimates are always relative, not absolute. Also, soil profiles can become complicated by the superimposition of a younger profile over an older one (Bown & Kraus 1987).

Some types of modern and fossil soils have been given particular names. For example, **seatearths** are histisols, argillisols or spodosols that are common in the coal measures of northwestern Europe and North America (Percival 1986). They are characterised by a bed of organic matter underlain by a leached horizon of white sandstone from which iron has been washed out. **Laterites** are oxisols that are the product of extensive weathering of bedrock to form a soil that consists mainly of iron and aluminium oxides: examples of laterites may be found in the stratigraphic record as strongly reddened layers between basalt lava flows and provide evidence that the eruption was subaerial. Iron-rich oxisols that become cemented are known as **ferricretes** and they are a type of hardened soil profile called a **duricrust**. Duricrusts are highly resistant surfaces that develop over very long time periods (e.g. they are found associated with major unconformities; Retallack 2001); as well as iron-rich forms there are records of **silcretes**, which are silica-rich.

Identification of a palaeosol profile is probably the most reliable indicator of a terrestrial environment. Channels are not unique to the fluvial regime because they also occur in deltas, tidal settings and deep marine environments, and thin sheets of sandstone are also common to many other depositional settings. However, sometimes the recognition of a palaeosol can be made difficult by diagenetic alteration (*18.2*), which can destroy the original pedogenic features.

9.8 FLUVIAL AND ALLUVIAL FAN DEPOSITION: SUMMARY

Fluvial environments are characterised by flow and deposition in river channels and associated overbank sedimentation. In the stratigraphic record the channel fills are represented by lenticular to sheet-like bodies with scoured bases and channel margins, although these margins are not always seen. The deposits of gravelly braided rivers are characterised by cross-bedded conglomerate representing deposition on channel bars. Both sandy braided river and meandering river deposits typically consist of fining-upward successions from a sharp scoured base through beds of trough and planar cross-bedded, laminated and cross-laminated sandstone. Lateral accretion surfaces characterise meandering rivers that are also often associated with a relatively high proportion of overbank facies. Floodplain deposits are mainly alternating thin sandstone sheets and mudstones with palaeosols; small lenticular bodies of sandstone may represent crevasse splay deposition. Palaeocurrent data from within channel deposits are unidirectional, with a wider spread about the mean in meandering river deposits; palaeocurrents in overbank facies are highly variable.

Alluvial fan deposits are located near to the margins of sedimentary basins and are limited in lateral extent to a few kilometres from the margin. The facies are dominantly conglomerates, and may include matrix-supported fabrics deposited by debris flows, well-stratified gravels and sands deposited by sheet-flood processes and in channels that migrate laterally across the fan surface. Alluvial and fluvial deposits will interfinger with lacustrine and/or aeolian facies, depending on the palaeoclimate, and many (but not all) river systems feed into marine environments via coasts, estuaries and deltas. Other characteristics of fluvial and alluvial facies include an absence of marine fauna, the presence of land plant fossils, trace fossils and palaeosol profiles in alluvial plain deposits.

Characteristics of fluvial and alluvial fan deposits

- lithologies – conglomerate, sandstone and mudstone
- mineralogy – variable, often compositionally immature
- texture – very poor in debris flows to moderate in river sands
- bed geometry – sheets on fans, lens shaped river channel units
- sedimentary structures – cross-bedding and lamination in channel deposits
- palaeocurrents – indicate direction of flow and depositional slope
- fossils – fauna uncommon, plant fossils may be common in floodplain facies
- colour – yellow, red and brown due to oxidising conditions
- facies associations – alluvial fan deposits may be associated with ephemeral lake and aeolian dunes, rivers may be associated with lake, delta or estuarine facies

FURTHER READING

Best, J.L. & Bristow, C.S. (Eds) (1993) *Braided Rivers*. Special Publication 75, Geological Society Publishing House, Bath.

Blum, M., Marriott, S. & Leclair, S. (Eds) (2005) *Fluvial Sedimentology VII*. Special Publication 35, International Association of Sedimentologists. Blackwell Science, Oxford.

Bridge, J.S. (2003) *Rivers and Floodplains: Forms, Processes, and Sedimentary Record.* Blackwell Science, Oxford.

Bridge, J.S. (2006) Fluvial facies models: recent developments. In: *Facies Models Revisited* (Eds Walker, R.G. & Posamentier, H.). Special Publication 84, Society of Economic Paleontologists and Mineralogists, Tulsa, OK; 85–170.

Collinson, J.D. (1996) Alluvial sediments. In: *Sedimentary Environments: Processes, Facies and Stratigraphy* (Ed. Reading, H.G.). Blackwell Science, Oxford; 37–82.

Harvey, A.M., Mather, A.E. & Stokes, M. (Eds) (2005) *Alluvial Fans: Geomorphology, Sedimentology, Dynamics.* Special Publication 251, Geological Society Publishing House, Bath.

Retallack, G.J. (2001) *Soils of the Past: an Introduction to Paleopedology* (2nd edition). Blackwell Science, Oxford.

Smith, N.D. & Rogers, J. (Eds) (1999) *Fluvial Sedimentology VI.* Special Publication 28, International Association of Sedimentologists. Blackwell Science, Oxford.

10

Lakes

Lakes form where there is a supply of water to a topographic low on the land surface. They are fed mainly by rivers and lose water by flow out into a river and/or evaporation from the surface. The balance between inflow and outflow and the rate at which evaporation occurs control the level of water in the lake and the water chemistry. Under conditions of high inflow the water level in the lake may be constant, governed by the spill point of the outflow, and the water remains fresh. Low water input coupled with high evaporation rates in an enclosed basin results in the concentration of dissolved ions, which may be precipitated as evaporites in a perennial saline lake or when an ephemeral lake dries out. Lakes are therefore very sensitive to climate and climate change. Many of the processes that occur in seas also occur in lakes: deltas form where rivers enter the lake, beaches form along the margins, density currents flow down to the water bottom and waves act on the surface. There are, however, important differences with marine settings: the fauna and flora are distinct, the chemistry of lake waters varies from lake to lake and certain physical processes of temperature and density stratification are unique to lacustrine environments.

10.1 LAKES AND LACUSTRINE ENVIRONMENTS

A lake is an inland body of water. Although some modern lakes may be referred to as 'inland seas', it is useful to draw a distinction between water bodies that have some exchange of water with the open ocean (such as lagoons – 13.3.2) and those that do not, which are true lakes. Lakes form where there is a depression on the land surface which is bounded by a sill such that water accumulating in the depression is retained. Lakes are typically fed by one or more streams that supply water and sediment from the surrounding hinterland. Groundwater may also feed water into a lake. The amount of sediment accumulated in lakes is small compared with marine basins, but they may be locally significant, resulting in strata hundreds of metres thick and covering hundreds to thousands of square kilometres. Sand and mud are the most common components of lake deposits, although almost any other type of sediment can accumulate in *lacustrine* (lake) environments, including limestones, evaporites and organic material. Plants and animals living in a lake may be preserved as fossils in lacustrine deposits, and concentrations of organic material can form beds of coal (*18.7.1*) or oil and gas source rocks (*18.7.3*). The study of modern lakes is referred to as **limnology**.

10.1.1 Lake formation

Large inland depressions that allow the accumulation of water to form a lake are usually the result of tectonic forces creating a sedimentary basin. The formation of different sedimentary basins is discussed in Chapter 24, and the most important processes for the creation of lake basins are those of continental extension to generate rifts (*24.2.1*), basins related to strike-slip within continental crust (*24.5.1*) and intracontinental sag basins (*24.2.3*). Rift and strike-slip basins are bounded by faults that cause parts of the land surface to subside relative to the surrounding area. Drainage will always follow a course to the lowest level, so rivers will feed into a subsiding area and may form a lake. With continued movement on the faults and hence continued subsidence, the lake may become hundreds of metres deep, and through time may accumulate hundreds or even thousands of metres of sediment. Depressions that are related to broad subsidence of the crust (sag basins) tend to be larger and shallower; lacustrine deposits in these settings are likely to be relatively thin (tens to hundreds of metres) but may be spread over a very large area. Lakes can also be created where thrust faults (*24.4*) locally uplift part of the land surface and create a dam across the path of a river.

A depression on the land surface can also form by erosion, but the erosive agent cannot be water alone because a stream will always follow a path down hill. Glaciers, on the other hand, can scour more deeply into a valley. Provided that the top surface of the glacier has an overall slope down-flow, the base of the ice flow can move down and up creating depressions in the valley floor. When the ice retreats these overdeepened parts of the valley floor will become areas where lakes form. Glacial processes can also create lakes by building up a natural dam of detritus across a valley floor through the formation of a terminal moraine (*7.4.1*). Lakes formed in glacial areas tend to be relatively small and the

chances of long-term preservation of deposits in glacial lakes is lower as they are typically in areas undergoing erosion (cf. continental glacial environments: *7.4*).

Other processes of dam building are by landslides (*6.5.1*) that block the path of a stream in a valley and large volumes of volcanic ash or lava that can create topography on the land surface and result in the formation of a lake. Volcanic activity can also create large lakes by caldera collapse and explosive eruptions that remove large quantities of material from the centre of a volcanic edifice, leaving a remnant rim within which a crater lake can form (*17.4.3*).

10.1.2 Lake hydrology

The supply of water to a lake is through streams, groundwater and by direct rainfall on the lake surface. If there is no loss of water from the lake, the level will rise through time until it reaches the spill point, which is the top of the sill or barrier around the lake basin (Figs 10.1 & 10.2). A lake is considered to be **hydrologically open** if it is filled to the spill point and there is a balance of water supply into and out of the basin. Under these circumstances the level of the water in the lake will be constant, and the constant supply from rivers will mean that the water in the lake will be fresh (i.e. with a low concentration of dissolved salts and hence low salinity).

The surface of a lake will be subject to evaporation of water vapour into the atmosphere, a process that becomes increasingly important at higher temperatures and where the air is dry. If the rate of evaporation exceeds or balances the rate of water supply there is no outflow from the lake and it is considered to be **hydrologically closed**. These types of lake basin are also sometimes referred to as endorheic and are basins of internal drainage. Soluble ions chemically weathered from bedrock are carried in solution in rivers to the lake. If the supply of dissolved ions is low the evaporation will have little effect on the concentration of ions in the lake water, but more commonly dissolved ions become concentrated by evaporation to make the waters saline. With sufficient evaporation and concentration evaporite minerals may precipitate (*3.2*) and under conditions of low water supply and high evaporation rate the lake may dry up completely.

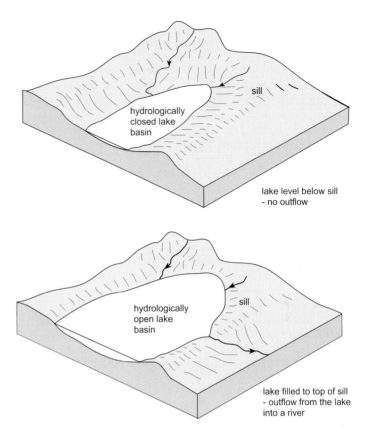

Fig. 10.1 Hydrological regimes of lakes.

Fig. 10.2 A lake basin supplied by a river in the foreground, with outflow through a sill to the sea in the distance.

From a sedimentological point of view, three types of lake can be considered, irrespective of their mode of formation or hydrology. *Freshwater lakes* have low salinity waters and are either hydrologically open, or are hydrologically closed with a low supply of dissolved ions allowing the water to remain fresh. *Saline lakes* are hydrologically closed and are perennial water bodies in which dissolved ions have become concentrated by evaporation. *Ephemeral lakes* mainly occur in arid climatic settings and are temporary bodies of water that exist for a few months or years after large rainstorms in the catchment area, but are otherwise dry.

10.2 FRESHWATER LAKES

The majority of large modern lakes are freshwater: they occur at latitudes ranging from the Equator to the polar regions (Bohacs et al. 2003) and include some of the largest and deepest in the world today. Lacustrine deposits from lakes of similar scales are known from the stratigraphic record, mainly from Devonian through to Neogene strata.

10.2.1 Hydrology of freshwater lakes

Lakes are relatively static bodies of water, with no currents driven by tidal processes or oceanic circulation (cf. seas). Waves form when a wind blows over the surface of the water, but the limited size of any lake means that there is not a large fetch (*4.4*) and hence the waves cannot grow to the sizes seen in the world's oceans. Wind-driven surface currents may reach velocities up to $30\,cm\ s^{-1}$ (Talbot & Allen 1996), especially in narrow valleys where the wind is funnelled by the topography. However, currents driven by the wind in lakes are too weak to move anything more than silt and fine sand and will not redistribute coarser sediment. These currents and the relatively small waves formed on a lake influence the upper part of the water body, and the effects of the water oscillation decrease with depth (*4.4.1*). Therefore, below about 10 or 20 m depth the lake waters are unaffected by any wave or current activity. This allows for the development of *lake water stratification*, which is seen as a contrast in the temperature, density and the chemistry of the waters in the upper and lower parts of the water body.

The surface of the lake is warmed by the Sun and the water retains the heat to acquire a steady temperature that varies gradually with the seasons. Due to the lack of circulation the water in the lower part of the lake remains at a constant, cooler temperature. These two divisions of the lake water are known as the ***epilimnion***, which is the upper, warmer lake water, and the ***hypolimnion***, the lower, colder part: they are separated by a surface across which the temperature changes, the ***thermocline*** (Fig. 10.3). The density of pure water is determined by the temperature and, above $4°C$, the density decreases as it becomes warmer. The stratification is therefore one of density as well as temperature, and, because the lower density warm water is above the higher density cold water, the situation is stable (Talbot & Allen 1996).

Agitation of the lake surface by waves and circulation in the epilimnion means that this part of the water body is oxygenated by contact with the air. In the hypolimnion, any oxygen is quickly used up by aerobic bacterial activity and, due to the lack of circulation, is not replenished. The bottom of the lake therefore becomes ***anaerobic*** (without air, and therefore oxygen) and this has two important consequences. First, any organic material that falls through the water column to the lake floor will not be subject to breakdown by the activity of the aerobic processes that normally cause decomposition of plant and animal tissue. If there is abundant plant material being swept into the lake, this has the potential to form a detrital coal layer (*18.7.1*), and the remains of algal or bacterial life within the lake may also accumulate to form a bed rich in organic matter, which may ultimately form a sapropelic coal or a

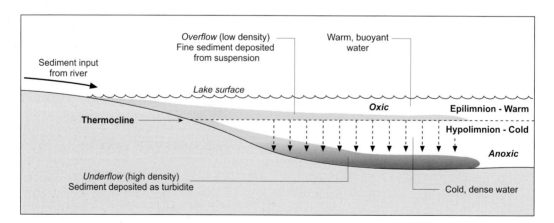

Fig. 10.3 The thermal stratification of fresh lake waters results in a more oxic, upper layer, the epilimnion, and a colder, anoxic lower layer, the hypolimnion. Sedimentation in the lake is controlled by this density stratification above and below the thermocline.

source rock for oil and gas (*18.7.3*). The second effect of anaerobic bottom conditions is that this is an environment that is unfavourable for life. Stratified lakes therefore have no animals living on the bottom or within the surface sediment and hence there is no bioturbation (*11.7.3*) to disturb the primary sedimentary layering.

10.2.2 Lake margin clastic deposits

Where a sediment-laden river enters a lake the water velocity drops abruptly and a delta forms as coarse material is deposited at the river mouth (Fig. 10.4). The form and processes on a lake delta will be similar to that seen in river-dominated deltas, with some wave reworking of sediment also occurring if the lake experiences strong winds. The character of the delta deposits will be largely controlled by the nature of the sediment supply from the river, and may range from fine-grained deposits to coarse, gravelly fan-deltas (*12.4.2*).

Away from the river mouth the nature of the lake shore deposits will depend on the strength of winds generating waves and currents in the lake basin. If winds are not strong, lake shore sediments will tend to be fine-grained but strong, wind-driven currents can redistribute sandy sediment around the edges of the lake where it can be reworked by waves into sandy beach deposits (Reid & Frostick 1985). These marginal lacustrine facies (Fig. 10.5) will be similar in character to beaches developed along marine shorelines (*13.2*).

In situations where the slope into the lake is very gentle the edge of the water body is poorly defined as the environment merges from wet alluvial plain into a lake margin setting. This lake margin marshy environment is sometimes referred to as a ***palustrine*** environment. Plants and animals living in this setting live in and on the sediment in a wet soil environment where sediments will be modified by soil (pedogenic) processes (*9.7*), resulting in a nodular texture that may sometimes be calcareous.

10.2.3 Deep lake facies

Away from the margins, clastic sedimentation occurs in the lake by two main mechanisms: dispersal as plumes of suspended sediment and transport by density currents (Fig. 10.3) (Sturm & Matter 1978). Plumes of water laden with suspended sediment may be brought into the lake by rivers: if the sediment–water mixture is a lower density than the hypolimnion the plume will remain above the thermocline and will be distributed around the lake by wind-driven surface currents. The suspended load will eventually start to settle out of the epilimnion and fall to the lake floor to form a layer of mud. Density currents (*4.5*) provide a mechanism for transporting coarser sediment across the lake floor. Mixtures of sediment and water brought in by a river or reworked from a lake delta may flow as a turbidity current (*4.5.2*), which can travel across the lake floor. The deposits will be layers of sediment that grade from coarse material deposited from the current first to finer sediment that settles out last. In lakes where sediment plumes and turbidity currents are the main transport mechanisms the deep lake facies will consist of very finely laminated muds deposited from suspension alternating with thin graded turbidites forming a characteristic, thinly bedded succession (Fig. 10.5).

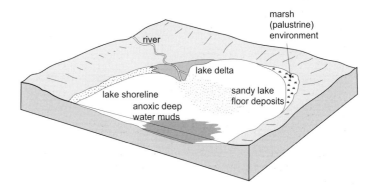

Fig. 10.4 Facies distribution in a freshwater lake with dominantly clastic deposition.

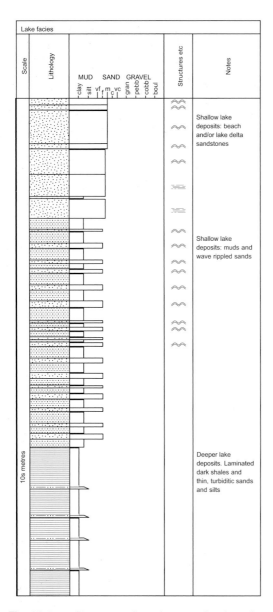

Fig. 10.5 A schematic graphic sedimentary log through clastic deposits in a freshwater lake.

supply. The spring thaw will result in an influx of sediment-laden cold water, which will form a layer of sediment on the lake floor. During the summer months organic productivity in and around the lake provides a supply of organic material that settles on the lake floor where it is preserved in the anaerobic conditions. This alternation of dark, organic-rich deposits formed in the summer months and paler, clastic sediment brought in by the spring thaw is a distinctive feature of many temperate lakes. The milli-metre-scale laminae formed in this way are known as **varves** and they have been used in chronostratigra-phy of Holocene deposits (*21.5.7*).

10.2.4 Lacustrine carbonates

Carbonates can form a significant proportion of the succession in any lake setting only if the terrigenous clastic input is reduced (Fig. 10.6). Direct chemical precipitation of carbonate minerals occurs in lakes with raised salinity, but in freshwater lakes the for-mation of calcium carbonates is predominantly asso-ciated with biological activity. The hard shells of animals such as bivalve molluscs, gastropods and ostracods can contribute some material to lake sedi-ments, and this coarse skeletal material may be depos-ited in shallow water or redistributed around the lake by wave-driven currents. However, the most abun-dant carbonate material in lakes is usually from algal and microbial sources.

The breakdown of calcareous algal filaments is an important source of lime mud, which may be deposited in shallow lake waters or redeposited by density currents into deeper parts of the lake. Cyano-bacteria and green algae form stromatolite bioherms and biostromes (*15.3.2*) in shallow (less than 10 m) lake waters: these carbonate build-ups may form mats centimetres to metres thick or form thick coatings of bedrock near lake margins. They form by the microbial and algal filaments trapping and binding carbonate (see also marine stromatolites, *3.1.3*). A common feature of lakes with areas of active carbo-nate deposition is coated grains. Green algae and cyanobacteria form **oncoids**, irregularly shaped, con-centrically layered bodies of calcium carbonate sev-eral millimetres or more across, formed around a nucleus. Oncoids form in shallow, gently wave-agitated zones: in these settings ooids may also form and build up oolite shoals in shallow water.

The absence of organisms living in deep lake environ-ments means that the fine lamination is preserved because it is not disrupted by biogenic activity.

In lakes formed in regions where there is an annual thaw of winter snow a distinctive stratification may develop due to seasonal variations in the sediment

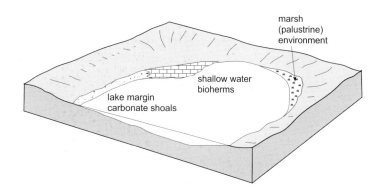

Fig. 10.6 Facies distributions in a freshwater lake with carbonate deposition.

Fig. 10.7 A saline lake, Mono Lake, California: the mineral deposit mounds are associated with underground spring waters.

Tufa (*travertine*) is an inorganic precipitate of calcium carbonate, which may form sheets or mounds in lakes. Springs along the margins or in the floor of the lake can be sites of quite spectacular build-ups of tufa (Fig. 10.7).

10.3 SALINE LAKES

Saline lakes are perennial, supplied by rivers containing dissolved ions weathered from bedrock and in a climatic setting where there are relatively high rates of evaporation. The salinity may vary from 5 g L^{-1} of solutes, which is **brackish** water, to **saline**, close to the concentration of salts in marine waters (*3.2*), to **hypersaline** waters, which have values well in excess of the concentrations in seawater. From a sedimentological point of view, brackish water lakes are similar to freshwater lakes because it is the high concentrations of salts that provide saline lakes with their distinctive character. The chemistry of saline lake waters

is determined by the nature of the salts dissolved from the bedrock of the catchment area of the river systems that supply the lake. The bedrock geology varies from place to place, so the chemical composition of every lake is therefore unique, unlike marine waters, which all have the same composition of salts. The types and proportions of evaporite minerals formed in saline lakes are therefore variable, and include minerals not found in marine evaporite successions.

The main ions present in modern saline lake waters are the cations sodium, calcium and magnesium and the carbonate, chloride and sulphate anions. The balance between the concentrations of different ions determines the minerals formed (Fig. 10.8) and three main saline lake types are recognised according to the composition of the **brines** (ion-rich waters) in them (Eugster & Hardie 1978). **Soda lakes** have brines with high concentrations of bicarbonate ions and sodium carbonate minerals such as **trona** and **natron**: these minerals are not precipitated from marine waters and are therefore exclusive indicators of non-marine evaporite deposition. **Sulphate lake** brines have lower concentrations of bicarbonate but are relatively enriched in magnesium and calcium: they precipitate mainly sulphate minerals such as gypsum and mirabilite (a sodium sulphate). Salt lakes or **chloride lakes** such as the Dead Sea are similar in mineral composition to marine evaporites.

Organisms in saline lakes are very restricted in variety but large quantities of blue-green algae and bacteria may bloom in the warm conditions. These form part of a food chain that includes higher plants, worms, specialised crustaceans and birds such as flamingoes which feed on them. Organic productivity may be high enough to result in sedimentary successions that contain both evaporite minerals and black,

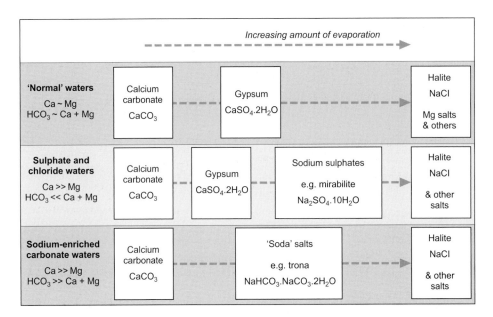

Fig. 10.8 Three general types of saline lake can be distinguished on the basis of their chemistry.

organic-rich shales. Seasonal temperature variations in permanent saline lakes result in a fine layering. This is due to direct precipitation of calcium carbonate as aragonite needles in the hot summer to form white laminae, which alternate with darker laminae formed by clastic input when sediment influx is greater in the winter.

10.4 EPHEMERAL LAKES

Large bodies of water that periodically dry out are probably best described as **ephemeral lakes**, although the term **playa lake** is also commonly used (Briere 2000). Terms such as 'saline pan' are also sometimes used to describe these temporary lake environments. It is perhaps simpler to just use the term 'ephemeral lakes' as this unambiguously implies that the water body is temporary. They occur in semi-arid and arid environments where the rainfall is low and the rate of evaporation is high.

Many desert areas are subject to highly irregular rainfall with long periods of dry conditions interrupted by intense rainfall that may occur only every few years or tens of years. After rainfall in the catchment area, the rivers become active (9.2.3) and flash

Fig. 10.9 A salt crust of minerals formed by evaporation in an ephemeral lake.

floods supply water and sediment to the basin centre where it ponds to form a lake. Once the lake has formed, particles suspended in the water will start to deposit and form a layer of fine-grained muddy sediment (Lowenstein & Hardie 1985). Evaporation of the water body gradually reduces its volume and the area of the lake starts to shrink, leaving areas of margin exposed where desiccation cracks (4.6) may form in the mud as it dries out. With further evaporation the ion concentration in the water starts to increase to the point where precipitation of minerals occurs. The

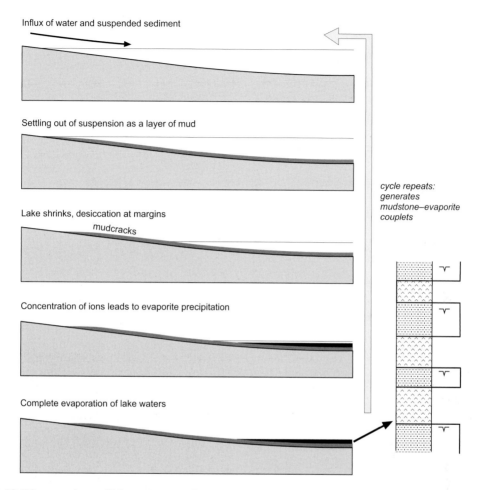

Influx of water and suspended sediment

Settling out of suspension as a layer of mud

Lake shrinks, desiccation at margins

mudcracks

Concentration of ions leads to evaporite precipitation

Complete evaporation of lake waters

cycle repeats:
generates
mudstone–evaporite
couplets

Fig. 10.10 When an ephemeral lake receives an influx of water and sediment, mud is deposited from suspension to form a thin bed that is overlain by evaporite minerals as the water evaporates. Repetitions of this process create a series of couplets of mudstone and evaporite.

least soluble minerals will precipitate first, followed by other evaporite minerals until the lake has dried up completely (Fig. 10.9). The resulting deposit consists of a layer of mud overlain by a layer of evaporite minerals (Fig. 10.10) The minerals formed by evaporation will be determined by the chemistry of the waters and show the same ranges of composition as is found in saline lakes. Subsequent flooding of the lake floor following another flood event does not necessarily result in solution of the evaporite minerals on the surface as they may become quickly blanketed by mud.

Repeated flooding and evaporation results in a series of **depositional couplets** of a layer of mud over-

lain by a layer of evaporite minerals: these couplets are typically a few millimetres to centimetres thick and are a characteristic signature of ephemeral lake deposition (Lowenstein & Hardie 1985). Ephemeral lake deposits occur in arid environments and are therefore likely to be associated with other facies formed in these settings: these will include aeolian sandflat and dune deposits, alluvial fan facies and material deposited by flash floods from ephemeral rivers. These facies are likely to be found interfingering with ephemeral lake deposits to form a facies association characteristic of arid depositional environments.

Evaporite minerals also form within the sediments surrounding ephemeral lakes. In these areas the sediment is saturated with saline groundwater, which evaporates at the ground surface, concentrating the dissolved minerals and leading to the crystallisation of evaporite minerals. These regions are sometimes referred to as *inland sabkhas* (cf. coastal sabkhas: 15.2.3) and the most common mineral to be formed is gypsum, which grows within the sediment in an interconnected mass of bladed crystals known as *desert rose*.

10.5 CONTROLS ON LACUSTRINE DEPOSITION

The characteristics of the deposits of lacustrine environments are controlled by factors that control the depth and size of the basin (which are largely determined by the tectonic setting), the sediment supply to the lake (which is a function of a combination of tectonics and climatic controls on relief and weathering) and the balance between water supply and loss through evaporation (which is principally related to the climate). If the climate is humid a lake will be hydrologically open, with water flowing both in and out of it. Such lakes can be considered to be *overfilled* (Bohacs et al. 2000, 2003), and their deposits are characterised by accumulation both at the margins, where sediment is supplied to deltas and beaches, and in the deep water from suspension and turbidity currents. The lake level remains constant, so there is no evidence of fluctuations in water depth under these conditions.

A *balanced fill* lake is one where the fluvial input is approximately balanced by the loss through evaporation. These lakes are sensitive to variations in the climate because a reduction in water input and/or an increase in evaporation (drier and/or warmer conditions) will result in a fall in the water level below the sill and the system becomes hydrologically closed. The area of the lake will contract, shifting the lake shoreline towards the basin centre and leaving a peripheral area exposed to subaerial conditions where desiccation cracks may form, plants colonise the surface and pedogenic processes modify the sediment. A fall in the water level will also bring parts of the lake floor that were previously below the wave base into shallower water where wave ripples rework the sediment. These changes will be reversed if the climate reverts to wetter and/or cooler conditions, as the lake level rises and the lake margin is reflooded. The deposits

of lakes subject to climatic and lake level fluctuations will exhibit frequent vertical changes in facies.

Saline and ephemeral lakes are *underfilled* (Bohacs et al. 2000, 2003). Some saline lakes may become relatively fresh if there is a change to wetter, cooler conditions, allowing the formation of a stratified water body and consequently the accumulation of organic-rich sediment on the lake floor. A return to a drier climate increases evaporation and concentration of ions leading to evaporite deposition. Cycles of climate change can be recognised in some lake deposits as alternations between dark, carbonaceous mudrock (sometimes oil shales: 3.6.3) and beds of gypsum and other evaporite minerals. The areas of shallow saline and ephemeral lakes can show considerable variations through time as a result of changes in climate.

The rate of sediment supply is significant in all lacustrine environments. If the rate of deposition of clastic, carbonate and evaporite deposits is greater than the rate of basin subsidence (Chapter 24) the lake basin will gradually fill. In overfilled lake settings this will result in a change from lacustrine to fluvial deposition as the river waters no longer pond in the lake but instead flow straight through the former lake area with channel and overbank deposits accumulating. Balanced fill and underfilled basins will also gradually fill with sediment, sometimes to the level of the sill such that they also become areas of fluvial deposition.

10.6 LIFE IN LAKES AND FOSSILS IN LACUSTRINE DEPOSITS

Palaeontological evidence is often a critical factor in the recognition of ancient lacustrine facies. Freshwater lakes may be rich in life with a large number of organisms, but they are of a limited number of species and genera when compared with an assemblage from a shallow marine environment. Fauna commonly found in lake deposits include gastropods, bivalves, ostracods and arthropods, sometimes occurring in *monospecific assemblages*, that is, all organisms belong to the same species. Some organisms, such as the brine shrimp arthropods, are tolerant of saline conditions and may flourish in perennial saline lake environments.

Algae and cyanobacteria are an important component of the ecology of lakes and also have sedimentological significance. A common organism found in lake deposits are ***charophytes***, algae belonging to the Chlorophyta (*3.1.3*), which are seen in many ancient lacustrine sediments in the form of calcareous encrusted stems and spherical reproductive bodies. Charophytes are considered to be intolerant of high salinities and the recognition of these millimetre-scale, often dark, spherical bodies in fine-grained sediment is a good indicator of fresh or possibly brackish water conditions.

Cold, sediment-starved lakes in mountainous or polar environments may be sites of deposition of siliceous oozes (*3.3*). The origin of the silica is diatom phytoplankton, which can be very abundant in glacial lakes. These deposits are typically bright white cherty beds that are called ***diatomites***, and they are basically made up entirely of the silica from diatoms.

10.7 RECOGNITION OF LACUSTRINE FACIES

If the succession is entirely terrigenous clastic material, it is not always easy to distinguish between the deposits of a lake and those of a low energy marine environment such as a lagoon (*13.3.2*), the outer part of a shelf (Chapter 14) or even the deep sea (Chapter 16). Shallow lake facies will have similar characteristics to lagoonal deposits, with wave ripple sands interbedded with muds deposited from suspension, while the deeper environments of a lake resemble those of seas with similar or greater depths, as they include deposits from suspension and turbidites. The main criteria for distinguishing between lacustrine and marine facies are often the differences between the organisms and habitats that exist in these environments.

There are a number of groups of organisms that are found only in fully marine environments: these include corals, echinoids, brachiopods, cephalopods, graptolites and foraminifers, amongst others (*3.1.3*). The occurrence of fossils of members of these groups therefore provides evidence of marine deposition. There are many genera of other phyla that can be used as indicators if found as fossils, in particular there are groups of bivalve and gastropod molluscs that are considered to be freshwater forms, and fish that are thought to be exclusively lake-dwellers. Some of the more reliable indicators of freshwater conditions are algal and bacterial (*3.1.3*) fossil forms. Reliance on fossils to provide indicators of

lacustrine environments becomes more difficult in rocks that are from further back in the stratigraphic record, and in Precambrian strata it may be almost impossible to be sure.

A feature of lakes that is much less commonly found in marine settings is the stratification of the water body (Fig. 10.3). The lack of mixing of the oxygenated surface water with the lower part of the water column results in anaerobic conditions at the bottom of a deep, stratified lake. Animals are unable to tolerate the anaerobic conditions so the lake floor is devoid of life, and therefore there is no bioturbation (*11.7*). Deep lake deposits may therefore preserve primary sedimentary lamination that in marine environments is typically destroyed by burrowing organisms. The anoxia also prevents the aerobic breakdown of organic material that settles on the lake floor, allowing the accumulation of organic-rich sediments. The deposits of saline and ephemeral lakes usually can be distinguished from marine facies by the chemistry of the evaporite minerals.

Characteristics of lake deposits
- lithologies – sandstone, mudstone, fine-grained limestones and evaporites
- mineralogy – variable
- texture – sands moderately well sorted
- bed geometry – often very thin-bedded
- sedimentary structures – wave ripples and very fine parallel lamination
- palaeocurrents – few with palaeoenvironmental significance
- fossils – algal and microbial plus uncommon shells
- colour – variable, but may be dark grey in deep lake deposits
- facies associations – commonly occur with fluvial deposits, evaporites and associated with aeolian facies

FURTHER READING

Anadón, P., Cabrera, L. & Kelts, K. (Eds) (1991) *Lacustrine Facies Analysis*. Special Publication 13, International Association of Sedimentologists. Blackwell Science, Oxford, 318 pp.

Bohacs, K.M., Carroll, A.R., Neal, J.E. & Mankiewicz, P.J. (2000) Lake-basin type, source potential, and hydrocarbon character: an integrated-sequence-stratigraphic–geochemical framework. In: *Lake Basins through Space and Time* (Eds Gierlowski-Kordesch, E.H. & Kelts, K.R.).

Studies in Geology 46, American Association of Petroleum Geologists, Tulsa, OK; 3–34.

Bohacs, K.M., Carroll, A.R. & Neal, J.E. (2003) Lessons from large lake systems – thresholds, nonlinearity, and strange attractors. In: *Extreme Depositional Environments: Mega End Members in Geologic Time* (Eds Chan, M.A. & Archer, A. W.). Geological Society of America Special Paper 370, Boulder, CO; 75–90.

Carroll, A.R. & Bohacs, K.M. (1999) Stratigraphic classification of ancient lakes: balancing tectonic and climatic controls: *Geology*, **27**, 99–102.

Matter, A. & Tucker, M.E. (Eds) (1978) *Modern and Ancient Lake Sediments*, Special Publication 2, International Association of Sedimentologists. Blackwell Scientific Publications, Oxford, 290 pp.

Talbot, M.R. & Allen, P.A. (1996) Lakes. In: *Sedimentary Environments: Processes, Facies and Stratigraphy* (Ed. Reading, H.G.). Blackwell Scientific Publications, Oxford; 83–124.

Tucker, M.E. & Wright, V.P. (1990) Lacustrine carbonates. In: *Carbonate Sedimentology*. Blackwell Scientific Publications, Oxford; 164–190.

11

The Marine Realm: Morphology and Processes

The oceans and seas of the world cover almost three-quarters of the surface of the planet and are very important areas of sediment accumulation. The oceans are underlain by oceanic crust, but at their margins are areas of continental crust that may be flooded by seawater: these are the continental shelves. The extent of marine flooding of these continental margins has varied through time due to plate movements and the rise and fall in global sea level related to climate changes. The sedimentary successions in these shallow shelf areas provide us with a record of global and local tectonic and climatic variations. There is considerable variety in the sedimentation that occurs in the marine realm, but there are a number of physical, chemical and biological processes that are common to many of the marine environments. Physical processes include the formation of currents driven by winds, water density, temperate and salinity variations and tidal forces: these have a strong effect on the transport and deposition of sediment in the seas. Chemical reactions in seawater lead to the formation of new minerals and the modification of detrital sediment. The seas also team with life: long before there was life on land organisms evolved in the marine realm and continue to occupy many habitats within the waters and on the sea floor. The remains of these organisms and the evidence for their existence provide important clues in the understanding of palaeoenvironments.

11.1 DIVISIONS OF THE MARINE REALM

The **bathymetry**, the shape and depth of the sea floor (Fig. 11.1), is fundamentally determined by the plate tectonic processes that create ocean basins by sea-floor spreading. The spreading ridges are areas of young, hot basaltic crust that is relatively buoyant and typically around 2500 m below sea level. Away from the ridges the water depth increases as the older crust cools and subsides, and most of the **ocean floor** is between about 4000 and 5000 m below sea level. The deepest parts of the oceans are the **ocean trenches** created by subduction zones, where water depths can be more than 10,000 m. At the ocean margins the transition from ocean crust to continental crust underlies the **continental rise** and the **continental slope**, which are the lower and upper

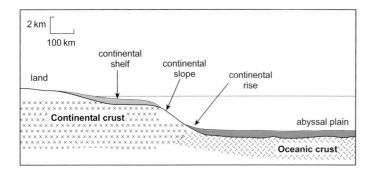

Fig. 11.1 A cross-section from the continental shelf through the continental slope and rise down to the abyssal plain.

parts of the bathymetric profile from the deep ocean to the shelf. The angle of the continental slope is relatively steep, usually between about 2° and 7°, while the continental rise is a lower angle slope down to the edge of the **abyssal plain**.

The **continental shelf** itself is underlain by continental crust, and the junction between the shelf and the slope usually occurs at about 200 m below sea level at present-day margins (the **shelf edge break**). Continental shelves are very gently sloping with gradients ranging from steep shelves of 1 in 40 to more typical gradients of 1 in 1000. They may extend for tens to hundreds of kilometres from the coastline to

the shelf edge break. Large areas of continental crust that are covered by seawater, which are mainly bordered by land masses and connected by straits to the oceans, are called **epicontinental seas** (sometimes called **epeiric seas**). The areas of epicontinental seas are greatest when relative sea levels are at the highest worldwide. A nomenclature for the division of the marine realm based on these depth zones is shown in Fig. 11.2. The shelf area, down to 200 m water depth, is called the **neritic zone**, the **bathyal zone** corresponds to the continental slope and extends from 200 m to 2000 m water depth, while the **abyssal zone** is the ocean floor below 2000 m. A depth limit

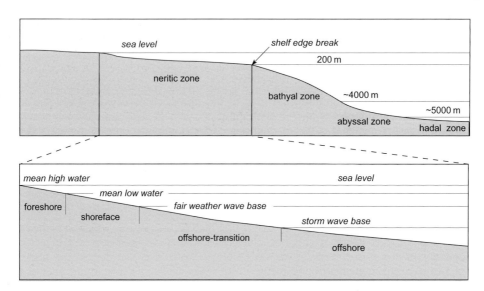

Fig. 11.2 Depth-related divisions of the marine realm: (a) broad divisions are defined by water depth; (b) the shelf is described in terms of the depth to which different processes interact with the sea floor, and the actual depths vary according to the characteristics of the shelf.

to this zone can also be applied at about 5000 m, below which the deepest parts of the oceans are called the **hadal zone**.

The shelf (neritic environment) can be usefully further divided into depth-controlled zones (Fig. 11.2), although in this case the divisions are not defined by absolute depths, but the depths to which certain processes operate. Their range therefore varies according to the conditions in a particular basin because the depths to which tidal processes, waves and storms affect the shelf vary considerably. The **foreshore** is the region between mean high water and mean low water marks of the tides. Depending on the tidal range (*11.2.2*) this may be a vertical distance of anything from a few tens of centimetres to many metres. The seaward extent of the foreshore is governed also by the slope and it may be anything from a few metres, if the shelf is steeply sloping and/or the tidal range is small, to over a kilometre in places where there is a high tidal range and a gently sloping shelf. The foreshore is part of the beach environment or **littoral zone** (*13.2*).

The **shoreface** is defined as the region of the shelf between the low-tide mark and the depth to which waves normally affect the sea bottom (*4.4.1*), and this is the **fair weather wave base**. The lower depth that the shoreface reaches depends on the energy of the waves in the area but is typically somewhere between 5 and 20 m. The width of the shoreface will be governed by the shelf slope as well as the depth of the fair weather wave base and may be hundreds of metres to kilometres across. In deeper water it is only the larger, higher energy waves generated by storms that affect the sea bed. The depth to which this occurs is the **storm wave base** and this is very variable on different

shelves. In some places it may be as little as 20 m water depth but can be 50 to 200 m water depth if the shelf borders an ocean with a large fetch for storm waves. This deeper shelf area between the fair weather and storm wave bases is called the **offshore-transition zone**. The **offshore zone** is the region below storm wave base and extends out to the shelf-edge break at around 200 m depth.

The activities of a number of physical, chemical and biological processes are determined by water depth, and in turn these influence the sediment accumulation on the different parts of the sea floor. The following sections consider some of these processes and how they affect depositional environments.

11.2 TIDES

11.2.1 Tidal cycles

According to Newton's Law of Gravitation, all objects exert gravitational forces on each other, the strength of which is related to their masses and their distances apart. The Moon exerts a gravitational force on the Earth and although ocean water is strongly attracted gravitationally to the Earth, it also experiences a small gravitational attraction from the Moon. The water that is closest to the Moon experiences the largest gravitational attraction and this creates a bulge of water, a **tidal bulge**, on that side of the Earth (Fig. 11.3). The bulge on the opposite side, facing away from the Moon, can be thought of as being the result of the Earth being pulled away from that water mass by the gravitational force of the Moon.

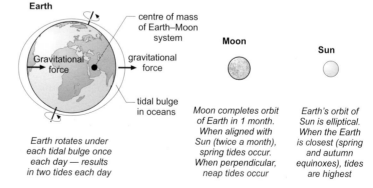

Fig. 11.3 The gravitational force of the Sun and Moon act on the Earth and on anything on the surface, including the water masses in oceans.

If the land areas are ignored the effect of these bulges is to create an ellipsoid of water with its long axis oriented towards the position of the Moon. As the Earth rotates about its axis the bulges move around the planet. At any point on the surface the level of the water will rise and fall twice a day as the two bulges are passed in each rotation. This creates the daily or **diurnal tides**. During the daily rotation, a point on the Earth will pass under one high bulge and a slightly lower bulge 12 hours or so later: this is referred to as the **diurnal tidal inequality**, the two high tides in a day are not of equal height. The two tides in the diurnal tidal cycle are just over twelve and a half hours because the Moon is orbiting the Earth as the planet is rotating, changing its relative position each day.

The Moon rotates around the Earth in the same plane as the Earth's orbit around the Sun. The Sun also creates a tide, but its strength is about half that of the Moon despite its greater mass, because the Sun is further away. When the Sun and Moon are in line with the Earth (an alignment known as **syzygy**) the gravitational effects of these two bodies are added together to increase the height of the tidal bulge. When the Moon is at $90°$ to the line joining the Sun and the Earth (the **quadratic alignment**), the gravitational effects of the two on the water tend to cancel each other. During the four weeks of the Moon's orbit, it is twice in line and twice perpendicular. This creates **neap–spring tidal cycles** with the highest tides in each month, the **spring tides**, occurring when the three bodies are in line. (The term 'spring' in this context is not referring to the season of the year.) A week either side of the spring tides are the **neap tides**, which occur when the Moon and Sun tend to cancel each other and the tidal effect is smallest.

Superimposed on the diurnal and neap–spring cycles is an **annual tidal cycle** caused by the elliptical nature of the Earth's orbit around the Sun. At the spring and autumn (Fall) equinoxes, the Earth is closest to the Sun and the gravitational effect is strongest. The highest tides of the year occur when there are spring tides in late March and late September. In mid-summer and mid-winter the Sun is at its furthest away and the tides are smaller. This pattern of three superimposed tidal cycles (diurnal, neap–spring and annual) is a fundamental feature of tidal processes that controls variations through time of the strength of tidal currents.

11.2.2 Tidal ranges

The tidal bulge created in the open ocean is only a few tens of centimetres, but of course the difference between high and low tide is many metres in some places, so there must be a mechanism to amplify the vertical change in sea level. The tidal bulge can be considered as a wave of water that passes over the surface of the Earth. In any waveform **resonance** effects are created by the shape of the boundaries of the 'vessel' the wave is moving through. In oceans and seas the shape of the continental shelf as it shallows towards land, indentations of the coastline and narrow straits between seas can all create resonance effects in the tidal wave. These can increase the amplitude of the tide and locally the tidal range is increased to several metres by tidal resonance effects. The highest tidal ranges in the world today are in bays on continental shelves, such as the Bay of Fundy, on the Atlantic seaboard of Canada, which has a tidal range of over 15 m (Dalrymple 1984).

In addition to the influence of land masses, the movement of water between high- and low-tide conditions is also affected by the Coriolis force (6.3): water masses moving in the northern hemisphere are deflected to the right of their path and in the southern hemisphere to the left. These effects break up the tidal wave into a series of **amphidromic cells** and at the centre of each cell there is an **amphidromic point** around which the tidal wave rotates (Fig. 11.4). At the amphidromic point there is no change in the water level during the tidal cycle. All oceans are divided into a number of major amphidromic cells and there are additional, smaller cells in shelf areas such as the North Sea and small seas such as the Gulf of Mexico. Tidal ranges are therefore very variable and within a body of water the pattern of tides can be very complex: in the North Sea, for example, the tidal range varies from less than a metre to over 6 m (Fig. 11.4). For sedimentological purposes it is useful to divide tidal ranges into the following categories: up to 2 m mean tidal range the regime is **microtidal**, between 2 and 4 m range it is **mesotidal** and over 4 m is **macrotidal**.

11.2.3 Characteristics of tidal currents

The horizontal movement of water induced by tides is a **tidal current**: tidal currents are weak in microtidal

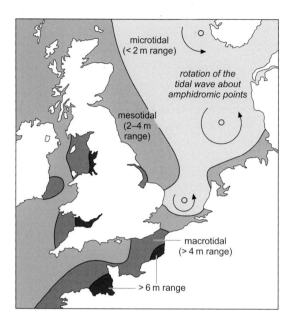

Fig. 11.4 The North Sea of northwest Europe has a variable tidal range along different parts of the bordering coasts. Amphidromic points mark the centres of cells of rotary tides that affect the shallow sea.

between the low and high tide, and the **ebb tide** current, which flows in the opposite direction as the water level returns to low tide. These are **bipolar currents** acting in two opposite directions. Second, the tidal flow varies in velocity in a cyclical manner. At times of high and low tide, the water is still, but as the tide turns, the water starts to move and increases in velocity up to a peak at the mid-tide point in each direction. Third, the strength of the flow is directly related to the difference between the levels of the high and low tides. As the tidal range varies according to the series of cycles (see above) the velocity of the current varies in the same pattern. The strongest tidal currents occur when there are the highest spring tides at the spring and autumn equinoxes.

The rotational pattern of the tidal wave within amphidromic cells results in a flow of water that follows a circular or elliptical pattern. These **rotary tides** can be important currents on shelves and in epicontinental seas. During the course of the tidal cycle the current varies in strength, but does not change direction and there may not be a period of slack water (Dalrymple 1992). These offshore tidal currents are important processes in the transport and deposition of sediment on some shelf areas (*14.3*).

11.2.4 Sedimentary structures generated by tidal currents

Bipolar cross-stratification

An analysis of current directions recorded by cross-bedding in sands deposited by tidal currents may

regimes, more pronounced if the range is mesotidal and are capable of carrying large quantities of sediment in macrotidal regimes. Nearshore tidal currents show a number of features that produce recognisable characteristics in sediment deposited by them (Fig. 11.5). First, tidal currents regularly change direction from the **flood tide** current, which moves water onshore

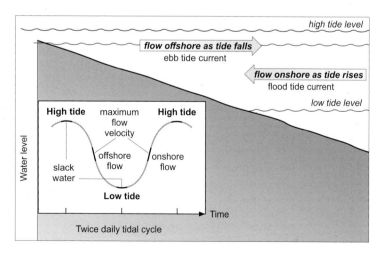

Fig. 11.5 During the diurnal tidal cycle the direction of flow reverses from ebb (offshore) to flood (onshore). The current velocity also varies from peaks at the mid points of ebb and flood flow, reducing to zero at high and low tide slack water before accelerating again.

show a **bimodal** (two main directions of flow) and **bipolar** (two opposite directions of flow) pattern. It should be noted, however, that these opposing palaeo-flow directions are not seen in all tidal sediments: first the flow in one direction (either the ebb or flood tide) may be much stronger than the other, and second the two flows might be widely separated and only one may have been active in the area examined (Dalrymple & Choi 2007). Under favourable circumstances, bipolar cross-stratification may be seen in a single vertical section produced by alternating directions of migration of ripples or dunes. This is known as **herringbone cross-stratification** (Figs 11.6 & 11.7) and it results from a tidal current flowing predominantly in one direction for a period of time, probably many years, followed by a change in the pattern of

Fig. 11.7 Herringbone cross-stratification in sandstone beds (width of view 1.5 m).

tidal flow that results in another period of opposite flow. This pattern of alternating directions should not be interpreted as a diurnal pattern as this would imply unrealistically high rates of sediment accumulation. The herringbone pattern is characteristic of tidal sedimentation, but is not found in all instances.

Mud drapes on cross-strata

At the time of high or low tide when the current is changing direction there is a short period when there is no flow. When the water is relatively still some of the suspended load may be deposited as a thin layer of mud. When the current becomes stronger during the next tide, the mud layer is not necessarily removed because the clay-rich sediment is cohesive and this makes it resistant to erosion (4.2.4). **Mud drapes** formed in this way can be seen in wave and current ripple laminated sands deposited in shallow water in places such as tidal mud flats (13.4): the heterolithic beds formed in this way display flaser or lenticular lamination depending on the proportion of sand and mud present (4.8). Mud drapes can also occur within cross-beds: a lamina of sand is deposited on the lee slope of the subaqueous dune during strong tidal flow but as the tide changes direction mud falls out of suspension and drapes the subaqueous dune (Dalrymple 1992). There are circumstances where mud drapes can form in other depositional regimes, for instance in rivers that have only seasonal flow, but they are most common in tidal settings: abundant, regular mud drapes are a good indicator of a tidally influenced environment (Figs 11.6 & 11.8).

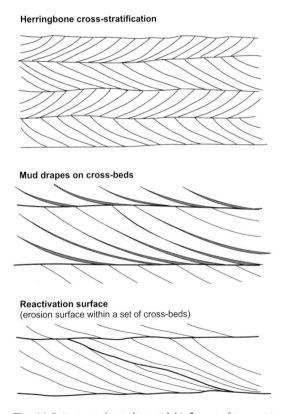

Herringbone cross-stratification

Mud drapes on cross-beds

Reactivation surface
(erosion surface within a set of cross-beds)

Fig. 11.6 Features that indicate tidal influence of transport and deposition: (a) herringbone cross-stratification; (b) mud drapes on cross-bedding formed during the slack water stages of tidal cycles; (c) reactivation surfaces formed by erosion of part of a bedform when a current is reversed.

Fig. 11.8 Cross-bedded sandstone in sets 35 cm thick with the surfaces of individual cross-beds picked out by thin layers of mud. Mud drapes on cross-beds are interpreted as forming during slack water stages in the tidal cycle.

Reactivation surfaces

In places where there is one dominant direction of tidal current the bedforms migrate in that direction producing unidirectional cross-stratification. These bedforms can be modified by the reverse current, principally by the removal of the crest of a subaqueous dune. When the bedform recommences migration in the direction of the dominant flow the cross-strata build out from the eroded surface. This leaves a minor erosion surface within the cross-stratification, which is termed a ***reactivation surface*** (Figs 11.6 & 11.9) (Dalrymple 1992).

Tidal bundles

The strength of the tidal current varies cyclically and hence its capacity to carry sediment varies in the

Fig. 11.9 A reactivation surface within cross-bedded sands is a minor erosion surface truncating some of the cross-beds.

same way. At the highest tides the current is strongest enabling more transport and deposition of sand on the bedforms in the flow. When the difference between high and low tide is smaller the current will transport a reduced bedload or there may be no sediment movement at all. A cyclical variation in the thickness of foreset laminae in cross-beds may therefore be attributed to variations in flow strength in the neap–spring cycle and these are called ***tidal bundles***. In an idealised case, the laminae would show thickness variations in cycle in multiples of 7 or 14 (Yang & Nio 1985), but there is often no sedimentation or bedform migration during the weaker parts of the tidal cycle, so this ideal pattern is rarely seen.

11.3 WAVE AND STORM PROCESSES

The depth to which surface waves affect a water body is referred to as the wave base (4.4.1) and on continental shelves two levels can be distinguished (11.1). The fair weather wave base is the depth to which there is wave-influenced motion under normal weather conditions. The storm wave base is the depth waves reach when the surface waves have a higher energy due to stronger winds driving them. Below the storm wave base the sea bed is not normally affected by surface waves.

11.3.1 Storms

Storms are weather systems that have associated strong surface winds, typically in excess of $100 \, km \, h^{-1}$, and they may affect both land and marine environments. In continental settings they are important in aeolian transport of material (8.1), which includes the transport of airborne sediment out into the oceans. Large storms have a very large impact in shallow marine environments and storm-related processes of sedimentation are dominant in most shelf and epicontinental seas. There are three components to the effects of storms on shelf environments. The strong winds drive currents in the oceans that move water and sediment across and along continental shelves. They also generate large waves that affect much deeper parts of the shelf than normal, fair weather waves: these waves rework the sediment on the sea floor generating characteristic sedimentary structures (14.2.1). Finally, the high-energy conditions bring a lot of sediment into suspension near the sea floor and the mixture of sediment and water moves as a gravity-driven underflow across the shelf, from shallower to deeper water. The deposits of these storm processes are referred to as *tempestites*: there is further discussion of the processes and products of storm-dominated shelves in Chapter 14.

11.3.2 Tsunami

Tsunami is the Japanese for 'harbour wave' and refers to waves with periods of 10^3 to 10^4 seconds that are generated by events such as subsea earthquakes, large volcanic eruptions and submarine landslides. In the past such waves were sometimes incorrectly called 'tidal waves', but their origins have no connection with tidal forces. These events can set up a surface wave a few tens of centimetres amplitude in deep ocean water and a wavelength of many kilometres. As the wave reaches the shallower waters of the continental shelf, the amplitude is increased to ten or more metres, producing a wave that can have a devastating effect on coastal areas (Scheffers & Kelletat 2003).

The effects of a tsunami are dramatic, with widespread destruction occurring near coasts, both near the source of the wave and also anywhere in the path of it, which can be thousands of kilometres across an ocean. They also have a serious impact on shallow marine environments causing disruption and redeposition of foreshore and shoreface sediments. It has been suggested that beds of poorly sorted debris containing a mixture of deposits and fauna from different coastal and shallow marine environments may form as a consequence of tsunami (Pilkey 1988). It may be possible to distinguish them from ordinary storm deposits by their larger size, but in practice it may be difficult to show that a deposit is generated by a specific mechanism.

11.4 THERMO-HALINE AND GEOSTROPHIC CURRENTS

Currents that are driven by contrasts in temperature and/or salinity are called *thermo-haline currents*. Cold water is dense and will sink relative to warmer water, and seawater is denser if the salinity is greater than normal: these temperature and salinity contrasts generate flow of the denser fluid beneath the less dense water. Cold surface water descends at high polar latitude, *sink points*, and these water masses then move around the oceans as thermo-haline bottom currents (Stow 1985). The water that is moved from the polar regions is replaced by warm surface waters and this sets up a circulation system that transports water thousands of kilometres in the world's oceans. *Geostrophic currents* are wind-driven currents related to the global wind systems, which result from differences in air mass temperatures combined with the Coriolis force (6.3). The pattern of ocean currents is shown in Fig. 11.10.

The effects of these currents on sedimentation are most noticeable in deeper waters (16.4) as their effects in shallower water are often masked by the influences of tides, waves and storms. Thermo-haline currents are typically weaker than storm and tidal currents but are of larger volume. They mainly move clay and silt in suspension and very fine sands as bedload. Thermo-haline currents are also important in the distribution of nutrients in the oceans. Bottom currents move nutrients from colder regions to areas where upwelling occurs and the nutrient-rich waters reach the surface. As a consequence, these areas of upwelling are regions of high organic productivity and can result in deposits rich in biogenic material.

11.5 CHEMICAL AND BIOCHEMICAL SEDIMENTATION IN OCEANS

The most important chemical and biochemical sediments in modern seas and ancient shelf deposits are carbonate sediments and evaporites, and in the oceans plankton generate large quantities of carbonate and siliceous sediment. In addition there are other, less abundant but significant chemical and biochemical deposits.

11.5.1 Glaucony and glauconite

The term glauconite is commonly used by geologists to refer to a dark green mineral that is found quite commonly in marine sediments. In correct usage the use of this term should be restricted to a potassium-rich mica, which has the mineral name *glauconite*, because this is in fact only one member of a group of potassium and iron-rich phyllosilicate minerals that

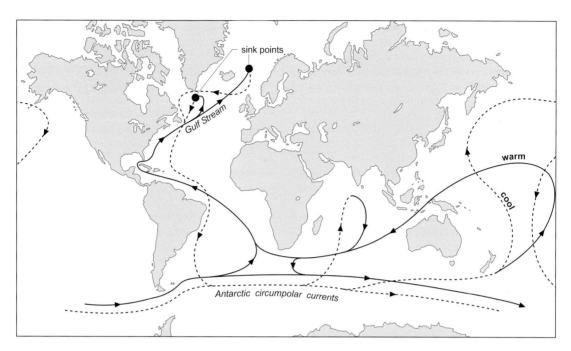

Fig. 11.10 The main geostrophic current pathways (thermo-haline circulation patterns) affecting the modern oceans. Sink points in the North Atlantic are due to input of cold glacial meltwater from the Greenland ice-cap.

are closely related (Amorosi 2003). Material made up of any of these distinctive, medium to dark green minerals is referred to as *glaucony*. Glaucony minerals are authigenic, that is, they crystallise within the sedimentary environment (*2.3.2*): this is in contrast to almost all other silicate minerals found within sediments that are detrital (*2.3.1*). The process of forming the mineral, glauconitisation, occurs at the sea floor on substrates such as the hard parts of foraminifers, other carbonate fragments, faecal pellets and lithic fragments. It appears the process requires a particular microenvironment at the interface between oxidising seawater and slightly reducing interstitial waters. This typically occurs at water depths of between about 50 and 500 m, on the outer parts of continental shelves and upper parts of continental slopes.

Glaucony/glauconite is important in sedimentology and stratigraphy for a number of reasons. Firstly, it is a reliable indicator of deposition in a shallow marine environment, although it can be reworked into deeper water and occasionally into shallower environments by currents. Secondly it is most abundant within shelf sediments under conditions where sedimentation of other material, terrigenous clastic or carbonate, is slow. It therefore commonly occurs in **condensed sections**, that is, strata which have been deposited at anomalously low sedimentation rates. The recognition of periods of low sedimentation rate on the shelf is important when assessing evidence of changes in sea level because outer shelf sedimentation tends to be slowest during periods of sea level rise (this is discussed further in Chapter 23). Thirdly, because the mineral is authigenic and also rich in potassium, it can be dated by radiometric methods and the age obtained corresponds to the time of deposition. As will be seen in Chapter 21, direct radiometric dating of sedimentary material is rarely possible, but glaucony/glauconite is the exception and consequently is very important in relating strata to the geological time scale (*19.1.2*).

11.5.2 Phosphorites

Phosphorites are sedimentary rocks that are enriched in phosphorus to a level where the bulk composition is over 15% P_2O_5. Phosphate may be present in sediment as primary bioclasts such as fish teeth and scales and vertebrate bones, but mostly it occurs as an authigenic precipitate, which coats grains, forms peloids and micronodules on the sea floor and may also occur as laminae encrusting the sea bed (Glenn & Garrison 2003). Accumulations of phosphorite are favoured by slow sedimentation rates of other materials and, like glaucony, are characteristic of condensed sections. Hardgrounds can be composed of laminated phosphorites, while the peloids and other grains are concentrated into phosphate-rich beds by reworking of the material by seafloor currents (Glenn & Garrison 2003).

Modern phosphorite concentrations occur on continental margins where there are regions of upwelling of nutrient-rich waters, such as off the west coast of South America and off west Africa where Antarctic water comes to the surface. These nutrient-rich cool waters coming up into warmer waters promote blooms of plankton, which are at the bottom of the food chain. Ancient phosphorites are thought to have formed in similar settings and it might also be expected that concentrations would be greatest at times of high sea level when supply of other sediment to the shelf is reduced. Phosphorite production is also related to the supply of phosphate, which ultimately comes from the weathering of continental rocks.

11.5.3 Organic-rich sediments: black shales

Organic material from dead plants, animals and microbial organisms is abundant in the oceans and becomes part of the material that falls to the sea floor. Where the sea floor is oxygenated by currents bringing water down from the surface the organic matter is oxidised or consumed by scavengers living on the sea bed. Poor circulation reduces the oxygen in the waters at the sea floor and the conditions become *anoxic*. Breakdown of the organic matter is slower or non-existent in the absence of oxygen and the conditions are not favourable for scavenging organisms. The organic matter accumulates under these anoxic conditions and contributes to the pelagic sediment to form **black shale**, a mudrock that typically contains 1–15% organic carbon (Wignall 1994; Stow et al. 1996). The black or dark grey colour is partly due to the presence of the organic matter and also because of finely disseminated pyrite (iron sulphide), which also forms under reducing conditions.

The conditions for the formation of black shales are therefore determined by the organic input, the efficiency of the breakdown of that material by microbial activity and the dilution effects of terrigenous clastic, biogenic carbonate or silica. The most favourable sites are therefore deep seas where there is poor circulation between the oxygenated surface water and the sea floor. Basins with restricted circulation, such as the modern Black Sea, provide optimal conditions (Wignall 1994), but not all black shales form in similar settings. Provided the supply of organic material is greater than the rate at which it can be broken down, black shales can form on shelves where circulation is moderately effective. They have considerable economic importance in sedimentology and stratigraphy as they are hydrocarbon source rocks (*18.7.3*).

11.6 MARINE FOSSILS

Shelves are areas of oxygenated waters periodically swept by currents to bring in nutrients. As such they are habitable environments for many organisms that may live swimming in water (***planktonic***) or on the sea floor (***benthic***), either on the surface or within the sediment. Plants and animals living in the marine realm contribute detritus, modify other sediments and create their own environments. Modern shelf environments team with life and it is rare to find an ancient shelf deposit that does not contain some evidence for the organisms that lived in the seas at the time.

In shallow seas with low clastic input the calcareous hard parts of dead organisms make up the bulk of the sediment, either as the loose detritus of mobile animals or as biogenic reefs, which are whole sediment bodies built up as a framework by organisms such as corals and algae. Terrigenous clastic sandy and muddy shelf deposits may also contain a rich flora and fauna, the type and diversity of which depends on the energy on the sea bed (fragmentation can occur in high-energy environments) and the post-depositional history (Chapter 18), which affects preservation of material.

Many plants and animals occupy ecological niches that are defined by such factors as water depth, temperature, nutrient supply, nature of substrate and so on. If the ecological niche of a fossil organism can be determined this can provide an excellent indication of the depositional environment. In the younger Cenozoic strata the fossils may be of organisms so similar to those alive today that determining the likely environment in which they lived is quite straightforward. Farther back in geological time this task becomes more difficult. Groups of organisms such as trilobites and graptolites, which were abundant in the Lower Palaeozoic seas, have no modern representatives for direct comparison of lifestyle. Clues as to the ecological niche occupied by a fossil organism are provided by considering the *functional morphology* of the body fossil. All organisms are in some way adapted to their environment so if these adaptations can be recognised the lifestyle of the organisms can be determined to some extent. In trilobites, for example, it has been recognised that some types had well-developed eyes whereas in others they were very poorly developed: one interpretation of this would be that the trilobites with eyes needed them to help move around on the sea floor but those that lived buried in the sediment had no need of sight.

Some organisms are thought to have occupied very specific niches and can provide quite precise information about the environment of deposition. Some algae and hermatypic corals require clear water and sunlight to thrive, so they are indicators of shallow, mud-free shelf environments. Other organisms (certain bivalves, for instance) are more tolerant of different environments and can live in a range of conditions and water depths provided that a supply of nutrients are available. In general, the abundance of benthic organisms decreases as the water depth increases. Shoreface environments usually have the most diverse assemblages of benthic fauna and flora due to the well-oxygenated conditions of the wave-agitated water and the availability of light (provided that it is not too muddy). The abundance of organisms living on the sea floor decreases in the offshore transition and offshore parts of the shelf. In the deep oceans only a few specialised organisms live on the sea floor adjacent to areas of hydrothermal activity.

The abundance of planktonic organisms is controlled by the supply of nutrients and the surface temperature of the water. The hard parts of planktonic organisms may be distributed in sediments of any water depth, although dissolution of calcium carbonate occurs in very deep water (*16.5.2*). One approach to the problem of determining the depth at which sediment was deposited is to consider the ratio of benthic to planktonic organisms present: if the proportion of benthic organisms is high the water was probably shallow, whereas a high count of planktonic organisms indicates deeper water. This method normally only provides a very rough guide to relative water depth but is applied in a semi-quantitative way in Cenozoic and Mesozoic strata by considering the proportions of benthic and planktonic forms of foraminifers.

11.7 TRACE FOSSILS

Although body fossils provide physical evidence of an organism having lived in the past, **trace fossils** are evidence of the activity of an organism. Traces include tracks of walking animals, trails of worms, burrows of molluscs and crustaceans, and are collectively called *ichnofauna*. Trace fossils are usually found on or within sediment that was unconsolidated but with sufficient strength to retain the shape of the animal's trace. Contrasts in sediment type between a burrow and the host sediment are a considerable aid to recognition. A distinction is made between **burrows** formed in soft sediment and **borings** made by organisms into hard substrate.

The different forms of trace fossils are given names similar to those used in the classification of animals and body fossils: so, for example, smaller vertical tubes in sands are called **Skolithos** and a crawling trail produced by a multilimbed organism is known as *Cruziana*. Comparison of the form of *Cruziana* traces with body fossils provides very strong evidence that trilobites formed these features, but this link between ichnofauna and body fossils is the exception rather than the rule. For the majority of trace fossils, we can only guess at the nature of the animal that formed them: other exceptions are **Ophiomorpha**, a pellet-lined burrow which has a morphology identical to burrows made by modern callianassid shrimps, and **Trypanites**, a boring made in rock or solid substrate that can be seen in modern seas as being made by bivalve molluscs such as *Lithophaga*.

Ichnofossils are classified according to the inferred manner in which they were formed, for example, by movement of an animal over a surface, feeding, creation of a shelter, and so on (Fig. 11.11) (Simpson 1975; Ekdale et al. 1984). However, there is considerable variation within these categories as dinosaur footprints and trilobite tracks classify as the same

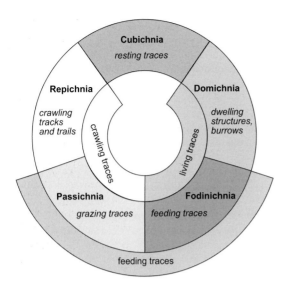

Fig. 11.11 Classification of trace fossils based on interpretation of the activity of the organism. (Adapted from Seilacher 2007.)

type of trace fossil. There is also a lot of overlap between categories, as an animal may have been walking and feeding at the same time. The most common trace fossils are some form of burrow made for dwelling or feeding or both. **Escape burrows**, formed by organisms moving up to the surface, are common in settings where there is rapid sedimentation by storms or turbidity currents.

11.7.1 Trace fossils in palaeoenvironmental analysis

Although we may not know the identity of many of the animals that produced trace fossils, their presence provides some very valuable information about the behaviour of organisms and the nature of the palaeoenvironments. From the perspective of the analysis of sedimentary rocks, ichnofossils will often be more useful than fossil shells or bones because they are conclusive evidence that an animal lived there. In contrast, a body fossil is, of course, a dead animal, and it is not always certain whether it lived in the place where the fossil is found. A coral may be preserved as part of the reef in which it lived, but a pelagic organism is not preserved where it lived, swimming in the open ocean, but on the ocean floor, where it ended up

after it died. In some cases, the environmental conditions might have actually caused the death of an animal, such as a skeleton of a mammal enclosed in volcanic ash. Most importantly, ichnofauna provide precise information about the environment where they were formed. For example, bird footprints are either evidence of a land surface, or of very shallow water where the bird may have been paddling, and a complex of burrows in sea-floor sediment is evidence of oxygenated conditions. Trace fossils are therefore a very powerful tool in palaeoenvironmental analysis, and we can use changes in trace fossil assemblages, known as **ichnofacies**, as evidence for changes in environment, such as rise and fall of sea level (23.8).

11.7.2 Trace fossil assemblages

The ecology of the sea floor and hence the ichnofauna found in the sediment is controlled by a number of interrelated factors (Pemberton et al. 1992). These factors are:

1 substrata type, whether it is hard or soft, sandy or muddy;
2 the strength of the currents that sweep the sea floor;
3 the rate at which sediment is being deposited;
4 turbidity, which is the amount of fine suspended sediment in the water;
5 oxygen levels in the water;
6 the salinity of the water;
7 the quality of the nutrient supply;
8 the quantity of nutrient supply.

These environmental variables can be simplified into a scheme based primarily on water depth (Fig. 11.12) and the hardness of the substrate (Fig. 11.13) (Pemberton et al. 1992; Pemberton & MacEachern 1995). Shallow marine environments tend to be higher energy and are richer in nutrients than deep water settings. There are, however, exceptions to this, as some shallow water settings (shelf seas with restricted water circulation and lagoons) can be low energy and relatively poorly supplied with nutrient, so these ichnofacies are not necessarily definitive indicators of water depth.

The conditions of the substrate may vary from loose sand in a foreshore setting to hard rocks in another beach environment: the ichnofacies that occur on hard or semiconsolidated shorelines (*Trypanites* and **Glossifungites** assemblages respectively) can also

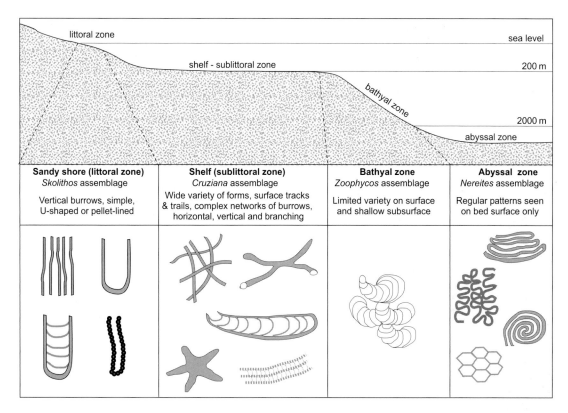

Fig. 11.12 Assemblages of trace fossil forms and their relationship to the major divisions of the marine realm. (Adapted from Pemberton et al. 1992.) The assemblages are named after characteristic ichnofauna and the 'type' ichnofossil does not need to be present in the assemblage.

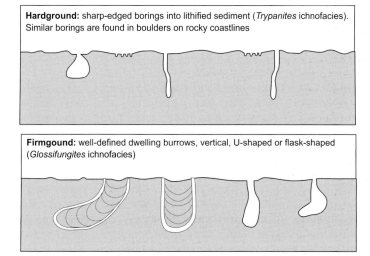

Fig. 11.13 The characteristics of trace fossils are influenced by the nature of the substrate. Boring organisms cut sharp-sided traces into solid rock or cemented sea floors (hardgrounds). Semiconsolidated surfaces (firm-grounds) result in well-defined burrows.

occur further out on the shelf if conditions result in a hard or firm sea floor (*11.7.4*). It should be noted that the names of the assemblages are taken from one particular ichnofossil which may be typical: the *Cruziana* assemblage does not necessarily include the actual form *Cruziana*, and is in fact unlikely to unless the deposits are Palaeozoic as they are thought to be formed by trilobites. Examples of trace fossils are shown in Fig. 11.14.

Trace fossil assemblages that occur along shorelines may be subdivided according to the degree of consolidation of the substrate. Along sandy shorelines *Skolithos* ichnofacies are characteristic. This facies is named after simple vertical tubes formed by organisms that lived in the high energy region of the foreshore. In this assemblage *Ophiomorpha* also occur, a larger, mainly vertical burrow lined with faecal pellets, and *Diplocraterion*, a U-shaped burrow. The animals that formed *Skolithos*, *Ophiomorpha* and *Diplocraterion* are thought to have moved up and down in the sediment with the changing water level of the foreshore. Where the sediment is semiconsolidated the *Glossifungites* ichnofacies assemblage occurs: the burrows are similar in form to those of the *Skolithos* assemblage but they tend to have sharp, well-defined margins to the tubes and may extend into excavated dwelling cavities. Some organisms (such as bivalves, echinoids and some sponges) are able to bore into rock to create dwelling traces: this assemblage is called *Trypanites*.

In the shoreface zone of the shelf, the *Cruziana* assemblage includes *Cruziana* itself, *Rhizocorallium*, an inclined U-shaped burrow, *Chondrites*, a vertically branching small burrow, *Planolites*, a horizontal branching burrow and *Thalassanoides*, larger (>10 mm diameter) burrows in a complex three-dimensional network. In the deeper waters of the outer bathyal zone the **Zoophycos** assemblage is the characteristic ichnofacies. *Zoophycos* has a rather variable, partly radial form that may be tens of centimetres across. Few other trace fossils are found in these depths. In the deeper bathyal to abyssal depths the **Nereites** ichnofacies assemblage traces are characteristically feeding traces showing regular patterns. These include *Helminthoidea*, which, like *Nereites* is a looping surface trace, and the enigmatic *Palaeodictyon* which has a regular hexagonal pattern. The regular structure of the traces of this ichnofacies is attributed to the scarcity of nutrients and the need to move efficiently; in shallower,

nutrient-rich sediment more random feeding structures are the norm.

11.7.3 Bioturbation

The presence of evidence of organisms disturbing sediment is known as **bioturbation**, and is a very common feature in sedimentary rocks. In fact, the absence of bioturbation in shallow marine deposits may be taken as an indicator of something unusual about conditions, such as an anoxic sea floor. The intensity of bioturbation in a body of sediment is an indication of the number of animals living there and the length of time over which they were active (Droser & Bottjer 1986). A scale of bioturbation intensity has been devised to allow comparison between deposits in different places.

Grade 1: a few discrete traces
Grade 2: bioturbation affects less than 30% of the sediment, bedding is distinct
Grade 3: between 30% and 60% of the sediment affected, bedding is distinct
Grade 4: 60% to 90% of the sediment bioturbated, bedding indistinct
Grade 5: over 90% of sediment bioturbated, and bedding is barely detectable
Grade 6: sediment is totally reworked by bioturbation

It should be noted that when a body of sediment is wholly bioturbated it can be difficult to recognise individual traces, and sometimes difficult to recognise that there is bioturbation present at all. The sediment will simply appear to be structureless, with the only evidence of trace fossils being that the sediment appears to be slightly mottled or with patches of different grain sizes.

11.7.4 Trace fossils and rates of sedimentation

Ichnofacies can be used as indicators of the degree of consolidation of the substrate (Fig. 11.13) and this can be a useful tool in the analysis of a stratigraphic succession. Where rates of sedimentation are high, the sea floor is covered by loose sandy or muddy material and a variety of ichnofacies occur according to the water depth. Sediment exposed on the sea floor starts to consolidate if the rate of sedimentation is

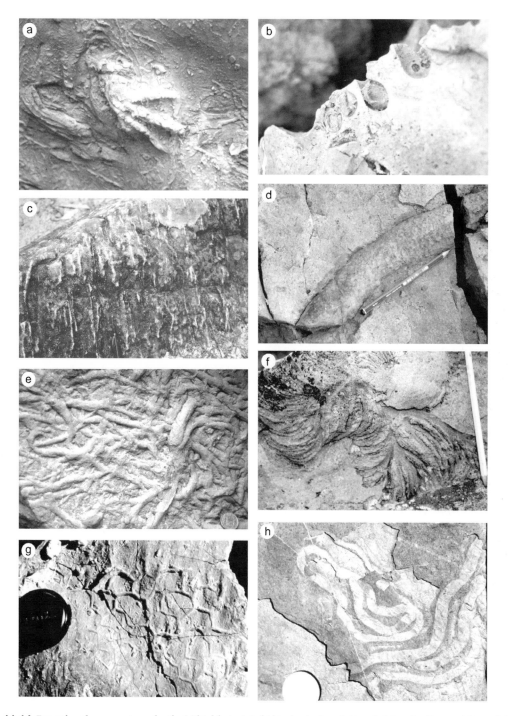

Fig. 11.14 Examples of common trace fossils: (a) bird footprint; (b) bivalve borings into rock; (c) vertical burrows in sandstone (*Skolithos*); (d) large crustacean burrow (*Ophiomorpha*); (e) complex burrows (*Thalassanoides*); (f) *Zoophycos*; (g) *Palaeodictyon*; (h) *Helmenthoides*.

relatively slow and a ***firmground*** forms. The characteristic ichnofacies of firmgrounds is *Glossifungites* (Ekdale et al. 1984). At even slower rates of sedimentation complete lithification (*18.2*) of the sea floor occurs with the formation of a ***hardground*** typified by the ichnofacies *Trypanites* (Ekdale et al. 1984). Recognition of hardgrounds and firmgrounds is particularly important in the sequence stratigraphic analysis of sedimentary successions (Chapter 23).

11.8 MARINE ENVIRONMENTS: SUMMARY

The physical processes of tides, waves and storms in the marine realm define regions bounded by water depth changes. The beach foreshore is the highest energy depositional environment where waves break and tides regularly expose and cover the sea bed. At this interface between the land and sea storms can periodically inundate low-lying coastal plains with seawater. Across the submerged shelf, waves, storms and tidal currents affect the sea bed to different depths, varying according to the range of the tides, the fetch of the waves and the intensity of the storms.

Sedimentary structures can be used as indicators of the effects of tidal currents, waves in shallow water and storms in the offshore transition zone. Further clues about the environment of deposition are available from body fossils and trace fossils found in shelf sediments. More details of the coastal, shelf and deep-water environments are presented in the following chapters.

FURTHER READING

Bromley, R.G. (1990) *Trace Fossils, Biology and Taphonomy.* Special Topics in Palaeontology 3, Unwin Hyman, London.

Johnson, H.D. & Baldwin, C.T. (1996) Shallow clastic seas. In: *Sedimentary Environments: Processes, Facies and Stratigraphy* (Ed. Reading H.G.). Blackwell Science, Oxford; 232–280.

Pemberton, S.G. & MacEachern, J.A. (1995) The sequence stratigraphic significance of trace fossils: examples from the Cretaceous foreland basin of Alberta, Canada. In: *Sequence Stratigraphy of Foreland Basin Deposits* (Eds Van Wagoner, J.C. & Bertram, G.T.). Memoir 64, American Association of Petroleum Geologists, Tulsa, OK; 429–476.

Seilacher, A. (2007) *Trace Fossil Analysis.* Springer, Berlin.

12

Deltas

The mouths of rivers may be places where the accumulation of detritus brought down by the flow forms a sediment body that builds out into the sea or a lake. In marine settings the interaction of subaerial processes with wave and tide action results in complex sedimentary environments that vary in form and deposition according to the relative importance of a range of factors. Delta form and facies are influenced by the size and discharge of the rivers, the energy associated with waves, tidal currents and longshore drift, the grain size of the sediment supplied and the depth of the water. They are almost exclusively sites of clastic deposition ranging from fine muds to coarse gravels. Deposits formed in deltaic environments are important in the stratigraphic record as sites for the formation and accumulation of fossil fuels.

12.1 RIVER MOUTHS, DELTAS AND ESTUARIES

The mouth of a river is the point where it reaches a standing body of water, which may be a lake or the sea. These are places where a delta may form (this chapter), an estuary may occur (next chapter) or where there is neither a delta nor an estuary. This variation depends on the morphology of the river mouth, the supply of sediment by the river and the processes acting in the lake or sea. A **delta** can be defined as a 'discrete shoreline protuberance formed at a point where a river enters the ocean or other body of water' (Fig. 12.1) (Elliott 1986; Bhattacharya & Walker 1992), and as such it is formed where sediment brought down by the river builds out as a

body into the lake or sea. In contrast, an **estuary** is a river mouth where there is a mixture of fresh water and seawater with accumulation of sediment within the confines of the estuary, but without any build-out into the sea. 'Ordinary' river mouths are settings where there is no significant mixing of waters and any sediment introduced by the river is reworked and carried away by processes such as waves and tides.

12.2 TYPES OF DELTA

Even a cursory survey of modern deltas reveals that they are widely variable in terms of scale, processes and the nature of the sediment deposited. A stream feeding into a lake may create a sediment body that is

Fig. 12.1 A delta fed by a river prograding into a body of water.

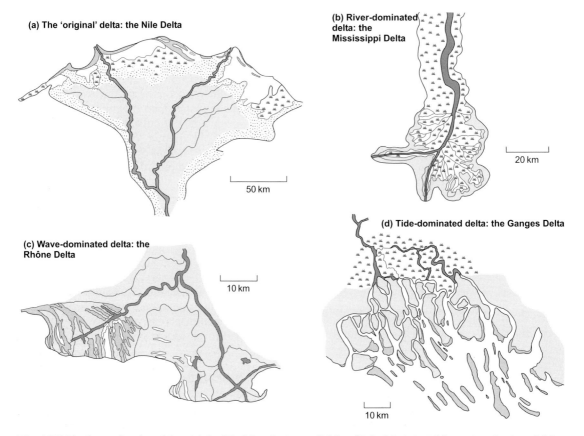

(a) The 'original' delta: the Nile Delta

50 km

(b) River-dominated delta: the Mississippi Delta

20 km

(c) Wave-dominated delta: the Rhône Delta

10 km

(d) Tide-dominated delta: the Ganges Delta

10 km

Fig. 12.2 The forms of modern deltas: (a) the Nile delta, the 'original' delta, (b) the Mississippi delta, a river-dominated delta, (c) the Rhone delta, a wave-dominated delta, (d) the Ganges delta, a tide-dominated delta.

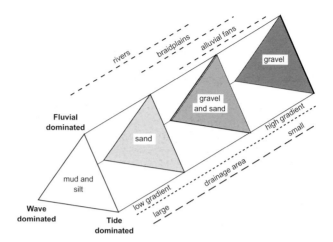

Fig. 12.3 Classification of deltas taking grain size, and hence sediment supply mechanisms, into account. (Modified from Orton & Reading 1993.)

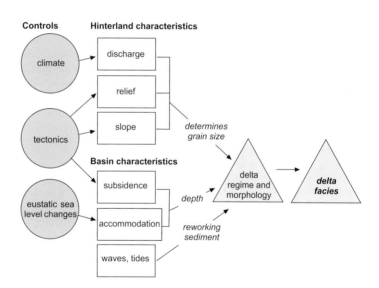

Fig. 12.4 Controls on delta environments and facies. (Adapted from Elliott 1986a.)

only a few tens to hundreds of metres across, while the largest deltas cover areas of thousands of square kilometres. The 'original' delta is at the mouth of the River Nile (Egypt), an area of flat land with river channels that had the triangular shape of the fourth letter of the Greek alphabet, Δ (Fig. 12.2). This shape, however, is not shared by many other deltas (Fig. 12.2) and the morphologies range from elongate 'fingers' building out into the sea (such as the Mississippi Delta, USA) to highly indented shapes

formed by multiple channels (e.g. the Ganges Delta, Bangladesh). The overall form is found to be related to the relative importance of three main processes: the current in the river, the action of waves and the action of tides. The sediments deposited by the Mississippi and Ganges deltas are mainly mud and silt, but others, such as the Rhone Delta (France) are much sandier, and deltas fed by pebbly streams can be made up of a high proportion of gravel (e.g. Skeidarsandur, Iceland). Gravelly deltas are not necessarily fed by a river: a debris-flow or sheetflood-dominated alluvial fan may build out into the lake or the sea to form a sediment body that is commonly referred to as a ***fan delta***, although it should be noted that the term 'fan delta' has also been applied to coarse-grained deltas fed by rivers (Nemec 1990a).

Deltas are now commonly classified in terms of the dominant grain size of the deposits and the relative importance of fluvial, wave and tide processes (Fig. 12.3, after Orton & Reading 1993). This scheme can be applied to modern deltas and is useful because the characteristics of the deposits formed by different deltas within it can be used as a basis for classifying strata that are interpreted as delta facies. The relationships between the controls, the form of the delta and the facies are summarised in Fig. 12.4. The supply of the sediment is determined by the nature of the hinterland, with the climate influencing the weathering and erosion processes and the discharge, the amount of water in the rivers, while there are tectonic controls on the topography, especially the gradient of the river and the effect this has on the grain size of the material carried. The relative importance of processes that rework the sediment in the basin is controlled by climatic and geomorphological factors: tidal range is determined by the local shape of the basin (*11.2*), while the wave activity is influenced by climate and the size of the water body (*11.3*).

One further factor needs to be added to the variables that are used to classify deltas in Fig. 12.3. The depth of the water in the basin is also important because it influences the effects of wave and tide processes and also controls the overall geometry of the delta body: if the delta is building out into shallow water it will spread further out into the basin than if the water is deeper (*12.4.3*). Water depth, variations in sea level and the formation of delta cycles are considered further in section *12.5*.

12.3 DELTA ENVIRONMENTS AND SUCCESSIONS

Marine deltas form at the interface of continental and marine environments. The processes associated with river channel and overbank settings occur alongside wave and tidal action of the shallow marine realm. Flora and fauna characteristic of land environments, such as the growth of plants and the development of soils, are found within a short distance of animals that are found exclusively in marine conditions. These spatial associations of characteristics seen in modern deltas occur as associations of facies in the stratigraphic record. Deltas can therefore be considered in terms of ***subenvironments***, divisions of the overall

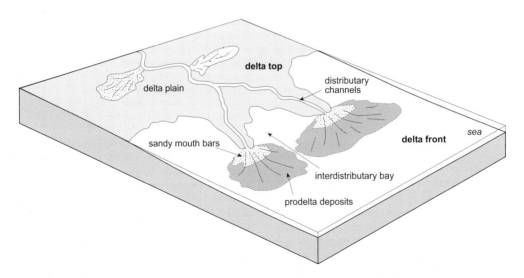

Fig. 12.5 Delta deposition can be divided into two subenvironments, the delta top and the delta front.

delta environment in which these combinations of processes occur.

12.3.1 Delta-top subenvironments

Deltas are fed by a river or an alluvial fan and there is a transition between the area that is considered part of the fluvial/alluvial environment and the region that is considered to be the **delta top** or **delta plain** (Fig. 12.5). Delta channels can be as variable in form as a river and may be meandering or braided, single or divided channels (9.2). Branching of the river channel into multiple courses is common, to create a distributary pattern of channels across the delta top. The coarsest delta-top facies are found in the channels, where the flow is strong enough to transport and deposit bedload material. Adjacent to the channels are subaerial overbank areas (9.3), which are sites of sedimentation of suspended load when the channels flood. These may be vegetated under appropriate climatic conditions and in wet tropical regions large, vegetated swamps may form on the delta top. These may be sites for the accumulation of peat (3.6.1), although if there is frequent overbank flow from the channel the deposit will be a mixture of organic and clastic material to form a carbonaceous mud. Crevasse splays (9.3) may result in lens-shaped sandy deposits on the delta top.

On deltas where the channels build out elongate lobes of sediment, sheltered areas of shallow water may be protected from strong waves and currents. These sheltered areas along the edge of the delta top are called **interdistributary bays** (Fig. 12.5) and they are regions of low-energy sedimentation between the lobes. The water may be brackish if there is sufficient influx of fresh water from the channel and overbank areas and the boundary between the floodplain and the interdistributary bays may be indistinct, especially if the delta top is swampy.

12.3.2 Delta-front subenvironments

At the mouth of the channels the flow velocity is abruptly reduced as the water enters the standing water of the lake or sea. The delta front immediately forward of the channel mouth is the site of deposition of bedload material as a **subaqueous mouth bar** (Fig. 12.5). The coarsest sediment is deposited

first, in shallow water close to the river mouth where it may be extensively reworked by wave and tide action. The current from the river is dissipated away from the channel mouth and wave energy decreases with depth, leading to a pattern of progressively finer material being deposited further away from the river mouth. This area, the **delta slope**, is often shown as a steep incline away from the delta top, but the slope varies from only $1°$ or $2°$ in many fine-grained deltas to as much as $30°$ in some coarse-grained deltas.

River-borne suspended load enters the relatively still water of the lake or sea to form a **sediment plume** in front of the delta. Fresh river water with a suspended load may have a lower density than saline seawater and the plume of suspended fine particles will be buoyant, spreading out away from the river mouth. As mixing occurs deposition out of suspension occurs, with the finest, more buoyant particles travelling furthest away from the delta front before being deposited in the **prodelta** region. Gravity currents may also bring coarser sediment down the delta front and deposit material as turbidites (4.5.2).

12.3.3 Deltaic successions

The definition of a delta includes the concept of **progradation**, that is, deposition results in the sediment body building out into the lake or sea. The sedimentary succession formed will therefore consist of progressively shallower facies as the prodelta is overlain by the delta front, which is in turn superposed by mouth-bar and delta-top sediments. The succession formed by the progradation of a delta therefore has a **shallowing-up** pattern, a series of strata that consistently shows evidence of the younger beds being deposited in shallower water than the older beds they overly (Fig. 12.6). In the delta-front subenvironment the deepest water facies, the prodelta deposits, are the finest grained as they are deposited in the lowest energy setting. In a shallowing-up succession they will be overlain by sediments of the delta slope, which will tend to be a little coarser, and the shallowest facies will be those of the mouth bars, which are typically sandy or even gravelly sediment. The beds formed by delta progradation will therefore show a coarsening-up pattern (4.2.5).

The shallowing-up, coarsening-up pattern is one of the distinctive characteristics of a deltaic succession,

(a)

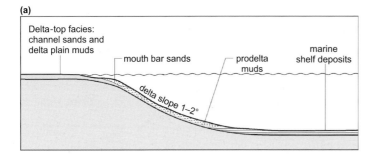

Delta-top facies:
channel sands and
delta plain muds

mouth bar sands

prodelta
muds

marine
shelf deposits

delta slope 1–2°

(b)

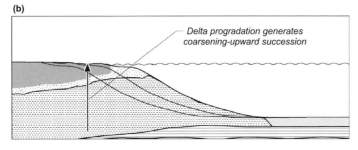

*Delta progradation generates
coarsening-upward succession*

Fig. 12.6 A cross-section across a delta
lobe: progradation results in a coarsening-
up succession.

but can be considered to be more diagnostic of a delta
only if the top of the succession shows a transition
from deposition in subaqueous to subaerial environments. Evidence of deposition on the delta top may be
the recognition of a river channel, signs of plant
growth and soil formation or other exclusively subaerial physical, chemical or biological processes, such
as desiccation cracks or tracks of land animals. The
sedimentary logs of different deltas illustrated in this
chapter all show this same, basic pattern, despite
differences in the processes and setting in each case.
One caveat must be added: a coarsening-up marine
succession capped by continental facies can be formed
at a coastline where sediment is supplied by marine
processes (*13.6.3*), so it is important to establish that
there is evidence of a river or alluvial fan supplying
sediment from the land if the succession is to be
reliably interpreted as a delta deposit.

12.4 VARIATIONS IN DELTA MORPHOLOGY AND FACIES

The combinations of factors that control delta
morphologies give rise to a wide spectrum of possible
delta characteristics. Modern deltas provide examples
of a number of positions within the 'toblerone plot' in
Fig. 12.3: the Mississippi Delta is fine-grained and

river dominated, the Rhone Delta is mixed sand and
mud and is wave-dominated, the Skeidarasandur is
mainly gravelly with river and wave influence, and
so on. Even with all the possible positions within that
plot, there is also the additional variable of water
depth to be added. Every modern delta will have
individual characteristics due to the different factors
controlling its form, and it may be expected that the
deposits of ancient deltas will be similarly variable.
A simple, neat classification of deltas into a small
number of types will represent only a small proportion of the possible forms, so it is more instructive to
consider the effects of different factors on the delta
morphology and consequently on delta facies. An
individual modern or ancient delta is likely to display
a combination of the features shown in the following
sections.

12.4.1 Effects of grain size: fine-grained deltas

The deposits on a delta will include a high proportion
of fine-grained material if the fluvial system supplying
it is a mixed-load river (*9.2.2*). Low gradient, mixed-
load river channels characterise the lower tracts of
large river systems. Large rivers like these carry sediment that is delivered to the delta as sandy bedload

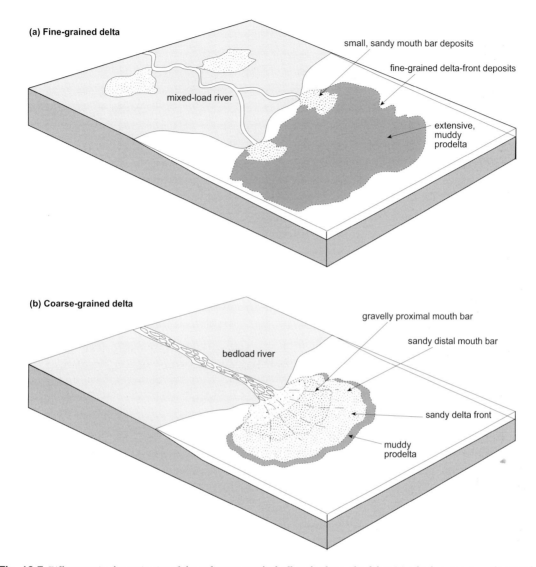

Fig. 12.7 Differences in the grain size of the sediment supplied affect the form of a delta: (a) a high proportion of suspended load results in a relatively small mouth bar deposited from bedload and extensive delta-front and prodelta deposits; (b) a higher proportion of bedload results in a delta with a higher proportion of mouth bar gravels and sands.

and a large suspended load of silt and clay. Sand deposition on the delta top is concentrated in the delta channels and on adjacent levees, while the bulk of the delta plain and any interdistributary bay areas are regions of mud accumulation (Fig. 12.7). The proximal mouth bars may also be sandy, but the rest of the delta slope and prodelta receive sediment fall-out from the plume of suspended sediment that issues from the river mouth (Orton & Reading 1993; Bhattacharya 2003). The delta front may also be the site of mass flows: the wet, muddy sediment brought down by the river may be transported by turbidity currents to deposit as turbidites on the lower part of the delta front, in the prodelta area and beyond. Deltas can be the supply systems for large submarine fan complexes (*16.2*). Muddy sediment deposited on

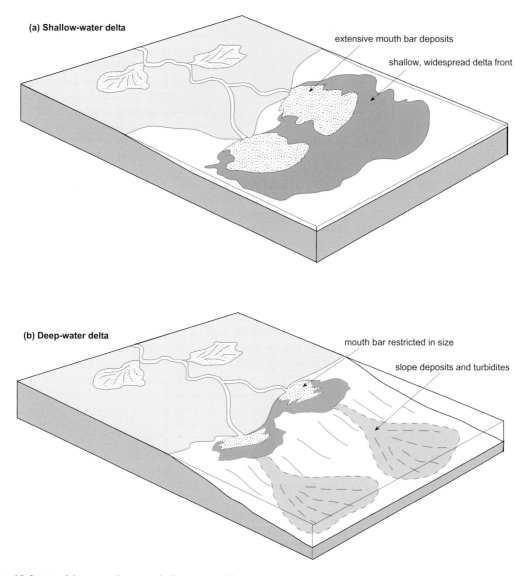

Fig. 12.8 (a) A delta prograding into shallow water will spread out as the sediment is redistributed by shallow-water processes to form extensive mouth-bar and delta-front facies. (b) In deeper water the mouth bar is restricted to an area close to the river mouth and much of the sediment is deposited by mass-flow processes in deeper water.

the delta slope is initially unstable and syndepositional deformation features (18.1) are common.

The proportion of sand in the delta deposits increases if the feeder river provides more bedload sediment. Sandy bedload rivers also transport material in suspension, but the delta environment becomes a setting for deposition of sand in channels, as overbank splays on the delta top and as shallow marine deposits in the upper part of the delta front. Extensive sand bodies form as mouth bars, perhaps reworked by wave and tide action. Fine-gained deposition occurs in parts of the delta plain away

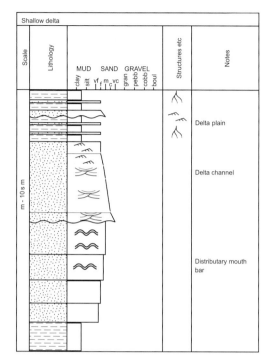

Fig. 12.9 A schematic sedimentary log of a sandy delta prograding into shallow water.

from the channels and in the lower parts of the delta slope and prodelta region.

12.4.2 Effects of grain size: coarse-grained deltas

Coarse-grained deltas, also referred to as fan deltas, are fed by pebbly braided rivers or alluvial fans. They form adjacent to areas of steep relief, where streams in the catchment areas of the rivers flow down steep slopes carrying coarse material into rivers or on to alluvial fans that prograde into a lake or the sea. Settings such as the faulted margins of rift basins (*24.2.1*) are typical sites for coarse-grained deltas to form.

The delta-top environment and hence the facies deposited are those of a coarse braided river (*9.2.1*) or an alluvial fan (*9.5*). Gravelly material is transported by fluvial or alluvial fan processes into the

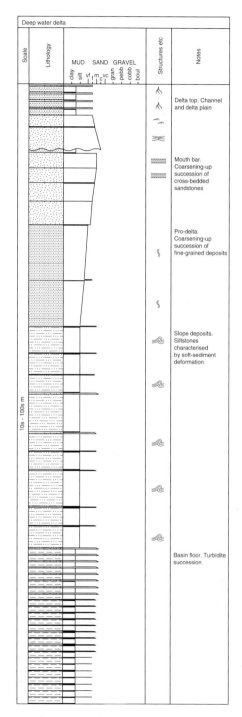

Fig. 12.10 A schematic sedimentary log of a sandy delta prograding into a deep-water basin.

Fig. 12.11 A modern Gilbert-type coarse-grained delta.

lake or sea, where they are reworked by waves or tidal currents. Documented modern examples are all from basins where the tidal range is small and wave action is the main mechanism for distribution of clasts in shallow water. The energy associated with waves is strongly depth-dependent (4.4) and so there is a sorting of the sediment into different grain sizes according to water depth. The largest clasts remain in the shallowest water where the wave action is strongest, while smaller clasts are carried by waves further offshore into slightly deeper water. Across a gently sloping shelf there will be a progressive fining of the clast size as the water depth increases and hence the energy of the waves decreases (Fig. 12.7).

Progradation of a coarse-grained delta across a shallow lake or sea floor results in a coarsening-up succession from finer sands deposited furthest offshore through coarser sands, granules, pebbles and even cobbles or boulders at the top of the delta-front succession, which is then overlain by coarse fluvial or alluvial fan facies of the delta top (Fig. 12.7). Coarse-grained deltas that display these characteristics have been classified as '***shelf-type fan deltas***' by Wescott & Ethridge (1990).

12.4.3 Water depth: shallow- and deep-water deltas

A delta progrades by sediment accumulating on the sea floor at the delta front where it builds up to sea level to increase the area of the delta top. For a given supply of sediment, the rate at which the delta progrades will depend on the thickness of the sediment pile that must be created to reach sea level. Delta progradation will hence occur at a greater rate if it is building into a shallow sea or lake (Fig. 12.8), and the area covered by a delta lobe will be greater because it forms a thin, widespread body of sediment. In contrast a delta building into deeper water will form a thicker deposit that progrades at a slower rate (Collinson et al. 1991).

A delta building into shallow water will tend to have a large delta-plain area. If the climate is suitable for abundant plant growth, peat mires may develop on parts of the plain away from the delta channels and delta successions that have developed in a shallow-water setting may therefore include coal beds. The delta-front facies will all be deposited in shallow water, and hence will be strongly influenced by processes such as wave action (Fig. 12.9). Sandy and gravelly deposits are therefore likely to be relatively well sorted.

In deeper water, a greater proportion of the sediment will be deposited in the lower part of the delta slope as a thicker coarsening-up succession is generated during delta progradation (Fig. 12.10). The area of the delta top will be relatively small, with less potential for the development of widespread fine-grained delta-plain facies and mires. Wave-reworked mouth-bar facies will be limited in extent because of the small area of shallow water where wave action is effective. The delta slope will be extensive and a potential site for gravity flows: coarser deposits may

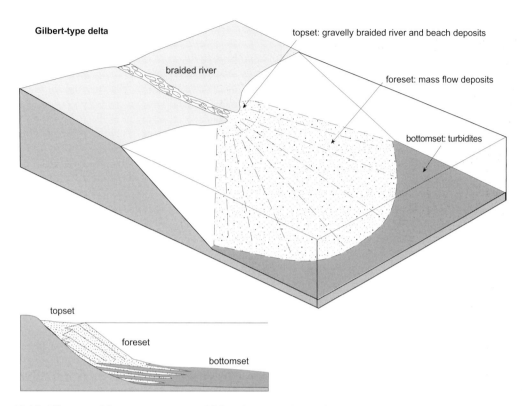

Fig. 12.12 Gilbert-type deltas are coarse-grained deltas that prograde into deep water. They display a distinctive pattern of steeply-dipping foreset beds sandwiched between horizontal topset and bottomset strata.

become remobilised to form debris flows and finer sediment mixed with more water will generate turbidity currents. The lower part of the delta front may therefore be a site of deposition of these mass-flow deposits, which may extend out on to the basin plain.

12.4.4 Coarse-grained deep-water deltas

The combination of a supply of coarse sediment and a steep basin margin results in a particular delta form that is unlike all other deltas and therefore merits a special mention (Fig. 12.11). They even have a special name '*Gilbert-type deltas*', named after the American geologist G.K. Gilbert who first described deposits of this type in 1895. Gilbert-type deltas have a characteristic three-part structure (Figs 12.12 & 12.13). The *topset* (the delta top) is a subaerial to shallow-marine environment, with gravels deposited by

braided rivers and, in some cases, reworked by wave processes at the shoreline. In front of the topset lies the *foreset* (the delta front), which is very distinctive because the beds are at a steep angle, typically up to around 30 degrees and close to the angle of rest of granular material. Deposition on a foreset occurs by two mechanisms (Nemec 1990b): debris flows of poorly sorted gravel mixed with sand and mud, and well-sorted gravels deposited by a grainflow (avalanche) process (*4.5.3*). Slumping (*18.1.1*) is often seen on the delta because the steep slopes of the foresets can become unstable. At the base of the foreset slope sediments are finer, comprising mud, sand and some gravel, which lie approximately horizontally and are the products of turbidites and suspension deposition in a prodelta setting, known in this context as the *bottomset*.

As a Gilbert-type delta progrades, the foreset builds out over the bottomset and in turn the foreset is overlain by topset facies: the resulting deposit is in

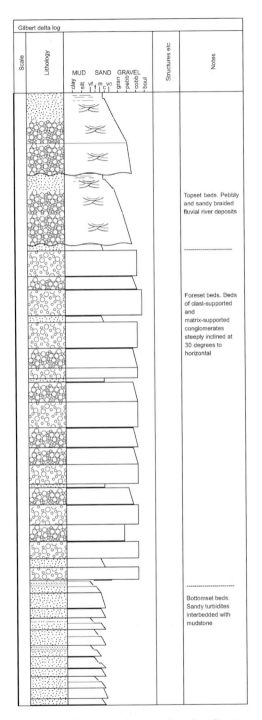

Gilbert delta log

Scale

Lithology

MUD   SAND   GRAVEL
clay silt vf f m c vc gran pebb cobb boul

Structures etc

Notes

Topset beds. Pebbly
and sandy braided
fluvial river deposits

Foreset beds. Beds
of clast-supported
and
matrix-supported
conglomerates
steeply inclined at
30 degrees to
horizontal

Bottomset beds.
Sandy turbidites
interbedded with
mudstone

Fig. 12.13 A schematic sedimentary log of a Gilbert-type coarse-grained delta deposit.

the form of a sandwich of steeply-dipping conglomeratic strata between layers of horizontal beds of conglomerate and sandstone (Fig. 12.14). The height of the foreset is determined by the depth of water the delta is building into, and ranges from a few tens of metres to over 500 ms (Fig. 12.14). These thick packages of steeply dipping strata are unique to Gilbert-type deltas: the only deposits that even approach this angle of deposition are on alluvial fans, and these are at a lower angle than the 30° recorded in Gilbert-type deltas. They are typically found at the edges of basins that have active faulted margins such as rift basins (24.2.1), where uplift of the land at the margin creates steep topography to supply the gravel and the basin is subsiding to form a deep, steep-sided basin.

12.4.5 Process controls: river-dominated deltas

A delta is regarded as river-dominated where the effects of tides and waves are minor. This requires a microtidal regime (11.2.2) and a setting where wave energy is effectively dissipated before the waves reach the coastline. Under these conditions, the form of the delta is largely controlled by fluvial processes of transport and sedimentation. The unidirectional fluvial current at the mouth of the river continues into the sea or lake as a subaqueous flow. The channel form is maintained, with well-defined subaqueous levees and overbank areas (Fig. 12.15). Bedload and suspended load carried by the river is deposited on the subaqueous levees, building up to sea level and extending the front of the delta basinwards as thin strips of land either side of the main channel to form the characteristic 'bird's foot' pattern of a river-dominated delta (Bhattacharya & Walker 1992). A common feature of fluvially dominated deltas is channel instability due to the very low gradient on the delta plain, resulting in frequent avulsion of the major and minor channels. The course of the river changes as one route to the sea becomes abandoned and a new channel is formed, leaving the former channel, its levees and overbank deposits abandoned. Repeated switching of the channels on the delta top builds up a pattern of overlapping abandoned lobes (Fig. 12.16).

The deposits of river-dominated deltas have well-developed delta-top facies, consisting of channel and

Fig. 12.14 A Gilbert-type coarse-grained delta exposed in a cliff over 500 m high. The exposure is made up mostly of foreset deposits dipping at around 30°: horizontal topset strata form the top of the cliff and the toes of the foreset beds pass into gently dipping bottomset facies.

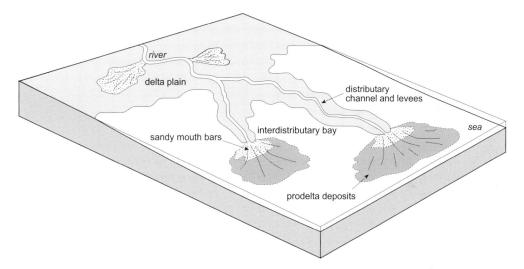

Fig. 12.15 A river-dominated delta with the distributary channels building out as extensive lobes due to the absence of reworking by wave and tide processes. Low-energy, interdistributary bays are a characteristic of river-dominated deltas.

overbank sediments. The characteristics of these facies will be essentially the same as those of a similar fluvial system. The overbank areas of a delta top may be sites of prolific growth of vegetation, leading to the formation of peat and eventually coal. The channels build out to form the 'toes' of the 'bird's foot', between which there are large interdistributary bays. These bays are relatively sheltered and are sites of fine-grained, subaqueous sedimentation. Crevasse splays from the distributary channels

supply sediment into these bays and they gradually fill to sea level to become the vegetated part of the delta plain. The filling of interdistributary bays results in small scale (a few metres thick) coarsening-up successions (Fig. 12.17). In front of the channels, mouth bars form and are localised to areas in front of the individual delta lobes. Little redistribution of mouth-bar sediments by wave or tidal processes occurs, so individual mouth-bar bodies are relatively small.

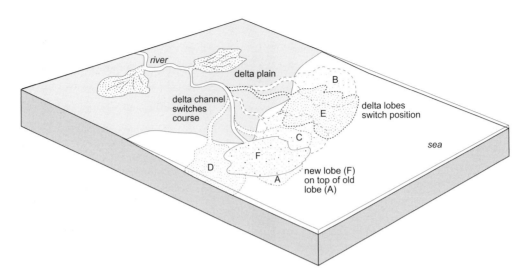

Fig. 12.16 When a delta channel avulses a new lobe starts to build out at the new location of the channel mouth. The abandoned lobe subsides by dewatering until completely submerged. Through time the channel will eventually switch back to a position overlapping the former delta lobe. This results in a series of delta-lobe successions, each coarsening-up.

12.4.6 Process controls: wave-dominated deltas

Waves driven by strong winds have the capacity to rework and redistribute any sediment deposited in shallow water, especially under storm conditions. The river mouth and mouth-bar areas of a delta are susceptible to the action of waves, resulting in a modification of the patterns seen in river-dominated deltas (Bhattacharya & Giosan 2003) (Fig. 12.18). Progradation of the channel outwards is limited because the subaqueous levees do not form and bedload is acted upon by waves as quickly as it is deposited. Any obliquity between the wind direction and the delta front causes a lateral migration of sediment as the waves wash material along the coast to form beach spits and mouth bars that build up as elongate bodies parallel to the coastline. Wave action is effective at sorting the bedload into different grain sizes and the mouth-bar deposits of a wave-influenced delta may be expected to be better sorted than those of a river-dominated delta.

Progradation of a wave-dominated delta occurs because the wave action does not transport all the material away from the region of the river mouth. A net supply of bedload by the river results in a series of shore-parallel sand ridges forming as mouth bars build up and out to form a new beach (Fig. 12.19). Wave-dominated delta deposits display well-developed mouth bar and beach sediments, occurring as elongate coarse sediment bodies approximately perpendicular to the orientation of the delta river channel. This is in contrast to the river-dominated delta deposits, which would be expected to show less continuous mouth bars, and a higher proportion of channel and overbank deposits forming the delta lobes. Delta-front and prodelta deposits may not significantly differ between these two delta types (Figs 12.17 & 12.20).

12.4.7 Process controls: tide-dominated deltas

Coastlines with high tidal ranges experience onshore and offshore tidal currents that move both bedload and suspended load. A delta building out into a region with strong tides will be modified into a pattern that is different to both river- and wave-dominated deltas (Fig. 12.21). First, the delta-top channel(s) are subject to tidal influence with reverses of flow and/or periods of stagnation as a flood tide balances the fluvial discharge. This may be seen in strata as reversals of palaeocurrent indicated by cross-stratification, and

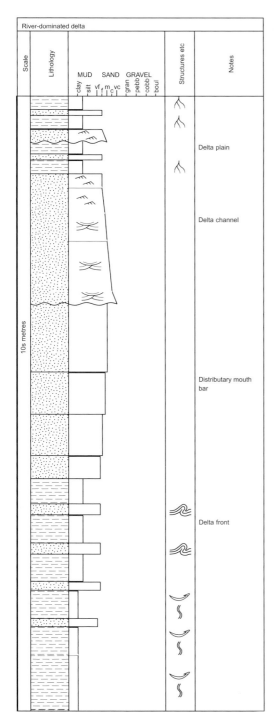

Fig. 12.17 A schematic graphic sedimentary log of river-dominated delta deposits.

the formation of mud drapes (*11.2.4*). Overbank areas on the delta top may be partially tidal flats, and all of the delta top will be susceptible to flooding during periods of high fluvial discharge coupled with high tides. The tidal currents rework sediments at the river mouth into elongate bars that are perpendicular to the shoreline. These are modified mouth bars, which may show bidirectional cross-stratification and mud drapes on the cross-bed foresets due to the reversing nature of the ebb and flood tidal currents (Willis et al. 1999) (*11.2.4*).

The deposits of a tidally influenced delta can be distinguished from other deltas by the presence of sedimentary structures and facies associations which indicate that tidal processes were active (reversals of palaeoflow, mud drapes, and so on), and subaqueous mouth bars will be elongate parallel to the river channels. The overall succession of strata will display the characteristic coarsening-up of a delta (Fig. 12.22), a feature that allows it to be distinguished from other tidally influenced environments such as estuaries, which have much in common in terms of depositional processes. The main distinguishing feature is that a delta is always a progradational feature, whereas an estuary commonly forms as part of a retrogradational, or transgressive, succession (*23.1.6*).

12.5 DELTAIC CYCLES AND STRATIGRAPHY

When the channel on the delta top changes course, the former lobe is abandoned as a new site of deposition is occupied. River-dominated deltas tend to have the most frequent changes in position of the active lobe, but avulsion of channel course also occurs in other delta types. The deposits of an abandoned lobe will gradually compact as water deposited with the fine-grained sediment escapes from the pore spaces and the bulk density increases. This compaction occurs without any additional load, and results in the abandoned lobe subsiding below sea level. The fall below sea level of the abandoned lobe will be accelerated if the delta is located in a region of overall subsidence or if there is a eustatic rise in sea level.

The beds that mark the end of sedimentation on a delta lobe are known as the ***abandonment facies*** (Reading & Collinson 1996). In the upper part of the

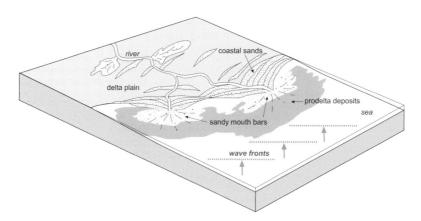

Fig. 12.18 A wave-dominated delta formed where wave activity reworks the sediment brought to the delta front to form coastal sand bars and extensive mouth-bar deposits.

Fig. 12.19 Sand bars at the mouth of a wave-dominated delta.

delta plain these will be peats or palaeosols, which represent a low clastic supply to this part of the plain now that active lobe progradation has moved elsewhere on the delta. The fringes of the delta lobe will be areas of slow, fine-grained deposition in shallow water, while further offshore, carbonate facies may form over the toe of the delta. Abandonment facies may show intense bioturbation because of the slow sedimentation rate.

After a number of changes in channel position the active delta lobe may reoccupy an earlier position and prograde over an older, compacted and submerged delta-lobe succession. In cross-section the result is one coarsening-up delta-lobe succession built up on top of another. Repetition of this pattern has been recognised in the stratigraphic record and are referred to as ***delta cycles***, each 'cycle' representing the progradation of an individual delta lobe. The thickness of

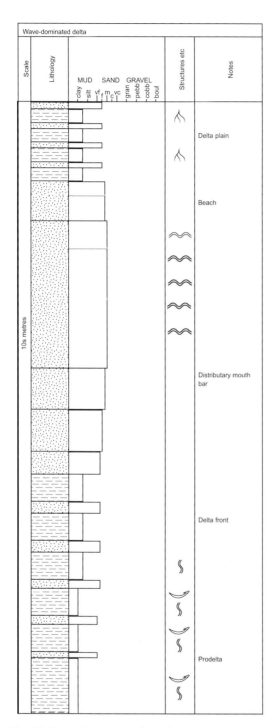

Fig. 12.20 A schematic graphic sedimentary log of wave-dominated delta deposits.

a delta 'cycle' will be controlled by the depth of water in the receiving basin (see above) and may range from a few metres to tens or hundreds of metres in thickness (Elliott 1986).

Variants on the idealised delta cycle are frequently encountered (Fig. 12.23). A complete succession from offshore fine-grained deposits up to the delta channel fill will be seen only at the point where the axis of the lobe has built out basinward. In other positions, the top of the cycle will vary from delta-plain carbonaceous mudstones, to interdistributary bay deposits or mouth-bar sands. In a hinterland direction, subsidence will not be great enough for fully marine conditions to develop at the base of each delta cycle, and only the upper parts of the typical succession may be seen (Elliott 1986).

In addition to the trends that represent the progradation of delta lobes, smaller scale grain-size patterns are also present. The filling of an interdistributary bay results in a coarsening-up succession, but this will normally be on a scale that is an order of magnitude smaller than the main delta cycle. Small-scale fining-up trends are formed by the filling of distributary channels when they are abandoned.

12.6 SYNDEPOSITIONAL DEFORMATION IN DELTAS

The delta front is a slope that can vary from about 1° in mud-rich settings to over 30° in coarse-grained deltas. Even the very low angle slopes are potentially unstable and mass movement of loose, soft sediment on the delta slope is common. Debris flows, slumps and slides (6.5.1) that consist of remobilised delta-front deposits reworked and remobilised occur and may be seen as part of the succession in deltaic facies. The slumps and slides can be large-scale, involving the movement of bodies of sediment tens of metres thick and hundreds of metres across. The surfaces on which the slides move are like faults, and these features are often regarded as growth faults, synsedimentary deformation structures (18.1.1) (Bhattacharya & Davies 2001). Further instabilities also arise as a result of the relatively rapid accumulation of sediment on a delta: coarser, and relatively denser sediment of the delta top is built up on top of muddy, wet and less dense delta-front facies and the result is the formation of mud diapirs (Hiscott 2003) (18.1.4).

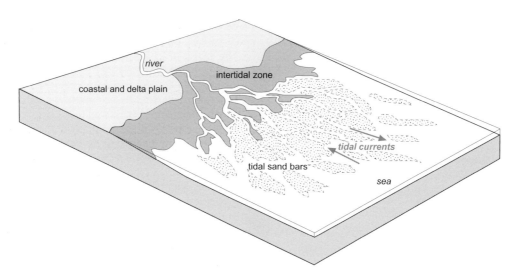

Fig. 12.21 A tide-dominated delta in a macrotidal regime will show extensive reworking of the delta front by tidal currents and the delta top will have a region of intertidal deposition.

These deformation processes occurring during the formation of the deltaic deposit give rise to some quite complicated sedimentary features within a delta succession. Beds may be deformed on a scale ranging from slump folds a few centimetres across to synsedimentary faults that rotate and displace packages of strata tens of metres thick (*18.1.1*). Similar features can occur in other depositional environments, but they are probably most common in deltaic facies, especially if the succession contains a high proportion of muddy sediments that deform relatively easily.

12.7 RECOGNITION OF DELTAIC DEPOSITS

A key feature of many deltas is the close association of marine and continental depositional environments. In delta deposits this association is seen in the vertical arrangement of facies. A single delta cycle may show a continuous vertical transition from fully marine conditions at the base to a subaerial setting at the top. This transition is typically within a coarsening-upwards succession from lower energy, finer grained deposits of the prodelta to the higher energy conditions of the delta mouth bar where coarser sediment accumulates.

The delta top contains both relatively coarse sediment of the distributary channel as well as finer grained material in overbank areas and interdistributary bays. The channel may be recognised by its scoured base, a fining-up pattern and evidence of flow, which will be unidirectional unless there is a strong tidal influence resulting in bidirectional currents. The delta top will show signs of subaerial conditions, including the development of a soil. Deposits in the sheltered interdistributary bays may show thin bedding resulting from influxes of sediment from the delta top and symmetrical ripples due to wave action. The shallower water deposits of the delta front may be extensively reworked by wave and/or tidal action resulting in cross-stratified mouth-bar facies. The geometry and extent of the mouth-bar sand bodies will be determined by the relative importance of river, tidal and wave processes. Deeper, lower delta slope deposits and prodelta facies are finer grained, deposited from plumes of suspended material disgorged by the river, or as turbidites that flowed down the delta front.

Deltaic deposits are almost exclusively composed of terrigenous clastic material supplied by rivers. However, there are examples of deltas formed by lavas and volcaniclastic material building out into the sea, and these are not fed by water, but by the volcanic processes: the term 'non-alluvial delta' may be applied to these deposits (Nemec 1990a). Limestone beds are

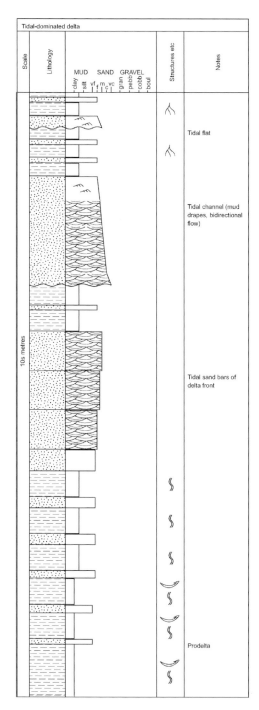

Fig. 12.22 A schematic graphic sedimentary log of tide-dominated delta deposits.

only rarely associated with deltas, occurring as accumulations on delta fronts where the supply of clastic detritus is low: examples in modern settings and from the stratigraphic record indicate that carbonates form on deltas in arid and semi-arid environments, where the supply of clastic sediment to the delta is highly ephemeral (Bosence 2005).

Palaeontological evidence from fauna and flora can be important in the recognition of the marine and continental subenvironments of a delta. A distinct fauna tolerant of brackish water may be found near the mouths of channels and in the interdistributary bays where fresh and marine water mix. The mixture of shallow-marine, brackish and freshwater fauna plus coastal vegetation is also characteristic of deltaic environments. The contrast between fresh and saline water is not present in deltas formed at the margins of freshwater lakes and in these settings the recognition of the delta must be based on the facies patterns.

A final point to emphasise is that the various models of different types of delta presented in this chapter are just a few examples of the possible combinations of the controls that determine the form and facies of a delta. Any modern or ancient delta should be considered in terms of the evidence for the effects of different factors – sediment grain size, basin water depth, the relative importance of river, tide and wave influences – and should not be expected to exactly match any of the models presented here or any other text books.

Characteristics of deltaic deposits
- lithologies – conglomerate, sandstone and mudstone
- mineralogy – variable, delta-front facies may be compositionally mature
- texture – moderately mature in delta-top sands and gravels, mature in wave-reworked delta-front deposits
- bed geometry – lens-shaped delta channels, mouth-bar lenses variably elongate, prodelta deposits thin bedded
- sedimentary structures – cross-bedding and lamination in delta-top and mouth-bar facies
- palaeocurrents – topset facies indicate direction of progradation, wave and tidal reworking variable on delta front
- fossils – association of terrestrial plants and animals of the delta top with marine fauna of the delta front
- colour – not diagnostic, delta-top deposits may be oxidised

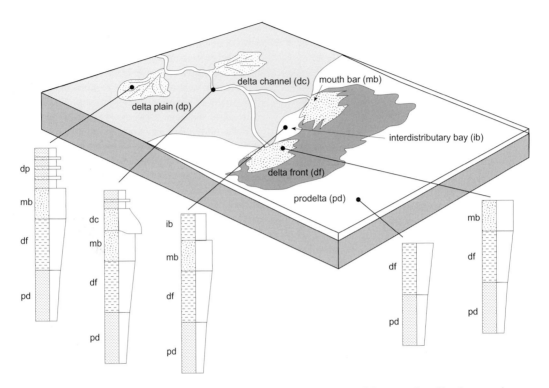

Fig. 12.23 Delta cycles: the facies succession preserved depends on the location of the vertical profile relative to the depositional lobe of a delta.

- facies associations – typically occur overlying shallow-marine facies and overlain by fluvial facies in an overall progradational pattern.

FURTHER READING

Bhattacharya, J.P. (2006) Deltas. In: *Facies Models Revisited* (Eds Walker, R.G. & Posamentier, H.). Special Publication 84, Society of Economic Paleontologists and Mineralogists, Tulsa, OK; 237–292.

Bhattacharya, J.P. & Giosan, L. (2003) Wave-influenced deltas: geomorphological implications for facies reconstruction. *Sedimentology*, **50**, 187–210.

Boyd, R., Dalrymple, R.W. & Zaitlin, B.A. (1992) Classification of clastic coastal depositional environments. *Sedimentary Geology*, **80**, 139–150.

Colella, A. & Prior, D.B. (Eds) (1990) *Coarse-Grained Deltas*. Special Publication 10, International Association of Sedimentologists. Blackwell Science, Oxford, 357 pp.

Reading, H.G. & Collinson, J.D. (1996) Clastic coasts. In: *Sedimentary Environments: Processes, Facies and Stratigraphy* (Ed. Reading, H.G.). Blackwell Science, Oxford; 154–231.

13

Clastic Coasts and Estuaries

The morphology of coastlines is very variable, ranging from cliffs of bedrock to gravelly or sandy beaches to lower energy settings where there are lagoons or tidal mudflats. Wave and tidal processes exert a strong control on the morphology of coastlines and the distribution of different depositional facies. Wave-dominated coasts have well-developed constructional beaches that may either fringe the coastal plain or form a barrier behind which lies a protected lagoon. Barrier systems are less well developed where there is a larger tidal range and the deposits of intertidal settings, such as tidal mudflats, become important. A very wide range of sediment types can be deposited in these coastal depositional systems and in this chapter only terrigenous clastic environments are considered: carbonate and evaporite coastal systems are covered in the following chapter. Estuaries are coastal features where water and sediment are supplied by a river, but, unlike deltas, the deposition is confined to a drowned river valley.

13.1 COASTS

Coasts are the areas of interface between the land and the sea, and the coastal environment can comprise a variety of zones, including coastal plains, beaches, barriers and lagoons. The **shoreline** is the actual margin between the land and the sea. Coastlines can be divided into two general categories on the basis of their morphology, wave energy and sediment budget. **Erosional coastlines** typically have relatively steep gradients where a lot of the wave energy is reflected back into the sea from the shoreline (a **reflective coast**, Fig. 13.1): both bedrock and loose material may be removed from the coast and redistributed by

wave, tide and current processes. At **depositional coastlines** the gradient is normally relatively gentle and a lot of the wave energy is dissipated in shallow water: provided that there is a supply of sediment, these **dissipative coasts** can be sites of accumulation of sediment (Woodroffe 2003).

13.1.1 Erosional coastlines

Exposure of bedrock in cliffs allows both physical and chemical processes of weathering: oxidation and hydration reactions are favoured in the wet environment, and the growth of salt crystals within cracks of

Dissipative coast

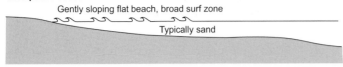

Reflective coast

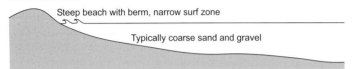

Fig. 13.1 Reflective coasts are usually erosional with steep beaches and a narrow surf zone. Dissipative coasts may be depositional, with sand deposited on a gently sloping foreshore.

Fig. 13.2 An erosional coastline: wave action has eroded the cliff and left a wave-cut platform of eroded rock on the beach.

rocks sprayed with seawater can play an important role in breaking up the material. Material accumulates at the foot of cliffs as loose clastic detritus and occasionally as large blocks when whole sections of the cliff face are removed. Cliff erosion may result in **wave cut platforms** (Fig. 13.2) of bedrock eroded subhorizontally at beach level. Wave action, storms and tidal currents will then remove the debris as bedload, as suspended load or in solution. This contributes to the supply of sediment to the marine environment, and away from river mouths can be an important source of clastic detritus to the shallow marine realm.

13.1.2 Depositional coastlines

A coastline that is a site of accumulation of sediment must have an adequate supply of material to build up a deposit. The sources of this sediment are from the marine realm, either terrigenous clastic detritus reworked from other sources or bioclastic debris. The terrigenous material ultimately comes from rivers, with a small proportion of wind-blown origin and from direct erosion of coastlines. This sediment is brought to a depositional coastline by tidal, wind-driven and geostrophic currents (11.4) that transport material parallel to shorelines or across shallow seas. Wind-driven waves acting obliquely to the shoreline are an important mechanism of transport, creating a shore-parallel current known as **longshore drift**. Shallow seas are generally rich in fauna, and the remains of the hard parts of these organisms provide an important source of bioclastic material to coastlines.

The form of a depositional coastline is determined by the supply of sediment, the wave energy, the tidal range and the climate. Climate exerts a strong control on coasts that are primarily sites of carbonate and evaporite deposition, and these environments are considered in Chapter 15. Along clastic coastlines a beach of sandy or gravelly material forms where there is a sufficient supply of clastic sediment and enough wave energy to transport the material on the foreshore. The form of the beach, and the development of barrier systems and lagoons, is dependent on whether the coastline is in a micro-, meso- or macrotidal regime. Sea-level changes also strongly influence coastal morphology. In the following sections the processes related to beach formation are first considered, followed by a description of the morphologies that can exist in wave-dominated and tidally influenced coasts.

13.2 BEACHES

The beach is the area washed by waves breaking on the coast. The seaward part of the beach is the foreshore (*11.1*), which is a flat surface where waves go back and forth and which is gently dipping towards the sea (Fig. 13.3). Where wave energy is sufficiently strong, sandy and gravelly material may be continuously reworked on the foreshore, abrading clasts of all sizes to a high degree of roundness, and effectively sorting sediment into different sizes (Hart & Plint 1995). Sandy sediment is deposited in layers parallel to the slope of the foreshore, dipping offshore at only a few degrees to the horizontal (much less than the angle of repose). This low-angle stratification of well-sorted, well-rounded sediment is particularly characteristic of wave-dominated sandy beach environments (Clifton 2003, 2006). Grains are typically compositionally mature as well as texturally mature (*2.5.3*) because the continued abrasion in the beach swash zone tends to break down the weaker clasts.

On gravel beaches the water washed up the beach by each wave tends to percolate down into the porous gravel, and the backwash of each wave is therefore weak. Clasts that are washed up the beach will therefore tend to build up to form a **storm ridge** at the top

of the foreshore, a back-beach gravel ridge that is a distinctive feature of gravelly beaches. The clast composition will vary according to local sediment supply, and may contain terrigenous clastic, volcaniclastic or bioclastic debris.

At the top of the beach, a ridge, known as a **berm**, marks the division between the foreshore and **backshore** area (Fig. 13.3). Water only washes over the top of the berm under storm-surge conditions. Sediment carried by the waves over the berm crest is deposited on the landward side forming layers in the backshore that dip gently landward. These low-angle strata are typically truncated by the foreshore stratification, to form a pattern of sedimentary structures that may be considered to be typical of the beach environment (Figs 13.3 & 13.4). The backshore area may become colonised by plants and loose sand can be reworked by aeolian processes.

Wave action in the lower part of the foreshore can rework sand and fine gravel into wave ripples that can be seen on the sediment surface at low tide and can be preserved as wave-ripple cross-lamination. However, wave-formed sedimentary structures on the beach may be obliterated by organisms living in the intertidal environment and burrowing into the sediment. This bioturbation may obscure any other sedimentary structures.

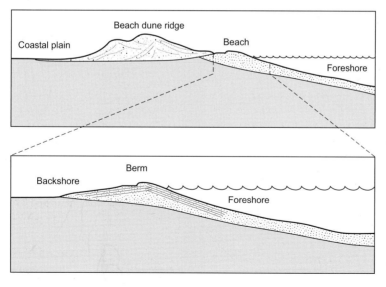

Fig. 13.3 Morphological features of a beach comprising a beach foreshore and backshore separated by a berm; beach dune ridges are aeolian deposits formed of sand reworked from the beach.

Fig. 13.4 Foreshore-dipping and backshore-dipping stratification in sands on a beach barrier bar.

13.2.1 Beach dune ridges

Aeolian processes can act on any loose sediment exposed to the air. Along coasts any sand that dries out on the upper part of the beach is subject to reworking by onshore winds that may redeposit it as aeolian dunes (*8.6.1*). Coastal dunes form as ridges that lie parallel to the shoreline and they may build up to form dune complexes over 10 m high and may stretch hundreds of metres inland (Wal & McManus 1993). Vegetation (grasses, shrubs and trees) plays an important role in stabilising and trapping sediment (Fig. 13.5). The limiting factor in beach dune ridge growth is the supply of sand from the beach. They commonly form along coasts with a barrier system, but can also be found along strand-plain coasts.

In a sedimentary succession these beach dune ridge deposits may be seen as well-sorted sand at the top of the beach succession (Fig. 13.6). Some preservation of the roots of shrubs and trees that colonised the dune field is possible, but the effect of the vegetation is often to disrupt the preservation of well-developed dune cross-bedding.

13.2.2 Coastal plains and strand plains

Coastal plains are low-lying areas adjacent to seas (Fig. 13.7). They are part of the continental environment where there are fluvial, alluvial or aeolian processes of sedimentation and pedogenic modification. Coastal plains are influenced by the adjacent marine environment when storm surges result in extensive flooding by seawater. A deposit related to storm flooding can be recognised by features such as the presence of bioclastic debris of a marine fauna amongst deposits that are otherwise wholly continental in character.

Sandy coastlines where an extensive area of beach deposits lies directly adjacent to the coastal plain are known as *strand plains* (Fig. 13.7). Along coasts supplied with sediment, beach ridges create strand plains that form sediment bodies tens to hundreds of metres across and tens to hundreds of kilometres long and progradation of strand plains can produce extensive sandstone bodies. The strand plain is composed of

Fig. 13.5 A beach dune ridge formed by sand blown by the wind from the shoreline onto the coast to form aeolian dunes, here stabilised by grass.

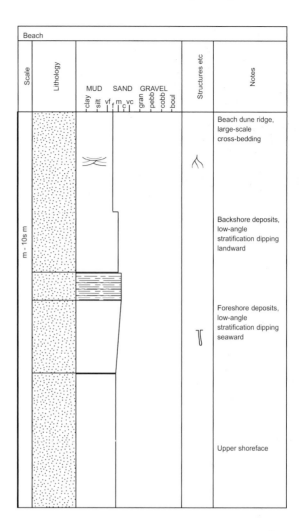

Fig. 13.6 A schematic graphic sedimentary log of sandy beach deposits.

the sediment deposited on the foreshore and backshore region. The backshore area merges into the coastal plain and may show evidence of subaerial conditions such as the formation of aeolian dunes and plant colonisation.

13.3 BARRIER AND LAGOON SYSTEMS

13.3.1 Barriers

Along some coastlines a barrier of sediment separates the open sea from a lagoon that lies between the barrier and the coastal plain (Fig. 13.8). Beach bar-

riers are composed of sand and/or gravel material and are largely built up by wave action. They may be partially attached to the land, forming a **beach spit**, or wholly attached as a **welded barrier** that completely encloses a lagoon, or can be isolated as a **barrier island** in front of a lagoon. In practice, the distinction between these three forms can be difficult to identify in ancient successions and their sedimentological characteristics are very similar. Barriers (Fig. 13.9) range in size from less than 100 m wide to several kilometres and their length ranges from a few hundred metres to many tens of kilometres (Davis & Fitzgerald 2004). The largest tend to form along the open coasts of large oceans where the wave energy is high and the tidal range is small.

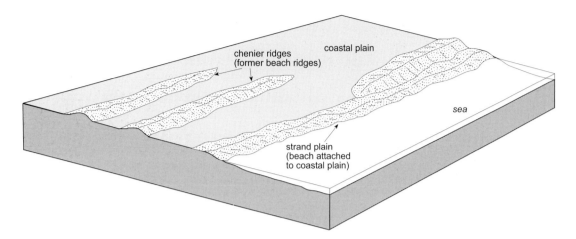

Fig. 13.7 A wave-dominated coastline with a coastal plain bordered by a sandy beach: chenier ridges are relics of former beach strand plains.

The seaward margin of a barrier island has a beach and commonly a beach dune ridge where aeolian processes rework the sand. Vegetation helps to stabilise the dunes. On the landward side of the island the layers of sand deposited during storms pinch out into the muddy marshes of the edge of the lagoon. During storm surges seawater may locally overtop the beach ridge and deposit **washovers** of sediment that has been reworked from the barrier and deposited in the lagoon (Fig. 13.8). Washover deposits are low-angle cones of stratified sands dipping landwards from the barrier into the lagoon.

The conditions required for a barrier to form are as follows. First, an abundant supply of sand or gravel-sized sediment is required and this must be sufficient to match or exceed any losses of sediment by erosion. The supply of the sediment is commonly by wave-driven longshore drift from the mouth of a river at some other point along the coast and there may also be some reworking of material from the sea bed further offshore. Second, the tidal range must be small. In macrotidal settings the exchange of water between a lagoon and the sea during each tidal cycle would prevent the formation of a barrier because a

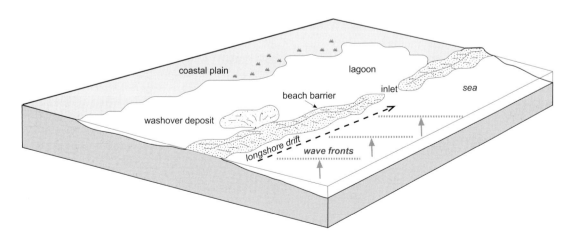

Fig. 13.8 A wave-dominated coastline with a beach-barrier bar protecting a lagoon.

Fig. 13.9 Beach-barrier bars along a wave-dominated coastline.

13.3.2 Lagoons

Lagoons are coastal bodies of water that have very limited connection to the open ocean. Seawater reaches a lagoon directly through a channel to the sea or via seepage through a barrier; fresh water is supplied by rainfall or by surface run-off from the adjacent coastal plain. If a lagoon is fed by a river it would be considered to be part of an estuary system (*13.6*). They are typically very shallow, reaching only a few metres in depth.

Lagoons generally develop along coasts where there is a wave-formed barrier and are largely protected from the power of open ocean waves (Reading & Collinson 1996). Waves are generated by wind blowing across the surface of the water, but the fetch of the waves (*4.4*) will be limited by the dimensions of the lagoon. Ripples formed by waves therefore affect the sediments only in very shallow water. The wind may also drive weak currents across the lagoon. Tidal effects are generally small because the barrier–lagoon morphology is only well developed along coasts with a small tidal range.

Fine-grained clastic sediment is supplied to lagoons as suspended material in seawater entering past the barrier and in overland flow from the adjacent coastal plain. Organic material may be abundant from vegetation which grows on the shores of the lagoon. In tropical climates, trees with aerial root systems (**mangroves**) colonise the shallow fringes of the lagoon. Mangroves cause the shoreline to prograde into the lagoon as they act as sites for accumulation of

restricted inlet would not be able to let the water pass through at a high enough rate. Barrier systems are therefore best developed in microtidal (Fig. 13.8) and, to some extent, mesotidal settings (Fig. 13.10). Third, barrier islands generally form under conditions of slow relative sea-level rise (Hoyt 1967; FitzGerald & Buynevich 2003). If there is a well-developed beach ridge, the coastal plain behind it may be lower than the top of the ridge. With a small sea-level rise, the coastal plain can become partially flooded to form a lagoon, and the beach ridge will remain subaerial, forming a barrier. For the barrier to remain subaerial as sea level rises further, sediment must be added to the beach to build it up, that is, the first condition of high sediment supply must be satisfied.

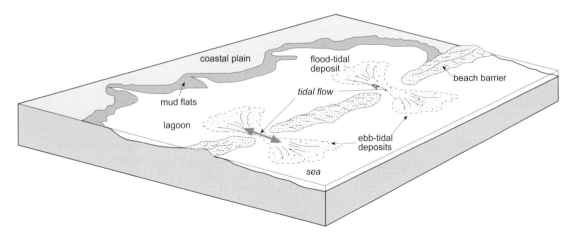

Fig. 13.10 Morphological features of a coastline influenced by wave processes and tidal currents.

sediment and organic matter along the water's edge. In more temperate climates, saline-tolerant grasses, shrubs and trees may play a similar role in trapping sediment. Coarser sediment may enter the lagoon when storms wash sediment over the barrier as **washover deposits**, which are thin layers of sand reworked by waves. Sand is also blown into the water by onshore winds picking up material from the dunes along the barrier.

An important characteristic of lagoons is their water chemistry. Due to the limited connection to open ocean, it is common for lagoon water to have either higher or lower salinity than seawater. Low salinity, brackish water (*10.3*) will be a feature of lagoons in areas of high rainfall, local run-off of fresh water from the coastal plain or small streams. Mixing of the lagoon water with the seawater is insufficient to maintain full salinity in these brackish lagoons. In more arid settings the evaporation from the surface of the lagoon may exceed the rate at which seawater exchanges with the lagoon water and the conditions become hypersaline (*10.3*), that is, with salinities higher than that of seawater. If salinities become very elevated, precipitation of evaporite minerals will occur (*3.2*).

A lagoonal succession is typically mudstone, often organic-rich, with thin, wave-rippled sand beds (Boggs 2006) (Fig. 13.11). The deposits of lagoons can be difficult to distinguish from those of lakes with similar dimensions and in similar climatic settings. The processes are almost identical in the two settings because they are both standing bodies of water. Two lines of evidence can be used to identify lagoonal facies. First, the fossil assemblage may indicate a marine influence, and specifically a restricted fauna may provide evidence of brackish or hypersaline water. Second, the association with other facies is also important: lagoonal deposits occur above or below beach/barrier island sediments and fully marine shoreface deposits.

13.4 TIDES AND COASTAL SYSTEMS

13.4.1 Microtidal coasts

Under microtidal conditions wave action can maintain a barrier system (Fig. 13.8) that can be more or less continuous for tens of kilometres. Exchange of water between the lagoon and the sea may be very

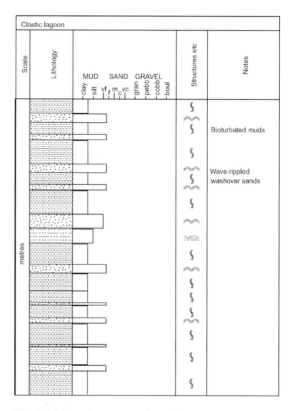

Fig. 13.11 A schematic graphic sedimentary log of clastic lagoon deposits.

limited, occurring through widely spaced inlets and as seepage through the barrier. Coarse sedimentation in the lagoon will be largely restricted to washovers that occur during storms. There is a strong likelihood of the lagoon waters becoming either brackish or hypersaline, depending upon the prevailing climate.

13.4.2 Mesotidal coasts

With the increased tidal range of mesotidal conditions, more exchange of water between the lagoon and the sea is required, resulting in more inlets forming, breaking up the barrier into a series of islands (Fig. 13.10). These inlets are the pathways for the tidal flows and the currents within them can be strong enough to redistribute sediment. On the lagoon side of the barrier sediment washed through the channel is deposited in a **flood-tidal delta** (Fig. 13.10). The water in the lagoon is shallow, so the sediment spreads out into a thin, low-angle cone of detritus dipping very gently landwards. Bedforms on the flood-tidal delta are typically subaqueous dunes migrating landwards, which result in cross-bedding with onshore palaeocurrent directions (Boothroyd 1985). **Ebb-tidal deltas** form on the seaward margin of the tidal channel as water flows out of the lagoon when the tide recedes. Building out into deeper water they are thicker bodies of sediment than flood-tidal deltas and the direction of bedform migration is seawards. The size and extent of an ebb-tidal delta is limited by the effects of reworking of the sediment by wave, storm and tidal current processes in the sea.

13.4.3 Macrotidal coasts

Coasts that have high tidal ranges do not develop barrier systems because the ebb and flood tidal currents are a stronger control on the distribution of sediment than wave action. A depositional coast in a macrotidal setting will be characterised by areas of intertidal **mudflats** that are covered at high tide and exposed at low tide. Water flooding over these areas with the rising tide spreads out and loses energy quickly: only suspended load is carried across the tidal flats, and this is deposited when the water becomes still at high tide. The upper parts of the tidal flats are only inundated at the highest tides. The incoming tide brings in nutrients and tidal flats are commonly areas of growth of salt-tolerant vegetation (**xerophytes**) and animal life is often abundant (worms, molluscs and crustaceans in particular). The deposits of this **salt marsh** environment (Belknap 2003) are therefore predominantly fine-grained clay and silt, highly carbonaceous because of all the organic material and the animal life results in extensive bioturbation. The vegetation on the tidal flats tends to trap sediment, and mud flats are commonly sites of net accumulation. Tidal flats are often cut by **tidal creeks**, small channels that act as conduits for flow during rising and falling tides: the stronger flow in these creeks allows them to transport and deposit sand, resulting in small channel sand bodies within the tidal-flat muds. Coarser sediment is also introduced onto the tidal flats during storms, forming thin layers of sand and bioclastic debris. Flaser and lenticular bedding (4.8) may occur on the lower parts of the tidal mudflats, where currents may be periodically strong enough to transport and form ripples. These ripple-laminated sands will occur interbedded with mud that drapes the ripple forms.

13.5 COASTAL SUCCESSIONS

The patterns of sedimentary successions built up at a coast are determined by a combination of sediment supply and relative sea-level change. (As will be seen in Chapter 23, these two factors are in fact dominant in controlling the large- and medium-scale stratigraphy in all shallow marine depositional environments.) **Prograding barriers** and strand plains are those that build out to sea through time as sediment is added to the beach from the sea. A barrier will become wider, and the inner margins may become more stabilised by vegetation growth. A prograding strand plain will result in a series of ridges parallel to the coastline, **chenier ridges** (Fig. 13.7), which are the relics of former beaches that have been left inland as the shoreline prograded (Augustinus 1989). **Retrograding barriers** form where the supply of sediment is too low to counteract losses from the beach by erosion. Removal of sediment from the front of the barrier reduces its width and, in turn, its height. This makes the coast more susceptible to washovers of sand occurring and the lagoon (or a marsh behind a strand plain) will become partly filled in. By this process the beach system will gradually move landward.

Under conditions of slow relative sea-level rise the beach may also move landward, but a lagoon will also expand, flooding the adjacent coastal plain in response to the sea-level rise. Through time these transgressive barrier systems will build up a succession from coastal plain deposits at the base, overlain by lagoon facies and capped by beach deposits of the barrier system (Fig. 13.12). A similar transgressive situation at a strand plain will result in coastal plain deposits overlain by beach deposits.

13.6 ESTUARIES

An estuary is the marine-influenced portion of a drowned valley (Dalrymple et al. 1992). A drowned valley is the seaward portion of a river valley that becomes flooded with seawater when there is a

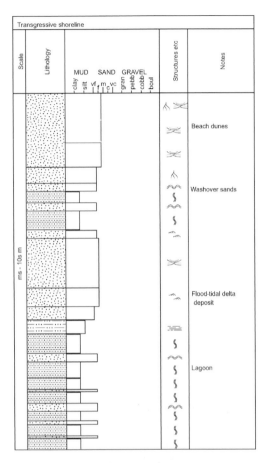

Fig. 13.12 A schematic graphic sedimentary log of a transgressive coastal succession.

relative rise in sea level (a transgression, *23.1.3*). They are regions of mixing of fresh and seawater. Sediment supply to the estuary is from both river and marine sources, and the processes that transport and deposit this sediment are a combination of river and wave and/or tidal processes. An estuary is different from a delta because in an estuary all the sedimentation occurs within the drowned valley, whereas deltas are progradational bodies of sediment that build out into the marine environment. A stretch of river near the mouth that does not have a marine influence would not be considered to be an estuary.

Estuaries are common features at the mouths of rivers in the present day because since the last glacial period there has been a relative rise in sea level. During this Holocene transgression many river

valleys have become flooded and these provide a spectrum of morphologies and process controls that can be used to construct models for estuarine sedimentation. Two end members are recognised (Dalrymple et al. 1992): ***wave-dominated estuaries*** and ***tide-dominated estuaries***, with a range of intermediate forms in between. In addition to these two basic process controls, the volume of the sediment supply and the relative importance of supply from marine and fluvial sources also play an important role in determining the facies distributions in an estuarine succession. The extent of estuarine deposits will depend upon the size of the valley and the depth to which it has been flooded. Modern estuaries range from a few kilometres to over 100 km long and from a few hundred metres to over 10 km wide. The thickness of the succession formed by filling an estuary is typically tens of metres.

Sedimentation in an estuary will eventually result in the drowned valley filling to sea level and, unless there is further sea-level rise, the area will cease to have an estuarine character. If there is a high rate of fluvial sediment supply, deposition will start to occur at the mouth of the river and a delta will start to form. Under conditions where the marine processes are dominant, the river mouth will become an area of tidal flats if tidal currents are strong, or the sediment will be reworked and redistributed by wave processes to form a strand plain. An estuary is therefore a temporary morphological feature, existing only during and immediately after transgression while sediment fills up the space created by the sea-level rise. The presence of estuarine deposits therefore can be used as an indicator of transgression – see Chapter 23 for further discussion of the relationship between sea-level changes and facies.

13.6.1 Wave-dominated estuaries

An estuary developed in an area with a small tidal range and strong wave energy will typically have three divisions (Figs 13.13 & 13.14): the bay-head delta, the central lagoon and the beach barrier.

Bay-head delta

The ***bay-head delta*** is the zone where fluvial processes are dominant. As the river flow enters the central lagoon it decelerates and sediment is

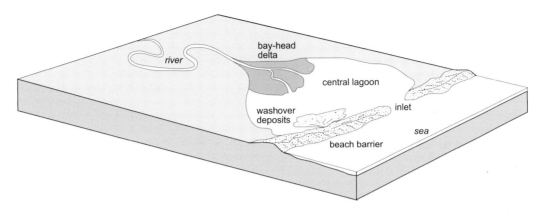

Fig. 13.13 Distribution of depositional settings in a wave-dominated estuary.

Fig. 13.14 A wave-dominated estuary, with an extensive beach barrier protecting a lagoonal area.

deposited. The form and processes of a bay-head delta will be those of a river-dominated delta (*12.4.5*) because the tidal effect is minimal and the barrier protects the central lagoon from strong wave energy. A coarsening-up, progradational succession will be formed, with channel and overbank facies building out over sands deposited at the channel mouth, which in turn overlies fine-grained deposits of the central lagoon.

Central lagoon

The lowest energy part of the estuarine system is the ***central lagoon***, where the river flow rapidly decreases and the wave energy is mainly concentrated

at the barrier bar. The central lagoon is therefore a region of fine-grained deposition, often rich in organic material, similar to normal lagoonal conditions (*13.3.2*). When the central lagoon becomes filled with sediment it becomes a region of salt-water marshes crossed by channels. In wave-dominated estuaries, parts of the lagoon that receive influxes of sand may be areas where wave-ripples form and these may also be draped with mud.

Beach barrier

The outer part of a wave-dominated estuary is a zone where wave action reworks marine sediment (bioclastic material and other sediment reworked by longshore drift) to form a barrier. The characteristics of the barrier will be the same as those found along clastic coasts (*13.3.1*). An inlet allows the exchange of water between the sea and the central lagoon, and if there is any tidal current, a flood-tidal delta of marine-derived sediment may prograde into the central lagoon.

Successions in wave-dominated estuaries

The sedimentary succession deposited in the estuary will reflect the three divisions of the system, although they may not all occur in a single vertical succession (Dalrymple et al. 1992). The relative thicknesses of each will depend on the balance between fluvial and marine supply of sediment: if fluvial supply is greatest, the bay-head delta facies will dominate, whereas the barrier deposits will be more important if the marine supply is higher. Many actual examples will show quite a lot of variation from the idealised successions in Fig. 13.15.

13.6.2 Tide-dominated estuaries

Tidal processes may dominate in mesotidal and macrotidal coastal regimes where tidal current energy exceeds wave energy at the estuary mouth. The funnel shape of an estuary tends to increase the flood-tidal current strength, but decreases to zero at the **tidal limit**, the landward extent of tidal effects in an estuary. The river flow strength decreases as it interacts with the tidal forces that are dominant. Three areas of deposition can be identified (Figs 13.16 & 13.17): tidal channel deposits, tidal flats and tidal sand bars.

Tidal channels

In the inner part of the estuary where the river channel is influenced by tidal processes, the low-gradient channel commonly adopts a meandering form (Dalrymple et al. 1992). Point bars form on the inner banks of meander bends in the same way as purely fluvial systems, but the tidal effects mean that there are considerable fluctuations in the strength of the flow during different stages of the tidal cycle: when a strong ebb tide and the river act together, the combined current may transport sand, but a strong flood tide may completely counteract the river flow,

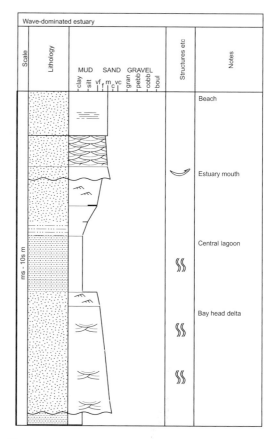

Fig. 13.15 A graphic sedimentary log of wave-dominated estuary deposits.

resulting in standing water, which allows deposition from suspension. The deposits in the point bar are therefore **heterolithic**, that is, they consist of more than one grain size, in this case alternating layers of sand and mud (Reineck & Singh 1972). This style of point-bar stratification has been called '**inclined heterolithic stratification**', sometimes abbreviated to 'IHS' (Thomas et al. 1987). These alternating layers of sand and mud dipping in to the axis of the channel (perpendicular to flow) are a distinctive feature of tidally influenced meandering channels.

Tidal flats

Adjacent to the channels and all along the sides of the estuary there are tidal flat areas that are variably covered with seawater at high tide and subaerially exposed at low tide. These are typically vegetated

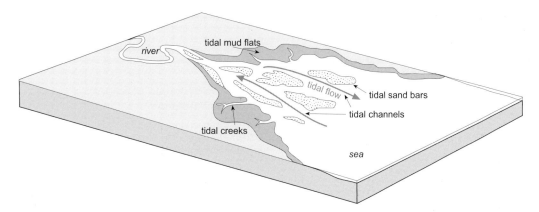

Fig. 13.16 Distribution of depositional settings in a tidally dominated estuary.

salt marsh areas cut by tidal creeks that act as the conduits for water flow during the tidal cycles. The processes and products of deposition in these settings are the same as found in macrotidal settings.

Tidal bars

The outer part of a tide-dominated estuary is the zone of strongest tidal currents, which transport and deposit both fluvially derived sediment and material brought in from the sea. In macrotidal regions the currents will be strong enough to cause local scouring and to move both sand and gravel: bioclastic debris is common amongst the gravelly detritus deposited as a lag on the channel floor (Reinson 1992). Dune

bedforms are created and migrate with the tidal currents to generate cross-bedded sandstone beds. Evidence for tidal conditions in these beds may include mud drapes, reactivation surfaces and herringbone cross-stratification (*11.2.4*). The mud drapes form as the current slows down when the tide turns, and the reactivation surfaces occur as opposing currents erode the tops of dune bedforms. Herringbone cross-bedding is relatively uncommon because the ebb and flood tidal flows tend to follow different pathways, with the flood tide going up one side of the estuary and the ebb tide following a different route down the other side. Dune bedforms that form on elongate banks (and hence the cross-beds) will be mainly oriented in either the flood tide direction or with the

Fig. 13.17 A tidally dominated estuarine environment with banks of sand covered with dune bedforms exposed at low tide.

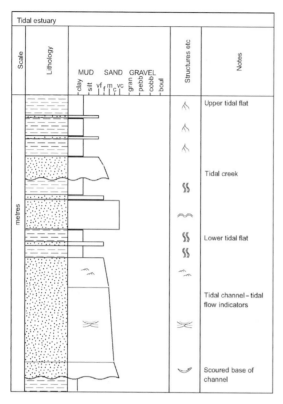

Fig. 13.18 A graphic sedimentary log of tidal estuary deposits.

indicators. Muddy tidal flat deposits rich in organic material may contain sandy sediment deposited within tidal creeks, at the highest tides and during storms.

13.6.3 Recognition of estuarine deposits: summary

There are many features in common between the deposits of deltas and estuaries in the stratigraphic record. Both are sedimentary bodies formed at the interface between marine and continental environments and consequently display evidence of physical, chemical and biological processes that are active in both settings (e.g. an association of beds containing a marine shelly fauna with other units containing rootlets). The key difference is that a delta is a progradational sediment body, that is, it builds out into the sea and will show a coarsening-up succession produced by this progradation. In contrast, estuaries are mainly aggradational, building up within a drowned river channel. The base of an estuarine succession is therefore commonly an erosion surface scoured at the mouth of the river, for example, in response to sea-level fall. It may be difficult to distinguish between the deposits of a tidal estuary and a tide-dominated delta if there is limited information and it is difficult to establish whether the succession is aggradational and valley-filling or progradational.

13.7 FOSSILS IN COASTAL AND ESTUARINE ENVIRONMENTS

Beaches are high-energy environments, continually washed by waves, which move the sediment around subjecting the clasts to abrasion. The supply of shelly material from the sea is often abundant, but much of it will be broken up into fragments that may be identifiable in only a general sense as pieces of mollusc, coral, echinoderm, etc. Only the most robust organisms remain intact, and among these are thick-shelled molluscs such as oysters, which are also found living in high-energy, shallow water of the shallow subtidal zone. The abundance of bioclastic debris in beach deposits will depend on the relative proportions of mineral grains and shelly material supplied to the beach.

For organisms living in a lagoon, both hypersaline and brackish conditions require adaptation that only a limited number of plants and animals achieve.

ebb tide. Herringbone cross-stratification will only form in areas of overlap between banks of cross-beds of different orientation, or if the currents change position. Where tidal currents are strongest the dune bedforms are replaced by upper flow regime plane beds that form horizontally laminated sands.

Successions in tide-dominated estuaries

A succession formed in a tide-dominated estuary will consist of a combination of tidal channel, tidal flat and tidal bar deposits. The proportions preserved of each will depend on the position in the estuary, the strength of the tidal currents and the amounts of mud, sand and gravel available for deposition (Fig. 13.18). The base of a tidal channel is marked by a scour and lag, and will typically be followed by a fining-upwards succession of cross-bedded sands, which may show mud drapes, inclined heterolithic stratification. Channel and bar deposits may also show bi-directional palaeocurrent

Lagoonal faunal and floral assemblages are therefore often limited in numbers of taxa, being dominated by those that are adapted to either brackish or hypersaline conditions. Although the diversity of fauna may be severely limited by brackish or hypersaline waters in the lagoon, those species that are tolerant flourish in the absence of competition in waters rich in nutrient from the surrounding vegetation. These specialised organisms may occur in very large numbers and fossil assemblages in lagoons are typically of very low diversity or even monospecific.

The traces of organisms can commonly be found and the ichnofacies (*11.7*) present will depend upon the energy of the environment and the nature of the substrate. In lagoons the fine, organic-rich sediment provides a favourable feeding area for organisms that are able to tolerate the reduced/enhanced salinity, and bioturbation may be common. In sandy intertidal areas the predominant style of trace is typically a vertical structure created by animals moving up to the surface when the area is covered by water and down within the sediment body when the water recedes. This form of trace fossil is known as the *Skolithos* assemblage, after the simple vertical tubes that are found in these settings. Other ichnofacies assemblages occur if the substrate is relatively firm (*Glossifungites* assemblage) or hard (*Trypanites* assemblage). Trypanites-type traces are borings made in solid rock (bedrock or loose boulders) by molluscs, and these are characteristic of rocky coastlines.

The association of marine and continental conditions is one of the characteristics of estuaries, and this is reflected in the fossil assemblages found in deposits in these environments. Some shelly debris may be brought in from the marine environment, but shelly fauna is also often abundant in estuarine settings. As well as body fossils, evidence for biogenic activity is also present in the form of trace fossils, which range from very abundant and diverse in tidal mudflats to sparse in the high energy, sandy environments of the outer parts of estuaries. Vegetation growth may be prolific in tidal mudflats, especially on the upper parts, and plant remains may be present as organic material or as root traces.

Characteristics of coastal and estuarine systems

These complex, heterogeneous depositional environments are divided into four elements for the purposes of summarising their characteristics.

Beach/barrier systems
- lithology – sand and conglomerate
- mineralogy – mature quartz sands and shelly sands
- texture – well sorted, well rounded clasts
- bed geometry – elongate lenses
- sedimentary structures – low-angle stratification and wave reworking
- palaeocurrents – mainly wave-formed structures
- fossils – robust shelly debris
- colour – not diagnostic
- facies associations – may be associated with coastal plain, lagoonal or shallow-marine facies

Lagoons
- lithology – mainly mud with some sand
- mineralogy – variable
- texture – fine-grained, moderately to poorly sorted
- bed geometry – thinly bedded mud with thin sheets and lenses of sand
- sedimentary structures – may be laminated and wave rippled
- palaeocurrents – rare, not diagnostic
- fossils – often monospecific assemblages of hypersaline or brackish tolerant organisms
- colour – may be dark due to anaerobic conditions
- facies associations – may be associated with coastal plain or beach barrier deposits

Tidal channel systems
- lithology – mud, sand and less commonly conglomerate
- mineralogy – variable
- texture – may be well sorted in high energy settings
- bed geometry – lenses with erosional bases
- sedimentary structures – cross-bedding and cross-lamination and inclined heterolithic stratification
- palaeocurrents – bimodal in tidal estuaries
- fossils – shallow marine
- colour – not diagnostic
- facies associations – may be overlain by fluvial, shallow marine, continental or delta facies

Tidal mudflats
- lithology – mud and sand
- mineralogy – clay and shelly sand
- texture – fine-grained, not diagnostic
- bed geometry – tabular muds with thin sheets and lenses of sand
- sedimentary structures – ripple cross-lamination and flaser/lenticular bedding
- palaeocurrents – bimodal in tidal estuaries

- fossils – shallow marine fauna and salt marsh vegetation
- colour – often dark due to anaerobic conditions
- facies associations – may be overlain by shallow marine or continental facies

FURTHER READING

Boyd, R., Dalrymple, R.W. & Zaitlin, B.A. (1992) Classification of clastic coastal depositional environments. *Sedimentary Geology*, **80**, 139–150.

Boyd, R., Dalrymple, R.W. & Zaitlin, B.A. (2006) Estuarine and incised valley facies models. In: *Facies Models Revisited* (Eds Walker, R.G. & Posamentier, H.). Special Publication 84, Society of Economic Paleontologists and Mineralogists, Tulsa, OK; 171–235.

Clifton, H.E. (2006) A re-examination of facies models for clastic shorelines. In: *Facies Models Revisited* (Eds Walker, R.G. & Posamentier, H.). Special Publication 84, Society of Economic Paleontologists and Mineralogists, Tulsa, OK; 293–337.

Dalrymple, R.W., Zaitlin, B.A. & Boyd, R. (1992) Estuarine facies models: conceptual basis and stratigraphic implications. *Journal of Sedimentary Petrology*, **62**, 1130–1146.

Davis, R.A. Jr & Fitzgerald, D.M. (2004) *Beaches and Coasts*. Blackwell Science, Oxford.

Reading, H.G. & Collinson, J.D. (1996) Clastic coasts. In: *Sedimentary Environments: Processes, Facies and Stratigraphy* (Ed. Reading, H.G.). Blackwell Science, Oxford; 154–231.

Woodroffe, C.D. (2003) *Coasts: Form, Process and Evolution*. Cambridge University Press, Cambridge.

14

Shallow Sandy Seas

Shallow marine environments are areas of accumulation of substantial amounts of terrigenous clastic material brought in by rivers from the continental realm. Offshore from most coastlines there is a region of shallow water, the continental shelf, which may stretch tens to hundreds of kilometres out to sea before the water deepens down to the abyssal depths of ocean basins. Not all land areas are separated by ocean basins, but instead have shallow, epicontinental seas between them. Terrigenous clastic material is distributed on shelves and epicontinental seas by tides, waves, storms and ocean currents: these processes sort the material by grain size and deposit areas of sand and mud, which form thick, extensive sandstone and mudstone bodies in the stratigraphic record. Characteristic facies can be recognised as the products of transport and deposition by tides and storm/wave processes. Deposition in shallow marine environments is sensitive to changes in sea level and the stratigraphic record of sea-level changes is recorded within sediments formed in these settings.

14.1 SHALLOW MARINE ENVIRONMENTS OF TERRIGENOUS CLASTIC DEPOSITION

The continental shelves and epicontinental seas (11.1) are important sites of deposition of sand and mud in the world's oceans and account for over half the volume of ocean sediments. These successions can be very thick, over 10,000 m, because deposition may be very long-lived and can continue uninterrupted for tens of millions of years. They occur as largely undeformed strata around the edges of continents and also in orogenic belts, where the collision of continental plates has forced beds deposited in shallow marine environments high up into mountain ranges. This chapter focuses on the terrigenous clastic deposits found in shallow seas; carbonate sedimentation, which is also important in these environments, is covered in Chapter 15.

14.1.1 Sediment supply to shallow seas

The supply of sediment to shelves is a fundamental control on shallow marine environments and depositional facies of shelves and epicontinental seas. If the area lies adjacent to an uplifted continental region and there is a drainage pattern of rivers delivering detritus to the coast, the shallow-marine sedimentation will be dominated by terrigenous clastic deposits. The highest concentrations of clastic sediment will be near the mouths of major rivers: adjacent coastal regions will also be supplied with sediment by longshore movement of material by waves, storms and tides. Shallow seas that are not supplied by much terrigenous material may be areas of carbonate sedimentation, especially if they are in lower latitudes where the climate is relatively warm. In cooler climates where carbonate production is slower, shelves and shallow seas with low terrigenous sediment supply are considered to be **starved**. The rate of sediment accumulation is slow and may be exceeded by the rate of subsidence of the sea floor such that the environment becomes gradually deeper with time.

14.1.2 Characteristics of shallow marine sands

Detritus that reaches a shallow sea is likely to have had a history of transport in rivers, may have passed through a delta or estuary, or could have been temporarily deposited along a coastline before it arrives at the shelf. If there is a long history of transport through these other environments the grain assemblage is likely to be mature (2.5.3). Texturally, the grains of sand will have suffered a degree of abrasion and the processes of turbulent flow during transport will separate the material into different grain sizes. The compositional maturity will probably be greater than the equivalent continental deposits, because the more labile minerals and grains (such as feldspar and lithic fragments) are broken down during transport: shallow marine sands are commonly dominated by quartz grains. In polar areas, the sediment supplied is much less mature because cold weather reduces chemical weathering of the grains and glacial transport does not result in much sorting or rounding of the clasts (7.3.4).

The detrital component is often complemented by material that orginates in the shallow marine environment. Shallow seas are rich in marine life, including many organisms that have calcareous shells and skeletons. The remains of these biogenic hard parts are a major component of shelf carbonate deposits (Chapter 15), but can also be very abundant in sands and muds deposited in these seas. Whole shells and skeletons may be preserved in mudrocks because they are low-energy deposits. In higher energy parts of the sea, currents move sand around and a lot of biogenic debris is broken up into bioclastic fragments ranging from sand-sized, unidentifiable pieces up to larger pieces of shelly material and bone. Bone is also the origin for phosphates that can form as authigenic deposits in shallow marine settings (3.4): these phosphates are relatively rare. However, another authigenic mineral, glauconite (11.5), is a common component of sandstones and mudrocks formed on shelves and epicontinental seas and is considered to be a reliable indicator of shallow marine conditions. The characteristic dark green colour of the mineral gives sediments rich in it a distinctly green tinge, although it is iron-rich and weathers to a rusty orange colour. 'Greensands' are shallow-marine deposits rich in glauconite that are particularly common in Cretaceous strata in the northern hemisphere.

Shallow seas are environments rich in animal life, particularly benthic organisms that can leave traces of their activity in the sediments. Bioturbation may form features that are recognisable of the activities of a particular type of organism (11.7), but also results in a general churning of the sediment, homogenising it into apparently structureless masses. Primary sedimentary structures (wave ripples, hummocky cross-stratification, trough cross-bedding, and so on) are not always preserved in shelf sediments because of the effects of bioturbation. Bioturbation is most intense in shallower water and is frequently more abundant in sandy sediment than in muddy deposits. This is because the currents that transport and deposit sand may also carry nutrients for benthic organisms living in the sand: many organisms also prefer to live on and within a sandy substrate.

The abundance of calcareous shell material in shallow-marine sandstones makes calcium carbonate available within the strata when the beds are buried. Groundwater moving through the sediments dissolves and reprecipitates the carbonate as cement (18.2.2). Shelly fossils within sandstones are therefore sometimes found only as **casts** of the original form, as the original calcite or aragonite shells have been dissolved away. Sandstone beds deposited in shallow marine settings also typically have a calcite cement.

14.1.3 Shallow marine clastic environments

The patterns and characteristics of deposition on shelves and epicontinental seas with abundant terrigenous clastic supply are controlled by the relative importance of wave, storm and tidal processes. The largest tidal ranges tend to be in epicontinental seas and restricted parts of shelves, although in some situations the tidal ranges in narrow or restricted seaways can be very small (*11.2.2*). Open shelf areas facing oceans are typically regions with a microtidal to mesotidal regime and are affected by ocean storms. Two main types, **storm-dominated shelves** and **tide-dominated shelves**, can be recognised in both modern environments and ancient facies: these are end-members of a continuum and many modern and ancient shelves and epicontinental seas show influence of both major processes (Johnson & Baldwin 1996). The majority of modern shelves are storm-dominated (80%): the remainder are mainly tide-dominated (17%), with just a small number (3%) of shelves influenced mainly by ocean currents (Johnson & Baldwin 1996). These ocean-current-dominated shelves are generally narrow (less than 10 km) and lie adjacent to strong geostrophic currents (*11.4*): sandwaves and sand ribbons form on them, and as such they are similar to tidal shelves, but the driving current is not of tidal origin.

The detailed characteristics of sands deposited on modern shelves can be determined directly only by taking shallow cores that provide a limited amount of information: indirect investigation by geophysical techniques, such as shallow seismic profiles (*22.2*), can also yield some information about the internal structures. Not all sandy deposits occurring on modern shelves have been formed by processes occurring in the present day: the sea-level rise in the past 10 kyr, the Holocene transgression, has drowned former strand plain and barrier island ridges, along with sands deposited in the shoreface, leaving them as inactive relics in deeper water.

14.2 STORM-DOMINATED SHALLOW CLASTIC SEAS

14.2.1 Facies distribution across a storm-dominated shelf

Shoreface

The shallower parts of the shelf and epicontinental sea are within the depth zone for wave action (*11.1*) and any sediment will be extensively reworked by wave processes. Sands deposited in these settings may preserve wave-ripple cross-lamination and horizontal stratification. Streaks of mud in flaser beds (*4.8*) deposited during intervals of lower wave energy become more common in the deposits of slightly deeper water further offshore (Fig. 14.1). Wave ripples are less common as the fair-weather wave base is approached in the lower part of the shoreface. Within the shoreface zone **sand ridges** may be formed by flows generated by eddy currents related to storms and/or wave-driven longshore drift (Stubblefield et al. 1984). These ridges occur in water depths of 5 to 15 m and are oriented obliquely to the coastline as oblique longshore bars. They are up to about 10 m

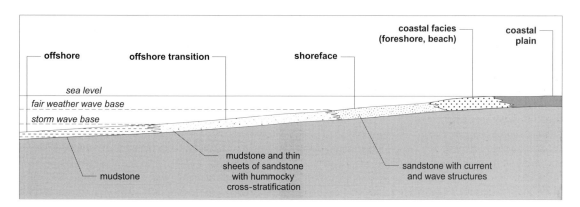

Fig. 14.1 Characteristics of a storm-dominated shelf environment.

high, a few kilometres wide and tens of kilometres in length, occurring spaced about 10 km apart. The sediments are typically well-sorted sands with a basal lag of gravel (Hart & Plint 1995).

Offshore transition zone

In the offshore transition zone, between the fair-weather and storm wave bases on storm-dominated shelves, sands are deposited and reworked by storms. A storm creates conditions for the formation of bedforms and sedimentary structures that seem to be exclusive to storm-influenced environments (Dott & Bourgeois 1982; Cheel & Leckie 1993). ***Hummocky cross-stratification*** (often abbreviated to ***HCS***) is distinctive in form, consisting of rounded mounds of sand on the sea floor a few centimetres high and tens of centimetres across. The crests of the hummocks are tens of centimetres to a metre apart. Internal stratification of these hummocks is convex upwards, dips in all directions at angles of up to 10° or 20°, and thickens laterally: these features are not seen in any other form of cross-stratification (Figs 14.2 & 14.3). Between the hummocks lie swales and where concave layers in them are preserved this is sometimes called ***swaley cross-stratification*** (abbreviated to ***SCS***).

Hummocky and swaley cross-stratification are believed to form as a result of ***combined flow***, that is, the action of both waves and a current. This occurs when a current is generated by a storm at the same time as high-amplitude waves reach deep below the surface. The strong current takes sand out into the deeper water in temporary suspension and as it is deposited the oscillatory motion caused by the waves results in deposition in the form of hummocks and swales. Swaley cross-stratification is mainly formed and preserved in shallow water where the hummocks have a lower preservation potential. One of the characteristics of HCS/SCS is that these structures are normally only seen in fine to medium grained sand, suggesting that there is some grain-size limitation involved in this process. Storm conditions affect the water to depths of 20 to 50 m or more so HCS/SCS may be expected in any sandy sediments on the shelf to depths of several tens of metres. These structures are not seen in shoreface deposits above fair-weather wave base due to reworking of the sediment by ordinary wave processes, so this characteristic form of cross-stratification is found only in sands deposited in the offshore transition zone (*11.1*).

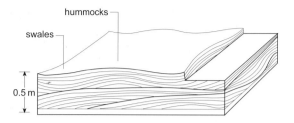

Fig. 14.2 Hummocky–swaley cross-stratification, a sedimentary structure that is thought to be characteristic of storm conditions on a shelf.

Fig. 14.3 An example of hummocky cross-stratified sandstone with very well-defined, undulating laminae. The bed is 30 cm thick.

Individual storm deposits, tempestites (*11.3.1*), deposited by single storm events typically taper in thickness from a few tens of centimetres to millimetre-thick beds in the outer parts of this zone several tens of kilometres offshore (Aigner 1985). Proximal tempestites have erosive bases and are composed of coarse detritus, whereas the distal parts of the bed are finer-grained laminated sands: hummocky and swaley cross-stratification occurs in the sandy parts of tempestites (Walker & Plint 1992). An idealised tempestite bed (Fig. 14.4) will have a sharp, possibly erosive base, overlain by structureless coarse sediment (coarse sand and/or gravel): the scouring and initial deposition occurs when the storm is at its peak strength. As the storm wanes, hummocky–swaley cross-stratification forms in finer sands and this is overlain by fine sand and silt that shows horizontal and wave-ripple lamination formed as the strength of the oscillation decreases. At the top of the bed the sediment grades into mud. The magnitude of the

Fig. 14.4 A bed deposited by storm processes. The base (bottom of the photograph) has a sharp erosional contact with underlying mudrocks. Planar lamination is overlain by hummocky cross-stratification and capped by wave-rippled sandstone and mudstone (just below the adhesive tape roll, 8 cm across).

storms that deposit beds tens of centimetres thick is not easy to estimate, because the availability of sand is probably of equal importance to the storm energy in determining the thickness of the bed.

In the periods between storm events this part of the shelf is an area of deposition of mud from suspension. This fine-grained clastic material is sourced from river mouths and is carried in suspension by geostrophic and wind-driven currents, and storms also rework a lot of fine sediment from the sea floor and carry it in suspension across the shelf. Storm deposits are therefore separated by layers of mud, except in cases where the mud is eroded away by the subsequent storm. The proportion of mud in the sediments increases offshore as the amount of sand deposited by storms decreases.

Offshore

The outer shelf area below storm wave base, the offshore zone, is predominantly a region of mud deposition. Exceptional storms may have some effect on this deeper part of the shelf, and will be represented by thin, fine sand deposits interbedded with the mudstone. Ichnofauna are typically less diverse and abundant than the associations found in the shoreface and offshore transition zone. The sediments are commonly grey because this part of the sea floor is relatively poorly oxygenated allowing some preservation of organic matter within the mud.

14.2.2 Characteristics of a storm-dominated shallow-marine succession

If there is a constant sediment supply to the shelf, continued deposition builds up the layers on the sea bed and the water becomes shallower. Shelf areas that were formerly below storm wave base experience the effects of storms and become part of the offshore transition zone. Similarly addition of sediment to the sea floor in the offshore transition zone brings the sea bed up into the shoreface zone above fair-weather wave base and a vertical succession of facies that progressively shallow upwards is constructed (Figs 14.5 & 14.6) (Walker & Plint 1992). The offshore facies mainly consists of mudstone beds with some bioturbation. This is overlain by offshore transition facies made up of sandy tempestite beds interbedded with bioturbated mudstone. The tempestite beds have erosional bases, are normally graded and show some hummocky–swaley cross-stratification. The thickness of the sandstone beds generally increases up through the succession, and the deposits of the shallower part of this zone show more SCS than HCS. The shoreface is characterised by sandy beds with symmetrical (wave) ripple lamination, horizontal stratification and SCS, although sedimentary structures may be obscured by intense bioturbation. Sandstone beds in the shoreface may show a broad lens shape if they were deposited as localised ridges on the shallow sea floor. The top of the succession may be capped by foreshore facies (13.2).

14.2.3 Mud-dominated shelves

Some shelf areas are wave- and storm-dominated, but receive large quantities of mud and relatively little sand. They occur close to rivers that have a high suspended load: the plumes of suspended sediment from the mouths of major rivers may extend for tens or hundreds of kilometres out to sea and then are reworked by wind-driven and geostrophic currents across the shelf (McCave 1984). Muddy deposits on the inner parts of the shelf are normally intensely bioturbated, except in cases where the rates of sedimentation of mud are so high that accumulation outpaces the rate at which the organisms can rework the sediment. High concentrations of organic matter may make these shelf muds very dark grey or black in colour.

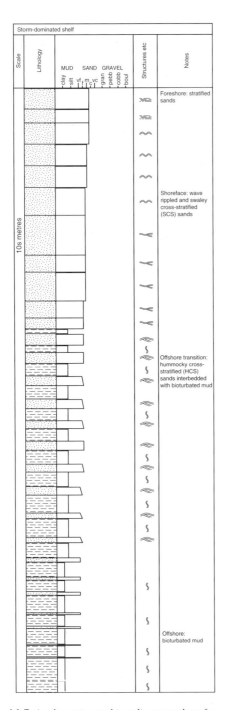

Fig. 14.5 A schematic graphic sedimentary log of a storm-dominated succession.

14.3 TIDE-DOMINATED CLASTIC SHALLOW SEAS

14.3.1 Deposition on tide-dominated shelves

Offshore sand ridges

Near shorelines that experience strong tidal currents large sand ridges (*14.2.1*) are found on modern shelves. The ridges form parallel to the shoreline in water depths of up to 50 m and may be tens of metres high, in places rising almost to sea level (Fig. 14.7). They are typically a few kilometres wide, similar distances apart and extend for tens of kilometres as straight or gently curving features, elongated parallel to the tidal current. Between the sandy ridges there may be thin layers of gravel on the sea floor, deposited during an earlier, probably shallower phase of sedimentation on the shelf, and left behind as a lag as sand has been winnowed away by the currents (Plint 1988; Hart & Plint 1995). The sands are moderately well sorted, medium grained but the deposits may include some mud occurring as clay laminae deposited during slack phases of the tidal flow. Internal sedimentary structures are cross-lamination and cross-bedding generated by the migration of ripples and subaqueous dune bedforms over the surface of the ridges. The resulting sandstone body preserved in the stratigraphic record is likely to have a basal lag and consist of stacks of cross-bedded and cross-laminated sandstone up to 10 m thick, or more: the primary sedimentary structures may be wholly or partly destroyed by bioturbation.

Tidal sandwaves and sand ribbons

Currents generated by tides influence large areas of shelves and epicontinental seas. These tidal currents affect the sea bed tens of metres below sea level and are strong enough to move large quantities of sand in shallow marine environments. The effects of waves and storms are largely removed by tidal currents reworking the material in macrotidal regimes and only the tidal signature is left in the stratigraphic record. In seas with moderate tidal effects the influence of tides is seen in shallower water, but storm deposits are preserved in the offshore transition zone in these mixed storm/tidal shelf settings.

Fig. 14.6 The strata in the hillside are a succession passing up from offshore mudstones (bottom left), to thin-bedded sandstone of the offshore transition zone up to the cliff-forming shoreface sandstones.

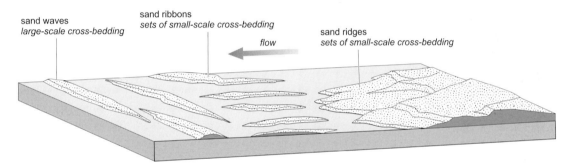

sand waves
large-scale cross-bedding

sand ribbons
sets of small-scale cross-bedding

flow

sand ridges
sets of small-scale cross-bedding

Fig. 14.7 Sandwaves, sand ridges and sand ribbons in shallow, tidally influenced shelves and epicontinental seas.

The form of tidal deposits in shallow marine environments depends on the velocity of the tidal current. In areas of low velocity currents (ca. $50\,\text{cm s}^{-1}$) sand occurs in low relief sheets and patches that are rippled on the surface. At low to moderate near-surface tidal current velocities (50 to $100\,\text{cm s}^{-1}$) **sandwaves** are typical: these bedforms are a class of large subaqueous dunes that have heights of at least 1.5 m and wavelengths ranging from 150 m to 500 m (Fig. 14.7). The crests are straight to moderately sinuous and the lee slope is a lower angle than most subaqueous bedforms at around $15°$ (Johnson & Baldwin 1996). Migration of sandwaves in the direction of the predominant tidal current generates cross-stratification with sets that may be many metres thick (Fig. 14.8). Cross-stratification on this scale is not generally seen in other marine environments and

is only matched in size by aeolian dunes and some large bar forms in rivers. Individual sandwaves are isolated on the sea floor if the supply of sediment is low, but form amalgamated banks of sandwaves if there is abundant sand supply to the shelf.

In shallow seas with higher velocity tidal currents (over $100\,\text{cm s}^{-1}$) sediment on the sea floor forms **sand ribbons** elongated parallel to the flow direction (Fig. 14.7). These ribbons are only a metre or so thick but are up to 200 m wide and stretch for over 10 km in the flow direction. Areas of low sand supply are characterised by isolated ribbons whereas higher sediment supply results in ribbons amalgamated into **sand ridges**. Very strong tidal currents (over $100\,\text{cm s}^{-1}$) can sweep sand off the sea floor leaving only patches of gravel and metre-deep furrows eroded into the sea bed.

Fig. 14.8 Large-scale cross-stratification formed by the migration of sandwaves in a tidally influenced shelf environment.

Fig. 14.9 Bioturbated, cross-bedded sandstones deposited on a tidally influenced shelf.

The offshore transition and offshore zones of shelves and epicontinental seas are too deep for the effects of the surface tidal currents to be felt and are sites for mud deposition and sands deposited by storm currents. Mud is also deposited in shallower areas that are not affected by tidal currents. Bioturbation is common in these fine-grained deposits (Fig. 14.9).

14.3.2 Characteristics of tide-dominated shallow-marine successions

Packages of cross-stratified sandstone that contain a fully marine fauna and lack evidence for any sub-aerial exposure are normally interpreted as the deposits of tidally dominated shallow seas (Fig. 14.10). In water depths of tens of metres tides are the only currents that can generate and maintain the large subaqueous dune or sandwave bedforms: geostrophic currents are generally too weak and storm-driven currents are too short-lived and infrequent to create these bedforms. Features of tidal sedimentation (*11.2.4*) that may be present in these offshore tidal facies include mud drapes on some of the smaller scale cross-bedding and reactivation surfaces within the sandwave cross-stratification (Allen 1982). There may be evidence of different directions of tidal currents from within a unit of tidally deposited sandstones, but herringbone cross-stratification is uncommon. Tidal currents on a shelf tend to follow regular patterns (rotary tides: *11.2.3*) that do not undergo the direct reversals seen in estuarine and coastal tidal settings. Erosion surfaces overlain by gravel or shelly lags are found, representing higher energy parts of the shelf or sea, but the distinct channels found in estuarine deposits are not seen. The packages of cross-bedded sandstone are typically tens of metres thick, sometimes amalgamated into even larger units, and are lens-shaped on a scale of kilometres.

14.4 RESPONSES TO CHANGE IN SEA LEVEL

The processes of waves, storms and tides on a shelf are related to the water depth and hence the characteristics of shelf sediments are largely controlled by relative positions of the sea floor and the sea level. Consequently, any change in the relative sea level is likely to have an effect on the sedimentation on a shallow shelf area. For example, an increase in relative sea level of 20 m in a nearshore area will result in a change from wave-influenced shoreface deposition to storm-influenced offshore-transition sedimentation. Conversely, a fall in relative sea level in the offshore transition area may have the opposite effect, resulting in shallower water over that part of the shelf that would now become part of the shoreface zone. The

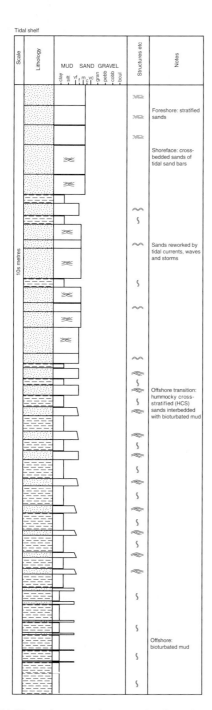

Fig. 14.10 A schematic sedimentary log through a tidally influenced shelf succession.

causes of sea-level changes and the responses to them recorded in sedimentary successions deposited on shelves are further discussed in Chapter 23.

Sand bodies formed as sand ridges may preserve the overall dimensions of the ridge if a relative sea-level rise occurs, leaving the sands in deeper water and therefore inactive. Mud deposited over the surface of the ridges will wholly enclose them, preserving them as large, elongate lenses of sandstone. These bodies make attractive oil and gas exploration targets (*18.7.4*) because they are made up of relatively well-sorted sandstone (a suitable reservoir rock) surrounded by mudstone (a suitable reservoir seal).

One particular response to sea-level change on sandy shelves is the deposition of a thin layer of gravel during sea-level rise. These **transgressive lags** form as coarse sediment deposited on the shelf during periods of low sea level is reworked by wave action: as the sea level rises (Plint 1988; Hart & Plint 1995), the gravel is moved by waves in a landward direction, resulting in a thin (usually only a few tens of centimetres) conglomerate bed within the succession. The clasts in the bed are likely to be well sorted and well rounded, and hence resemble pebbly beach deposits (*13.2*): the context will, however, be different, as transgressive lag deposits will be associated with deeper water facies of the shoreface in contrast to the foreshore associations of a beach deposit.

14.5 CRITERIA FOR THE RECOGNITION OF SANDY SHALLOW-MARINE SEDIMENTS

The environments of deposition on continental shelves vary according to water depth, sediment supply, climate and the relative importance of wave, tide and storm processes. The products of these interacting processes are extremely variable in terms of facies character, sediment body geometry and stratigraphic succession. There are, however, certain features that can be considered to be reliable indicators of shallow marine environments. First, the physical processes are generally distinctive: for example, extensive sheets and ridges of cross-bedded sand deposited by strong currents are easily recognised and cannot be the deposits of any other environment, especially if there is evidence that the currents were tidal; hummocky and swaley cross-stratification are distinctive sedimentary structures that are believed to be unique to storm-deposited sands. Second, the organisms that occur in shelf deposits are distinctive of shallow marine conditions, either as body fossils, specifically benthic organisms that are only abundant in shelf environments, or as trace fossils that display diverse morphologies (*11.7*). Third, successions of shelf sandstones and mudstones may also be associated with limestones deposited during periods of low supply of terrigenous clastic detritus.

Tempestite beds can superficially resemble turbidites (*4.5.2*) because they are also normally graded sandstone beds, with sharp bases and interbedded with mudrocks. Turbidites are more commonly found in deep basin environments (Chapter 16), so distinguishing between them and tempestites provides information about the water depth. The presence of HCS–SCS in tempestites provides evidence of deposition on a shelf, and the ichnofacies association will typically be more diverse than that found in deeper water environments (*11.7*).

Characteristics of deposits of shallow sandy seas
- lithology – mainly sand and mud, with some gravel
- mineralogy: – mature quartz sands, shelly sands
- texture – generally moderately to well sorted
- bed geometry – sheets of variable thickness, large lenses formed by ridges and bars
- sedimentary structures – cross-bedding, cross- and horizontal lamination, hummocky and swaley cross-stratification

- palaeocurrents – flow directions very variable, reflecting tidal currents, longshore drift, etc.
- fossils – often diverse and abundant, benthic forms are characteristic
- colour – often pale yellow-brown sands or grey sands and muds
- facies associations – may be overlain or underlain by coastal, deltaic, estuarine or deeper marine facies.

FURTHER READING

DeBatist, M. & Jacobs, P. (Eds) (1996) *Geology of Siliciclastic Shelf Seas*. Special Publication 117, Geological Society Publishing House, Bath.

Fleming, B.W. & Bartholomä, A. (Eds) (1995) *Tidal Signatures in Modern and Ancient Sediments*. Special Publication 24, International Association of Sedimentologists. Blackwell Science, Oxford.

Johnson, H.D. & Baldwin, C.T. (1996) Shallow clastic seas. In: *Sedimentary Environments: Processes, Facies and Stratigraphy* (Ed. Reading, H.G.). Blackwell Science, Oxford; 232–280.

Suter, J.R. (2006) Facies models revisited: clastic shelves. In: *Facies Models Revisited* (Eds Walker, R.G. & Posamentier, H.). Special Publication 84, Society of Economic Paleontologists and Mineralogists, Tulsa, OK; 331–397.

Swift, D.J.P., Oertel, G.F., Tillman, R.W. & Thorne, J.A. (Eds) (1991) *Shelf Sand and Sandstone Bodies: Geometry, Facies and Sequence Stratigraphy*. Special Publication 14, International Association of Sedimentologists. Blackwell Science, Oxford.

15

Shallow Marine Carbonate and Evaporite Environments

Limestones are common and widespread sedimentary rocks that are mainly formed in shallow marine depositional environments. Most of the calcium carbonate that makes up limestone comes from biological sources, ranging from the hard, shelly parts of invertebrates such as molluscs to very fine particles of calcite and aragonite formed by algae. The accumulation of sediment in carbonate-forming environments is largely controlled by factors that influence the types and abundances of organisms that live in them. Water depth, temperature, salinity, nutrient availability and the supply of terrigenous clastic material all influence carbonate deposition and the build up of successions of limestones. Some depositional environments are created by organisms, for example, reefs built up by sedentary colonial organisms such as corals. Changes in biota through geological time have also played an important role in determining the characteristics of shallow-marine sediments through the stratigraphic record. In arid settings carbonate sedimentation may be associated with evaporite successions formed by the chemical precipitation of gypsum, anhydrite and halite from the evaporation of seawater. Shallow marine environments can be sites for the formation of exceptionally thick evaporite successions, so-called 'saline giants', that have no modern equivalents.

15.1 CARBONATE AND EVAPORITE DEPOSITIONAL ENVIRONMENTS

There are a number of features of shallow marine carbonate environments that are distinctive when compared with the terrigenous clastic depositional settings considered in Chapter 14. First, they are largely composed of sedimentary material that has formed *in situ* (in place), mainly by biological processes: they are therefore not affected by external processes influencing the supply of detritus, except where increased terrigenous clastic supply reduces **carbonate productivity**, i.e. the rate of formation of calcium carbonate by biological processes. Second, the grain size of the material deposited is largely determined by the biological processes that generate the material, not by the strength of wave or current action, although these processes may result in break-up of clasts during reworking. Third, the biological processes can determine the characteristics of the environment, principally in places where reef formation strongly controls the distribution of energy regimes. Finally, the production of carbonate material by organisms is rapid in geological terms, and occurs at rates that can commonly keep pace with changes in water depth due to tectonic subsidence or eustatic sea-level rises: this has important consequences for the formation of depositional sequences (Chapter 23).

15.1.1 Controls on carbonate sedimentation

Areas of shallow marine carbonate sedimentation are known as **carbonate platforms**. They can occur in a wide variety of climatic and tectonic settings provided that two main conditions are met: (a) isolation from clastic supply and (b) shallow marine waters. The types of carbonate grains deposited and the facies they form are mainly controlled by climatic conditions and they have varied through time with the evolution of different groups of organisms. The places where carbonate platforms occur are determined by tectonic controls on the shape and depth of sedimentary basins: tectonic subsidence factors also strongly influence the stratigraphy of successions on carbonate platforms (Bosence 2005). Patterns of depositional sequences are also affected by sea-level fluctuations (Chapter 23).

Isolation from clastic supply

The primary requirement for the formation of carbonate platforms is an environment where the supply of terrigenous clastic and volcaniclastic detritus is very low and where there is a supply of calcium carbonate. Clastic supply to shallow marine environments can be limited by both tectonic and climatic factors. Most terrigenous sediment is supplied to shallow seas by rivers, and the pathways of fluvial systems are controlled by the distribution of areas of uplift and subsidence on the continents. On most continents the bulk of the drainage is concentrated into a small number of very large rivers that funnel sediment to coastal deltas. Along coastlines distant from these deltas the clastic supply is generally low, with only relatively small river systems providing detritus. This allows for quite extensive stretches of continent to be areas that receive little or no terrigenous sandy or muddy sediment. The climate of the continent adjacent to the shelf also has an important effect. In desert

regions the rainfall, and hence the run-off, is very low, which means that there is little transport of sediment to the sea by rivers.

Shallow marine waters

Biogenic carbonate production is inhibited by the presence of clastic material so the areas of low input of detritus are potential sites for carbonate deposition. Under favourable conditions, the amount of biogenic carbonate produced in shallow seas is determined by the productivity within the food chain. Photosynthetic plants and algae at the bottom of the food chain are dependent on the availability of light, and penetration by sunlight is controlled by the water depth and the amount of suspended material in the sea. Relatively shallow waters with low amounts of suspended terrigenous clastic material are therefore most favourable and in bright tropical regions with clear waters this **photic zone** may extend up to 100 m water depth (Fig. 15.1) (Bosscher & Schlager 1992). Photosynthetic organisms typically flourish in the upper 10 to 20 metres of the sea and it is in this zone that the greatest abundance of calcareous organisms is found. This shallow region of high biogenic productivity is referred to as the **carbonate factory** (Tucker & Wright 1990). Increased or reduced salinity inhibits production and the optimum temperature is around 20 to 25°C. Hermatypic corals dependent on symbiotic algae are most productive in shallow clear water with strong currents, while most other benthic marine organisms prefer quieter waters.

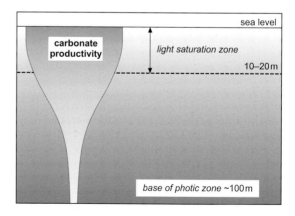

Fig. 15.1 The relationship between water depth and biogenic carbonate productivity, which is greatest in the photic zone.

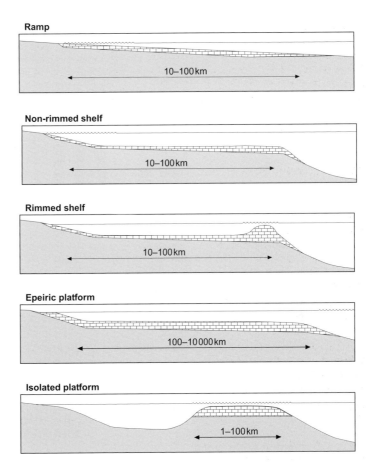

Fig. 15.2 The types of carbonate platform in shallow marine environments.

15.1.2 Controls on evaporite sedimentation

Precipitation of evaporate minerals, principally calcium sulphates (gypsum and anhydrite) and sodium chloride (halite) (*3.2*), occurs where bodies of seawater become wholly or partially isolated from the open ocean under arid conditions. The fundamental controlling factor in the formation of evaporite deposits is climate, because the seawater can become sufficiently concentrated for precipitation to occur only if the rate of loss through evaporation exceeds the input of water. These arid environments are principally found in subtropical regions where the mean annual temperatures are relatively high but the rainfall is low. Modern marine evaporite deposits are all found in coastal settings where precipitation occurs in semi-isolated water bodies such as lagoons or directly within sediments of the coastal plain, places where

recharge by seawater is limited. In the past, larger areas of evaporate precipitation resulted from the isolation from the open ocean of epicontinental seas and small ocean basins (*16.1*).

15.1.3 Morphologies of shallow marine carbonate-forming environments

The term 'carbonate platform' can be generally applied to any shallow marine environment where there is an accumulation of carbonate sediment. If the platform is attached to a continental landmass it is called a ***carbonate shelf*** (Fig. 15.2), a region of sedimentation that is analogous to shelf environments for terrigenous clastic deposition. A carbonate shelf may receive some supply of material from the adjacent landmass. ***Carbonate banks*** are isolated platforms that are

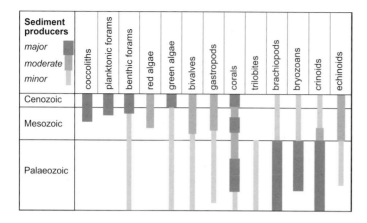

Fig. 15.3 Different groups of organisms have been important producers of carbonate sedimentary material through the Phanerozoic; limestones of different ages therefore tend to have different biogenic components.

completely surrounded by deep water and therefore do not receive any terrigenous clastic supply. A *carbonate atoll* is a particular class of carbonate bank formed above a subsiding volcanic island. Three morphologies of carbonate platform are recognised: they may be flat-topped with a sharp change in slope at the edge forming a steep margin, either as a *rimmed* or *non-rimmed shelf*, or they may have a *ramp* morphology, a gentle (typically less than 1°) slope down to deeper water with no break in slope (James 2003).

15.1.4 Carbonate grain types and assemblages

The range of types of carbonate grain is reviewed in Chapter 3. The relative abundance of the different carbonate-forming organisms has varied considerably though time (Fig. 15.3) (Walker & James 1992), so, in contrast to terrigenous clastic facies in shallow marine environments, the characteristics of shallow-marine carbonate facies depend on the time period in which they were deposited. Most significantly, the absence of abundant shelly organisms in the Pre-cambrian means that carbonate facies from this time are markedly different from Phanerozoic deposits in that they lack bioclastic components.

The skeletal grain associations that occur on carbonate platforms are temperature and salinity dependent. In low latitudes where the shallow sea is always over 15°C and the salinity is normal, corals and calcareous green algae are common and along with numerous other organisms form a *chlorozoan* assemblage. In restricted seas where the salinities are higher only green algae flourish, and form a *chloralgal* association (Lees 1975). Temperate carbonates formed in cooler waters are dominated by the remains of benthic foraminifers and molluscs, a *foramol* assemblage (Wilson & Vecsei 2005). Ooids are most commonly associated with chlorozoan and chloralgal assemblages.

15.2 COASTAL CARBONATE AND EVAPORITE ENVIRONMENTS

15.2.1 Beaches

The patterns of sedimentation along high-energy coastlines with carbonate sedimentation are very similar to those of clastic, wave-dominated coastlines (*13.3*). Carbonate material in the form of bioclastic debris and ooids is reworked by wave action into ridges that form strand plains along the coast or barrier islands separated from the shore by a lagoon (Tucker & Wright 1990; Braithwaite 2005). The texture of carbonate sediments deposited on barrier island and strand plain beaches is typically well-sorted and with a low mud matrix content (grainstone and packstone). Few organisms live in the high-energy foreshore zone, so almost all of the carbonate detritus is reworked from the shoreface. Sedimentary structures are low angle (3° to 13°) cross-stratification dipping seaward on the foreshore and landwards in the backshore area. Barrier islands formed of carbonate sediment form in

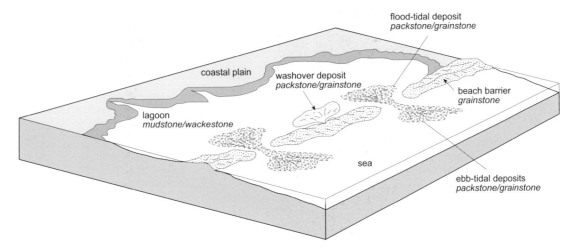

Fig. 15.4 Morphological features of a carbonate coastal environment with a barrier protecting a lagoon.

microtidal regimes, where they occur as laterally continuous barriers parallel to the shoreline. In common with barriers made up of terrigenous clastic material (*13.3.1*) they form in response to a slow rise in sea level.

An important difference between beaches made up of terrigenous clastic material and carbonate-rich beaches is the formation of **beachrock** in the latter. Carbonate in solution precipitates between sand and gravel material deposited on the beach and cement the beach sediments into fully lithified rock. Beachrock along the foreshore may act as a host for organisms that bore into the hard substrate (*11.7.2*), a feature that may make it possible to recognise early cementation of a beachrock in the stratigraphic record. A prograding strandplain or barrier island generates a coarsening-upwards succession of well-sorted, stratified grainstone and packstone. The deposits are typically associated with lagoonal, supratidal and inner shelf/ramp facies.

At the top of the beach sands composed of bioclastic and other carbonate detritus may be reworked by wind to form aeolian dunes (*8.4.2*). When these dune sands become wet calcium carbonate is locally dissolved and reprecipitated to cement the material at the surface into a rock, which is often referred to as an **aeolianite** (Tucker & Wright 1990). Carbonate also precipitates around the roots of vegetation growing in the dune sands and may be preserved as nodular rhizocretions (*9.7.2*) (McKee & Ward 1983).

15.2.2 Beach barrier lagoons

Lagoons form along carbonate coastlines where a beach barrier wholly or partly encloses an area of shallow water (Fig. 15.4). The character of the lagoon deposits depends on the salinity of the water and this in turn is determined by two factors: the degree of connection with the open ocean and the aridity of the climate.

Carbonate lagoons

Carbonate lagoons are sites of fine-grained sedimentation forming layers of carbonate mudstone and wackestone with some grainstone and packstone beds deposited as washovers near the beach barrier. Where a barrier island ridge is cut by tidal channels in a mesotidal regime, the tidal currents passing through form flood- and ebb-tidal deltas in much the same way as in clastic barrier island systems (*13.3*). The shape and internal sedimentary structures of these deposits are also similar on both clastic and carbonate coastlines, with lenses of cross-bedded oolitic and bioclastic packstone and grainstone formed by subaqueous dunes on flood-tidal deltas. The nature of the carbonate material deposited on ebb- and flood-tidal deltas depends on the type of material being generated in the shallow marine waters: it may be bioclastic debris or oolitic sediment forming beds of grainstone and packstone (Fig. 15.4).

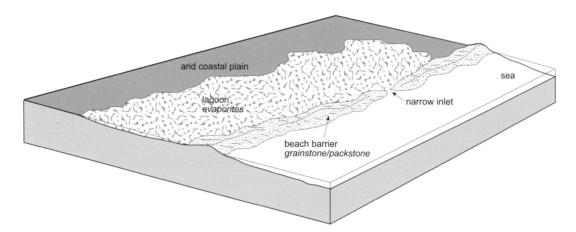

Fig. 15.5 A carbonate-dominated coast with a barrier island in an arid climatic setting: evaporation in the protected lagoon results in increased salinity and the precipitation of evaporite minerals in the lagoon.

The source of the fine-grained carbonate sediment in lagoons is largely calcareous algae living in the lagoon, with coarser bioclastic detritus from molluscs. Pellets formed by molluscs and crustaceans are abundant in lagoon sediments. The nature and diversity of the plant and animal communities in a carbonate lagoon is determined by the salinity. Lagoons in mesotidal coastlines tend to have better exchange of seawater through tidal channels than more isolated lagoons in microtidal regimes. Where the climate is relatively humid evaporation is lower, and as the lagoon has near-normal salinities a diverse marine fauna is present. In more arid regions the lagoon may become hypersaline and there will be a restricted fauna, with organisms such as stromatolites and marine grasses (*Thalassia*) abundant.

Arid lagoons

In hot, dry climates the loss of water by evaporation from the surface of a lagoon is high. If it is not balanced by influx of fresh water from the land or exchange of water with the ocean the salinity of the lagoon will rise and it will become hypersaline (*10.3*), more concentrated in salts than normal seawater (Fig. 15.5). An area of hypersaline shallow water that precipitates evaporite minerals is known as a *saltern*. Deposits are typically layered gypsum and/ or halite occurring in units metres to tens of metres thick. In the restricted circulation of a lagoon conditions are right for large crystals of selenitic gypsum

(*3.2.1*) to form by growing upwards from the lagoon bed (Warren & Kendall 1985). Connection with the ocean may be via gaps in the barrier or by seepage through it. Variations in the salinity within the lagoon may be because of climatically related changes in the freshwater influx from the land or increased exchange with open seawater during periods of higher sea level. The extent of the lagoon and the minerals precipitated in it are therefore likely to be variable, resulting in cycles of sedimentation, including layers of carbonate deposited during periods when the salinity was closer to normal marine values. An alternation between laminated gypsum deposited subaqueously in a lagoon and nodular gypsum formed in a supratidal sabkha (see below) around the edges of the water body may represent fluctuations in the area of the water body.

15.2.3 Supratidal carbonates and evaporites

Supratidal carbonate flats

The **supratidal zone** lies above the mean high water mark and is only inundated by seawater under exceptional circumstances, such as very high tides and storm conditions. Where the gradient to the shoreline is very low the supratidal zone is a marshy area where microbial (algal and bacterial) mats form (Fig. 15.6). Aeolian action may also bring in carbonate sand and dust that is bound by the microbes and, as

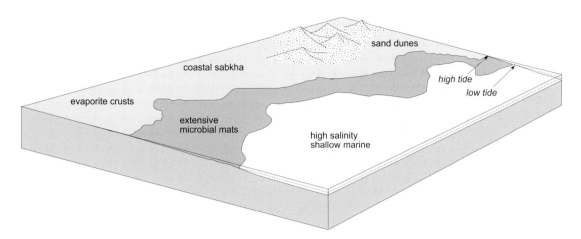

Fig. 15.6 In arid coastal settings a sabkha environment may develop. Evaporation in the supratidal zone results in saline water being drawn up through the coastal sediments and the precipitation of evaporite minerals within and on the sediment surface.

syndepositional cementation (*18.2.2*) occurs, a hard **carbonate pavement** is formed. Desiccation breaks up the pavement, but the broken pieces of crust are reincorporated into the sediment again as further cementation occurs. The fabric created shows some primary lamination, but also appears to be brecciated (note, however, that these breccias form *in situ* and do not involve transport of the clasts).

Arid sabkha flats

Arid shorelines are found today in places such as the Arabian Gulf, where they are sites of evaporite formation within the coastal sediments. These arid coasts are called **sabkhas** (Kendall 1992); they typically have a very low relief and there is not always a well-defined beach (Fig. 15.6). The coastal plain of a sabkha is occasionally wetted by seawater during very high tides or during onshore storm winds, but more important is also a supply of water through groundwater seepage from the sea (Yechieli & Wood 2002). The surface of the coastal plain is an area of evaporation and water is drawn up through the sediment to the surface. As the water rises it becomes more concentrated in salts that precipitate within the coastal plain sediments, and a dense, highly concentrated brine is formed. Gypsum and anhydrite grow within the sediment while a crust of halite forms at the surface.

In general, anhydrite forms in the hotter, drier sabkhas and gypsum where the temperatures are lower or where there is a supply of fresh, continental water to the sabkha. Both gypsum and anhydrite are formed in some sabkhas: close to the shore in the intertidal and near-supratidal zone gypsum crystals grow in the relatively high flux of seawater through the sediment, whereas further up in the supratidal area conditions are drier and nodules of anhydrite form in the sediment. The gypsum and anhydrite grow by displacement within the sediment, with the gypsum in clusters and the anhydrite forming amorphous coalesced nodules with little original sediment in between. These layers of anhydrite with remnants of other sediment have a characteristic **chicken-wire structure**. Halite crusts are rarely preserved because they are removed by any surface water flows. The terrigenous sediment of the sabkha is often strongly reddened by the oxidising conditions.

The succession formed by sedimentation along an arid coast starts with beds deposited in a wave-reworked shallow subtidal setting and overlain by intertidal microbial limestone beds. Gypsum formed in the upper intertidal and lower supratidal area occurs next in the succession, overlain by anhydrite with a chicken-wire structure. Coalesced beds of anhydrite formed in the uppermost part of the sabkha form layers, contorted as the minerals have grown, known as an **enterolithic** bedding structure. This

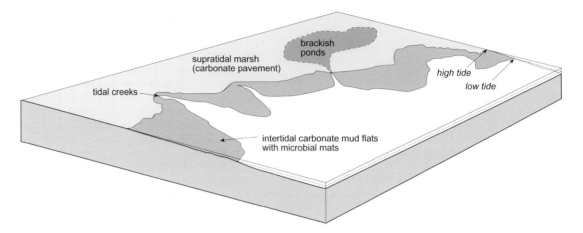

Fig. 15.7 Tide-influenced coastal carbonate environments.

cycle may be repeated many times if there is continued subsidence along the coastal plain (Kendall & Harwood 1996). The displacive growth of the gypsum within the sediment is a distinctive feature of sabkhas, which allows the deposits of these arid coasts to be distinguished from other marine evaporite deposits (15.5). Similar evaporite growth occurs within continental sediments in arid regions (10.3).

15.2.4 Intertidal carbonate deposits

Tidal currents along carbonate-dominated coastlines transport and deposit coarse sediment in tidal channels and finer carbonate mud on tidal flats. The tidal channel sediments are similar in character to those found in tidal channels in clastic estuarine deposits (13.6). The base of the channel succession is marked by an erosive base overlain by a lag of coarse debris: this may consist of broken shells and intraclasts of lithified carbonate sediment. Carbonate sands deposited on migrating bars in the tidal channels form cross-bedded grainstone and packstone beds (Pratt et al. 1992). As a channel migrates or is abandoned the sands are overlain by finer sediment, forming beds of carbonate mudstone and wackestone. Bioturbation is normally common throughout.

In the intertidal zones deposits of lime mud and shelly mud are subject to subaerial desiccation at low tide (Fig. 15.7). Terrigenous clastic mud remains relatively wet when exposed between tidal cycles, but carbonate mud in warm climates tends to dry out and form a crust by syndepositional cementation. Repeated precipitation of cements in this crust causes the surface layer to expand and form a polygonal pattern of ridges, called **tepee structures** or **pseudoanticlines**, a few tens of centimetres across. As the pseudoanticlines grow they leave cavities beneath them, which are sites for the growth of sparry calcite cements. Smaller isolated cavities and vertically elongate hollow tubes also form in lime muds in intertidal areas due to air and water being trapped in the sediment during the wetting and drying process. Patches of calcite cement that grow in the cavities in the host of lime mud give rise to a fabric generally called **fenestrae** or **fenestral cavities**. Vertical fenestrae may also result from roots and burrows. Lime mudstones with small cavities filled with calcite are sometimes called a '**birds-eye limestone**'. The lithified crusts can be reworked by storms and redeposited elsewhere.

A common feature of carbonate tidal flats is the formation of algal and bacterial mats, which trap fine-grained sediment in thin layers to form the well-developed, fine lamination of a stromatolite (3.1.3). Stromatolites may form horizontal layers or irregular mounds on the tidal flats. Their distribution is partly controlled by the activity of organisms, which either feed on the microbial mats or disrupt it by bioturbation. Stromatolites tend to be better developed in the higher parts of the intertidal area that are less favourable for other organisms that may graze on the mats.

15.3 SHALLOW MARINE CARBONATE ENVIRONMENTS

The character of deposits in shallow marine carbonate environments is determined by the types of organisms present and the energy from waves and tidal currents. The sources of the carbonate material are predominantly biogenic, including mud from algae and bacteria, sand-sized bioclasts, ooids and peloids and gravelly debris that is skeletal or formed from intraclasts. Bioturbation is usually very common and faecal pellets contribute to the sediment. A number of different carbonate deposits are characteristic of many shallow marine environments, for example shoals of sand-sized material, reefs and mud mounds.

15.3.1 Carbonate sand shoals

Sediment composed of sand to granule-sized, loose carbonate material occurs in shallow, high energy areas. These **carbonate shoals** may be made up of ooids (*3.1.4*), mixtures of broken shelly debris or may be accumulations of benthic foraminifers (*3.1.3*). Reworking by wave and tidal currents results in deposits made up of well-sorted, well-rounded material: when lithified these form beds of grainstone, or sometimes packstone. Sedimentary structures may be similar to those found in sand bodies on clastic shelves, including planar and trough cross-bedding generated by the migration of subaqueous dune bedforms. However, the degree of reworking is often limited by early carbonate cementation (*18.2.2*). Extensive wave action tends to build up shoals that form banks parallel to the coastline, whereas tidal currents in coastal regions result in bodies of sediment elongated perpendicular to the shoreline.

15.3.2 Reefs

Reefs are carbonate bodies built up mainly by framework-building benthic organisms such as corals (Figs 15.8 & 15.9) (Kiessling 2003). They are wave-resistant structures that form in shallow waters on carbonate platforms. The term 'reef' is used by mariners to indicate shallow rocky areas at sea, but in geological terms they are exclusively biological features. Reef build-ups are sometimes referred to as **bioherms**: carbonate build-ups that do not form dome-shaped reefs but are instead tabular forms known as **biostromes**.

Reef-forming organisms

Scleractinian corals are the main reef builders in modern oceans, as they have been for much of the Mesozoic and Cenozoic (Fig. 15.10), These corals are successful because many of the taxa are hermatypic,

Fig. 15.8 Modern coral atolls.

Fig. 15.9 Modern corals in a fringing reef. The hard parts of the coral and other organisms form a boundstone deposit.

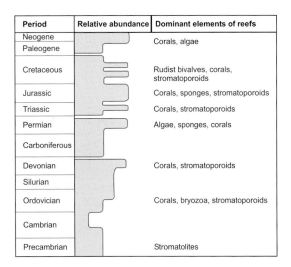

Period	Relative abundance	Dominant elements of reefs
Neogene		Corals, algae
Paleogene		
Cretaceous		Rudist bivalves, corals, stromatoporoids
Jurassic		Corals, sponges, stromatoporoids
Triassic		Corals, stromatoporoids
Permian		Algae, sponges, corals
Carboniferous		
Devonian		Corals, stromatoporoids
Silurian		
Ordovician		Corals, bryozoa, stromatoporoids
Cambrian		
Precambrian		Stromatolites

Fig. 15.10 The type and abundance of carbonate reefs has varied through the Phanerozoic (data from Tucker, 1992).

that is, they have a symbiotic relationship with algae, which allows the corals to grow rapidly in relatively nutrient-poor water. The other main modern reefs builders are calcareous algae. However, over the past 2500 Myr a number of different types of organisms have performed this role (Tucker 1992). The earliest reef-builders were cyanobacteria, which created stromatolites, followed in the Palaeozoic by rugose and tabulate corals and calcareous sponges (including stromatoporoids, which were particularly important in the Devonian – Fig. 15.11). The most

unusual reef-forming organism was a type of bivalve, the **rudists**: the shells of these molluscs were thick and conical, forming massive colonies, which are characteristic of many Cretaceous reefs (Ross & Skelton 1993). Not only has the type of organism forming reefs varied through time, but also the relative importance of reefs as depositional systems has changed, with four peaks of dominance in the Phanerozoic (Fig. 15.10) separated by times when mud mounds were the more common bioherms.

Reef structures

Modern reefs can be divided into a number of distinct subenvironments (Fig. 15.12). The reef crest is the site of growth of the corals that build the most robust structures, encrusting and massive forms capable of withstanding the force of waves in very shallow water. Going down the reef front these massive and encrusting forms of coral are replaced by branching and more delicate plate-like forms in the lower energy, deeper water. Behind the reef crest is a reef flat, also comprising relatively robust forms, but conditions become quieter close to the back-reef area and globular coral forms are common in this region (Tucker & Wright 1990; Wright & Burchette 1996).

In addition to the main reef builders that form the framework, other organisms play an important role too: encrusting organisms such as bryozoa and calcareous algae also help to stabilise the framework and the remains of a wide variety of organisms that live within the reef provide additional mass to it. There are also many organisms that

Fig. 15.11 The core of a Devonian reef flanked by steeply dipping forereef deposits on the right-hand side of the exposure.

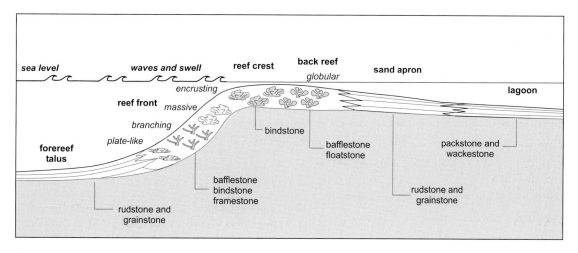

Fig. 15.12 Facies distribution in a reef complex.

remove mass from the reef structure, a process of **bioerosion** carried out by some types of fish and molluscs that bore into the reef. The voids between the framework structures may be filled with the remains of organisms, debris formed by the mechanical breakdown and bioerosion of the framework and by carbonate mud. If burial occurs before the voids have been filled with sediment, crystalline calcite cement may subsequently be precipitated during diagenesis (*18.2*).

Break-up of the reef core material by wave and storm action leads to the formation of a talus slope of reefal debris. This **forereef** setting is a region of accumulation of carbonate breccia to form bioclastic rudstone and grainstone facies. As these are gravity deposits formed by material falling down from the reef crest they build out as steeply sloping depositional units inclined at 10° to 30° to the horizontal. Behind the reef crest the **back reef** is sheltered from the highest energy conditions and is the site of deposition of debris removed from the reef core and washed towards the lagoon. A gradation from rudstone to grainstone deposits of broken reef material, shells and occasionally ooids forms a fringe along the margin of the lagoon.

Reef settings

Three main forms of reef have been recognised in modern oceans, and in fact were recognised by Charles Darwin in the middle of the 19th century

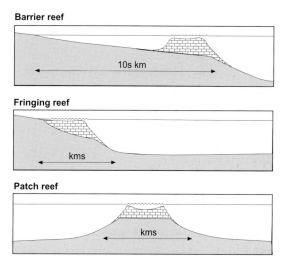

Fig. 15.13 Reefs can be recognised as occurring in three settings: (a) barrier reefs form offshore on the shelf and protect a lagoon behind them, (b) fringing reefs build at the coastline and (c) patch reefs or atolls are found isolated offshore, for instance on a seamount.

(Fig. 15.13). **Fringing reefs** are built out directly from the shoreline and lack an extensive back-reef lagoonal area. **Barrier reefs**, of which the Great Barrier Reef of eastern Australia is a distinctive example, are linear reef forms that parallel the shoreline, but lie at a distance of kilometres to tens of kilometres offshore: they create a back-reef lagoon area which is a large area of shallow, low-energy sea, which is itself an important ecosystem and depositional setting. In open ocean areas *coral atolls* develop on localised areas of shallow water, such as seamounts, which are the submerged remains of volcanic islands. In addition to these settings of reef formation, evidence from the stratigraphic record indicates that there are many examples of *patch reefs*, localised build-ups in shallow water areas such as epicontinental seas, carbonate platforms and lagoons.

Reefs as palaeoenvironmental indicators

Present-day reefs are mainly in tropical seas, occurring up to 35° latitude either side of the Equator. It is therefore tempting to apply this observation to the sedimentary record and conclude that if a carbonate reef body is found it indicates an environment of deposition in warm tropical waters. This assumption can be made only with certain caveats. First, other reef-builders live in different environments: in modern seas coralline algae build reefs at higher latitudes and the environmental tolerances of extinct taxa are not fully known. Second, even if the reef is made of corals, then it must be remembered that pre-Mesozoic groups, the Rugosa and Tabulata, may not have had the symbiotic relationship with algae that is such a distinctive feature of the Mesozoic to present-day Scleractinia corals, and their distribution in the seas was more controlled by the availability of nutrients.

Cessation of reef development

The growth of coral reefs can normally keep pace with both tectonic subsidence and global sea-level rise. The cessation of reef development is therefore usually due to changes in environmental conditions, such as an increase in the flux of terrigenous clastic material or a change in the nutrient supply. When this occurs the dominant facies formed is fine-grained pelagic material, which is similar in character to deep-sea pelagic carbonates (*16.5.1*). Pelagic carbonate sedimentation is considerably slower than shallow-marine accumulation rates resulting in much thinner layers in a given period of time. Successions deposited under these conditions are known as *condensed sections* and they may have as many millions of years of accumulation in them as a shallow-water deposit

two or three orders of magnitude thicker (Bernoulli & Jenkyns 1974).

15.3.3 Carbonate mud mounds

A *carbonate mud mound* is a sediment body consisting of structureless or crudely bedded fine crystalline carbonate. Modern examples of carbonate mud mounds are rare, so much less is known about the controls on their formation than is the case for reefs. From studies of mounds of fine-grained carbonate in the rock record (e.g. Monty et al. 1995) it appears that there are two, possibly three types. Many mounds are made of the remains of microbes that had calcareous structures and these microbes grew in place to build up the body of sediment. Others have a large component of detrital material, again mainly the remains of algae and bacteria, which have been piled up into a mound of loose material. It is also possible that some skeletal organisms such as calcareous sponges and bryozoans are responsible for building carbonate mud mounds. They appear to form in deeper parts of the shelf than reefs, but within the photic zone. Cementation of the mud requires circulation of large amounts of water rich in calcium carbonate, a process that is not well understood.

15.3.4 Outer shelf and ramp carbonates

On the outer parts of shelves carbonate sedimentation is dominated by fine-grained deposits. These carbonate mudstones are composed of the calcareous remains of planktonic algae (*3.1.3*) and other fine-grained biogenic carbonate. This facies is found in both modern and ancient outer platform settings and when lithified the fine-grained carbonate sediment is called *chalk*. Similar facies also occur in deeper water settings (*16.5.1*). Chalk deposited in shallower water may contain the shelly remains of benthic and planktonic organisms and there is extensive evidence of bioturbation in some units (Ekdale & Bromley 1991). Chert nodules within the beds are common in places, the result of the redistribution of silica from the skeletons of siliceous organisms. Bedding is picked out by slight variations in the proportions of clay minerals, which occur in most chalk deposits, or by variations in the degree of cementation. Deposits of this type may be found in strata of various ages, particularly

Fig. 15.14 Cliffs of Cretaceous Chalk.

from the Mesozoic and Cenozoic, but are most commonly found in the Late Cretaceous in the northern hemisphere in a stratigraphic unit which is called *The Chalk* (capitalised) (Fig. 15.14).

15.3.5 Platform margins and slopes

The edge of a carbonate platform may be marked by an abrupt change in slope or there may be a lower angle transition to deeper water facies. The front of a reef can form a vertical 'wall' and along with other slopes too steep for sediment accumulation are *by-pass margins*. Sediment accumulates at the base of the slope, brought in by processes ranging from large blocks fallen from the reef front to submarine talus slopes, slumps, debris flows and turbidites (Mullins & Cook 1986). The most proximal material forms rudstone deposits, which are sometimes called megabreccias if they contain very large blocks, passing distally to redeposited packstones, to turbiditic wackestones and mudstones. *Depositional margins* form on more gentle slopes with a continuous spectrum of sediments from the reef boundstones or shoal grainstones of the shelf margin to packstones, wackestones and mudstones further down the slope. Finer grained sediments tend to be unstable on slopes and slumping of the mudstones and wackestones may occur, resulting in contorted, redeposited beds.

15.4 TYPES OF CARBONATE PLATFORM

A number of different morphologies of carbonate platform are recognised (Fig. 15.15), the most widely documented being *carbonate ramps*, which are gently sloping platforms, and *rimmed shelves*, which are flat-topped platforms bordered by a rim formed by a reef or carbonate sand shoal. The tectonic setting influences the characteristics of carbonate platforms (Bosence 2005), with the largest occurring on passive continental margins (*24.2.4*) while smaller platforms form on localised submarine highs such as fault blocks in extensional settings (*24.2*) and on salt diapirs (*18.1.4*). The different types of carbonate platform can sometimes occur associated with each other: an isolated platform may be a carbonate ramp on one side and a rimmed shelf on the other and one form may evolve into another, for example, a ramp may evolve into a rimmed shelf as a fringing reef develops (Bosence 2005).

15.4.1 Carbonate ramps

The bathymetric profile of a carbonate ramp (Fig. 15.15) and the physical processes within the sea and on the sea floor are very similar to an open shelf with clastic deposition. The term 'ramp' may give the impression of a significant slope but in fact the slope is a gentle one of less than a degree in most instances (Wright & Burchette 1996), in contrast to slope environments associated with rimmed shelves, which are much steeper. Modern ramps are in places where reefs are not developed, such as regions of cooler waters, increased salinity or relatively high input of terrigenous clastic material. However, in the past carbonate ramps formed in a wider range of climatic and environmental settings, especially during periods when reef development was not so widespread. In macro- to mesotidal regimes tidal currents distribute carbonate sediment and strongly influence the coastal facies. Wave and storm processes are dominant in microtidal shelves and seas. The effects of tides, waves and storms are all depth-dependent and ramps can be divided into three depth-related zones: inner, mid- and outer ramp.

Distribution of facies on a carbonate ramp

The *inner ramp* is the shallow zone that is most affected by wave and/or tidal action. Coastal facies along tidally influenced shorelines are characterised by deposition of coarser material in channels and carbonate muds on tidal flats (Tucker & Wright

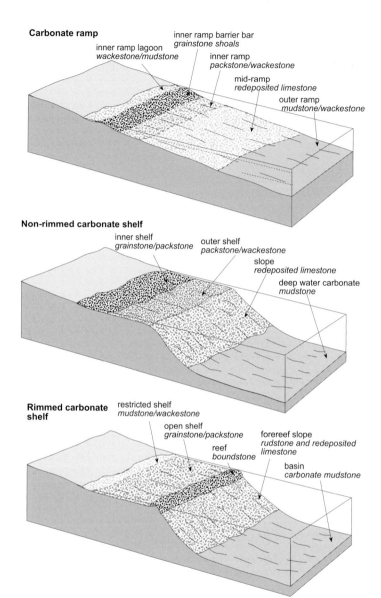

Fig. 15.15 Generalised facies distributions on carbonate platforms: (a) ramps, (b) non-rimmed shelves and (c) rimmed shelves.

1990; Jones & Desrochers 1992). Wave-dominated shorelines may have a beach ridge that confines a lagoon or a linear strand plain attached to the coastal plain. Ramps with mesotidal regimes will show a mixture of beach barrier, tidal inlet, lagoon and tidal-flat deposition. Agitation of carbonate sediment in shallow nearshore water results in a shoreface facies of carbonate sand bodies. Skeletal debris and ooids formed in the

shallow water form bioclastic and oolitic carbonate sand shoals. Benthic foraminifers are the principal components of some Tertiary carbonate ramp successions.

The **_mid-ramp_** area lies below fair-weather wave base and the extent of reworking by shallow-marine processes is reduced. Storm processes transport bioclastic debris out on to the shelf to form deposits of wackestone and packstone, which may include hummocky and swaley cross-stratification (*14.2.1*). In deeper water below storm wave base the **outer ramp** deposits are principally redeposited carbonate mudstone and wackestone, often with the characteristics of turbidites. Redeposition of carbonate sediments is common in situations where the outer edge of the ramp merges into a steeper slope at a continental margin as a **distally steepened ramp**. **Homoclinal ramps** have a consistent gentle slope on which little reworking of material by mass-flow processes occurs (Read 1985). In contrast to rimmed shelves reefal build-ups are relatively rare in ramp settings. Isolated patch reefs may occur in the more proximal parts of a ramp and mud mounds are known from Palaeozoic ramp environments.

Carbonate ramp succession

A succession built up by the progradation of a carbonate ramp is characterised by an overall coarsening-up from carbonate mudstone and wackestone deposited in the outer ramp environment to wackestones and packstones of the mid-ramp to packstone and grainstone beds of the inner ramp (Fig. 15.16) (Wright 1986). The degree of sorting typically increases upwards, reflecting the higher energy conditions in shallow water. Inner ramp carbonate sand deposits are typically oolitic and bioclastic grainstone beds that exhibit decimetre to metre-scale cross-bedding and horizontal stratification. The top of the succession may include fine-grained tidal flat and lagoonal sediments. Ooids, broken shelly debris, algal material and benthic foraminifers may all be components of ramp carbonates. Locally mud mounds and patch reefs may occur within carbonate ramp successions.

On shelves and epicontinental seas where there are fluctuations in relative sea level, cycles of carbonate deposits are formed on a carbonate ramp. A sea-level rise results in a shallowing-up cycle a few metres to tens of metres thick that coarsens up from beds of mudstone and wackestone to grainstone and packstone. A fall in sea level may expose the inner ramp deposits to dissolution in karstic subaerial weathering (*6.6.3*) (Emery & Myers 1996).

15.4.2 Non-rimmed carbonate shelves

Non-rimmed carbonate shelves are flat-topped shallow marine platforms that are more-or-less

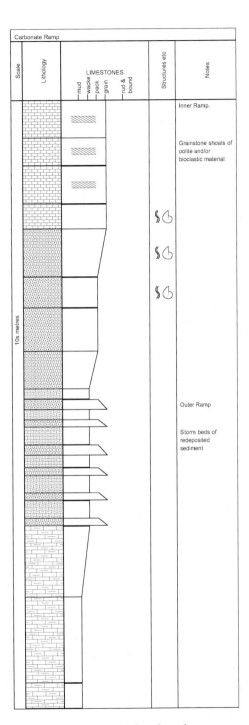

Fig. 15.16 Schematic graphic log of a carbonate ramp succession.

horizontal (Fig. 15.16), in contrast to the gently dipping morphology of a carbonate ramp. They lack any barrier at the outer margin of the shelf (cf. rimmed shelves) and as a consequence the shallow waters are exposed to the full force of oceanic conditions. These are therefore high-energy environments where carbonate sediments are repeatedly reworked by wave action in the inner part of the shelf and where redeposition by storms affects the outer shelf area (Fig. 15.17) (James 2003). They therefore resemble storm-dominated clastic shelves (14.2), but the deposits are predominantly carbonate grains. Extensive reworking in shallow waters may result in grainstones and packstones, whereas wackestones and mudstones are likely to occur in the outer shelf area. Coastal facies are typically low energy tidal-flat deposits but a beach barrier may develop if the wave energy is high enough.

15.4.3 Rimmed carbonate shelves

A *rimmed carbonate shelf* is a flat-topped platform that has a rim of reefs or carbonate sand shoals along the seaward margin (Fig. 15.16). The reef or shoal forms a barrier that absorbs most of the wave energy from the open ocean. Modern examples of rimmed shelves all have a coral reef barrier because of the relative abundance of hermatypic scleractinian corals in the modern oceans. Landward of the barrier lies a low-energy shallow platform or shelf lagoon that is sheltered from the open ocean and may be from a few kilometres to hundreds of kilometres wide and vary in depth from a few metres to several tens of metres deep.

Distribution of facies on a carbonate rimmed shelf

In cases where the barrier is a reef, the edge of the shelf is made up of an association of reef-core, fore-reef and back-reef facies (15.3.2): the reef itself forms a bioherm hundreds of metres to kilometres across. Sand shoals may be of similar extent where they form the shelf-rim barrier. Progradation of a barrier results in steepening of the slope at the edge of the shelf and the slope facies are dominated by redeposited material in the form of debris flows in the upper part and turbidites on the lower part of the slope. These pass laterally into pelagic deposits of the deep

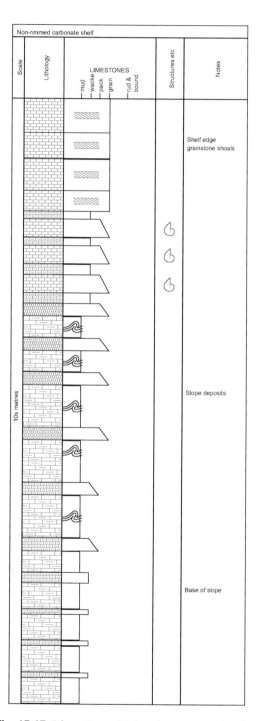

Fig. 15.17 Schematic graphic log of a non-rimmed carbonate shelf succession.

basin. The back-reef facies near to the barrier may experience relatively high wave energy resulting in the formation of grainstones of carbonate sand and skeletal debris reworked from the reef. Further inshore the energy is lower and the deposits are mainly wackestones and mudstones. However, ooidal and peloidal complexes may also occur in the shelf lagoon and patch reefs can also form. In inner shelf areas with very limited circulation and under conditions of raised salinities the fauna tends to be very restricted. In arid regions evaporite precipitation may become prominent in the shelf lagoon if the barrier provides an effective restriction to the circulation of seawater.

Rimmed carbonate shelf successions

As deposition occurs on the rimmed shelf under conditions of static or slowly rising sea level the whole complex progrades. The reef core builds out over the fore reef and back-reef to lagoon facies overlie the reef bioherm (Fig. 15.18). Distally the slope deposits of the fore reef prograde over deeper water facies comprising pelagic carbonate mud and calcareous turbidite deposits. The steep depositional slope of the fore reef creates a clinoform bedding geometry, which may be seen in exposures of rimmed shelf carbonates. This distinctive geometry can also be recognised in seismic reflection profiles of the subsurface (22.2) (Emery & Myers 1996). The association of reef-core boundstone facies overlying fore-reef rudstone deposits and overlain by finer grained sediments of the shelf lagoon forms a distinctive facies association. Under conditions of sea-level fall the reef core may be subaerially exposed and develop karstic weathering, and a distinctive surface showing evidence of erosion and solution may be preserved in the stratigraphic succession if subsequent sea-level rise results in further carbonate deposition on top (Bosence & Wilson 2003).

15.4.4 Epicontinental (epeiric) platforms

There are no modern examples of large epicontinental seas dominated by carbonate sedimentation but facies distributions in limestones in the stratigraphic record indicate that such conditions have existed in the past, particularly during the Jurassic and Cretaceous when large parts of the

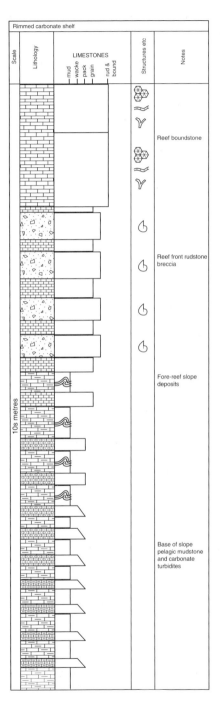

Fig. 15.18 Schematic graphic log of a rimmed carbonate shelf succession.

continents were covered by shallow seas (Tucker & Wright 1990). The water depth across an epicontinental platform would be expected to be variable up to a few tens to hundreds of metres. Both tidal and storm processes may be expected, with the latter more significant on platforms with small tidal ranges. Currents in broad shallow seas would build shoals of oolitic and bioclastic debris that may become stabilised into low-relief islands. Deposition in intertidal zones around these islands and the margins of the sea would result in the progradation of tidal flats. The facies successions developed in these settings would therefore be cycles displaying a shallowing-up trend, which may be traceable over large areas of the platform.

15.4.5 Carbonate banks and atolls

Isolated platforms in areas of shallow sea surrounded on all sides by deeper water are commonly sites of carbonate sedimentation because there is no source of terrigenous detritus. They are found in a number of different settings ranging from small atolls above extinct volcanoes to horst blocks in extensional basins and within larger areas of shallow seas (Wright & Burchette 1996; Bosence 2005). All sides are exposed to open seas and the distribution of facies on an isolated platform is controlled by the direction of the prevailing wind. The characteristics of the deposits resemble those of a rimmed shelf and result in similar facies associations. The best developed marginal reef facies occurs on the windward side of the platform, which experiences the highest energy waves. Carbonate sand bodies may also form part of the rim of the platform. The platform interior is a region of low energy where islands of carbonate sand may develop and deposition occurs on tidal flats.

15.5 MARINE EVAPORITES

Evaporite deposits in modern marine environments are largely restricted to coastal regions, such as evaporite lagoons and sabkha mudflats (*15.2.2 & 15.2.3*). However, evaporite successions in the stratigraphic record indicate that precipitation of evaporite minerals has at times occurred in more extensive marine settings.

15.5.1 Platform evaporites

In arid regions the restriction of the circulation on the inner ramp/shelf can lead to the formation of extensive platform evaporites. On a gently sloping ramp a sand shoal can partially isolate a zone of very shallow water that may be an area of evaporite precipitation; the subtidal zone here often merges into a low-energy mudflat coastline. Shelf lagoons behind rims formed by reefs or sand shoals can create similar areas of evaporite deposition, although the barrier formed by a reef usually allows too much water circulation. Evaporite units deposited on these platforms can be tens of kilometres across (Warren 1999).

15.5.2 Evaporitic basins (saline giants)

Evaporite sedimentation occurs only in situations where a body of water becomes partly isolated from the ocean realm and salinity increases to supersaturation point and there is chemical precipitation of minerals. This can occur in epicontinental seas or small ocean basins that are connected to the open ocean by a strait that may become blocked by a fall in sea level or by tectonic uplift of a barrier such as a fault block. These are called **barred basins** and they are distinguished from lagoons in that they are basins capable of accumulating hundreds of metres of evaporite sediment. To produce just a metre bed of halite a column of seawater over 75 m deep must be evaporated, and to generate thick succession of evaporite minerals the seawater must be repeatedly replenished (Warren 1999).

Deposition of the thick succession can be produced in three ways (Warren 1999) each of which will produce characteristic patterns of deposits (Fig. 15.19).

1 A shallow-water to deep-basin setting exists where a basin is well below sea level but is only partly filled with evaporating seawater, which is periodically replenished. The deep-water setting will be evident if the basin subsequently fills with seawater and the deposits overlying the evaporites show deep marine characteristics such as turbidites.

2 A shallow-water to shallow-basin setting is one in which evaporites are deposited in salterns but continued subsidence of the basin allows a thick succession to be built up. The deposits will show the characteristics of shallow-water deposition throughout.

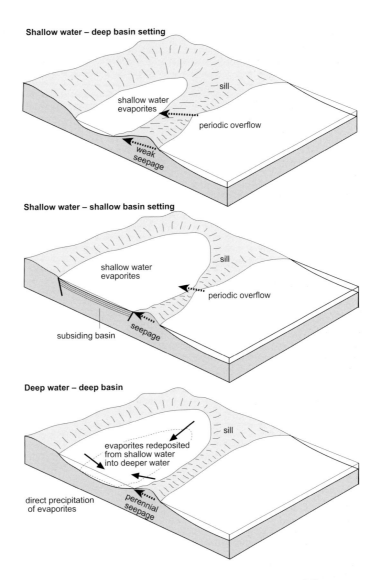

Fig. 15.19 Settings where barred basins can result in thick successions of evaporites.

3 A deep-water to deep-basin setting is a basin filled with hypersaline water in which evaporite sediments are formed at the shallow margins and are redeposited by gravity flows into deeper parts of the basin. Normally graded beds generated by turbidites and poorly sorted deposits resulting from debris flows are evidence of redeposition. Other deep-water facies are laminated deposits produced by settling of crystals of evaporite minerals out of the water body. As a basin fills up, the lower part of the succession will be deeper water facies and the overlying succession will show characteristics of shallow-water deposition.

Deep-basin succession can show two different patterns of deposition (Einsele 2000). If the barred basin is completely enclosed the water body will gradually shrink in volume and area and the deposits that result will show a ***bulls-eye pattern*** with the most soluble salts in the basin centre (Fig. 15.20). In circumstances where there is a more permanent connection a gradient of increasing salinity from the connection with the ocean to the furthest point into the basin will exist. The minerals precipitated at any point across the basin will depend on the salinity at the point and may range from highly soluble sylvite (potassium

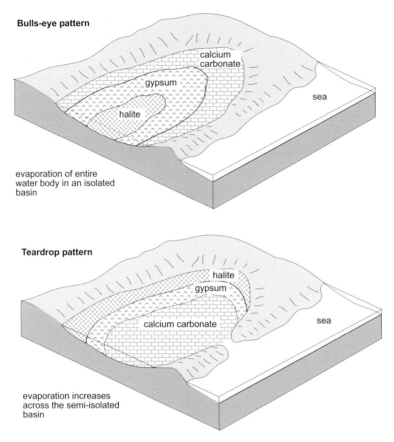

Fig. 15.20 (a) A barred basin, 'bulls-eye' pattern model of evaporite deposition; (b) a barred basin 'teardrop' pattern model of evaporite deposition.

chloride) at one extreme to carbonates deposited in normal salinities at the other. If equilibrium is reached between the inflow and the evaporative loss then stable conditions will exist across the basin and tens to hundreds of metres of a single mineral can be deposited in one place. This produces a ***teardrop pattern*** of evaporite basin facies (Fig. 15.20).

Changes in the salinity and amount of seawater in the basins result in variations in the types of evaporite minerals deposited. For example, a global sea-level rise will reduce the salinity in the basin and may lead to widespread carbonate deposition. Cycles in the deposits of barred basins may be related to global sea-level fluctuations or possibly due to local tectonics affecting the width and depth of the seaway connection to the open ocean. Organic material brought into the basin during

periods of lower salinity can accumulate within the basin deposits and be preserved when the salinity increases because hypersaline basins are anoxic.

There are no modern examples of very large, barred evaporitic basins but evidence for seas precipitating evaporite minerals over hundreds of thousands of square kilometres exist in the geological record (e.g. Nurmi & Friedman 1977; Taylor 1990). These ***saline giants*** have over 1000 m thickness of evaporite sediments in them and represent the products of the evaporation of vast quantities of seawater. Evaporite deposits of latest Miocene (Messinian) age in the Mediterranean Sea are evidence of evaporative conditions produced by partial closure of the connection to the Atlantic Ocean. This period of hypersaline conditions in the Mediterranean is sometimes referred to as the ***Messinian salinity crisis*** (Hsü 1972).

15.6 MIXED CARBONATE–CLASTIC ENVIRONMENTS

The depositional environments described in this chapter are made up of 'pure' carbonate and evaporite deposits that do not contain terrigenous clastic or volcaniclastic material. There are, however, modern environments where the sediments are mixtures of carbonate and other clastic materials, and in the stratigraphic record many successions consist of mixtures of limestones, sandstones and mudstones. These typically occur in shallow-marine settings. The changes from carbonate to non-carbonate deposition and vice versa are the result of variations in the supply of terrigenous clastic material and this is in turn determined by tectonic or climatic factors, or fluctuations in sea level.

Climate plays an important role in determining the supply of sand and mud to shallow marine environments. Under more humid conditions, the increased run-off on the land surface results in more sediment being carried by rivers, which are themselves more vigorous and hence deliver more sediment to the adjacent seas. A change to a wetter climate on an adjacent landmass will therefore result in increased deposition of sand and mud, which will suppress carbonate production on a shelf. Alternation of beds of limestone with beds of mudstone or sandstone may therefore be due to periodic climatic fluctuations of alternating drier and wetter conditions. However, other mechanisms can also cause fluctuations in the supply of detritus from the continent to parts of the shelf. Tectonic uplift of the landmass can also increase the sediment supply by increasing relief and hence the rate of erosion. Tectonic activity can also result in subsidence of the shelf, which will make the water deeper across the shelf area: a relative sea-level rise will have the same effect. With increased water depth, more of the shelf area will be 'starved' of mud and sand, allowing carbonate sedimentation to occur in place of clastic deposition. Fluctuations in sea level (which are described in more detail in Chapter 23) may therefore result in alternations between limestone and mudstone/sandstone deposition.

Carbonate deposits can co-exist with terrigenous clastic and volcaniclastic sediments under certain conditions. Deltas built by ephemeral rivers in arid environments may experience long periods without supply of debris and during these intervals carbonates may develop on the delta front (Wilson 2005), for example, in the form of small reefs that build up in the shallow marine parts of ephemeral fan-deltas (Chapter 17). Time intervals between eruption episodes in island arc volcanoes (12.4.2) may be long enough for small carbonate platforms to develop in the shallow water around an island volcano, giving rise to an association between volcanic and carbonate deposition (Wilson & Lokier 2002).

Characteristics of shallow marine carbonates
- lithology – limestone
- mineralogy – calcite and aragonite
- texture – variable, biogenic structures in reefs, well sorted in shallow water
- bed geometry – massive reef build-ups on rimmed shelves and extensive sheet units on ramps
- sedimentary structures – cross-bedding in oolite shoals
- palaeocurrents – not usually diagnostic, with tide, wave and storm driven currents
- fossils – usually abundant, shallow marine fauna most common
- colour – usually pale white, cream or grey
- facies associations – may occur with evaporites, associations with terrigenous clastic material may occur

Characteristics of marine evaporites
- lithology – gypsum, anhydrite and halite
- mineralogy – evaporite minerals
- texture – crystalline or amorphous
- bed geometry – sheets in lagoons and barred basins, nodular in sabkhas
- sedimentary structures – intrastratal solution breccias and deformation
- palaeocurrents – rare
- fossils – rare
- colour – typically white, but may be coloured by impurities
- facies associations – often with shallow marine carbonates

FURTHER READING

Braithwaite, C. (2005) *Carbonate Sediments and Rocks*. Whittles Publishing, Dunbeath.

Kendall, A.C. & Harwood, G.M. (1996) Marine evaporites: arid shorelines and basins. In: *Sedimentary Environments: Processes, Facies and Stratigraphy* (Ed. Reading, H.G.). Blackwell Science, Oxford; 281–324.

Tucker, M.E. & Wright, V.P. (1990) *Carbonate Sedimentology*. Blackwell Scientific Publications, Oxford, 482 pp.

Warren, J. (1999) *Evaporites: their Evolution and Economics.* Blackwell Science, Oxford.

Wright, V.P. & Burchette, T.P. (1996) Shallow-water carbonate environments. In: *Sedimentary Environments: Processes, Facies and Stratigraphy* (Ed. Reading, H.G.). Blackwell Science, Oxford; 325–394.

16

Deep Marine Environments

The deep oceans are the largest areas of sediment accumulation on Earth but they are also the least understood. Around the edges of ocean basins sediment shed from land areas and the continental shelves is carried tens to hundreds of kilometres out into the basin by gravity-driven mass flows. Turbidity currents and debris flows transport sediment down the continental slope and out on to the ocean floor to form aprons and fans of deposits. Towards the basin centre terrigenous clastic detritus is limited to wind-blown dust, including volcanic ash and fine particulate matter held in temporary suspension in ocean currents. The surface waters are rich in life but below the photic zone organisms are rarer and on the deep sea floor life is relatively sparse, apart from strange creatures around hydrothermal vents. Organisms that live floating or swimming in the oceans provide a source of sediment in the form of their shells and skeletons when they die. These sources of pelagic detritus are present throughout the oceans, varying in quantity according to the surface climate and related biogenic productivity.

16.1 OCEAN BASINS

Altogether 71% of the area of the globe is occupied by ocean basins that have formed by sea-floor spreading and are floored by basaltic oceanic crust. The mid-ocean ridge spreading centres are typically at 2000 to 2500 m depth in the oceans. Along them the crust is actively forming by the injection of basic magmas from below to form dykes as the molten rock solidifies and the extrusion of basaltic lava at the surface in the form of pillows (17.11). This igneous activity within the crust makes it relatively hot. As further injection occurs and new crust is formed, previously formed material gradually moves away from the spreading centre and as it does so it cools, contracts and the density increases. The older, denser oceanic crust sinks relative to the younger, hotter crust at the spreading centre and a profile of increasing water depth away from the mid-ocean ridge results (Fig. 16.1) down to around 4000 to 5000 m where the crust is more than a few tens of millions of years old.

The ocean basins are bordered by continental margins that are important areas of terrigenous clastic and carbonate deposition. Sediment supplied to the ocean basins may be reworked from the shallow marine shelf areas, or is supplied directly

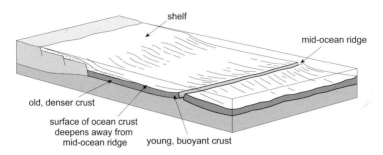

Fig. 16.1 Deep water environments are floored by ocean crust and are the most widespread areas of deposition worldwide.

from river and delta systems and bypasses the shelf. There is also intrabasinal material available in ocean basins, comprising mainly the hard part of plants and animals that live in the open oceans, and airborne dust that is blown into the oceans. These sources of sediment all contribute to oceanic deposits (Douglas 2003). The large clastic depositional systems are mainly found near the margins of the ocean basin, although large systems may extend a thousand kilometres or more out onto the basin plain, and the ocean basin plains provide the largest depositional environments on Earth.

The problem with these deep-water depositional systems, however, is the difficulty of observing and measuring processes and products in the present day. The deep seas are profoundly inaccessible places. Our knowledge is largely limited to evidence from remote sensing: detailed bathymetric surveys, side-scan sonar images of the sea floor and seismic reflection surveys (22.2) of the sediments. There are also extremely localised samples from boreholes, shallow cores and dredge samples. Our database of the modern ocean floors is comparable to that of the surface of the Moon and understanding the sea floor is rather like trying to interpret all processes on land from satellite images and a limited number of hand specimens of rocks collected over a large area. However, our knowledge of deep-water systems is rapidly growing, partly through technical advances, but also because hydrocarbon exploration has been gradually moving into deeper water and looking for reserves in deep-water deposits.

16.1.1 Morphology of ocean basins

Continental slopes typically have slope angles of between 2° and 10° and the continental rise is even

less (*11.1*). Nevertheless, they are physiographically significant, as they contrast with the very low gradients of continental shelves and the flat ocean floor. Continental slopes extend from the shelf edge, about 200 m below sea level, to the basin floor at 4000 or 5000 m depth and may be up to a hundred kilometres across in a downslope direction. Continental slopes are commonly cut by **submarine canyons**, which, like their counterparts on land, are steep-sided erosional features. Submarine canyons are deeply incised, sometimes into the bedrock of the shelf, and may stretch all the way back from the shelf edge to the shoreline. They act as conduits for the transfer of water and sediment from the shelf, sometimes feeding material directly from a river mouth. The presence of canyons controls the formation and position of submarine fans.

The generally flat surface of the ocean floor is interrupted in places by **seamounts**, underwater volcanoes located over isolated hotspots. Seamounts may be wholly submarine or may build up above water as volcanic islands, such as the Hawaiian island chain in the central Pacific. As subaerial volcanoes they can be important sources of volcaniclastic sediment to ocean basins. The flanks of the volcanoes are commonly unstable and give rise to very large-scale submarine slides and slumps that can involve several cubic kilometres of material. Bathymetric mapping and sonar images of the ocean floor around volcanic islands such as Hawaii in the Pacific and the Canary Islands in the Atlantic have revealed the existence of very large-scale slump features. Mass movements on this scale would generate tsunami (*11.3.2*) around the edges of the ocean, inundating coastal areas.

The deepest parts of the oceans are the trenches formed in regions where subduction of an oceanic plate is occurring. Trenches can be up to 10,000 m deep. Where they occur adjacent to continental margins (e.g. the Peru–Chile Trench west of South America) they are filled with sediment supplied from the continent, but mid-ocean trenches, such as the Mariana Trench in the west Pacific, are far from any source of material and are unfilled, starved of sediment.

16.1.2 Depositional processes in deep seas

Deposition of most clastic material in the deep seas is by mass-flow processes (4.5). The most common are debris flows and turbidity currents, and these form part of a spectrum within which there can be flows with intermediate characteristics.

Debris-flow deposits

Remobilisation of a mass of poorly sorted, sediment-rich mixture from the edge of the shelf or the top of the slope results in a debris flow, which travels down the slope and out onto the basin plain. Unlike a debris flow on land an underwater flow has the opportunity to mix with water and in doing so it becomes more dilute and this can lead to a change in the flow mechanism and a transition to a turbidity current. The top surface of a submarine debris flow deposit will typically grade up into finer deposits due to dilution of the upper part of the flow. Large debris flows of material are known from the Atlantic off northwest Africa (Masson et al. 1992) and examples of thick, extensive debris-flow deposits are also known from the stratigraphic record (Johns et al. 1981; Pauley 1995). Debris-flow deposits tens of metres thick and extending for tens of kilometres are often referred to as **megabeds**.

Turbidites

Dilute mixtures of sediment and water moving as mass flows under gravity are the most important mechanism for moving coarse clastic material in deep marine environments. These turbidity currents (4.5.2) carry variable amounts of mud, sand and gravel tens, hundreds and even over a thousand kilometres out onto the basin plain. The turbidites deposited can range in thickness from a few millimetres to tens of metres and are carried by flows with sediment concentrations of a few parts per thousand to 10%.

Denser mixtures result in high-density turbidites that have different characteristics to the 'Bouma Sequences' seen in low- and medium-density turbidites. Direct observation of turbidity currents on the ocean floor is very difficult but their effects have been monitored on a small number of occasions. In November 1929 an earthquake in the Grand Banks area off the coast of Newfoundland initiated a turbidity current. The passage of the current was recorded by the severing of telegraph cables on the sea floor, which were cut at different times as the flow advanced. Interpretation of the data indicates that the turbidity current travelled at speeds of between 60 and $100 \, \text{km h}^{-1}$ (Fine et al. 2005). Also, the deposits of recent turbidity flows have been mapped out, for example, in the east Atlantic off the Canary Islands a single turbidite deposit has been shown to have a volume of $125 \, \text{km}^3$ (Masson 1994).

High- and low-efficiency systems

A deep marine depositional system is considered to be a ***low-efficiency system*** if sandy sediment is carried only short distances (tens of kilometres) out onto the basin plain and a ***high-efficiency system*** if the transport distances for sandy material are hundreds of kilometres (Mutti 1992). High-volume flows are more efficient than small-volume flows and the efficiency is also increased by the presence of fines that tend to increase the density of the flow and hence the density contrast with the seawater. The deposits of low-efficiency systems are therefore concentrated near the edge of the basin, whereas muddier, more efficient flows carry sediment out on to the basin plain. The high-efficiency systems will tend to have an area near the basin margin called a bypass zone where sediment is not deposited, and there may be scouring of the underlying surface, with all the deposition concentrated further out in the basin.

Initiation of mass flows

Turbidity currents and mass flows require some form of trigger to start the mixture of sediment and water moving under gravity. This may be provided by an earthquake as the shaking generated by a seismic shock can temporarily liquefy sediment and cause it to move. The impact of large storm waves on shelf sediments may also act as a trigger. Accumulation of sediment on the edge of the shelf may reach the point where it becomes unstable, for example where a delta front approaches the edge of a continental shelf. High river discharge that results in increased sediment supply can result in prolonged turbidity current flow as sediment-laden water from the river mouth flows as a hyperpycnal flow across the shelf and down onto the basin plain. Such ***quasi-steady flows*** may last for much longer periods than the instantaneous triggers that result in flows lasting just a few hours. A fall in sea level exposes shelf sediments to erosion, more storm effects and sediment instability that result in increased frequency of turbidity currents.

16.1.3 Composition of deep marine deposits

The detrital material in deep-water deposits is highly variable and directly reflects the sediment source area. Sand, mud and gravel from a terrigenous source are most common, occurring offshore continental margins that have a high supply from fluvial sources. Material that has had a short residence time on the shelf will be similar to the composition of the river but extensive reworking by wave and tide processes can modify both the texture and the composition of the sediment before it is redeposited as a turbidite. A sandstone deposited by a turbidity current can therefore be anything from a very immature, lithic wacke to a very mature quartz arenite. Turbidites composed wholly or partly of volcaniclastic material occur in seas offshore of volcanic provinces. The deep seas near to carbonate shelves may receive large amounts of reworked shallow-marine carbonate sediment, redeposited by turbidity currents and debris flows into deeper water: recognition of the redeposition process

is particularly important in these cases because the sediment will contain bioclastic material that is characteristic of shallow water environments. Because there is this broad spectrum of sandstone compositions in deep-water sediments, the use of the term 'greywacke' to describe the character of a deposit is best avoided: it has been used historically as a description of lithic wackes (*2.3.3*) that were deposited as turbidites and the distinction between composition and process became confused as the terms turbidite and greywacke came to be used almost as synonyms. 'Greywacke' is not part of the Pettijohn classification of sandstones and it no longer has any widely accepted meaning in sedimentology.

16.2 SUBMARINE FANS

A ***submarine fan*** is a body of sediment on the sea floor deposited by mass-flow processes that may be fan-shaped, but more elongate, lobate geometries are also common (Fig. 16.2). They vary in size from a few kilometres radius to depositional systems covering over a million square kilometres and forming some of the largest geomorphological features on Earth. The morphology and depositional character of submarine fan systems are strongly controlled by the composition of the material supplied, particularly the proportions of gravel, sand and mud present. In this sense submarine fans are very much like other depositional systems such as deltas (Chapter 12), which also show considerable variability depending on the grain-size distribution in the material supplied. Note that although coarse-grained deltas are sometimes referred to as fan deltas and are largely submarine, the term submarine fan is restricted to fan-shaped bodies that

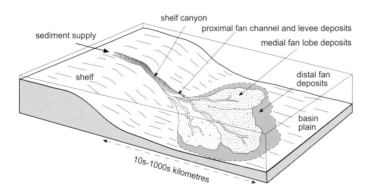

Fig. 16.2 Depositional environments on a submarine fan.

are deposited by mass-flow, mainly turbidity current, processes.

A submarine fan could form of any clastic material, but the larger fans are all composed of terrigenous clastic material supplied by large river systems. Carbonate shelves can be important sources of sediment redeposited in the ocean basins by turbidites, but the supply of carbonate sediment is rarely focused at discrete points along the continental slope: submarine fans composed of carbonate material are therefore rarely formed, and most carbonate turbidites are associated with slope-apron systems (*16.3*).

16.2.1 Architectural elements of submarine fan systems

A submarine fan can be divided into a number of '*architectural elements*', components of the depositional system that are the products of different processes and subenvironments of deposition (Fig. 16.3). *Submarine fan channels* form distinct elements on the fan surface and may have levees associated with them: these channels may incise into, or pass distally

into, *depositional lobes*, which are broad, slightly convex bodies of sediment.

Submarine fan channels and levees

The canyons that incise into the shelf edge funnel sediment and water to discrete points at the edge of the ocean basin where turbidity currents flowing down the canyons pass into channels. Unlike the canyons, the channels are not incised into bedrock, but may scour into underlying submarine fan deposits (Fig. 16.4). Submarine fan channels are variable in size: some of the larger modern examples are several tens of kilometres wide and over a thousand metres deep, and in the stratigraphic record there are submarine fan channels with thicknesses of up to 170 m and 20 km across (Macdonald & Butterworth 1990). The deposits in the channel are typically coarse sands and gravel that form thick, structureless or crudely graded beds characterised by T_{ab} of the Bouma sequence and S_{1-3} of the 'Lowe-type' high-density turbidite model (*4.5.2*). The lateral extent of these turbidite beds is limited by the width of the channel, which, when it is filled, forms a lenticular body made up of stacked coarse-grained turbidites.

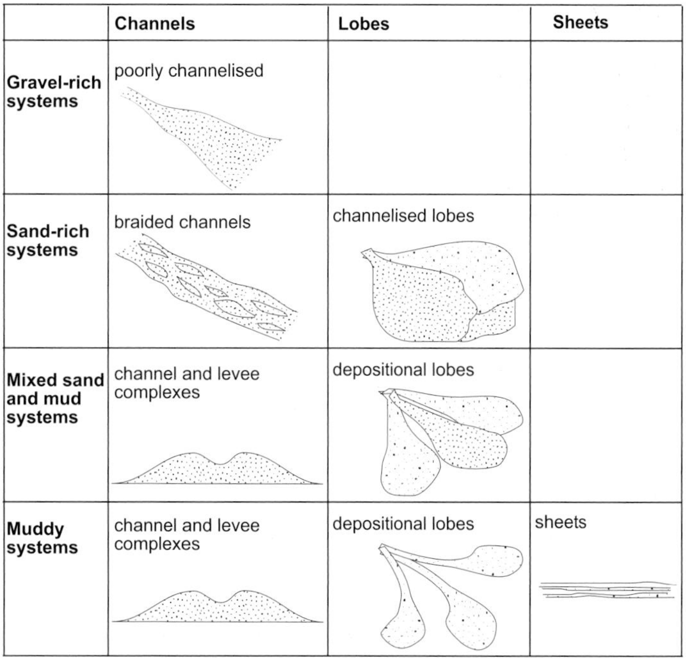

Fig. 16.3 The proportions of different architectural elements on submarine fans are determined by the dominant grain size deposited on the fan.

	Channels	Lobes	Sheets
Gravel-rich systems	poorly channelised		
Sand-rich systems	braided channels	channelised lobes	
Mixed sand and mud systems	channel and levee complexes	depositional lobes	
Muddy systems	channel and levee complexes	depositional lobes	sheets

Fig. 16.4 Thick sandstone beds deposited in a channel in the proximal part of a submarine fan complex.

Most of an individual turbidity flow is confined to the channel but the upper, more dilute part of the flow may spill out of the channel laterally. This is analogous to the channel and overbank setting familiar from fluvial environments (Chapter 9). The overbank flow from the channel contains fine sand, silt and mud and this spreads out as a fine-grained turbidity current away from the channel to form a ***submarine channel levee***. The levee turbidites consist of the upper parts of Bouma sequences (T_{c-e} and T_{de}) and they thin away from the channel margin with a low-angle, wedge-shaped geometry. Levee successions can build up to form units hundreds of metres thick, especially if the channel is aggrading, that is, filling up with sediment and building up its banks at the same time. Channel and levee complexes are also preserved when the channel migrates laterally or avulses, to leave its former position abandoned.

Depositional lobes

At the distal ends of channels the turbidity currents spread out to form a lobe of turbidite deposits that occupies a portion of the fan surface. An individual lobe is constructed by a succession of turbidity currents that tend to deposit further and further out on the lobe through time. A simple progradational geometry results if fan deposition is very ordered, with each turbidity current event of approximately the same magnitude and each depositing progressively further from the mouth of the channel. However, turbidity currents are of varying magnitude and so the pattern tends to be more complex. As the lobe builds out the flow in the more proximal part tends to become channelised. Lobe progradation continues until the channel avulses to another part of the fan. Avulsion occurs because an individual lobe will start to build up above the surrounding fan surface and eventually flows start to follow the slightly steeper gradient on to a lower area of the fan.

The succession built up by depositional lobe progradation is ideally a coarsening-up succession capped by a channelised unit (Fig. 16.5). Individual turbidites will show normal grading but as the lobe progrades currents will carry coarser sediment further out on the fan surface. Successive deposits therefore should contain coarser sediment and hence generate an overall coarsening-up pattern. A thickening-up of the beds should accompany the coarsening-up pattern (Fig. 16.5). Commonly this overall coarsening-up and thickening-up is not seen because of the complex, often random pattern of deposition on depositional lobes (Anderton 1995). Therefore there may not be any consistent vertical pattern of beds deposited on a submarine fan lobe. Depositional lobe deposits often contain the most complete Bouma sequences (T_{a-e} and T_{b-e}). The whole lobe succession may be tens to hundreds of metres thick and an individual lobe may be kilometres or tens of kilometres across. Lobes will be stacked both vertically and laterally against each other, although the lateral limits of an individual lobe may be difficult to identify.

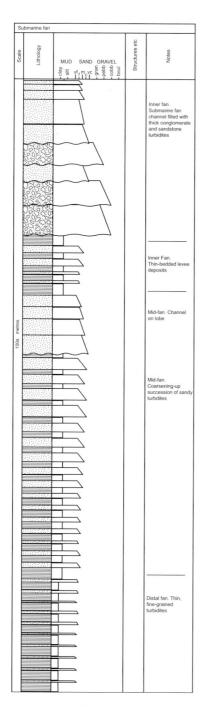

Fig. 16.5 Schematic graphic sedimentary logs through submarine fan deposits: proximal, mid-fan lobe deposits and lower fan deposits.

Turbidite sheets

Turbidite sheets are deposits of turbidity currents that are not restricted to deposition on a lobe but have spread out over a larger area of the fan. They are thin, fine-grained turbidites characterised by Bouma divisions T_{c-e} and T_{de} with little or no organisation into patterns or trends in grain size and bed thickness (Fig. 16.6). Interbedding with hemipelagic mudstones (*16.5.3*) is common.

16.2.2 Submarine fan systems

The architectural elements described are found in various proportions and are made up of different grain sizes of material depending on the characteristics and volume of the sediment supplied to the submarine fan. Any combination is possible, but it is convenient to consider four representative models: gravel-dominated, sandy, mixed sand and mud, and muddy, with the usual caveat that any intermediate form can exist. The examples shown in Figs 16.7–16.10 are for systems that have a single entry point supplying a fan-shaped body of sediment, but for each case there are also scenarios of multiple supply points, which form coalescing bodies of sediment that do not form an overall fan-shape (Reading & Richards 1994; Stow et al. 1996). Submarine fan systems are commonly divided into upper fan (inner fan), mid-fan and lower fan (outer fan) areas: in these schemes the upper fan is dominated by channel and levee complexes, the mid-fan by depositional lobes and the lower fan by sheets. Although this works well for some examples (e.g. sandy and mixed systems) the divisions are not so appropriate for gravelly and muddy systems (see below).

Gravel-rich systems

Coarse sediment may be deposited at the edge of a basin in coarse-grained deltas supplied by braided river or alluvial fans. The deeper parts of these deltas can merge into small submarine fans of material forming wedge-shaped bodies at the base of the slope (Fig. 16.7). The gravel is mainly deposited by debris flows and sands are rapidly deposited by high-density turbidity currents. These fan bodies tend to pass abruptly into thin-bedded distal turbidites and hemipelagic mudstones.

Fig. 16.6 A succession of sandy and muddy turbidite beds deposited on the distal part of a submarine fan complex.

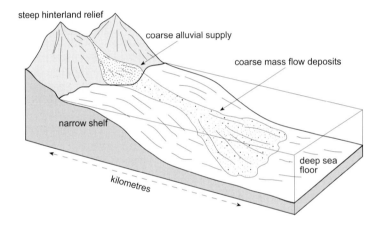

Fig. 16.7 Facies model for a gravel-rich submarine fan: typically found in front of coarse fan deltas, the fan is small and consists mainly of debris flows.

Sand-rich systems

A submarine fan system is considered to be sand-rich if at least 70% of the deposits in the whole system are sandy material (Fig. 16.8). They are usually sourced from sand-rich shelves where waves, storms and tidal currents have sorted the material, removing most of the mud and leaving a sand-rich deposit that is reworked by turbidity currents. Sand-rich turbidity currents have a low efficiency and do not travel very far, so the fan body is likely to be relatively small, less than 50 km in radius (Reading & Richards 1994). Deposition is largely by high-density turbidity currents and the fan is characterised by sandy channels and lobes. The inner fan area is dominated by chan-

nels with some lobes, while the mid-fan area is mainly coalesced lobes, often channelised. Due to the low transport efficiency the transition to finer grained sheet deposits of the lower fan is abrupt. Inactive areas of the fan (abandoned lobes) become blanketed by mud. Strata formed by these systems consist of thick, moderately extensive packages of sandy high-density turbidites separated by mud layers that represent the periods of lobe abandonment.

Mixed sand–mud systems

Where a river/delta system provides large quantities of both sandy and muddy material, a mixed sand–mud

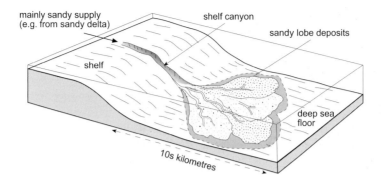

Fig. 16.8 Facies model for a sand-rich submarine fan: sand-rich turbidites form lobes of sediment that build out on the basin floor, with switching of the locus of deposition occurring through time.

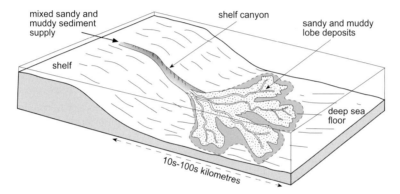

Fig. 16.9 Facies model for a mixed sand–mud submarine fan: the lobes are a mixture of sand and mud and build further out as the turbidites travel longer distances.

depositional system results; these systems are defined as consisting of between 30% and 70% sand (Fig. 16.9). These higher efficiency systems are tens to hundreds of kilometres in diameter and consist of well-developed channel levee systems and depositional lobes. Deposits in the channels in the inner and mid-fan areas include lags of coarse sandstone, sandy, high-density turbidite beds and channel abandonment facies that are muddy turbidites (Reading & Richards 1994). They form lenticular units flanked by levee deposits of thin, fine-grained turbidites and muds. The depositional lobes of the mid-fan are very variable in composition, including both high- and lower density turbidites, becoming muddier in the lower fan area. In a sedimentary succession, the lobe deposits form very broad lenses encased in thin sheets of the lower fan and muds of the basin plain.

Muddy systems

The largest submarine fan systems in modern oceans are mud-rich (Stow et al. 1996), and are fed by very large rivers. These large mud-rich fans include the Bengal Fan fed by the Ganges and Brahmaputra rivers and the large submarine fan beyond the mouth of the Mississippi. These submarine fan systems are over 1000 km in radius and consist of less than 30% sand (Fig. 16.10). Channels are the dominant architectural element of these systems and when some modern submarine fan channels are viewed in plan they are seen to follow a strongly sinuous course that looks like a meandering river pattern (Reading & Richards 1994). The channels deposits are sandy while some sand and more mud are deposited on the channel margins as well-developed levees.

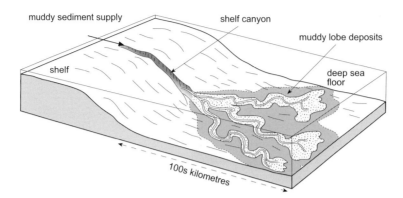

Fig. 16.10 Facies model for a muddy submarine fan: lobes are very elongate and most of the sand is deposited close to the channels.

Depositional lobes are rather poorly developed and thin: the outer fan area consists mainly of thin sandstone sheets interbedded with mudrocks of the basin plain. In a succession of mud-rich fan deposits the sandstone occurrences are limited to lenticular channel units and isolated, thin lobes and sheets in the lower fan.

16.2.3 Ancient submarine fan systems

Successions of turbidites are found in places where deep-sea deposits have been uplifted by tectonic forces and are now exposed on land: this occurs at ocean margins where accretionary prisms form (*24.3.2*) and around mountain belts where foreland basin deposits are thrust to the surface (*24.4.1*). The beds are commonly quite deformed, having been folded and faulted during the process of uplift, so interpretation of the successions, which may be many thousands of metres thick, is not always easy. The type of depositional system can be assessed by considering the ranges of the grain sizes of the material and the distribution of channel, levee, lobe and sheet facies. Because of the size of most submarine fan systems, the beds exposed will often represent only a very small part of a whole system, even if the outcrop extends for tens of kilometres or more.

The palaeogeography of the system can be established by using the distribution of the different facies, and by using indicators of palaeoflow. Scouring during the flow of a turbidity current leaves marks on the underlying surface that are filled in as casts when deposition subsequently occurs. These scour and tool marks (*4.7*) can be very abundant on the bases of turbidite sandstone beds and measurement of their orientation can be used to determine the direction of flow of the turbidity current: flute marks indicate the flow direction while groove marks show the orientation of the axis of the flow. Cross-lamination in the Bouma 'c' division can also be measured and used as a palaeocurrent indicator. Palaeoflow indicators in turbidites provide reliable information about where the source area was (back-tracking along the flow directions), except where a turbidity current encounters an obstacle and is diverted or in small basins where they may flow all the way across to the opposite margin and rebound back again.

Through time a deep-water basin may be wholly or partially filled up with the deposits of a submarine fan. During this process the fan system will prograde as it builds out into the basin. This means that the deposits of the lower fan will be overlain by mid-fan deposits and capped by upper fan facies (Fig. 16.5), but the succession is unlikely to be as simple as presented in this diagram.

16.3 SLOPE APRONS

Slope aprons are depositional systems on continental slopes and adjacent parts of the basin floor that are not fed by discrete point sources but instead have a linear supply from a stretch of the shelf. Deposition is by mass flow processes ranging from submarine slides and slumps to debris flows and turbidity currents (Fig. 16.11). Coarser debris tends to move as avalanches of detritus, including blocks of rock metres

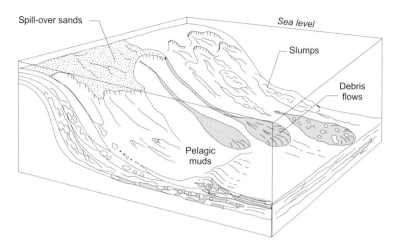

Fig. 16.11 Slope apron deposits include pelagic sediment, slumps, debris flows and sands from the shelf edge. (From Stow 1986.)

to tens of metres across known as **_olistoliths_**, and as debris flows of material that accumulate on the slope. Sand reworked from the edge of the shelf is transported as high-density turbidity currents down the slope to form **_spill-over sand_** deposits (Stow 1986). Mud and mixtures of sand and mud are redistributed further down the slope and onto the adjacent plain by turbidity currents. Finer deposits on the slope are reworked by debris flows and as more coherent slides and slumps of material.

The proportions of gravel, sand and mud will be determined by the nature of the sediment supply from the shelf edge and there is additionally hemipelagic deposition on the slope. The mass-flow deposits are therefore often interbedded with hemipelagic muds, although these are sometimes remobilised and deformed in slump units. The character of slope basin deposits therefore tends to be heterogeneous and generally chaotic. At the edges of carbonate platforms **_carbonate slopes_** may develop at angles ranging from a few degrees to slopes in excess of 30° (Wright & Burchette 1996). The steeper slopes are sites of slumping and redeposition of material by debris flows, while at the base of the slope an apron of carbonate turbidites is deposited.

16.4 CONTOURITES

Ocean currents that are geostrophic and/or thermohaline in origin (*11.4*) are mainly currents that flow along the sea floor parallel to, or nearly parallel to, the bathymetric contours of the continental margin. The deposits of bottom currents are hence called **_contourites_**. The contour currents are significant in the deeper marine environments and therefore contourites are not considered to accumulate on shelf areas. Contourites may be sheets of fine muddy and silty sediment, known as 'drifts', that cover hundreds of thousands to millions of square kilometres on the abyssal plain and may be tens to hundreds of metres thick (Stow et al. 1998). Sometimes they form more elongate bodies parallel to the basin margin, formed by bottom (contour) currents that may be strong enough to transport sand. Their composition depends on the material available to the current and may be terrigenous, calcareous, or volcaniclastic: the grain size and sedimentary structures formed depend on the flow velocity.

In contrast to turbidites, which are single event deposits, contourites are the products of continuous flow, and their characteristics are determined by variations in the current velocity, the amount and type of sediment supply and the degree of modification by processes such as bioturbation. A cycle of increase followed by decrease in velocity would produce a reverse and then normally graded pattern (Fig. 16.12), but other patterns are possible. Contourites can be difficult to recognise in the stratigraphic record, and a fine-grained deposit can often only be confidently interpreted as the product of a bottom current if it can be shown that it was not

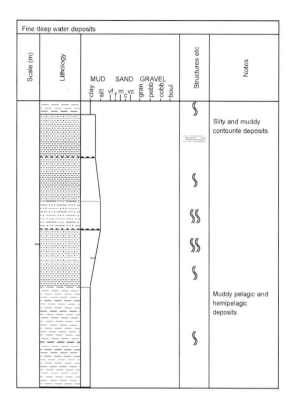

Fig. 16.12 Schematic graphic sedimentary log through contourite deposits.

deposited by a turbidite: this may be possible where there is a difference in composition between the sediment being transported down the continental slope by turbidite currents and sediment being transported parallel to the slope by bottom currents (Stow 1979; Stow & Lovell 1979).

16.5 OCEANIC SEDIMENTS

16.5.1 Pelagic sediments

The term *pelagic* refers to the open ocean, and in the context of sedimentology, *pelagic sediments* are made up of suspended material that was floating in the ocean, away from shorelines, and has settled on the sea floor. This sediment comprises terrigenous dust, mainly clay and some silt-sized particles blown from land areas by winds, very fine volcanic ash, particularly from major eruptions that send fine ejecta high into the atmosphere, and airborne particulates

from fires, mainly black carbon. It also includes bioclastic material that may be the remains of calcareous organisms, such as foraminifers and coccoliths, and the siliceous skeletons of Radiolaria and diatoms (*3.3*). All of these particles reside in the ocean water in suspension, moved around by currents near to the surface, but when they reach quieter, deeper water they gradually fall down through the water column to settle on the seabed.

The origin of the terrigenous clastic material is airborne dust (it is aeolian, *8.6.2*), and much of this is likely to have come from desert areas. The particles are therefore oxidised and the resulting sediments are usually a dark red-brown colour. These 'red clays' are made up of 75% to 90% clay minerals and they are relatively rich in iron and manganese. They lithify to form red or red-brown mudstones. These pelagic red mudrocks are a good example of how the colour of a sedimentary rock should be interpreted with caution: it is tempting to think of all red beds as continental deposits, but these deep-sea facies are red too. The accumulation rate of pelagic clays is very slow, typically only 1 to 5 mm kyr^{-1}, which means it could take up to a million years of continuous sedimentation to form just a metre of sediment.

Pelagic sediments with a biogenic origin are the most abundant type in modern oceans, and two groups of organisms are particularly common in modern seas and are very commonly found in strata of Mesozoic and Cenozoic age as well. Foraminifera are single-celled animals that include a planktonic form with a calcareous shell about a millimetre or a fraction of a millimetre across. Algae belonging to the group chrysophyta include coccoliths that have spherical bodies of calcium carbonate a few tens of microns across (*3.1.3*); organisms this size are commonly referred to as nanoplankton. The hard parts of these organisms are the main contributors to fine-grained deposits that form *calcareous ooze* on the sea bed: where one group is dominant the deposits may be called a *nanoplankton ooze* or *foraminiferal ooze*. Calcareous oozes accumulate at rates ten times that of pelagic clays, around 3 to 50 mm kyr^{-1} (Einsele 2000). This sediment consolidates to form a fine-grained limestone, which is a lime mudstone using the Dunham Classification (*3.1.6*), although these deposits are often called *pelagic limestones*. The foraminifers are normally too small to be seen with the naked eye, and the coccoliths are only recognisable using an electron microscope.

An electron microscope is also required to see any details of the siliceous biogenic material: diatoms are only 5 to 50 μm across while Radiolaria are 50 to 500 μm, so the larger ones can be seen with the naked eye. They are made of *opal*, a hydrated amorphous form of silica that is relatively soluble, and diatoms in particular are often dissolved. Accumulations of this material on the sea floor are known as *siliceous ooze* and they form more slowly than calcareous oozes, at between 2 and 10 mm kyr^{-1}. Upon lithification siliceous oozes form chert beds (*3.3*). The opal is not stable and readily alters to another form of silica such as chalcedony, which makes up the chert rock. *Deep sea cherts* are distinctive, thinly bedded hard rocks that may be black due to the presence of fine organic carbon, or red if there are terrigenous clays present (Fig. 16.13). The Radiolaria can often be seen as very fine white spots within the rock and where this is the case they are referred to as *radiolarian chert*. These beds formed from the lithification of a siliceous ooze deposited in deep water (*primary chert*) should be distinguished from chert formed as nodules due to a diagenetic silicification of a rock (secondary chert: *18.2.3*). Secondary cherts are developed in a host sediment (usually limestone) and have an irregular nodular shape: they do not provide information about the depositional environment but may be important indicators of the diagenetic history.

16.5.2 Distribution of pelagic deposits

Pelagic sediments form a significant proportion of the succession only in places that do not receive sediment from other sources, so any ocean areas close to margins tend to be dominated by sediment derived from the land areas, swamping out the pelagic deposits. The distribution of terrigenous and bioclastic material on the ocean floors away from the margins is determined by the supply of the airborne dust, the biogenic productivity of carbonate-forming organisms, the productivity of siliceous organisms, the water depth and the ocean water circulation (Einsele 2000). The highest productivity of the biogenic material is in the warmer waters near the Equator and also in areas where there is a good supply of nutrients provided by ocean currents. In these regions there is a continuous 'rain' of calcareous and, to a much lesser extent, siliceous biogenic material down towards the sea floor: this 'rain' is

Fig. 16.13 Thin-bedded cherts deposited in a deep marine environment.

less intense in cooler regions or areas with less nutrient supply.

The solubility of calcium carbonate is partly dependent on pressure as well as temperature. At higher pressures and lower temperatures the amount of calcium carbonate that can be dissolved in a given mass of water increases. In oceans the pressure becomes greater with depth of water and the temperature drops so the solubility of calcium carbonate also increases. Near the surface most ocean waters are near to saturation with respect to calcium carbonate: animals and plants are able to extract it from seawater and precipitate either aragonite or calcite in shells and skeletons. As biogenic calcium carbonate in the form of calcite falls through the water column it starts to dissolve at depths of around 3000 m and in most modern oceans will have been completely dissolved once depths of around 4000 m are reached (Fig. 16.14). This is the *calcite compensation depth* (*CCD*) (Wise 2003). Aragonite is more soluble than calcite and an aragonite compensation depth can be defined at a higher level in the water column than a calcite compensation depth (Scholle et al. 1983). The calcite compensation depth is not a constant level throughout the world's oceans today. The capacity for seawater to dissolve calcium carbonate depends on the amount that is already in solution, so in areas of high biogenic productivity the water becomes saturated with calcium carbonate to greater depths and higher pressures are required to put the excess of ions into solution. The depth of the CCD is also known to vary with the temperature of the water and the degree of deep water circulation that is present.

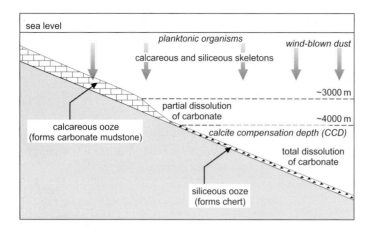

Fig. 16.14 The distribution of pelagic sediment in the oceans is strongly influenced by the effects of depth-related pressure on the solubility of carbonate minerals. Below the calcite compensation depth particles of the mineral dissolve resulting in concentrations of silica, which is less soluble, and clay minerals.

Above the CCD the remains of the less abundant siliceous organisms are swamped out by the carbonate material; below the CCD the skeletons of Radiolaria can form the main biogenic component of a pelagic sediment (Stow et al. 1996). High concentrations of siliceous organisms need not always indicate deep waters. The cold waters of polar regions favour diatoms over calcareous plankton and in pre-Mesozoic strata calcareous foraminifers and nanoplankton are not present. At water depths of around 6000 m the opaline silica that makes up radiolarians and diatoms is subject to dissolution because of the pressure and an **opal compensation depth** (or silica compensation depth) can be recognised.

In the deepest ocean waters it may be expected that only pelagic clays would be deposited. In some parts of the world's oceans this is the case, and there are successions of red-brown mudrocks in the stratigraphic record that are interpreted as hadal (very deep water) deposits. In some instances, these deep-water mudstones include thin beds of limestone and chert: radiolarian chert beds also sometimes include thin limestone beds. The occurrence of these beds might be explained in terms of fluctuations in the compensation depths, but a simpler explanation is that these deposits are actually turbidites and this can be verified by the presence of a very subtle normal grading within the beds. Carbonate, for example, can be deposited at depths below the CCD if it is introduced by a mechanism other than settling through the water column. If the material is brought into deep water by turbidity currents it will pass through the CCD quickly and will be deposited rapidly. The top of a calcareous turbidite may

subsequently start to dissolve at the sea floor, but the waters close to the sea floor will soon become saturated with the mineral and little dissolution of a calcareous turbidite deposit occurs.

16.5.3 Hemipelagic deposits

Fine-grained sediment in the ocean water that has been directly derived from a nearby continent is referred to as **hemipelagic**. It consists of at least 25% non-biogenic material. Hemipelagic deposits are classified as calcareous if more than 30% of the material is carbonate, terrigenous if more than half is detritus weathered from the land and there is less than 30% carbonate, or volcanigenic if more than half is of volcanic origin, with less than 30% of the material carbonate (Einsele 2000). Most of the material is brought into the oceans by currents from the adjacent landmass and is deposited at much higher rates than pelagic deposits (between 10 and over 100 mm kyr^{-1}) (Einsele 2000). Storm events may cause a lot of shelf sediment to be reworked and redistributed by both geostrophic currents and sediment gravity underflows. A lot of hemipelagic material is also associated with turbidity currents: mixing of the density current with the ocean water results in the temporary suspension of fine material and this remains in suspension for long after the turbidite has been deposited. The provenance and hence the general composition of the hemipelagic deposit will be the same as that of the turbidite.

Consolidated hemipelagic sediments are mudrocks that may be shaly and can have a varying proportion of fine silt along with dominantly clay-sized material. The provenance of the material forms a basis for distinguishing hemipelagic and pelagic deposits: the former will be compositionally similar to other material derived from the adjacent continent, whereas pelagic sediments will have a different composition. Clay mineral and geochemical analyses can be used to establish the composition in these cases. Mudrocks interbedded with turbidites are commonly of hemipelagic origin, representing a long period of settling from suspension after the short event of deposition directly from the turbidity current.

16.5.4 Chemogenic sediments

A variety of minerals precipitate directly on the sea floor. These **chemogenic oceanic deposits** include zeolites (silicates), sulphates, sulphides and metal oxides. The oxides are mainly of iron and manganese, and manganese nodules can be common amongst deep-sea deposits (Calvert 2003). The manganese ions are derived from hydrothermal sources or the weathering of continental rocks, including volcanic material, and become concentrated into nodules a few millimetres to 10 or 20 cm across by chemical and biochemical reactions that involve bacteria. This process is believed to be very slow, and manganese nodules may grow at a rate of only a millimetre every million years. They occur in modern sediments and in sedimentary rocks as rounded, hard, black nodules.

At volcanic vents on the sea floor, especially in the region of ocean spreading centres, there are specialised microenvironments where chemical and biological activity result in distinctive deposits. The volcanic activity is responsible for **hydrothermal deposits** precipitated from water heated by the magmas close to the surface (Oberhänsli & Stoffers 1988). Seawater circulates through the upper layers of the crust and at elevated temperatures it dissolves ions from the igneous rocks. Upon reaching the sea floor, the water cools and precipitates minerals to form deposits localised around the hydrothermal vents: these are **black smokers** rich in iron sulphide and **white smokers** composed of silicates of calcium and barium that form chimneys above the vent several metres high. The communities of organism that live around the vents are unusual and highly specialised: they include bacteria, tubeworms, giant clams and blind shrimps. Ancient examples of mid-ocean

hydrothermal deposits have been found in ophiolite suites (*24.2.6*) but fossil fauna are sparse (Oudin & Constantinou 1984).

16.6 FOSSILS IN DEEP OCEAN SEDIMENTS

The most abundant fossils in Mesozoic and Tertiary deep-ocean deposits are the skeletons of planktonic microscopic organisms such as foraminifers, coccoliths and Radiolaria. Foraminifera can be used as a relative depth indicator: both planktonic to benthic forms exist and the ratio of the two provides an approximate measure of water depth because deeper water sediments tend to contain a higher proportion of planktonic forms. Most of this biogenic pelagic material is very fine grained, but any floating or free-swimming organisms can contribute to pelagic deposits on death. These include the shells of large free-swimming organisms such as cephalopods, bones and teeth of fish or aquatic reptiles and mammals. Life in the open oceans in the early Palaeozoic was apparently dominated by graptolites, a hemichordate colonial organism with a free-swimming or floating lifestyle that had a 'backbone'. The compressed remains of graptolites are found in large quantities in Lower Palaeozoic mudrocks deposited in oceanic settings and are important in biostratigraphic correlation (Chapter 20).

Trace fossils in deep-water sediments typically belong to the *Zoophycos* and *Nereites* assemblages (*11.7.2*). The latter are bed-surface traces such as spirals and closely spaced loops made by organisms grazing the sea floor for the sparse nutrients that reach abyssal depths. *Zoophycos* is a shallow subsurface helical form found in the bathyal zone of the continental slope and rise. The occurrence of these ichnofauna is a good, but not infallible indicator of deep-water conditions: they can occur in shallower water if nutrient supply is low and water circulation is poor.

16.7 RECOGNITION OF DEEP OCEAN DEPOSITS: SUMMARY

Our knowledge of the deep oceans today is very poor compared with other depositional environments and considering the sizes of these areas of sediment accumulation. Much of the direct information on deep-water processes and products comes from sea-floor surveys and drilling as part of international collaborative research programmes, such as the Deep Sea Drilling Project, the Ocean Drilling Program and its successor the Integrated Ocean Drilling Program. Submersibles have also allowed direct observation of the sea floor and revealed features such as black and white smokers. The rest of our knowledge of deep-sea sedimentary processes comes from analysis of ancient successions of strata that are rather more conveniently exposed on land, but are somewhat fragmentary.

Evidence in sedimentary rocks for deposition in deep seas is as much based on the absence of signs of shallow water as positive indicators of deep water. Sedimentary structures, such as trough cross-bedding, formed by strong currents are normally absent from sediments deposited in depths greater than a hundred metres or so, as are wave ripples and any evidence of tidal activity. The main sedimentary structures in deep-water deposits are likely to be parallel and cross-lamination formed by deposition from turbidity currents and contour currents. Some authigenic minerals can provide some clues: glauconite does not form anywhere other than shelf environments, but is by no means ubiquitous there, and manganese nodules are characteristically formed at abyssal depths, but are not widespread. Absence of pelagic carbonate deposits may indicate deposition below the calcite compensation depth, although care must be taken not to mistake fine-grained redeposited limestones for pelagic sediments.

Establishing what the water depth was at the time of deposition is problematic beyond certain upper and lower limits. The effects of waves, tides and storm currents usually can be recognised in sediments deposited on the shelf and are absent below about 200 m depth. There are almost no reliable palaeowater-depth indicators between that point and the depths at which carbonate dissolution becomes a recognisable process at several thousand metres water depth and even then, establishing that deposition took place below the CCD is not always straightforward. Some of the most reliable indicators of water depth are to be found from an analysis of body fossils and trace fossils, because many benthic organisms can only exist in shelf environments, although body fossils may be redeposited into deep water by turbidity currents. When describing a facies

as 'deep water' it should be remembered that the actual palaeowater depth of deposition might have been anything below 200 m.

Characteristics of deep marine deposits

- lithology – mud, sand and gravel, fine-grained limestones
- mineralogy – arenites may be lithic or arkosic; carbonate and chert
- texture – variable, some turbidites poorly sorted
- bed geometry – mainly thin sheet beds, except in submarine fan channels
- sedimentary structures – graded turbidite beds with some horizontal and ripple lamination
- palaeocurrents – bottom structures and ripple lamination in turbidites show flow direction
- fossils – pelagic, free swimming and floating organisms
- colour – variable with red pelagic clays, typically dark turbidites and pale pelagic limestones
- facies associations – may be overlain or underlain by shelf facies.

FURTHER READING

Hartley, A.J. & Prosser, D.J. (Eds) (1995) *Characterization of Deep Marine Clastic Systems*. Special Publication 94, Geological Society Publishing House, Bath.

Nittrouer, C.A., Austin, J.A., Field, M.E., Kravitz, J.H., Syvitski, J.P.M. and Wiberg, P.L. (Eds) (2007) *Continental Margin Sedimentation: from Sediment Transport to Sequence Stratigraphy*. Special Publication 37, International Association of Sedimentologists. Blackwell Science, Oxford, 549 pp.

Pickering, K.T., Hiscott, R.N. & Hein, F.J. (1989) *Deep Marine Environments; Clastic Sedimentation and Tectonics*. Unwin Hyman, London.

Posamentier, H.W. & Walker, R.G. (2006) Deep-water turbidites and submarine fans. In: *Facies Models Revisited* (Eds Walker, R.G. & Posamentier, H.). Special Publication 84, Society of Economic Paleontologists and Mineralogists, Tulsa, OK; 399–520.

Stow, D.A.V. (1985) Deep-sea clastics: where are we and where are we going? In: *Sedimentology, Recent Developments and Applied Aspects* (Eds Brenchley, P.J & Williams, B.P.J.). Blackwell Scientific Publications, Oxford; 67–94.

Stow, D.A.V., Faugères, J-C., Viana, A & Gonthier, E. (1998) Fossil contourites: a critical review. *Sedimentary Geology*, **115**, 3–31.

Stow, D.A.V., Reading, H.G. & Collinson, J.D. (1996) Deep Seas. In: *Sedimentary Environments: Processes, Facies and Stratigraphy* (Ed. Reading, H.G.). Blackwell Science, Oxford; 395–453.

17

Volcanic Rocks and Sediments

The study of volcanic processes is normally considered to lie within the realm of igneous geology as the origins of the magmatism lie within the crust and mantle. However, the volcanic material is transported and deposited by sedimentary processes when it is particulate matter ejected from a vent as volcanic ash or coarser debris. Furthermore, both ashes and lavas can contribute to sedimentary successions, and in some places the stratigraphic record is dominated by the products of volcanism. Transport and deposition by primary volcanic mechanisms involve processes that are not encountered in other settings, including air fall of large quantities of ash particles that have been ejected into the atmosphere by explosive volcanic activity, and flows made up of mixtures of hot particulate matter and gases that may travel at very high velocities away from the vent and rapidly form a layer of volcanic detritus. Volcanic activity can create depositional environments and it can also contribute material to all other settings, both on land and in the oceans. The record of volcanic activity preserved within stratigraphic successions provides important information about the history of the Earth and the presence of volcanic rocks in strata offers a means for radiometric dating of these successions.

17.1 VOLCANIC ROCKS AND SEDIMENT

Volcanic rocks are formed by the extrusion of molten magma at the Earth's surface. Molten rock is erupted from fissures on land or under the sea and where volcanic material builds up a hill or mountain a volcano is formed. The products of volcanic activity occur as **lava** that flows across the land surface or sea floor before solidifying, or as volcaniclastic material (*3.7*) consisting of solid fragments of the cooled

magma that are transported and deposited by processes associated with eruption, gravity, air, water or debris flows. Close to the site of the volcanic activity the eruption products dominate the depositional environments and hence the stratigraphic succession: particles ejected by explosive volcanism can be carried high into the atmosphere and distributed around the whole globe, contributing some material to all depositional environments worldwide (Einsele 2000). The nature of the products of volcanism is determined by the chemistry of the magma and the physical setting where the eruptions occur, and a number of different eruption styles are recognised (*17.3*), each producing a characteristic suite of volcanic rocks.

17.1.1 Lavas

Molten magma flowing from fissures normally has a high viscosity and hence lava flows are laminar (*4.2.1*). This may result in a banding within the flow that is preserved when the lava cools and may be seen in some lava flows with relatively high silica compositions. On land, evidence for laminar flow may often be seen near the edges of lava flows between a margin of solidified lava, which forms a sort of levee, and the central part of the flow that moves as a simple plug with no internal deformation. Very fluid lavas may develop a *pahoehoe* texture, a ropy pattern on the surface (Fig. 17.1), whereas more viscous flows have a blocky surface texture, known as *aa*: these features may be preserved in the top parts of ancient flows. If an eruption occurs under water the lava cools rapidly to form *pillow lava* structures that are typically tens of centimetres in diameter and provide a reliable indicator of subaqueous eruption.

17.1.2 Formation of volcaniclastic material

Volcaniclastic material may be divided into fragments that result from primary volcanic processes, that is, those that are related to events during eruption and movement of the material, and those that are a result of secondary processes of weathering and erosion on the land surface. Primary processes can be further divided into those that are a part of the eruption, producing **pyroclastic** material, and those that are not related to the eruption event and are known as **autoclastic** processes. The products of these processes are volcanic blocks/bombs, lapilli and ash depending on their size (*3.7.2*) and they solidify to form agglomerate, lapillistone or tuff respectively (Fig. 17.2).

Pyroclastic material

Fragmentation of volcanic material during eruption can occur in a number of ways. *Magmatic explosions* occur when gases dissolved in the magma come out of solution as the melt rises to the surface and decompresses (Orton 1996). The solubility of volatile components decreases as the confining pressure falls to reach the point where the vapour pressure equals or exceeds the confining pressure. The sudden release of the gases to form bubbles within the magma causes both the gas bubbles and the melt to be violently ejected through a fissure or vent. The expanding bubbles fragment the cooling magma and generate clasts of pyroclastic material. Where this process occurs underground in a shallow magma chamber explosive failure of the roof of the chamber occurs when the pressure within the magma exceeds the strength of the rock above. The force of the explosion will then incorporate the overlying rock that is fragmented in the process. Explosive eruption also occurs when ascending magma reacts with water: these *phreatomagmatic explosions* happen when molten rock interacts with groundwater, wet sediment with

Fig. 17.2 Beds of volcaniclastic sediments: the lower layers are coarse lapillistones while the upper beds are finer ash forming tuff beds.

Fig. 17.1 The ropy surface texture of a pahoehoe lava.

shallow water in a lake or sea or under ice. They also occur when a subaerial lava flow or hot pyroclastic flow enters the water at the shoreline of the sea or a lake (Cas & Wright 1987). Fragmentation occurs as the water expands upon being heated and forming steam interacting with the rapidly cooling magma. Heating of water by volcanic processes to form steam can also create enough pressure to fragment surrounding and overlying rock, generating a **phreatic explosion**. Unlike phreatomagmatic explosions, these phreatic explosions do not involve the formation of fragments from molten magma. Phreatic and phreatomagmatic eruptions are both types of **hydrovolcanic processes**, occurring as a consequence of the interaction of volcanic activity and water.

Autoclastic material

Fragmentation also occurs as a consequence of non-explosive hydrovolcanic processes. The rapid cooling of the surface of a lava flow in contact with water results in **quench-shattering** and the creation of glassy fragments of rock of various shapes and sizes. This process can occur in shallow water but is found often in lavas formed in deeper water where the pressure of the overlying water column inhibits explosive reactions (Cas & Wright 1987). These autoclastic products are referred to as **hydroclastites** or more specifically **hyaloclastites**, which are poorly sorted breccias made up of fragments of volcanic glass formed by the rapid quenching of a molten lava. They often occur associated with pillow lavas, filling in the gaps between the pillows. A second autoclastic mechanism of fragmentation occurs during flow as the surface of a viscous lava flow partially solidifies and is then fractured and deformed as flow continues. This **flow fragmentation** process is also referred to as **autobrecciation**.

Epiclastic material

Epiclastic fragmentation of lava or ash deposits occurs after the episode of eruption has finished. Weathering processes (6.4) attack volcanic rocks very quickly, particularly if it is of basaltic composition and made up of minerals that readily oxidise and hydrolyse on contact with air and water. The surface of an ash layer or lava flow is therefore susceptible to breakdown and the formation of detritus that may be subsequently reworked and redeposited to form a bed of volcaniclastic sediment. There may be evidence of the weathering processes in the form of alteration around the edges of the clasts, and a degree of rounding of the clasts will indicate that the debris has been transported by water. Other indications of an epiclastic origin of a deposit may be the presence of clasts of non-volcanic origin within the deposit, although it is possible for pre-existing sediment to be included with primary volcaniclastic deposits during eruption and initial transport.

17.2 TRANSPORT AND DEPOSITION OF VOLCANICLASTIC MATERIAL

There are some important differences between the way that primary volcaniclastic material behaves during transport and deposition and the terrigenous clastic detritus considered in earlier chapters. An important physical control on sedimentation is that the settling velocity is proportional to fragment size, shape and density (4.2.5). Unlike terrigenous clastic material, the density of pyroclastic particles is very variable. In particular pumice pyroclasts may have a very low density and can float until they become waterlogged (Whitham & Sparks 1986). Grading in pyroclastic deposits may show both normal and reverse grading of different components in the same bed. Lithic fragments and crystals will be normally graded, with the coarsest material at the base. Pumice pyroclasts deposited in water may be reverse graded because the larger fragments will take longer to become waterlogged and hence will be the last to be deposited, resulting in reverse grading. Three primary modes of transport and deposition are recognised: falls, flows and surges, but it should be noted that all three can occur associated with each other in a single deposit.

17.2.1 Pyroclastic fall deposits

When an explosive volcanic eruption sends a cloud of debris into the air the pyroclastic fragments may return to the ground under gravity as a shower of **pyroclastic fall deposits**. Volcanic blocks and bombs travel only a matter of hundreds of metres to kilometres from the vent, depending on the force with which they were ejected. Finer lapilli and ash may be sent kilometres into the atmosphere and be

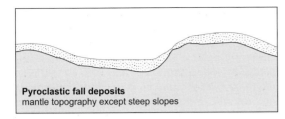

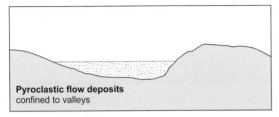

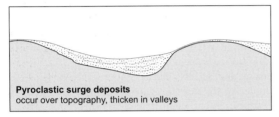

Fig. 17.3 Distribution of ash over topography from pyroclastic falls, pyroclastic flows and pyroclastic surges.

distributed by wind, and large eruptions can result in ash distributed thousands of kilometres from the volcano. A distinctive feature of air-fall deposits is that they mantle the topography forming an even layer over all but the steepest ground surface (Fig. 17.3). The deposits become thinner and are composed of finer grained material with increasing distance from the volcanic vent. Pyroclastic falls range in size from small cinder cones to large volumes mantling topography over large areas.

17.2.2 Pyroclastic flows

Mixtures of volcanic particles and gases can form masses of material that move in the same way as other sediment–fluid mixtures, as sediment gravity flows (4.5), and if they have a high concentration of particles they are referred to as **pyroclastic flows** (Fig. 17.3) (cf. pyroclastic surges, which are lower density mixtures). Pyroclastic flows can originate in a number of ways, including the collapse of a vertical

eruption column of ash, lateral or inclined blasts from the volcano, and the collapse of part of the volcanic edifice. They may move at very high velocities, up to $300\,\mathrm{m\ s^{-1}}$, and can have temperatures of over $1000°C$: a pyroclastic flow made up of a hot mixture of gas and tephra is sometimes referred to as a **nuée ardente**, a 'glowing cloud' (Cas & Wright 1987).

Flows that contain a high proportion of large clasts form **block- and ash-flow** deposits: these poorly sorted agglomerates have a monomict clast composition and cooling cracks in the blocks may indicate that they were hot when deposited. **Scoria-flow** deposits are a mixture of basaltic to andesitic ash, lapilli and blocks that are poorly sorted and commonly show reverse grading. An **ignimbrite** is the deposit of a pyroclastic flow composed of pumiceous material that is a poorly sorted mixture of blocks, lapilli and ash. Ignimbrites commonly contain fragments that are hot enough to fuse together when deposited and form a **welded tuff**, but it should be noted that not all pumice-rich flow deposits are welded. In general pyroclastic flow deposits do not show sedimentary structures other than normal or reverse grading and the poorly sorted character reflects their deposition from relatively dense flows.

17.2.3 Pyroclastic surges

Low concentrations of particles in a sediment gravity flow made up of volcanic particles and gas are known as **pyroclastic surges** (Fig. 17.3), and are distinct from pyroclastic flows because of their dilute nature and turbulent flow characteristics (Sparks 1976; Carey 1991). Phreatic and phreatomagmatic eruptions commonly generate a low cloud made up of a low-density mixture of volcanic debris and fluids, known as a **base surge**: both 'wet' and 'dry' base surges are recognised, depending on the amount of water that is involved in the flow. They travel at high velocity in a horizontal direction away from the eruption site. The deposits of base surges are typically stratified and laminated with low angle cross-stratification formed by the migration of dune and antidune bedforms. Accretionary lapilli (*3.7.2*) are a feature of 'wet' base surges and near to the vent large volcanic bombs may occur within the deposit. The thickness of a base surge varies from as much as a hundred metres close to a phreatomagmatic vent to units only a few centimetres thick further away.

It is common for low-density surges to occur associated with a high-density pyroclastic flow, either as a precursor to the main flow, and hence forming a deposit underlying the flow unit (a **_ground surge_** deposit), or (and) as an **_ash-cloud surge_** that forms at the same time as a flow but above it and depositing a surge deposit on top of the flow unit. Ground-surge deposits at the base of flow units are normally less than a metre thick and are typically stratified, including cross-stratification. At the tops of pyroclastic flow units ash-cloud surges also form thin stratified and cross-stratified beds of ash-size material. They form by dilution by mixing with air at the top of a flow and hence contain the same clast types as the underlying flow. An ash-cloud surge has similar characteristics to a turbidity current but instead of the clasts mixing with water, the ash is in a turbulent suspension of gas.

17.2.4 Pyroclastic flow, surge and fall deposits

A single eruption event may result in a combination of surge, flow and fall deposits (Fig. 17.4). Block- and ash-flow deposits lack the ground-surge unit that may be seen at the bottom of scoria-flow and ignimbrite deposits. Pyroclastic flow units are typically structureless, although they may display some grading, with reverse grading occurring in the lower density pumice and vesiculated scoria fragments and normal grading in the more dense lithic clasts. The process of **_elutriation_**, the mixing of the upper part of the sediment gravity flow with the surrounding air and volcanic gases, leads to a dilution and formation of a turbulent ash-cloud surge. Bedforms created by the flow result in cross-stratification as well as horizontal lamination in the deposits. Flow units are commonly capped by air-fall deposits that show no depositional structures.

A depositional feature that is quite commonly found in pyroclastic deposits but is very rare in terrigenous clastic sediments is the presence of antidune cross-bedding (4.3.4). Antidunes may form in many high velocity flows, but are normally destroyed as the flow velocity decreases and the sediment is reworked to form lower flow regime bedforms (4.3.6). Preservation of antidunes occurs when the rate of sedimentation from the flow is high enough to mantle the bedform before it can be reworked, and this occurs where there is volcaniclastic material entrained in a turbulent gravity flow in air (Schminke et al. 1973). The cross-stratification of antidunes dips in the opposite direction to dune cross-stratification, that is, it is directed in an up-flow direction.

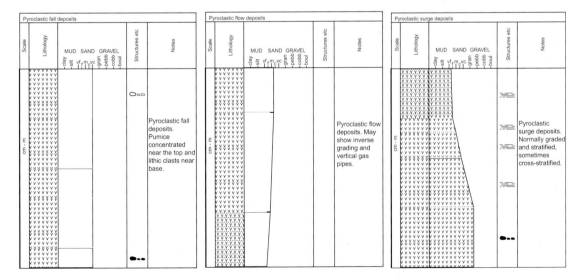

Fig. 17.4 Sketch graphic sedimentary logs of pyroclastic fall, flow and surge deposits.

17.2.5 Volcanic debris-flow avalanches

Structural collapse of part of a volcano can result in catastrophic avalanches of material downslope as a **debris-flow avalanche**. They may be triggered by explosive eruptions, volcanic earthquakes or by the oversteepening of the side of a volcanic edifice due to addition of material during eruption, such that part of it fails under gravity. Large amounts of unstable volcanic material move downslope under gravity including blocks that may be tens to hundreds of metres across in a matrix of finer-grained volcanic ash (Urgeles et al. 1997). The deposits of these vents are extremely poorly sorted, chaotic masses of detritus that may be tens to hundreds of metres thick and cover hundreds of square kilometres. Where water is involved in the debris-flow avalanche it may pass into a lahar (see below).

17.2.6 Lahars

A **lahar** is a debris flow (4.5.1) that contains a significant proportion of material of volcanic origin. They form as a result of mixing of unconsolidated volcanic material with water and the subsequent movement of the dense mixture as a sediment gravity flow (Smith & Lowe 1991). Lahars can form during or immediately after an eruption where pyroclastic material is erupted into or onto water, snow or ice and when heavy rains contemporaneous with the eruption fall on freshly deposited ash. Mobilisation of wet ash can also result in a lahar in circumstances where the ground is disturbed by an earthquake or there is a failure of a temporary lake formed by erupted material. Remobilisation of wet volcanic detritus can occur at any time after eruption, and some lahars may be unrelated to volcanic activity, including cases where epiclastic volcaniclastic debris is involved.

The characteristics of a lahar are essentially the same as those of other debris flows, with the distinction being the composition of the material deposited. The deposits are very poorly sorted and often matrix-supported with no sedimentary structures. Lahars can be readily distinguished from primary volcaniclastic deposits where there is a mixture of terrigenous clastic and volcaniclastic detritus, but where all the material is of volcanic origin, there can be similarities between lahars and pyroclastic flow deposits.

17.3 ERUPTION STYLES

17.3.1 Plinian eruptions

Plinian eruptions are large, explosive eruptions involving high-viscosity magmas of andesitic to rhyolitic composition. They involve large quantities of pumice that are ejected to form extensive pumiceous pyroclastic fall deposits over hundreds of square kilometres. Close to a vent the deposits of a single eruption may be 10 or 20 m thick: the distribution of material depends on the magnitude of the eruption, but deposits a metre thick can be found tens of kilometres away from the vent (Cas & Wright 1987). The deposits are typically clast-supported, angular, fragmented, pumice or scoria clasts with subordinate crystals and lithic fragments. The fabric may be massive or stratified, the former resulting from sustained eruptions, whereas stratification may result from fluctuations of eruption intensity or wind direction (Fig. 17.5). The bedding of Plinian falls tends to mantle

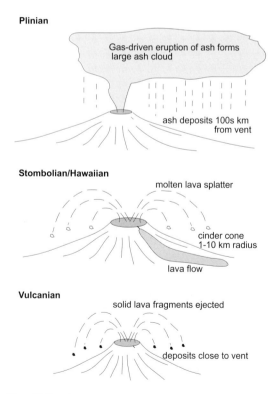

Fig. 17.5 Pelée, Merapi and St Vincent types of pyroclastic flow.

topography except where it is reworked into depressions by secondary processes. The distribution of material from a Plinian eruption is very strongly influenced by the strength and direction of the prevailing wind.

17.3.2 Strombolian (Hawaiian) eruptions

Strombolian or **Hawaiian eruptions** are characterised by a spatter of molten lava that solidifies to form glassy, vesicular fragments of basaltic composition known as **scoria**. The deposits of poorly sorted, coarse lapilli, blocks and bombs are commonly interbedded with lava to form small cones close to the vent (Fig. 17.5). The scoria is largely air-fall material that may be remobilised in grain flows if steep slopes are built up on the sides of the cone. A characteristic feature of some of the scoria in these eruptions is the presence of **Pelée's tears**, which are small, pear-shaped blobs of molten basaltic lava that solidify as they fall through the air, and **Pelée's hair**, which are filaments of solidified lava (Cas & Wright 1987).

17.3.3 Vulcanian eruptions

Eruptions of pyroclastic material from basaltic to andesitic stratovolcanoes typically consist of relatively small volumes of tephra ejected in a series of explosions from a vent. These **vulcanian eruptions** result from periodic breaches of the material that is plugging the vent and involve blocks of both magma and country rock along with ash and lapilli (Cas & Wright 1987). The air-fall deposits of these eruptions are characteristically stratified, due to the episodic character of the eruption, and poorly sorted, with bombs and blocks commonly occurring with the finer grained material (Figs 17.5 & 17.6).

17.4 FACIES ASSOCIATIONS IN VOLCANIC SUCCESSIONS

The facies approach used in the analysis of terrigenous clastic and carbonate sediments can also be applied to volcanic successions. Different processes of transport and deposition of volcanic material have been considered in the previous sections and these can be recognised as having variable importance in

Fig. 17.6 A small ash cone formed by a pyroclastic eruption.

the different environments where volcanic activity is dominant. As with all other depositional environments, a general division can be made between those on land and others that are marine: further subdivision in volcanic successions is largely determined by the characteristics of the magma (Cas & Wright 1987).

17.4.1 Continental basalt provinces

In continental areas associated with a mantle hot spot there may be eruption of large amounts of lava and pyroclastic material from multiple vents and fissures forming a **flood basalt** province. Valleys become filled and pre-existing landforms completely enveloped when flood basalts cover many thousands of square kilometres, with successions that can be several thousand metres thick. Where individual vents build up a volcano by the repeated eruption of basaltic magmas they tend to have relatively gentle slopes and are known as **shield volcanoes**. Associated eruptions of basaltic pyroclastic material form **scoria cones**, circular landforms that may be only a few hundred metres across but with steep sides. Other morphological types of crater composed of volcaniclastic material are **maars**, which have steep-sided craters and gentle outer slopes, **tuff rings** that have roughly equal slopes either side of the rim and **tuff cones** that have steep outer cones and small craters. These all have relatively low preservation potential because they are composed of loose material and are hence readily reworked. Weathering processes acting on basaltic material rapidly lead to breakdown and the formation of pedogenic profiles (9.7) that may be recognised, often as reddened units within the succession. Deposition of volcanic material over wide areas affects the fluvial systems and rivers tend to incise to form valleys within the succession. The fluvial deposits within these valleys can be preserved by overlying volcanic units.

17.4.2 Continental stratovolcanoes

The classic volcanoes forming steep conical mountains with a vent in a crater near the summit are *stratovolcanoes*. These volcanic landforms are composite bodies resulting from repeated eruptions of pyroclastic falls, pyroclastic flows and relatively short lava flows and they typically result from the eruption of intermediate to acidic magmas. The deposits preserved in the stratigraphic record close to the volcanic centre are likely to be ash fall products of large Plinian eruptions and welded pumiceous tuffs resulting from ignimbrites. Further away from the vent the pyroclastic fall ashes are reworked to form lahars and become mixed with terrigenous clastic material in rivers, lakes and on shorelines.

17.4.3 Continental silicic volcanoes

Eruptions involving silicic material are typically explosive resulting in the ejection of large amounts of magma. This can result in the formation of a *caldera*, an approximately circular depression with steep walls formed by collapse associated with the eruption of pyroclastic materials. The caldera itself will be the site of accumulation of lavas and ignimbrites along with epiclastic products of reworking by mass flows, rivers and into lakes. Beyond the rim of the caldera pumiceous pyroclastic flows and fall deposits will be subject to epiclastic reworking by fluvial processes that may result in large-scale redeposition, especially of unwelded pyroclastic deposits.

17.4.4 Mid-ocean ridge basalts

The mid-ocean spreading ridges are sites of voluminous extrusion of basaltic magma. Most of the extrusive material is in the form of pillowed and non-pillowed lavas, with hydroclastites/hyaloclastites

forming as a result of the rapid quenching of the lavas in contact with the seawater. Non-volcanic material can occur between pillows where eruption occurs on a sea floor of soft sediment and during periods of volcanic quiescence pelagic material is deposited between units of lava. The succession will therefore consist of basaltic lava flows with variable amounts of pillow structures, autobrecciated basaltic material and either fine-grained limestones or cherty mudrocks occurring between pillows and interbedded with the basalts. Pyroclastic material only occurs associated with the lavas in places where the eruption occurs in shallow water. These strata are preserved where pieces of ocean floor are tectonically emplaced on continental margins as ophiolites (24.2.6).

17.4.5 Seamounts

Seamounts are sites of volcanism within areas of oceanic lithosphere that develop into volcanic edifices that are close to or above sea level. They form where there is localised magmatism, for example over hot spots in the mantle, and may be isolated from any plate boundaries. The succession of volcanic rocks is similar to that built up by mid-ocean ridge volcanism but may be capped by shallow-marine facies such as limestone reefs that form atolls on tops of the seamounts. Preservation of a seamount in ancient successions is only likely where there has been obduction (24.2.6) of oceanic crust containing these edifices.

17.4.6 Marine stratovolcanoes

Large volcanoes build up from the sea floor to above sea level in island arcs where subduction-related magmatism results in the extrusion of large amounts of basaltic to andesitic magma. The subaerial parts of the volcanic edifice will resemble continental stratovolcanoes, consisting of assemblages of lavas, pyroclastic falls and flows. The associated epiclastic deposits will be different in this setting because of the effects of reworking by shallow marine processes and redeposition of sediment by mass flows to form an apron around the volcano. The redeposited facies include the products of slumps, slides, avalanches, debris flows and turbidity currents of volcaniclastic material: they will occur in the marine succession associated with shallow to deep-marine carbonate facies, pelagic deposits and pyroclastic air-fall ash which is spread by wind from the volcano.

17.4.7 Submarine silicic volcanoes

Eruptions within continental crust in marine settings result in the extrusion of magmas of silicic composition. The confining pressure of the water column above the magma means that underwater eruptions are less explosive because gases are less able to come out of solution in the magma. Submarine silicic volcanism therefore gives rise to much more extensive lavas of rhyolitic composition than is seen in continental successions. Hyaloclastites are extensively developed as lava is rapidly chilled in contact with seawater or soft, wet sediment, but pyroclastic products only form where the eruptions occur in shallow water, where the processes are much more likely to be explosive in character. Hydrothermal activity associated with silicic volcanism results in the formation of sulphide deposits.

17.5 VOLCANIC MATERIAL IN OTHER ENVIRONMENTS

Volcaniclastic clasts have a poor preservation potential in aeolian environments because the processes of abrasion and attrition during wind transport are too severe for the relatively fragile grains to survive. Preservation of ashes is favoured by low-energy continental environments where there is active sediment accumulation, such as lakes, which provide ideal conditions for the preservation of an ash fall deposit in a stratigraphic succession. Floodplain environments may also be suitable but pedogenic and weathering processes will rapidly alter the volcanic material. Ash bands also occur bedded with coals in swamps and mires where organic detritus is rapidly accumulating.

Volcanic deposits that can be classified as deltas are rare, but have been documented as either cones of lava and hyaloclastite advancing into the sea (Porebski & Gradzinski 1990) or aprons of volcaniclastic material (Nemec 1990a). Beach sands composed of grains of basalt can be found along the coasts of many volcanic islands; they exhibit a high degree of textural maturity, but their compositional maturity is very low, consisting mainly of unstable lithic fragments. An association with carbonate sediments is quite common and the fringes of volcanic islands are ideal locations for carbonate sedimentation because of the absence of terrigenous clastic detritus: in the periods between volcanic eruptions faunal communities are able to develop and provide a source of carbonate as bioclastic sands or reef build-ups.

17.6 VOLCANIC ROCKS IN EARTH HISTORY

17.6.1 Volcanic rocks in stratigraphy

Lavas and volcaniclastic deposits within sedimentary successions play a key role in stratigraphy because, unlike almost all other sedimentary rocks, they can be dated by radiometric isotope analysis (*21.2*). The absolute dates that can be determined from volcanic rocks provide the time framework for the calibration of stratigraphic schemes based on other criteria, particularly the fossils within the succession (*20.1*). In situations where dating is based on an igneous rock occurring as a layer within strata it is important to distinguish between a unit that formed as a surface flow, a lava, and a layer that was an intrusive body, a **sill**. The date for a lava provides a date for that part of the sedimentary succession, whereas the date for a sill is some time after the sediments were deposited. Sills may be identified by features that are not seen in lava flows such as a **baked margin**, which provides evidence of heating at the contact with both the beds below and the beds above. Additionally, when tracing the sill laterally it may be found to locally cut through beds up or down stratigraphy, behaving as a dyke at these points. Lava flows, on the other hand, may display characteristics that would not occur in sills such as a pillow structure if the eruption occurred under water, or a weathered top surface of the flow in the case of subaerial eruptions. Severe weathering or alteration of volcanic rocks causes problems for radiometric dating because it can make the ages obtained unreliable.

17.6.2 Magnitude of volcanic events

Most eruptions are relatively small, producing a steady stream of lava and/or the ejection of small quantities of ash. However, periodically there are more violent eruptions that eject many cubic kilometres of ash and volcanic gases. These larger events are recognisable in the stratigraphic record as thicker and more widespread deposits of volcanic ash sometimes occurring in depositional environments many hundreds of kilometres from the site of the eruption. These ash bands are very useful marker horizons as distinctive beds and have the additional benefit of being potentially datable. The effects of a large volcanic eruption may be experienced all over the world. Airborne ash and aerosols (droplets of water containing dissolved sulphates and nitrates) from an eruption can be carried high into the atmosphere where they affect the penetration of radiation from the Sun and can result in temporary global cooling. In comparison to volcanic eruption in geological history the recent large eruptions of Mount St Helens in 1980 and Mount Pinatubo in 1991 were relatively small events, involving between 1 and $10\,km^3$ of eruptive material. In comparison there were eruptions in Yellowstone, northwest USA, in the Quaternary that are thought to have produced up to $2500\,km^3$ of ash (Smith & Braile 1994) and a late Pleistocene event on Sumatra in western Indonesia deposited a layer of ash, the Toba Tuff, that can be traced across large areas of the Bay of Bengal and India (Ninkovitch et al. 1978).

17.6.3 Volcanicity and plate tectonics

The recognition of volcanigenic deposits in the stratigraphic record and analysis of their chemistry provides important clues to the plate tectonics of the past, making it possible to recognise ancient plate boundaries a long way back through Earth history. Most volcanicity around the world is associated with plate margins, with chains of volcanic islands related to subduction of oceanic plates. Volcanism also occurs in extensional tectonic regimes along all the oceanic spreading ridges and in intracontinental rifts: strike-slip plate boundaries may also be sites of volcanism. Exceptions to this pattern of association with plate boundaries are volcanoes situated above 'hot spots', sites around the surface of the Earth where *mantle plumes* provide exceptional amounts of heat to the crust. Geochemical work has shown distinct chemical signatures for the volcanic rocks associated with each of these different tectonic settings and hence the occurrence of volcanic

successions in the stratigraphic record provides evidence of the plate setting.

17.7 RECOGNITION OF VOLCANIC DEPOSITS: SUMMARY

The single most important criterion for the recognition of volcanigenic deposits is the composition of the material. Lavas and primary volcaniclastic detritus rarely contain any material other than the products of the eruption, the nature of which depends on the chemical composition of the magma and the nature of the eruption. Recognition of the volcaniclastic origin of rocks in the stratigraphic record becomes more difficult if the material is fine-grained, altered or both. In hand specimen a fine-grained volcaniclastic rock can be confused with a terrigenous clastic rock of similar grain size. Microscopic examination of a thin-section usually resolves the problem by making it possible to distinguish the crystalline forms within the volcaniclastic deposit from the eroded, detrital grains of terrigenous clastic material. Alteration can destroy the original volcanic fabric of the rock principally by breakdown of feldspars and other minerals to clays: rocks of basaltic composition are particularly susceptible to alteration. Complete alteration may mean that the original nature of the material can be determined only from relict fabrics, such as the outlines of the shapes of feldspar crystals remaining despite total alteration to clay minerals, and the chemistry of the clays as determined by XRD analysis (*2.4.4*).

Characteristics of volcaniclastic deposits
- lithology – basaltic to rhyolitic composition with lithic, crystal and glass fragments
- mineralogy – feldspar, other silicate minerals, some quartz
- texture – poorly to moderately sorted
- bed geometry – may mantle or fill topography
- sedimentary structures – parallel bedding, dune and antidune cross-bedding in pyroclastic flows
- palaeocurrents – cross-bedding may indicate pyroclastic flow direction
- fossils – rare except for plants and animals trapped during ash falls and flows
- colour – from black in basaltic deposits to pale grey rhyolitic material
- facies associations – pyroclastic deposits may occur associated with any continental and shallow-marine facies.

FURTHER READING

Cas, R.A.F., & Wright, J.V. (1987) *Volcanic Successions: Modern and Ancient*. Unwin Hyman, London.

Fisher, R.V. & Schmincke, H-U. (1994) Volcaniclastic sediment transport and deposition. In: *Sediment Transport and Depositional Processes* (Ed. Pye, K.). Blackwell Scientific Publications, Oxford; 351–388.

Fisher, R.V. & Smith, G.A. (1991) *Sedimentation in Volcanic Settings*. Special Publication 45, Society of Economic Paleontologists and Mineralogists, Tulsa, OK.

Orton, G.J. (1995) Facies models in volcanic terrains: time's arrow versus time's cycle. In: *Sedimentary Facies Analysis: a Tribute to the Teaching and Research of Harold G. Reading* (Ed. Plint, A.G.) Special Publication 22, International Association of Sedimentologists. Blackwell Science, Oxford; 157–193.

Orton, G.J. (1996) Volcanic Environments. In: *Sedimentary Environments: Processes, Facies and Stratigraphy* (Ed. Reading, H.G.). Blackwell Science, Oxford; 485–567.

18

Post-depositional Structures and Diagenesis

A sediment body deposited on land or in the sea normally undergoes significant modification before it becomes a sedimentary rock. Physical, chemical and, to some extent, biological processes act on the sediment at scales that range from the molecular to basin-wide. Generally these processes change sediment into sedimentary rocks by compacting loose detritus and adding material to create cements that bind the sediment together. Chemical changes occur to form new minerals and organic substances, and physical processes affect the layers on large and small scales. An important product of these post-depositional processes is the formation and concentration of fossil fuels: coal, oil and natural gas are all products of processes within sedimentary strata that occur after deposition.

18.1 POST-DEPOSITIONAL MODIFICATION OF SEDIMENTARY LAYERS

Sediment is generally deposited as layers that may contain features such as cross-bedding, wave ripples or horizontal lamination formed during deposition: these are referred to as **primary sedimentary structures**. The original layering and these sedimentary structures may be subject to modification by fluid movement and gravitational effects if the sediment remains soft. Disruption of the sedimentary layers may occur within minutes of deposition or may happen at any time up to the point when the material becomes lithified. **Soft-sediment deformation** is the general term for changes to the fabric and layering of

beds of recently deposited sediment. The deformation structures are mostly formed as a result of sediment instabilities caused by density contrasts and by movement of pore fluids through the sediment.

When sediment is deposited in marine environments it is saturated with water, and many continental deposits are also saturated by groundwater. Burial by more sediment usually leads to gradual expulsion of the pore waters, except where the water gets trapped within a layer by an impermeable bed above. This trapped water becomes **overpressured**, and when a crack in the overlying layer allows fluid to be released it travels at high velocity upwards (Leeder 1999). Rapidly moving pore water causes **fluidisation** of the sediment, which is carried upwards with the moving water. Finer sediment can

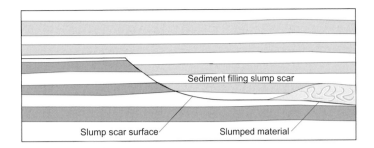

Fig. 18.1 Instabilities within the beds result in parts of the succession slumping to form deformed masses of material: slump scars are the surfaces on which movement occurs.

be more easily carried upwards, so the process of elutriation occurs, as fine sand is carried away by the fluid, leaving behind coarser, and more cohesive, material. *Liquefaction* is a shorter-term process that happens when a mass of saturated sediment is affected by a shock, such as an earthquake, and becomes momentarily liquid, behaving like a viscous fluid. There is usually only very localised movement of sediment and fluid during liquefaction.

Soft-sediment deformation takes a variety of forms at various scales and can occur in any sediment deposited subaqueously that retains some water after deposition. They can be loosely grouped into structures due to sediment instabilities, liquefaction, fluidisation and loading, although these are not mutually exclusive categories.

18.1.1 Structures due to sediment instabilities

Slumps and slump scars

Slumps and slump scars (Figs 18.1 & 18.2) form as a result of gravitational instabilities in sediment piles. When a mass of sediment is deposited on a slope it is often unstable even if the slope is only a matter of a degree or so. If subjected to a shock from an earthquake or sudden addition of more sediment failure may occur on surfaces within the sediment body and this leads to slumping of material. *Slumped beds* are deformed into layers that will typically show a fold structure with the noses of the anticlines oriented in the downslope direction. The surface left as the slumped material is removed is a *slump scar*, which is preserved when later sedimentation subsequently fills in the scar. Slump scars can be recognised in the stratigraphic record as spoon-shaped surfaces

in three dimensions and they range from a few metres to hundreds of metres across. They are common in deltaic sequences but may also occur within any material deposited on a slope.

Growth faults

There is a continuum of process and scale between slump scars and *growth faults*, which are surfaces within sedimentary succession along which there is relative displacement. Growth faults are considered to be *synsedimentary structures*, that is, they form during the deposition of a package of strata. They are most commonly found in delta-front successions (*12.6*), where the depositional slope and the superposition of mouth-bar sands on top of delta-front and prodelta muds results in gravitational instabilities within the succession (Collinson 2003; Collinson et al. 2006). Failure occurs on weak horizons and propagates upwards to form a spoon-shaped fault (a *listric fault*) within the sedimentary succession

Fig. 18.2 The layers of strata at different angles are a result of slumps rotating the strata.

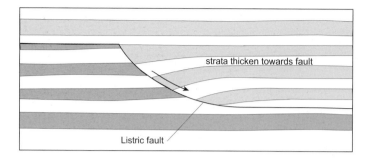

Fig. 18.3 Faulting during sedimentation results in the formation of a growth fault: the layers to the right thickening towards the fault are evidence of movement on the fault during deposition.

(Fig. 18.3). Movement of the beds above the fault over the curved fault surface results in a characteristic rotation of the beds. Growth faults can be distinguished from post-depositional faulting because a single fault affects only part of the succession, with overlying beds unaffected by that fault.

18.1.2 Structures due to liquefaction

Convolute bedding and convolute lamination

The layering within sediments can be disrupted during or after deposition by localised and small-scale liquefaction of the material. The structures range from slight oversteepening of cross-strata, to the development of highly folded and contorted layers called *convolute lamination* and *convolute bedding* (Figs 18.4 & 18.5). These structures form where the sediment is either deposited on a slight slope or where there is a shear stress on the material due to flow of overlying fluid (Leeder 1999; Collinson et al. 2006). The folds in the layering tend to be asymmetric, with the noses of the anticlines pointing downslope or in the direction of the flow. Convolute lamination is particularly common in turbidites, where it can be seen within the laminated and cross-laminated parts of the beds.

Fig. 18.5 Convolute lamination in thinly bedded sandstone and mudstone formed as a result of slumping.

Overturned cross-stratification

Sands deposited by avalanching down the lee slope of subaqueous dunes are loosely packed and saturated with water. They are easily liquefied and can be deformed by the shear stress caused by a strong current over a set of cross-beds. Shearing of the upper part of the cross-beds creates a characteristic form called *recumbent cross-bedding* or *overturned cross-stratification* (Fig. 18.6).

18.1.3 Structures due to fluidisation

Dish and pillar structures

Soft-sediment deformation structures formed by fluidisation processes are often called *dewatering structures* (Fig. 18.7) as they result from the expulsion of pore water from a bed. *Dish structures* are concave disruptions to the layering in sediments a few centimetres to tens of centimetres across formed by the

Convolute lamination, and convolute bedding

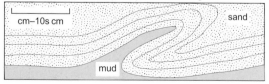

Fig. 18.4 Convolute lamination and convolute bedding form as a result of local liquefaction of deposits.

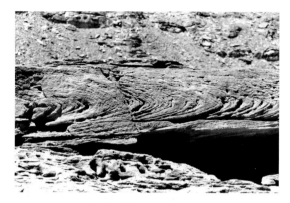

Fig. 18.6 Overturned cross-stratification in sandstone beds 60 cm thick: these would have been originally deposited as simple cross-beds by the migration of a subaqueous dune bedform and subsequently the upper part of the cross-bed set was deformed by the shear stress of a flow over the top.

Dewatering structures

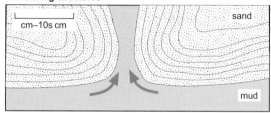

Fig. 18.7 Movement of fluid up from lower layers results in the formation of dewatering structures.

upward movement of fluid (Leeder 1999; Collinson et al. 2006). They are often picked out by fine clay laminae that are the cause of local barriers to fluid flow within the sediment. In plan view the dish structures form polygonal shapes. **Pillar structures**, also known as **elutriation pipes**, are vertical water-escape channels that can be simple tubes or have a vertical sheet-like form. Dish and pillar structures often occur together, although they can form separately.

Clastic dykes

Fluidisation of a large body of sediment in the subsurface can result in elutriation of sediment and the formation of vertical **clastic dykes** centimetres to tens of centimetres across. These sheet-like vertical bodies are typically made of fine sand and they cross-cut other beds. They form when a fracture occurs above an overpressured bed and the upward rush of pore waters carries sediment with it into the crack. The sand may show some layering parallel to the walls of the dyke but is otherwise structureless.

A distinction must be drawn between clastic dykes, which are injected from below, and **fissure fills** formed by the passive infill from above of fissures and cracks in the underlying layers. Fissure fills form where cracks occur at the surface due to earthquake activity or where solution opens cracks in the process of karstic weathering (6.6.3). They can usually be distinguished from clastic dykes because they taper downwards, can be filled with any size of clast (breccia is common) and can show multiple phases of opening and filling where they are earthquake-related. The term '**Neptunian dyke**' has been used in the past for these fissure fills.

Sand volcanoes and extruded sheets

Liquified sediment brought to the surface in isolated pipes emerges to form small **sand volcanoes** a few tens of centimetres to metres across (Fig. 18.8) (Leeder 1999; Collinson et al. 2006). These eruptions of sand on the surface can be preserved only if low-energy conditions prevent the sand being reworked by currents. Sand brought to the surface through clastic dykes can also spread out on the surface, usually as an **extruded sheet** of sandy sediment. These sheets can be difficult to recognise if the connection with an underlying dyke cannot be established. Intrusions

Fig. 18.8 Movement of fluid up from lower layers incorporates sand that reaches the sediment surface to form a sand volcano.

forming 'sills' of sand can form, but can also be difficult to identify.

18.1.4 Structures related to loading

Load casts

If a body of material of relatively low density is overlain by a mass of higher density, the result is an unstable situation. If both layers are relatively wet, the lower density mass will be under pressure and will try to move upwards by exploiting weaknesses in the overlying unit, forcing it to deform. **Load casts** form where the higher density sand has partially sunk into the underlying mud to form downward-facing, bulbous structures (Fig. 18.9): the mud may also become forced up into the overlying sand bed to form a ***flame structure*** (Collinson et al. 2006). As sand is forced downwards and the mud upwards, **load balls** of sand may become completely isolated within the muddy bed. These load-cast features are sometimes referred to as '***ball-and-pillow structures***' (Owen 2003). They are common at the bases of sandy turbidite beds and other situations where sand is deposited directly on wet muds.

Diapirism

In cases where the instability due to density differences between layers of unconsolidated sediment results in movements of material on a large scale, the process is known as ***diapirism***. This process can occur in a range of rock and sediment types in a variety of geological settings, but it is most commonly observed where the density contrast is large and the low-density material is relatively mobile. The bulk density of a layer of rock or sediment is determined by two factors: (a) the density of the minerals and (b) the proportion of the material that is occupied by pore spaces filled with gas or liquid. Two types of diapirism are commonly seen in sedimentary successions, ***salt diapirism*** and ***mud diapirism***, and they have two important implications for sedimentology and stratigraphy: first, diapiric structures can create local highs on the sea floor that may become the locus for carbonate development (Chapter 15), and second, diapirism can create subsurface structures that can be traps for hydrocarbons (*18.7.4*).

Halite (NaCl) has a mineral density of $2.17\,\text{g cm}^{-3}$, which is considerably lower than most sandstones and limestones, even if they are moderately porous. Halite is solid, but in common with all geological materials it will behave in a plastic manner and deform if put under sufficient heat and/or pressure. The pressure required to cause halite to behave plastically can be generated by only a few hundred metres thickness of overlying strata (***overburden***) and, due to its lower density, the halite mass will start to move up in areas where the overburden is thinner or weakened by faults. The diapiric movement of salt deforms the overlying strata, a phenomenon that is known as '***salt tectonics***'. The effects range from creating swells in the layer of salt, to creating dome-like bodies that intrude into the overlying strata (Fig. 18.10), to places where the salt mass breaks through to the surface (Alsop et al. 1996). In very arid regions the extruded salt may form a mass of halite in the landscape like a very viscous volcanic flow.

Load structures

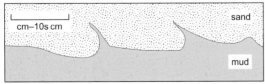

Fig. 18.9 Load casts and ball and pillow structures form where denser sediment, typically sand, is deposited on top of soft mud.

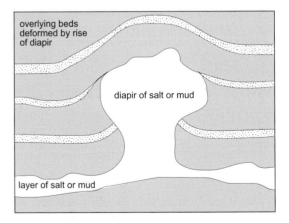

Fig. 18.10 Diapiric structures form where low-density material such as salt or water-saturated mud is overlain by denser sediments.

The second main form of diapirism occurs where a layer of sediment has a high porosity and its density is reduced due to the presence of a high proportion of water mixed with the sediment. This tends to occur where muddy sediment is deposited rapidly. Mud freshly deposited on the sea floor has about 75% of its mass composed of water. As more sediment is deposited on top, the water is gradually squeezed out, but clay-rich deposits, although they may be porous, have a low permeability because the plate-like clay minerals inhibit the passage of fluids through the material. Therefore water tends to become trapped within muddy layers if there is insufficient time for the water to escape. This creates a layer of water-rich, low-density material that may be overlain by denser sediment. This situation most commonly occurs in deltas where fine-grained prodelta facies are overlain by sands of the delta front and delta top as the delta progrades. Mud diapirism (also sometimes called *shale diapirism*) is therefore a common feature of muddy deltaic successions (Hiscott 2003).

18.2 DIAGENETIC PROCESSES

The physical and chemical changes that alter the characteristics of sediment after deposition are referred to as *diagenesis* (Milliken 2003). These processes occur at relatively low temperatures, typically below about 250°C, and at depths of up to about 5000 m (Fig. 18.11). There is a continuum between diagenesis and metamorphism, the latter being considered to be those processes that occur at higher temperatures (typically above 250°C to 300°C) and

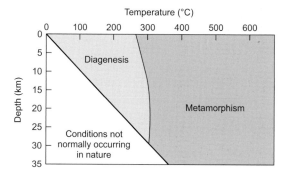

Fig. 18.11 Depth and temperature ranges of diagenetic processes.

pressures: metamorphism involves the destruction of the original sedimentary fabric.

Sediments are generally unconsolidated material at the time of deposition and are in the form of loose sand or gravel, soft mud or accumulations of the body parts of dead organisms. **Lithification** is the process of transforming sediment into sedimentary rock, and involves both chemical and physical changes that take place at any time after initial deposition. Some sediments are lithified immediately, others may take millions of years: there are sediments that never become consolidated, remaining as loose material millions of years after deposition. Lithification upon deposition occurs in some limestones, evaporite deposits and volcaniclastic sediments, which may all form rocks at the time of deposition. Boundstones are formed from the frameworks of organisms that build up solid masses of calcium carbonate as bioherms, for example, coral reefs (*15.3.2*); loose material between the coral mass may be subsequently lithified but the main framework of the rock is formed *in situ*. Chemical precipitation out of water results in beds of solid crystalline evaporite minerals. A further example is that of pyroclastic deposits deposited from hot clouds of ash and gases (nuée ardentes): the temperatures may be high enough for the ash particles to fuse together on deposition as a welded tuff (*17.2.2*).

18.2.1 Burial diagenesis: compaction

The accumulation of sediment results in the earlier deposits being overlain by younger material, which exerts an **overburden pressure** that acts vertically on a body of sediment and increases as more sediment, and hence more mass, is added on top. Loose aggregates initially respond to overburden pressure by changing the packing of the particles; clasts move past each other into positions that take up less volume for the sediment body as a whole (Fig. 18.12). This is one of the processes of **compaction** that increases the density of the sediment and it occurs in all loose aggregates as the clasts rearrange themselves under moderate pressure. Pore water in the voids between the grains is expelled in the process and compaction by particle repacking may reduce the volume of a body of sand by around 10%. During compaction weaker grains, such as mica flakes or mud clasts in sandstone, may be deformed plastically by the pressure from stronger grains such as quartz: fracturing,

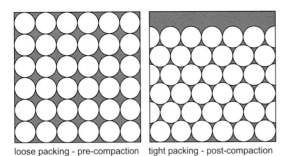

loose packing - pre-compaction   tight packing - post-compaction

Fig. 18.12 Changes to the packing of spheres can lead to a reduction in porosity and an overall reduction in volume.

or ***cataclasis***, of grains can also occur under pressure.

When muds are deposited they may contain up to 80% of water by volume: this is reduced to around 30% under burial of a thousand metres, representing a considerable compaction of the material. Certain sediments, for example boundstones formed as a coral reef, may not compact at all under initial burial. Compaction has little effect on horizontal layers of sediment except to reduce the thickness. Internal sedimentary structures such as cross-stratification may be slightly modified by compaction and the angle of the cross-strata with respect to the horizontal may be decreased slightly.

Differential compaction

Where there is a lateral change in sediment type ***differential compaction*** occurs as one part of a sediment pile compacts more than the part adjacent to it. Possible examples are lime muds deposited around an isolated patch reef, a sand bar surrounded by mud and a submarine channel cut into muds and filled with sand. In each case the degree to which the finer material will compact under overburden pressure will be greater than the sand body or reef. A 'draping' of the finer sediments around the isolated body will occur under compaction (Fig. 18.13). This can occur on all scales from bodies a few metres across to masses hundreds of metres wide. The differential compaction effect is less marked in fluvial successions where sand-filled channel bodies are surrounded by overbank mudstones. This is because the fine sediment on the floodplain dries out between flood events and loses most of its pore waters at that stage. As a consequence the effect of overburden

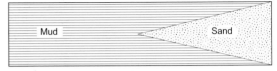

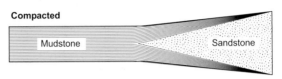

Fig. 18.13 Differential compaction between sandstone and mudstone results in draping of layers around a sandstone lens.

pressure on overbank muds and channel sands may be the same. Differential compaction effects can also be seen on the scale of millimetres and centimetres where there are contrasts in sediment type. Mud layers may become draped around lenses of sand formed by ripple and dune bedforms. Local compaction effects also occur around nodules and concretions where there is early cementation (Fig. 18.14).

Pressure solution/dissolution

The burial of sediment under layers of more strata results in overburden pressures that cause more extreme physical and chemical changes. In sandstones and conglomerates the pressure is concentrated at the contacts between grains or larger clasts, creating concentrations of stress at these points (Fig. 18.15). In the presence of pore waters, diffusion takes place moving

Fig. 18.14 Compaction of layers within a mudrock around a concretion.

Fig. 18.15 Pressure solution has occurred at the contact between two limestone pebbles.

some of the mineral material away from the contact and reprecipitating it on free surfaces of the mineral grains. This process is called **pressure solution** or **pressure dissolution** (Renard & Dysthe 2003) and it results in grains becoming interlocked, providing a rigidity to the sediment, that is, it becomes lithified. These effects may be seen at grain contacts when a rock is examined in thin-section using a petrographic microscope. Beds of limestone may show extensive effects of pressure dissolution (*18.4.1*).

Compaction effects

The degree of compaction in an aggregate can be determined by looking at the nature of the grain contacts (Fig. 18.16). If the sediment has been subjected to very little overburden pressure the clasts will be in contact mainly at the point where they touch, **point contacts**. Reduction in porosity by changes in the packing will bring the edges of more grains together as **long contacts**. Pressure solution between grains results in **concavo-convex contacts** where one grain has dissolved at the point of contact with another (Fig. 18.16). Under very high overburden pressures the boundaries between grains become complex **sutured contacts**, a pattern more commonly seen under the more extreme conditions of metamorphism.

18.2.2 Chemical processes of diagenesis: cementation

A certain amount of modification of the sediment occurs at the sediment–water and sediment–air interfaces:

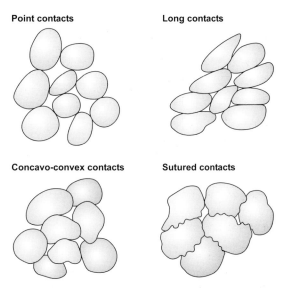

Fig. 18.16 Types of grain contact: there is generally a progressive amount of compaction from point, to long contacts (involving a re-orientation of grains), to concavo-convex and to sutured contacts (which both involve a degree of pressure dissolution.

cements formed at this stage are referred to as **eogenetic cements** and they are essentially synsedimentary, or very soon after deposition (Scholle & Ulmer-Scholle 2003). Most chemical changes occur in sediment that is buried and saturated with pore waters, and cements formed at this stage are called **mesogenetic**. Rarely cement formation occurs during uplift, known as **telogenetic cementation**. During these diagenetic stages, chemical reactions take place between the grains, the water and ions dissolved in the pore waters: these reactions take place at low temperatures and are generally very slow. They involve dissolution of some mineral grains, the precipitation of new minerals, the recrystallisation of minerals and the replacement of one mineral by another.

Dissolution

The processes of grain dissolution are determined by the composition of the grain minerals and the chemistry of the pore waters. Carbonate solubility increases with decreasing temperature and increasing acidity (decreasing pH): the presence of carbon dioxide in solution will increase the acidity of pore waters and leaching of compounds from organic matter may also reduce the pH. It is therefore common for calcareous shelly debris within terrigenous clastic sediment to be dissolved, and if this happens before any lithification occurs then all traces of the fossil may be lost. Dissolution of a fossil after cementation may leave the mould of it, which may either remain as a void or may subsequently be filled by cement to create a cast of the fossil. Silica solubility in water is very low compared with calcium carbonate, so large-scale dissolution of quartz is very uncommon. Silica is, however, more soluble in warmer water and under more alkaline (higher pH) conditions, and opaline silica is more soluble than crystalline quartz. Most quartz dissolution occurs at grain boundaries as a pressure dissolution effect, but the silica released is usually precipitated in adjacent pore spaces.

Precipitation of cements

The nucleation and growth of crystals within pore spaces in sediments is the process of **cementation**. A distinction must be made between matrix (*2.3*), which is fine-grained material deposited with the larger grains, and **cements**, which are minerals precipitated within pore spaces during diagenesis. A number of different minerals can form cements, the most common being silica, usually as quartz but occasionally as chalcedony, carbonates, typically calcite but aragonite, dolomite and siderite cements are also known, and clay minerals. The type of cement formed in a sediment body depends on the availability of different minerals in pore waters, the temperature and the acidity of the pore waters. Carbonate minerals may precipitate as cements if the temperature rises or the acidity decreases, and silica cementation occurs under increased acidity or cooler conditions.

Growth of cement preferentially takes place on a grain of the same composition, so, for example, silica cement more readily forms on a quartz grain than on grains of a different mineral. Where the crystal in the cement grows on an existing grain it creates an **overgrowth** with the grain and the cement forms a continuous mineral crystal (Scholle & Ulmer-Scholle 2003). These are referred to as **syntaxial overgrowths** (Fig. 18.17). Overgrowths are commonly seen in silica-cemented quartz sands; thin-section examination reveals the shape of a quartz crystal formed around a detrital quartz grain, with the shape of the original grain picked out by a slightly darker rim within the new crystal. In carbonate rocks overgrowths of sparry calcite form over biogenic fragments of organisms such as crinoids and echi-

noids that are made up of single calcite crystals (Scholle 1978).

Cementation lithifies the sediment into a rock and as it does so it reduces both the porosity and the permeability. The **porosity** of a rock is the proportion of its volume that is not occupied by solid material but is instead filled with a gas or liquid. **Primary porosity** is formed at the time of deposition and is made up mainly of the spaces between grains, or **interparticle porosity**, with some sediments also possessing **intraparticle porosity** formed by voids within grains, usually within the structures of shelly organisms. Cements form around the edges of grains and grow out into the pore spaces reducing the porosity. **Secondary porosity** forms after deposition and is a result of diagenetic processes: most commonly this occurs as pore waters selectively dissolve parts of the rock such as shells made of calcium carbonate. **Permeability** is the ease with which a fluid can pass through a volume of a rock and is only partly related to porosity. It is possible for a rock to have a high porosity but a low permeability if most of the pore spaces are not connected to each other: this can occur in a porous sandstone which develops a partial cement that blocks the 'throats' between interparticle pore spaces, or a limestone that has porosity sealed inside the chambers of shelly fossils. A rock can also have relatively low porosity but be very permeable if it contains large numbers of interconnected cracks. Cement growth tends to block up the gaps between the grains reducing the permeability. Pore spaces can be completely filled by cement resulting in a complete lithification of the sediment and a reduction of the porosity and permeability to zero.

Recrystallisation

The *in situ* formation of new crystal structures while retaining the basic chemical composition is the process of recrystallisation. This is common in carbonates of biogenic origin because the mineral forms created by an organism, such as aragonite or high magnesium calcite, are not stable under diagenetic conditions and they recrystallise to form grains of low magnesium calcite (Mackenzie 2003). The recrystallised grains will commonly have the same external morphology as the original shell or skeletal material, but the internal microstructure may be lost in the process. Recrystallisation occurs in many molluscs, but does not occur under diagenetic conditions in

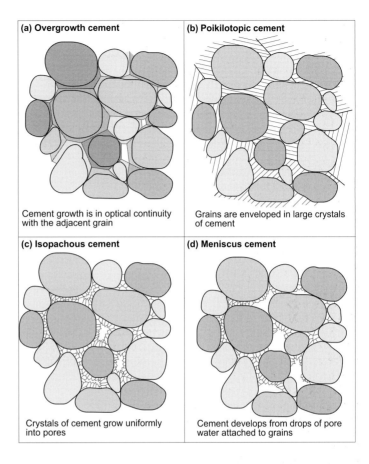

Fig. 18.17 Cement fabrics: (a) overgrowths formed by precipitation of the same mineral (such as quartz or calcite) are in optical continuity with the grain; (b) a poikilotopic fabric is the result of cement minerals completely enveloping grains; (c) an isopachous cement grows on all surfaces within pores, a pattern commonly seen in sparry calcite cements; (d) a meniscus fabric forms when cement precipitation occurs from water flowing down through the sediment.

groups such as crinoids, echinoids and most brachiopods, all of which have hard parts composed of low-magnesium calcite. Recrystallisation of the siliceous hard parts of organisms such as sponges and radiolaria occurs because the original structures are in the form of amorphous opaline silica, which recrystallises to microcrystalline quartz.

Replacement

The *replacement* of a grain by a different mineral occurs with grains of biogenic origin and also detrital mineral grains. For example, feldspars are common detrital grains and to varying degrees all types of feldspar undergo breakdown during diagenesis. The chemical reactions involve the formation of new clay minerals that may completely replace the volume of the original feldspar grain. Feldspars rich in

calcium are the most susceptible to alteration and replacement by clay minerals, whereas sodium-rich, and particularly potassium-rich, feldspars are more resistant. These reactions may take millions of years to complete. Silicification is a replacement process that occurs in carbonate rocks: differences between the mineralogy of a shelly fossil and the surrounding carbonate rock may allow the calcium carbonate of the fossil to be partly or completely replaced by silica if there are silica-rich pore waters present in the rock.

18.2.3 Nodules and concretions

Most sedimentary deposits are heterogeneous, with variations in the concentrations of different gain sizes and grain compositions occurring at all scales. The passage of pore waters through the sediment will be affected by variations in the porosity and permeability due to the distribution of clay particles that inhibit the flow. The presence of the remains of plants and animals creates localised concentrations of organic material that influence biochemical reactions within the sediment. These heterogeneities in the body mean that the processes that cause cementation are unevenly dispersed and hence some parts become cemented more quickly than others (Collinson et al. 2006). Where the distinction between well-indurated patches of sediment and the surrounding body of material is very marked the cementation forms nodules and concretions. Irregular cemented patches are normally referred to as **nodules** and more symmetrical, round or discoid features are called **concretions**.

Nodules and concretions can form in any sediment that is porous and permeable. They are commonly seen in sand beds (where large nodules are sometimes referred to as **doggers**), mudrock and limestone. Sometimes they may be seen to have nucleated around a specific feature, such as the body of a dead animal or plant debris, but in other cases there is no obvious reason for the localised cementation. Concretions formed at particular levels within a succession may coalesce to form bands of well-cemented rock. A variety of different minerals can be the cementing medium, including calcite, siderite, pyrite and silica. In places there is clear evidence that concretions in mudrocks form very soon after deposition: if the layering within the mudstones drapes around the concretion (Fig. 18.14) this is evidence that the cementation occurred locally before the rock as a whole underwent compaction.

Septarian concretions

The interiors of some carbonate concretions in mudstones display an array of cracks that are often filled with sparry calcite. These are known as **septarian structures**, and they are believed to form during the early stages of burial of the sediment. The precise mechanism of formation of the cracks is unclear, but is believed to be either the result of shrinkage (in a process similar to syneresis, 4.6), or related to excess pore fluid pressure in the concretion during compaction, or a combination of the two (Hounslow 1997).

Flints and other secondary cherts

Chert can form directly from siliceous ooze deposited on the sea floor (16.5.1): these primary cherts occur in layers associated with other deep-water sediments. Chert may also form in concretions or nodules as a result of the concentration of silica during diagenesis. These **secondary cherts** are diagenetically formed and are common in sedimentary rocks, particularly limestones (Knauth 1979, 1994). They are generally in the form of nodules that are sometimes coalesced to form layers. The diagenetic origin to these cherts can be seen in **replacement fabrics**, where the structures of organisms that originally had carbonate hard parts can be seen within the chert nodules. The edges of a chert nodule may also cut across sedimentary layering. The nodules form by the very fine-scale dissolution of the original material and precipitation of silica, often allowing detailed original biogenic structures to be seen.

The source of the silica is generally the remains of siliceous organisms deposited with the calcareous sediment. These organisms are sponges, diatoms and radiolarians that originally have silica in a hydrated, opaline form, and in shelf sediments sponge spicules are the most important sources of silica. The **opaline silica** is relatively soluble and it is transported through pore waters to places where it precipitates, usually around fossils, or burrows as microcrystalline or chalcedonic quartz in the form of a nodule. **Flint** is the specific name given to nodules of chert formed in the Cretaceous Chalk.

18.2.4 Colour changes during diagenesis

The colour of a sedimentary rock can be very misleading when interpretation of the depositional environment is being attempted. It is very tempting to assume, for example, that all strongly reddened sandstone beds have been deposited in a strongly oxidising environment such as a desert. Although an arid continental setting will result in oxidation of iron oxides in the sediment, changes in the oxidation state of iron minerals, the main contributors to sediment colour, can occur during diagenesis (Turner 1980). A body of sediment may be deposited in a reducing environment but if the pore waters passing through the rock long after deposition are oxidising then any iron minerals are likely to be altered to iron oxides. Conversely, reducing pore waters may change the colour of the sediment from red to green.

Diagenetic colour changes are obvious where the boundaries between the areas of different colour are not related to primary bedding structures. In fine-grained sediments **reduction spots** may form around particles of organic matter: the breakdown of the organic matter draws oxygen ions from the surrounding material and results in a localised reduction of oxides from a red or purple colour to grey or green. Bands of colour formed by concentrations of iron oxides in irregular layers within a rock are called **liesegangen bands** (Mozley 2003). The bands are millimetre-scale and can look very much like sedimentary laminae. They can be distinguished from primary structures as they cut across bedding planes or cross-strata and there is no grain-size variation between the layers of liesegangen bands. They form by precipitation of iron oxides out of pore waters. Other colour changes may result from the formation of minerals such as zeolites, which are much paler than the dark volcanic rocks within which they form.

18.3 CLASTIC DIAGENESIS

The early stage of diagenesis in terrigenous clastic rocks is mainly characterised by burial compaction as overburden pressure expels water from between grains and reduces the porosity. There is usually little eogenetic cementation, although there are a number of particular environments in which early cementation is important. These are principally beaches where there is precipitation of calcite from seawater washing over the upper parts of a gravelly beach, calcrete, silcrete and ferricrete formation in soils (*9.7.2*), carbonate cementation forming hardgrounds (*11.7*) where there are very low rates of sea-floor sedimentation and gypsum cementation in the coastal plains of arid coasts (*15.2.3*).

Mesogenetic cements are much more extensive than early-stage cements in most clastic rocks and mainly involve the growth of authigenic minerals such as quartz, calcite and clays. Calcite is an important cement in many sandstones and conglomerates (Fig. 18.18) deposited in marine environments: the calcium carbonate commonly originates from aragonitic shelly material deposited along with the sand or gravel. These cements will nucleate on any carbonate grains within the sediment and if these are sparse then the calcite crystals may grow to completely envelop a number of grains (Fig. 18.17b): this **poikilotopic** cement fabric can sometimes be seen in hand specimen as a shiny surface on parts of sandstone.

Fig. 18.18 An isopachous, sparry calcite cement formed in the pore spaces between pebbles lines the surfaces of the pebbles.

Quartz cements in sandstones commonly occur as syntaxial overgrowths: they form adjacent to pressure solution contacts by diffusion along grain boundaries or where there are waters rich in silica derived from dissolution of volcanic glass, extremely fine quartz dust or skeletal material from sponges, diatoms and radiolaria. Silica cements are commonly found only in circumstances where there is an absence of calcium carbonate, for example in quartz-rich sands deposited in a continental environment. The breakdown of volcanic and other lithic fragments in sands leads to the formation of clay mineral cements: these can form either early or late in the diagenetic history as direct precipitation from pore waters or by the recrystallisation of other clay minerals.

18.3.1 Diagenesis and sandstone petrography

Diagenetic features can be difficult to see in hand specimen, and investigation of the post-depositional features in a sedimentary rock normally requires examination of a thin-section. Petrographic description and interpretation of the diagenetic features in a rock will normally follow on from the analysis of the clasts discussed in Chapter 2. The first step is usually to distinguish between pore spaces, the areas between the grains that are voids, matrix, which is fine sediment (usually silt and clay) deposited with the grains, and cement, which is made of minerals precipitated within the pore spaces.

To make the recognition of pore spaces easier it is common practice to fill them with a dyed resin. This can be achieved by using a vacuum or pressure system to force a liquid resin into a sample of the rock before it is cut into a thin-section, and allowing the resin to harden. The resin also strengthens the rock and makes it easier to cut samples that are very friable. A blue dye is usually added to the resin so that it can be readily distinguished from mineral grains or cement, as there are very few blue minerals. Once the thin-section is cut, the porosity can then be identified as the areas of blue on the microscope slide. A matrix composed of detrital fine grains of clay and silt usually can be distinguished from a cement that is crystalline, except in cases where the cement is formed of clay minerals (see below).

Silica cement

Silica cement in quartz sandstone is usually seen as overgrowths of silica on the surfaces of some of the quartz grains. As the silica precipitates out of the pore fluids it nucleates on the surfaces of quartz grains and results in growth of the quartz crystal. The new growth of the crystal will be an extension to the crystal structure, so the orientation of the crystal lattice in the overgrowth will be the same as the host quartz grain (Fig. 18.17a). The overgrowth will hence be in 'optical continuity' with the adjacent part of the grain, i.e. it will show the same birefringence colour and will go into extinction at the same angle (2.3.5). If there is space between grains, the overgrowth will show clear crystal faces. In thin-section a quartz overgrowth usually can be recognised by switching between plane polarised light, which allows the edge of the original grain to be seen, and crossed polars, which will show that the original grain appears to have been extended.

Carbonate cement

If a porous sandstone is cemented by calcite, all of the spaces between the grains may be filled by a mosaic of interlocking crystals of sparry calcite (Fig. 18.17c). The size of the crystals may vary from very small crystals between the grains to large poikilotopic crystals that may completely envelop a number of grains. The calcite cement is identifiable by its high relief compared with the grains and high birefringence colours (2.3.5). Stages in cementation can sometimes be recognised by staining the thin-section with potassium ferricyanide (3.1.2): early stage calcite cements formed under relatively oxidising conditions will not be stained by the dye, but if the calcite forms under reducing conditions, which is typically the case in later diagenesis (Tucker & Wright 1990), then iron incorporated in the crystal lattices will make the calcite ferroan and hence will be stained blue by the potassium ferricyanide dye. Zoning of cements due to changes in chemical conditions during diagenesis can also be identified using specialist optical analysis techniques such as cathodoluminescence.

Clay mineral cements

Direct precipitation of clay minerals belonging to the illite and smectite groups can occur from pore waters and form a cement in sandstone beds. Illite formed in this way has a distinctive filamentous structure that can be recognised in scanning electron microscope images of the rock (3.4.4), making it possible to distinguish these cements from detrital clay minerals that tend to have a more platy structure. Using a petrographic microscope, clay mineral cements can be difficult to recognise although sometimes an even, brown rim around grains may be interpreted as a clay cement layer. However, care must be taken to distinguish such features from iron oxide coatings (which will be very thin) and a clay matrix (which will be randomly distributed clays).

Compaction effects

Evidence for compaction of sediment during burial diagenesis can be recognised in thin-section by considering the spatial arrangement of the grains and the nature of the contacts between them. Grains that are discoid or elongate tend to become reoriented, with their longer axes parallel to the bedding and long grain contacts (Fig. 18.16) will be common. With increasing overburden pressure dissolution starts to occur at grain boundaries leading to the formation of concavo-convex grain boundaries. Sediment that consists of a mixture of different clast types will also show evidence of deformation of the weaker grains under compaction: for example, mica flakes can be deformed between harder quartz grains and lithic clasts of mudrock may be very deformed and squeezed between stronger mineral grains.

18.3.2 Clay mineral diagenesis

Mud deposited on the sea floor contains up to 80% water by volume, so the first diagenetic process affecting muddy sediment is compaction and a considerable reduction in volume. During burial and increasing temperature, clay minerals undergo a range of changes in mineralogy that are determined by the original composition of the material and the temperatures reached during burial diagenesis. Illite is a very common clay mineral, but it is uncommon as a weathering product formed at the surface: most illite minerals are formed diagenetically from other clay minerals such as smectite and kaolinite at temperatures in excess of 70°C (Einsele 2000). With increasing burial the degree of crystallisation increases and it is possible to use an **illite crystallinity** index as a measure of burial temperature. Once formed illite is very stable, and is easily reworked into other sediments. Smectite is formed at lower temperatures (typically less than 50°C) by the weathering or diagenetic alteration of volcanic glass, feldspar and other silicate minerals, but is not stable at higher burial temperatures and tends to transform into illite. Chlorite is less common as a diagenetic mineral, occurring as part of the formation of illite under deep burial conditions. Kaolinite forms as a weathering product above the water table and tends to alter to illite upon burial.

18.3.3 Diagenesis of organic matter in marine muds

Shallow marine environments are regions of high biogenic productivity and the mud deposited on the sea floor of the shelf is rich in the remains of organisms. The diagenesis of this organic material takes place in a series of depth-defined zones (Burley et al. 1985). Organic material at the surface is subject to bacterial oxidation, a process that dominates the upper few centimetres of the sediment, where it is oxygenated by diffusion and bioturbation. Below this surface layer **sulphate reduction** takes place down to about 10 m (Fig. 18.19). Bacteria are involved in reducing sulphate ions to sulphide ions and in the same region ferric iron is reduced to ferrous iron. Under these reducing conditions calcite is precipitated and ferrous iron reacts with the sulphide ions to form pyrite (iron sulphide). At deeper levels within the

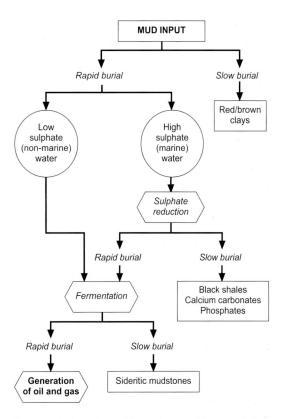

Fig. 18.19 Flow-chart of the pathways of diagenesis of organic matter in sediments.

sediment pile no sulphate ions remain and the dominant reactions are bacterial fermentation processes that break down organic material into carbon dioxide and methane. Carbonate minerals such as calcite and siderite are precipitated in this zone, which extends down to about 1000 m. At deeper levels any remaining organic matter is broken down inorganically.

18.4 CARBONATE DIAGENESIS

Cements in carbonate rocks are mainly made up of calcium carbonate derived from the host sediment. Lithification of aggregates of carbonate material can occur as eogenetic cementation contemporaneously with deposition in any settings where there is either a lot of seawater being circulated through the sediment or where sedimentation rates are low. Beachrock (15.2.1) may be formed of carbonate debris deposited on the beach that is cemented by calcium carbonate from seawater washing through it in the intertidal to supratidal zone (Gischler & Lomando 1997). In warm tropical shallow marine environments the seawater is often saturated with respect to calcium carbonate and cementation can take place on the sea floor forming a hardground or firmground if sedimentation rates are low. The cementation can be localised and related to microbial activity within the sediment, for example, it may be associated with burrows. Colder seawater is undersaturated with calcium carbonate and dissolution of carbonate material can occur.

In non-marine environments calcite cementation occurs in both the **vadose zone** (above the water table) and in the **phreatic zone** (below the water table). In the vadose environment, for example in caves and in streams, the precipitation of the calcite to form these cements is due to the degassing of water: the resulting deposits are stalactites and stalagmites in caves (or **speleothems**, the general term for cave deposits), and **travertine** deposited from surface waters in places such as waterfalls. In soils calcite precipitation forms cements as rhizoliths and calcrete as a result of the evaporation of groundwater and the addition of calcium carbonate as wind-blown dust. Synsedimentary precipitation of siderite can occur where there is mixing of seawater and fresh water under reducing conditions: this can happen in coastal marshes.

Burial stage (mesogenetic) cementation by calcite largely involves carbonate derived from the dissolution of carbonate grains. These cements are low-magnesium calcite and are in the form of bladed crystals that grow out from the grain margins into the pore spaces or as overgrowths, particularly on crystalline fragments of echinoids and crinoids, from which they may develop a poikilotopic fabric.

18.4.1 Compaction effects in limestones: stylolites and bedding planes

Calcite undergoes pressure dissolution under the pressure of a few hundred metres of overburden, forming solution surfaces within the rock known as **stylolites** (Bathurst 1987). At a small scale (millimetres to centimetres), stylolites are usually highly irregular solution surfaces that are picked out by concentrations of clay, iron oxides or other insol-

Fig. 18.20 Stylolites are surfaces of pressure dissolution, in this case marked by an irregular band of insoluble residue in a limestone.

uble components of the rock (Fig. 18.20). Where a stylolite cuts through a fossil it may be possible to determine the amount of calcium carbonate that has been dissolved at the surface. They normally form horizontally in response to overburden pressure, but can also form in response to tectonic pressures at high angles to the bedding. At a larger scale, horizontal pressure solution surfaces within a limestone succession create apparent bedding surfaces that may be very sharply defined by the higher concentration of clay along the surface, but do not necessarily represent a break in sedimentation. This apparent bedding, which is diagenetic in origin, may be more sharply defined in outcrop than true bedding surfaces representing primary changes and breaks in deposition. Pressure solution can result in the removal of large amounts of calcium carbonate and concentrate the clay component of an impure, muddy limestone to leave nodules of limestone in a wavy-bedded mudstone.

18.4.2 Dolomitisation

Dolomite is a calcium magnesium carbonate ($CaMg(CO_3)_2$) mineral that is found in carbonate sedimentary rocks of all ages and when the mineral forms more than 75% of the rock it is called a dolostone (Machel 2003), although the term dolomite is also often used for the rock as well as for the mineral (3.1). The mineral is relatively uncommon in modern depositional environments: it is known to occur in small quantities in arid coastal settings (15.2.3), where its formation may be related to microbial activity (Burns et al. 2000; Mazullo 2000). However, these modern examples do not provide an explanation for the thick successions of dolostone that are known from the stratigraphic record and most dolomite is believed to form diagenetically, a process known as **dolomitisation**. Many dolostones in the stratigraphic record contain fossils that indicate normal marine environments of deposition and show replacement fabrics where material that was clearly originally made up of calcite or aragonite has been wholly or partially replaced by dolomite. The mechanism of formation of dolomite by reaction of seawater and pore water with calcite and aragonite has been the subject of much debate and a number of different models have been proposed, all of which may be applicable in different circumstances (Machel 2003). All models have certain things in common: the original rock must be limestone, the water that reacts with it must be marine, or pore water derived from seawater, and there must be abundant, long-term supply of those waters for large-scale dolomitisation to take place. The process of dolomitisation also seems to be favoured by elevated temperatures and by either enhanced or reduced salinities compared with seawater.

The **mixing-zone model** for dolomitisation proposes that where fresh water, which is undersaturated with respect to calcite but oversaturated with respect to dolomite, mixes with marine waters then dolomitisation would occur (Fig. 18.21) (Humphrey & Quinn 1989). Although there may be a theoretical basis for this model, the process has not been observed in any of the many coastal regions around the world where conditions should be favourable. Arid coastal regions where concentrated brines promote dolomitisation have been suggested in the **reflux model**, but although this may result in formation of dolomite in the sediment within 1 or 2 m of the surface, this mechanism does not seem to be capable of generating large volumes of dolomite (Patterson & Kinsman 1982). It seems more likely that large-scale dolomitisation occurs at some point after burial and hence a number of **burial models** (Morrow 1999) or **seawater models** (Purser et al. 1994) have been proposed. Thick successions of platform limestone can be transformed wholly or partly into dolostone if seawater, or pore-water brines that originated as seawater, can be made to pass through the rock in large quantities for long periods of time. Compaction has been suggested as a potential driving force for

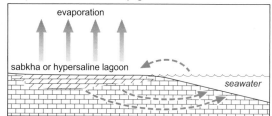

Evaporite brine residue/seepage reflux model

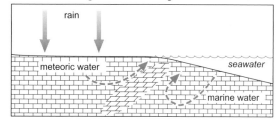

Meteoric–marine/groundwater mixing model

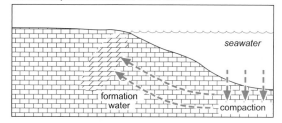

Burial compaction/formation water model

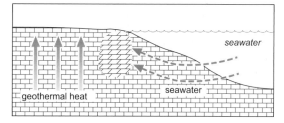

Seawater/convection model

Fig. 18.21 Four of the models proposed for the processes of dolomitisation. (From Tucker & Wright 1990.)

fluid transport, but seems unlikely to be capable of producing the quantities of fluids required. Thermally driven circulation, either by a geothermal heat source or by temperature differences between the interior of a platform and seawater, is the most likely candidate for generating long-term flow of the large quantities of fluid required (Qing & Mountjoy 1992). Topography can also provide a means of forcing water flow through rocks, but although **meteoric waters** (i.e. derived from rainfall) may provide an abundant flux of fluids, they rarely contain sufficient magnesium to promote dolomitisation.

A reversal of the process that causes dolomitisation in association with evaporites can result in dolomite being replaced by calcite. This **dedolomitisation** occurs where beds of gypsum are dissolved enriching groundwaters in calcium sulphate. The sulphate-rich waters passing through dolostone result in the replacement of dolomite by calcite.

18.4.3 Diagenesis and carbonate petrography

Most carbonate sediments become lithified during diagenesis and can readily be cut to make thin-sections: injection of blue resin into the pore spaces is nevertheless commonly carried out in order to make any voids within the rock visible. The blue-dyed resin shows up porosity in carbonate rocks that can either be between the grains (interparticle porosity) or within grains as intraparticle porosity, usually chambers within fossils such as foraminifers, cephalopods and gastropods. Distinguishing between cement and matrix and even between grains and cement is not always straightforward in carbonate rocks because all have the same, or very similar, mineralogy: the morphology of the carbonate material therefore provides most of the important clues as to its origin.

Grains within limestone that are biogenic in origin usually have distinctive shapes that reflect the structure of the organism, even if they are only small fragments (*3.1.5*). Similarly, ooids and peloids are easily recognised in thin-sections. Lithic clasts of limestone and intraclasts have more variable shapes and structures and, because they are in fact pieces of rock, may include areas of cement: distinguishing between the cement within intraclasts and the later cement of the whole rock can sometimes be difficult. Peloids are typically made up of carbonate mud, and must therefore be distinguished from a muddy matrix on the basis of their shape.

Neomorphism

Carbonate mud is the main constituent of carbonate mudstones and wackestones, and can occur as a matrix in packstones, grainstones and boundstones. Individual grains are clay-sized and therefore cannot be individually seen with a petrographic microscope. **Neomorphism** (replacement by recrystallisation) of carbonate mud to form microcrystalline sparry calcite commonly occurs, and as this results in an increase in crystal size, it may then be possible to see the crystalline form under the microscope: although it may be difficult to resolve individual crystals, the microspar appears as a mass of fine crystalline materials showing different birefringence colours under crossed-polars. The birefringence colours of carbonates are high-order pink and green, which may appear to merge into a pale brown if the individual crystals are very small or the magnification is low.

Shelly or skeletal material composed of aragonite undergoes replacement by calcite, either by the solution of the aragonite to create a void later filled by calcite, or by a direct mineral replacement. In the former case the internal structure is completely lost, but where the aragonite is transformed into calcite some relics of the original internal structure may be retained, seen as inclusions of organic matter. The neomorphic calcite crystals are larger than the original aragonite crystals, are often slightly brown due to the presence of the organic material and occur as an irregular mosaic occupying the external form of the skeletal material.

Carbonate cements

Cementation of carbonate sediment to form a limestone can involve a number of stages of cement formation. The form of eogenetic cements is determined by the position of the sediment relative to the groundwater level. In the phreatic zone, in which all the pore spaces are filled with water, the first stage is the formation of a thin fringe of calcite or aragonite growing perpendicular to the grain boundary out into the pore space: these crystals form a thin layer of approximately equal thickness over the grains and are hence known as **isopachous cement** (Fig.18.17c). Above the water level, in the intertidal and supratidal zones, the sediment is in the vadose zone and is only periodically saturated with water: the cement forms only where grains are close together within water held by surface tension to form a meniscus, and hence they are called **meniscus cements** (Fig. 18.17d). A bladed, fibrous or acicular morphology is characteristic of these early cements, with the long axes of the crystals oriented perpendicular to the grain edge. Very fine-grained, micritic, cements can also form at this stage. Recrystallisation of these eogenetic cements commonly occurs because if their original mineralogy was either aragonite or high-magnesium calcite they undergo change to low-magnesium calcite through time.

Many limestones have a cement of sparry calcite that fills in any pore space that is not occupied by an early cement. The interlocking crystals of clear calcite are believed to form during burial diagenesis (meso-genetic cement) from pore waters rich in calcium carbonate. If there are fragments of echinoids or cri-noids present in the sediment the sparry cement pre-cipitates as a syntaxial overgrowth (*18.2.2*) and can form poikilotopic fabric as the cement crystals com-pletely envelop a number of grains. The source of the calcium carbonate for these sparry cements may be from the dissolution of aragonite from shelly material or it may come from pressure solution at grain con-tacts and along stylolites.

Dolomite

Most dolomite occurring in sedimentary rocks is diagenetic in origin, occurring as a replacement of calcite. Although the optical properties of calcite and dolomite are very similar, dolomite commonly occurs as distinctive, small rhomb-shaped crystals that replace the original calcite fabric. Staining the thin-section with Alizarin Red-S (*3.1.2*) provides confirma-tion that the mineral is dolomite (which does not stain pink) as opposed to calcite (which does). Extensive dolomitisation may completely obliterate the primary fabric of the limestone, resulting in a rock that appears in thin-section as a mass of rhombic crystals. The transformation of calcite into dolomite results in a decrease in mineral volume and consequently an increase in porosity

18.5 POST-DEPOSITIONAL CHANGES TO EVAPORITES

Evaporite minerals may either be dissolved out of beds in the subsurface or be replaced by other, less soluble minerals such as calcite and silica (Warren 1999). Dissolution by pore waters passing through the beds leaves vugs and caverns that collapse under the weight of the overburden forming a **dissolution breccia** (Fig. 18.22). Breccias formed in this way consist of angular pieces of the strata bedded with or

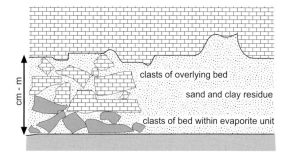

Fig. 18.22 Dissolution of evaporite minerals within a stratigraphic succession results in the formation of a breccia due to collapse of the beds.

immediately overlying the evaporite, with no sign of transport of the clasts: voids between the clasts may be filled with cement.

Initial burial and heating of gypsum leads to dehy-dration and replacement by anhydrite. Conversely, if anhydrite beds are uplifted to a hydrous, near-surface environment a change to gypsum may occur. Volume changes associated with these transitions may result in local deformation and disruption of the bedding. Replacement of halite, gypsum and anhydrite by cal-cite and silica may occur at any stage in diagenesis. The original cubic crystal form of halite may be pre-served as a **pseudomorph**, a cast made up of fine-grained sediment; pseudomorphs of selenite, a form of gypsum, can also occur. Anhydrite may be replaced by microcrystalline or chalcedonic quartz.

18.6 DIAGENESIS OF VOLCANICLASTIC SEDIMENTS

All the crystal, lithic and vitric particulate materials in volcaniclastic deposits are susceptible to diagenetic alteration. Crystals of minerals such as hornblende, pyroxene and plagioclase feldspar all readily react with pore waters to form clay minerals and lithic fragments that contain these minerals will similarly undergo alteration. Volcanic glass changes form in the absence of any other medium because it is meta-stable and **devitrifies** (changes from glass to mineral form) to form very finely crystalline minerals (Cas & Wright 1987). Devitrification can also result in dis-solution of silica in pore waters and the formation of siliceous cements. Clay minerals are also common cements. In some cases the original depositional fabric of the volcaniclastic sediment may be completely lost as a result of alteration during diagenesis. **Tonsteins** are kaolinite-rich mudrocks formed from volcanic, and **bentonites** are composed mainly of smectite clays that are alteration products of basaltic rocks (Spears 2003). The interaction of volcaniclastic material and alkaline waters results in the formation of members of the zeolite group of silicate minerals that may occur as replacements or cements in volcaniclastic successions. Where the original volcanic material has been largely altered during diagenesis the only clues to the origin of the sediment may be the mineralogy of the clays in a mudrock, such as the presence of a high proportion of smectite, and the relics of glass shards and mineral crystals preserved in the sediment.

18.7 FORMATION OF COAL, OIL AND GAS

The branch of geology that has the greatest economic importance worldwide is the study of **fossil fuels** (coal, oil and natural gas): they form by diagenetic processes that alter material made up of the remains of organisms. The places where the original organic material forms can be understood by studying depositional processes, but the formation of coal from plant material and the migration of volatile hydrocarbons

as oil and gas require an understanding of the diagenetic history of the sedimentary rocks where they are found.

18.7.1 Coal-forming environments

Vegetation on the land surface is usually broken down either by grazing animals or by microbial activity. Preservation of the plant material is only likely if the availability of oxygen is restricted, as this will slow down microbial decomposition and allow the formation of peat, which is material produced by the decay of land vegetation (3.6.1). In areas of standing or slowly flowing water conditions can become anaerobic if the oxygen dissolved in the water is used up as part of the decay process. These waterlogged areas of accumulation of organic material are called **mires**, and are the principal sites for the formation of thick layers of peat (3.6.1).

Mires can be divided into two types: areas where most of the input of water is from rainfall are known as **ombotrophic mires** or **bogs**; places where there is a through-flow of groundwater are called **rheotrophic mires** or **swamps**. In addition there are also rheotrophic mires that have an input of clastic sediment, and these are referred to as **marshes**, or **salt marshes** if the water input is saline (Fig. 18.23).

Peat (coal)-forming environments

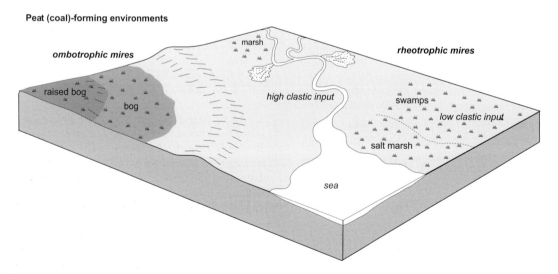

Fig. 18.23 Peat-forming environments: waterlogged areas where organic material can accumulate may either be regions of stagnant water (ombotrophic mires or bogs) or places where there is a through-flow of fresh or saline water (rheotrophic mires or marshes).

The significance of these different settings for peat formation is that these environmental factors have a strong influence on the quality and economic potential of a coal that might subsequently be formed (McCabe 1984; Bohacs & Suter 1997). Bogs tend to have little clastic input, so the peat (and hence coal) is almost pure plant material: the peat can be many metres thick, but is usually of limited lateral extent. Swamp environments can be more extensive, but the through-flowing water may bring in clay, silt and sand particles that make the coal impure (it will have a high **ash content** – *3.6.2*). Also, if the water is saline, it will contain sulphates and these lead to the formation of sulphides (typically iron pyrite) in the coal and give the deposit a high sulphur content: this is not desirable because it results in sulphur dioxide emissions when the coal is burnt. The ash and sulphur content are the two factors that are considered when assessing the **coal grade**, as the lower they are, the higher the grade.

A wet environment is required to form a mire and therefore a peat, so environments of their formation tend to be concentrated in the wetter climatic belts around the Equator and in temperate, higher latitudes. In warmer climates plant productivity is greater, but the microbial activity that breaks down tissue is also more efficient. Both plant growth and microbial breakdown processes are slower in cooler environments, but nevertheless the fastest rates of peat accumulation (over $2 \, \text{mm} \, \text{yr}^{-1}$) are from tropical environments and are ten times the rate of peat accumulation in cooler climes.

Coals that originate as peat deposits are known as **humic coals**, but not all coals have this origin. **Sapropelic coals** are deposits of aquatic algae that accumulate in the bottoms of lakes and although they are less common, they are significant because they can be source rocks for oil: humic coals do not yield oil, but can be the origins of natural gas.

18.7.2 Formation of coal from peat

The first stage of peat formation is the aerobic, biochemical breakdown of plant tissue at the surface that produces a brownish mass of material. This initially formed peat is used as a fuel in places, but has a low calorific value. The calorific value is increased as the peat is buried under hundreds of metres of other sediment and subjected to an increase in temperature and

pressure. Temperature is in fact the more important factor, and as this increases with depth (the geothermal gradient) the peat goes through a series of changes. Volatile compounds such as carbon dioxide and methane are expelled, and the water content is also reduced as the peat goes through a series of geochemical changes. As oxygen, hydrogen and nitrogen are lost, the proportion of carbon present increases from 60% to over 90%, and hence the calorific value of the coal increases.

Differences in the degree to which the original peat has been coalified are described in terms of **coal rank**. Transitional between peat and true coal is **lignite** or **brown coal**, which is exploited as an energy source in places. Going on through the series, low-rank coal is referred to as **sub-bituminous** coal, middle rank is **bituminous** and the highest rank coals are known as **anthracite**. In the process of these reactions, the original layer of peat is reduced considerably in thickness (Fig. 18.24) and a bed of bituminous coal may be only a tenth of the thickness of the original layer of peat.

18.7.3 Formation of oil and gas

Naturally occurring oil and gas are principally made up of **hydrocarbons**, compounds of carbon and hydrogen: **petroleum** is an alternative collective term for these materials. The hydrocarbon compounds originate from organic matter that has accumulated within sedimentary rocks and are transformed into petroleum by the processes of **hydrocarbon maturation**. This takes place in a series of stages dependent upon both temperature and time (Fig. 18.25).

The first stage is biochemical degradation of proteins and carbohydrates in organic matter by processes such as bacterial oxidation and fermentation. This **eogenesis** eliminates oxygen from kerogen, the solid part of the organic matter that is insoluble in organic solvents (Bustin & Wüst 2003; Wüst & Bustin 2003). Three main types of kerogen are recognised: Type I is derived from planktonic algae and amorphous organic material and is the most important in terms of generating oil; Type II consists of mixed marine and continental organic material (algae, spores, cuticles) which forms gas and waxy oils; Type III originates from terrestrial woody matter and is a source of gas only. Eogenesis occurs at

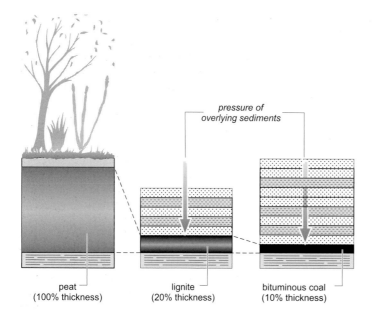

Fig. 18.24 The formation of coal from peat involves a considerable amount of compaction, initially converting peat into brown coal (lignite) before forming bituminous coal.

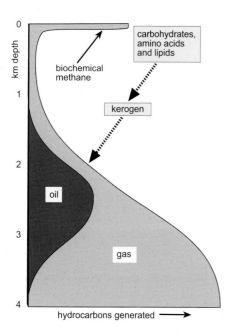

Fig. 18.25 With increased burial the maturation of kerogen results in the formation initially of oil and later gas: greater heating results in the complete breakdown of the hydrocarbons.

temperatures of up to 40°C and at up to depths of just over 1000 m.

At burial depths of between about 1000 and 4000 m and at temperatures of between 40°C and 150°C, the phase of diagenesis known as **catagenesis** further transforms the kerogen. This stage of thermal maturation is also known as the 'oil window' because liquid petroleum forms from Type I kerogen under these conditions. With increasing temperature the proportion of gas generated increases. Generation of oil by organic maturation of kerogen is a process that requires millions of years, during which time the strata containing the organic matter must remain within the oil window of depth and temperature. At higher temperatures and burial depths only methane is produced from all kerogen types, a stage known as **metagenesis**.

Formation of oil, which is made up of relatively long-chain hydrocarbons that are liquid at surface temperatures, from sedimentary organic matter requires a particular set of conditions. First, the organic matter must include the remains of plank-tonic algae that will form Type I kerogen: this material normally accumulates in anaerobic conditions in anoxic marine environments and in lakes. Second, the organic material must be buried in order that catagenesis can generate liquid hydrocarbons within the correct temperature window: if buried too far too quickly only methane gas will be formed. The third factor is time, because the kerogen source rock has to lie within the oil window for millions of years to generate significant quantities of petroleum.

Gas consisting of short-chain hydrocarbons, principally methane, is formed from Type III kerogen and at higher maturation temperatures. Burial of coal also generates natural gas (principally methane) and no oil. The methane generated from coal may become stored in fractures in the coal seam as **coal bed methane**, which is a hazard in underground coal mining, but can also be exploited economically.

18.7.4 Oil and gas reservoirs

The hydrocarbons generated from kerogen are compounds that have a lower density than the formation water present in most sedimentary successions. They are also immiscible with water and droplets of oil or bubbles of gas tend to move upwards through the pile of sedimentary rocks due to their buoyancy. This **hydrocarbon migration** proceeds through any permeable rock until the oil or gas reaches an impermeable barrier.

Hydrocarbon traps

Oil and gas become trapped in the subsurface where there is a barrier formed by impermeable rocks, such as well-cemented lithologies, mudrock and evaporite beds. These impermeable lithologies are known as **cap rocks**. The hydrocarbons will find their way around the cap-rock barrier unless there is some form of **hydrocarbon trap** that prevents further upward migration. **Structural traps** are formed by folds, such as anticlines, especially if they are dome-shaped in three dimensions, and by faults that seal a porous reservoir rock against an impermeable unit (Fig. 18.26). Other traps are **stratigraphic traps**, formed beneath unconformities and in places where the reservoir rock pinches out laterally: porous rocks such as limestone reefs that pass laterally into finer grained deposits and where sand bodies are laterally limited and enclosed by mudrocks are examples of stratigraphic traps. The size and shape of the trap determines the volume of oil and/or gas that is contained by the structure, and hence is an important factor in assessing the economics of a potential oil field. In the absence of traps and caps the hydrocarbon reaches the surface and leaks to the atmosphere. Partial release of hydrocarbons from the subsurface as

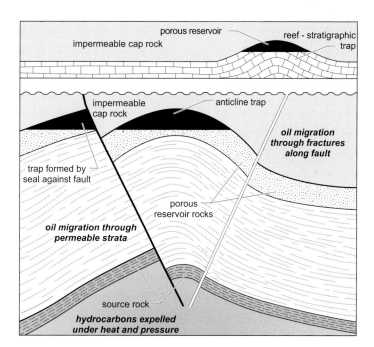

Fig. 18.26 Cartoon of the relationships between the source rock, migration pathway, reservoir, trap and cap rocks required for the accumulation of oil and gas in the subsurface.

oil seeps and *gas seeps* can be important indicators of the presence of hydrocarbons.

Reservoir rocks

Almost all oil and gas accumulations occur underground within the pore spaces of beds of sedimentary rocks. In a few rare cases there are accumulations of hydrocarbons in subterranean caverns formed by dissolution of limestone, but the vast majority of reserves are known hosted between grains in sandstones or within the structures of limestones. For a sedimentary rock to be a suitable reservoir unit, it must be both porous and permeable. Porosity is presented as a percentage of the rock volume. Permeability is expressed in *darcy units*, with a value of 1 darcy representing a very good permeability for a hydrocarbon reservoir.

Some of the best reservoir facies are beds of well-sorted sands deposited in sandy deserts and shallow seas, because these contain a high primary porosity. For similar reasons oolitic grainstones can be good reservoirs, and boundstones formed in reefs have a lot of void spaces within the original structure. There are examples of hydrocarbon reservoirs in deposits of many other environments, including rivers, deltas and submarine fans. Limestones may also have important secondary porosity due to dissolution and diagenetic changes. The reservoir quality of a rock is reduced by two main factors. First, the presence of mud reduces both porosity and permeability because clay minerals fill the spaces between grains and block the throats between them. Second, cementation reduces porosity and permeability by crystallising minerals in the pore spaces, sometimes to the extent of reducing the porosity to zero.

Economic oil and gas accumulations

Exploration for economic reserves of hydrocarbons requires knowledge of the depositional history of an area to determine whether suitable source rocks are likely to have formed and if there are any suitable reservoir and cap lithologies in the overlying succession. This analysis of the sedimentology is an essential part of oil and gas exploration. Knowledge of post-depositional events is also important to provide an assessment of the thermal and burial history that controls the generation of hydrocarbons.

FURTHER READING

Burley, S. & Worden, R. (Eds) (2003) *Sandstone Diagenesis: Recent and Ancient*. Reprint Series Vol. 4, International Association of Sedimentologists. Blackwell Science, Oxford.

Collinson, J.D., Mountney, N. & Thompson, D. (2006) *Sedimentary Structures*. Terra Publishing, London.

Gluyas, J. & Swarbrick, R.E. (2003) *Petroleum Geoscience*. Blackwell Science, Oxford.

Leeder, M.R. (1999) *Sedimentology and Sedimentary Basins: from Turbulence to Tectonics*. Blackwell Science, Oxford.

Scholle, P.A. & Ulmer-Scholle, D.S. (2003) *A Color Guide to the Petrography of Carbonate Rocks: Grains, Textures, Porosity, Diagenesis*. American Association of Petroleum Geologists, Tulsa, OK.

Tucker, M.E. & Wright, V.P. (1990) *Carbonate Sedimentology*. Blackwell Scientific Publications, Oxford, 482 pp.

19

Stratigraphy: Concepts and Lithostratigraphy

Our observations about rocks need to be set in the context of a time framework if we are to use them to understand Earth processes and history. That framework is provided by stratigraphy, and it is one of the oldest disciplines of the geological sciences. Stratigraphy is primarily concerned with the following issues: the recognition of distinct bodies of rock and their spatial relationships with each other; the definition of lithostratigraphic units and the correlation of lithostratigraphic units with each other; the correlation of rock units with a chronostratigraphic standard, which is a formal time-framework to which all of Earth geology can be related. Lithostratigraphy forms the basis for making geological maps and by correlating lithostratigraphic units it is possible to reconstruct the changing palaeogeography of an area through time.

19.1 GEOLOGICAL TIME

Time in geology is a bit like distance in astronomy: the numbers are so vast that it is difficult to make much sense of them. Periods of tens of years are easy to comprehend, because we experience them, and centuries are not so difficult, but once we start dealing with thousands of years our concept of the passage of these amounts of time becomes increasingly divorced from our life experience. So when a geologist refers to a million years, and then tens and hundreds of millions and ultimately billions of years, we have no reference points with which to gauge the passing of those lengths of time. However, a million years is a relatively short period in the history of the Earth, which is about 4.5 billion years. As we go further

and further back in geological time, dating something to within a million years becomes more and more difficult. When considering a geological 'event', such as the position of a succession of sandstones or limestones, we may refer to it as having happened over, for example, 4 million years, but in doing so we are talking about something which occurred over a period which is longer than we can realistically imagine. The geologist therefore has to develop a peculiar sense of time, and may consider 100,000 years as a 'short' period, even though it is unimaginably long when compared with our everyday life.

The passage of time since the formation of the Earth is divided into *geochronological units* and these are divisions of time that may be referred to in terms of years or by name. The Permian Period, for example, was the time between 299 and 251 million years before present. This is analogous to historians referring to the time between 1837 and 1901 as 'the Victorian period'. Geological time is normally expressed in millions of years or thousands of years before present ('present' is commonly defined as 1950, although this distinction is not necessary on a scale of millions of years!). Geological time units are abstract concepts, they do not exist in any physical sense.

The abbreviations used for dates are 'Ma' for millions of years before present and 'ka' for thousands of years before present. The time thousands of millions of years before present is abbreviated to 'Ga' (Gigayears). The North American Stratigraphic Code (North American Commission on Stratigraphic Nomenclature 1983) suggests that to express an interval of time of millions of years in the past abbreviations such as 'my', 'm.y.' or 'm.yr' could be used. This convention has the advantage of distinguishing 'dates' from 'intervals of time' but it is not universally applied.

19.1.1 Geological time units

It has commonly been the practice to distinguish between *geochronology*, which is concerned with geological time units and *chronostratigraphy*, which refers to material stratigraphic units. The difference between these is that the former is an interval of time that is expressed in years, whereas the latter is a unit of rock: for example, the Chalk strata in northwest Europe form a part of the Cretaceous System, a unit of rock, and they were deposited in shallow seas which existed in the area during a period of time that we call the Cretaceous Period, an interval of time. There is a hierarchical set of terms for geochronological units that has an exact parallel in chronostratigraphic units (Fig. 19.1), but the distinction between the two sets of terminologies is not made by all geologists and some (e.g. Zalasiewicz et al. 2004) question whether it is either useful or necessary to employ this dual stratigraphic terminology. The argument for maintaining both is that it provides a distinction between the physical reality of the strata themselves, the rocks of, say, the Silurian System, and the more abstract concept of the time interval during which they were deposited, which would be the Silurian Period. However, as Zalasiewicz et al. (2004) point

Geochronological units (intervals of time measured in years)	Chronostratigraphic units (material units defined by the ages of the rocks within them)
Eons (e.g. Phanerozoic Eon)	
Eras (e.g. Cenozoic Era)	
Periods (e.g. Neogene Period)	Systems (e.g. Neogene System)
Epochs (e.g. Miocene Epoch)	Series (e.g. Miocene Series)
Ages	Stages (e.g. Messinian Stage)
Divisions into 'Early', 'Middle' and 'Late'	Divisions into 'Lower', 'Middle' and 'Upper'

Fig. 19.1 Nomenclature used for geochronological and chronostratigraphic units.

out, the use of 'golden spikes' (see below) for stratigraphic correlation means that the beginning and end of the period of time are now defined by a physical point in a succession of strata, and thus there is no real need to distinguish between the 'time unit' (geochronology) and the 'rock unit' (chronostratigraphy) as they amount to the same thing. The terms for the geochronological units are described below, with the equivalent chronostratigraphic units also noted where they are also in common use.

Eons

These are the longest periods of time within the history of the Earth, which are now commonly divided into three *eons*: the Archaean Eon up to 2.5 Ga, the Proterozoic Eon from 2.5 Ga to 542 Ma (together these constitute the Precambrian), and the Phanerozoic Eon from 542 Ma up to the present.

Eras

Eras are the three time divisions of the Phanerozoic: the Palaeozoic Era up to 251 Ma, the Mesozoic Era from then until 65.5 Ma and finally the Cenozoic Era up to the present. Precambrian eras have also been defined, for example dividing the Proterozoic into the Palaeoproterozoic, the Mesoproterozoic and the Neoproterozoic.

Periods/Systems

The basic unit of geological time is the **period** and these are the most commonly used terms when referring to Earth history. The Mesozoic Era, for example, is divided into three periods, the Triassic Period, the Jurassic Period and the Cretaceous Period. The term **system** is used for the rocks deposited in this time, e.g. the Jurassic System.

Epochs/Series

Epochs are the major divisions of periods: some have names, for example the Llandovery, Wenlock, Ludlow and Pridoli in the Silurian, while others are simply Early, (Mid-) and Late divisions of the period (e.g. Early Cretaceous and Late Cretaceous). The chronostratigraphic equivalent is the **series**, but it is important to note that the terms Lower, Middle and Upper are used instead of Early, Middle and Late. As an example, rocks that belong to the Lower Triassic Series were deposited in the Early Triassic Epoch. Logically a body of rock cannot be 'Early', nor can a period of time be considered 'Lower' so it is important to employ the correct adjective and use, for example, 'Early Jurassic' when referring to events which took place during that time interval.

Ages/Stages

The smallest commonly used divisions of geological time are **ages**, and the chronostratigraphic equivalent is the **stage**. They are typically a few million years in duration. For example, the Oligocene Epoch is divided into the Rupelian and Chattian Ages (the Rupelian and Chattian Stages of the Oligocene Series of rocks).

Chrons are short periods of time that are sometimes determined from palaeomagnetic information, but these units do not have widespread usage outside of magnetostratigraphy (21.4). The Quaternary can be divided into short time units of only thousands to tens of thousands of years using a range of techniques available for dating the recent past, such as marine isotope stages (21.5).

19.1.2 Stratigraphic charts

The division of rocks into stratigraphic units had been carried out long before any method of determining the

geological time periods had been developed. The main systems had been established and partly divided into series and stages by the beginning of the 20th century by using stratigraphic relations and biostratigraphic methods. Radiometric dating has provided a time scale for the chronostratigraphic division of rocks. The published geological time scales (Fig. 19.2) have been constructed by integrating information from biostratigraphy, magnetostratigraphy and data from radiometric dating to determine the chronostratigraphy of rock units throughout the Phanerozoic.

A simplified version of the most recent version of the international stratigraphic chart published by Gradstein et al. (2004) is shown in Fig. 19.2 (Gradstein & Ogg 2004). This shows the names of the stratigraphic units that have been agreed by the International Commission on Stratigraphy and the ages, in millions of years, of the boundaries between each unit. The ages shown are based on the best available evidence and are not definitive. For reasons that will be discussed in Chapter 21, it is often difficult to directly measure the ages of a body of sedimentary rocks in terms of millions of years. Strata are normally defined stratigraphically as being, for example, 'Oxfordian' on the basis of the fossils that they contain (Chapter 20) or the physical relationships that they have with other rock units (see Lithostratigraphy). The time interval of the Oxfordian, 161.2 Ma to 155.0 Ma, that is shown on the chart is subject to change as new information from radiometric dating becomes available, or a recalibration is carried out. Older versions of these stratigraphic charts show different ages for boundaries, and no doubt future charts will also contain modifications to these dates. A unit of sedimentary rocks is therefore never referred to as being, say, 160 Ma old unless there has been a direct radiometric measurement made of that unit: instead it might be referred to as Oxfordian on the basis of its fossils, and this will not change, whatever happens in future versions of these charts.

19.1.3 Golden spikes

From the foregoing it should be clear that the Cambrian, for example, is not defined as the interval of time between 542 Ma and 488.3 Ma, but those numbers are the ages that are currently thought to be the times when the Cambrian Period started and ended. It is

Eon	Era	Period/System	Epoch/Series		Age (Ma)
Phanerozoic	Cenozoic	Neogene	Holocene		0.01
			Pleistocene		1.8
			Pliocene		5.3
			Miocene		23.0
		Paleogene	Oligocene		33.9
			Eocene		55.8
			Paleocene		65.5
	Mesozoic	Cretaceous	Upper		99.6
			Lower		145.5
		Jurassic	Upper		161.2
			Middle		175.8
			Lower		199.6
		Triassic	Upper		228.0
			Middle		245.0
			Lower		251.0
	Paleozoic	Permian	Lopingian		260.4
			Guadalupian		270.6
			Cisuralian		299.0
		Carboniferous	Pennsylvanian	Upper	306.3
				Middle	311.7
				Lower	318.1
			Mississippian	Upper	326.4
				Middle	345.3
				Lower	359.2
		Devonian	Upper		385.2
			Middle		397.5
			Lower		416.0
		Silurian	Pridoli		418.7
			Ludlow		422.9
			Wenlock		428.2
			Llandovery		443.7
		Ordovician	Upper		460.9
			Middle		471.8
			Lower		488.3
		Cambrian	Furongian		501.0
			Middle		513.0
			Lower		542.0
Proterozoic	Neoproterozoic				1000
	Mesoproterozoic				1600
	Paleoproterozoic				2500
Archean	Neoarchean				2800
	Meosarchean				3200
	Paleoarchean				3800
	Eoarchean				4560

Fig. 19.2 A stratigraphic chart with the ages of the different divisions of geological time. (Data from Gradstein et al. 2004.)

therefore necessary to have some other means of defining all of the divisions of the geological record, and the internationally accepted approach is to use the '**Global Standard Section and Point**' (**GSSP**) scheme, otherwise known as the process of establishing '**golden spikes**'.

Some of the periods of the Phanerozoic were originally named after the areas where the rocks were first described in the 18th and 19th centuries: the Cambrian from Wales (the Roman name of which was Cambria), the Devonian from Devon, England, the Permian after an area in Russia and the Jurassic from the Jura mountains of France. (Others were given names associated with a region, such as the Ordovician and Silurian Periods that have their origins in the names of ancient Welsh tribes, and some have names related to the character of the rocks, such as the Carboniferous, coal-bearing, and Cretaceous, from the Latin for chalk). This effectively established the principle of a 'type area', a region where the rocks of that age occur that could act as a reference for other occurrences of similar rocks. It was, in fact, mainly the fossil content that provided the means of correlating: if strata from two different places contain the same fossils, they are considered to be from the same period – this is the basis of biostratigraphic correlation (*20.6*).

The GSSP scheme takes the 'type area' concept further by defining the base of a period or epoch as a particular point, in a particular succession of strata, in a particular place. A 'golden spike' is metaphorically hammered into the rocks at that point, and all beds above it are defined as belonging to one epoch/period and all below it to another (Walsh et al. 2004). All other beds of similar age around the world are then correlated with the strata that contain the 'golden spike', using any of the correlation techniques that are described in this and the following chapters (lithostratigraphy, biostratigraphy, magnetostratigraphy, and so on). The locations chosen are typically ones with fossiliferous strata, because the fossils can be used for biostratigraphic correlation. Successions where there appears to be continuous sedimentation are also preferred so that all of the time interval is represented by beds of material: if there is a gap in the record at the GSSP location due to a break in sedimentation there is the possibility that there are rocks elsewhere which represent a time interval that has no equivalent at the GSSP site, and these beds could therefore not be defined as being of one unit or the other. The exact choice

of horizon is usually made on the basis of fossil content: the base of the Devonian, for example, is defined by a golden spike in a succession of marine strata in the Czech Republic at a point where a certain graptolite is found in higher beds, but not in the lower beds.

Golden Spikes have been established for about half the Age/Stage boundaries in the Phanerozoic, with the remainder awaiting the location of a suitable site and international agreement. The procedure of defining GSSPs cannot easily be applied in older rocks because it is essentially a biostratigraphic approach. The scarcity of stratigraphically useful fossils in Precambrian strata means that only one pre-Phanerozoic system has been defined so far: this is the Ediacaran Period/System, the youngest part of the Neoproterozoic Era. Other Precambrian boundaries have been ascribed with ages, a Global Standard Stratigraphic Age, or 'GSSA'. Therefore, in contrast to the Phanerozoic, the Precambrian is largely defined in terms of the age of the rocks in millions of years: for example, the Palaeoproterozoic is an era that is defined as being between 2500 Ma and 1600 Ma.

19.2 STRATIGRAPHIC UNITS

The *International Stratigraphic Chart* and the *Geologic Time Scale* that it shows provides an overall framework within which we can place all the rocks on Earth and the events that have taken place since the planet formed. It is, however, of only limited relevance when faced with the practical problems of determining the stratigraphic relationships between rocks in the field. Strata do not have labels on them which immediately tell us that they were deposited in a particular epoch or period, but they do contain information that allows us to establish an order of formation of units and for us to work out where they fit in the overall scheme. There are a number of different approaches that can be used, each based on different aspects of the rocks, and each of which is of some value individually, but are most profitably used in combinations.

First, a body of rock can be distinguished and defined by its lithological characteristics and its stratigraphic position relative to other bodies of rock: these are **lithostratigraphic units** and they can readily be defined in layered sedimentary rocks. Second, a body of rock can be defined and characterised by its fossil content, and this would be considered to be a

'Layer-cake' stratigraphy

Stratigraphic relations around a
reef or similar structure

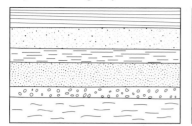

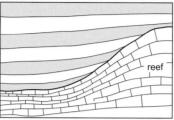

reef

Fig. 19.3 Principles of superposition: (a) a 'layer-cake' stratigraphy; (b) stratigraphic relations around a reef or similar feature with a depositional topography.

biostratigraphic unit (*20.6*). Third, where the age of the rock can be directly or indirectly determined, a *chronostratigraphic unit* can be defined (*21.3.3*): chronostratigraphic units have upper and lower boundaries that are each *isochronous surfaces*, that is, a surface that formed at one time. The fourth type of stratigraphic unit is a *magnetostratigraphic unit*, a body of rock which exhibits magnetic properties that are different to adjacent bodies of rock in the stratigraphic succession (*21.4.3*). Finally, bodies of rock can be defined by their position relative to unconformities or other correlatable surfaces: these are sometimes called *allostratigraphic units*, but this approach is now generally referred to as 'Sequence Stratigraphy', which is the subject of Chapter 23. Each of these approaches to stratigraphy are covered in this and the following chapters.

19.3 LITHOSTRATIGRAPHY

In *lithostratigraphy* rock units are considered in terms of the lithological characteristics of the strata and their relative stratigraphic positions. The relative stratigraphic positions of rock units can be determined by considering geometric and physical relationships that indicate which beds are older and which ones are younger. The units can be classified into a hierarchical system of members, formations and groups that provide a basis for categorising and describing rocks in lithostratigraphic terms.

19.3.1 Stratigraphic relationships

Superposition

Provided the rocks are the right way up (see below) the beds higher in the stratigraphic sequence of depos-

its will be younger than the lower beds. This rule can be simply applied to a layer-cake stratigraphy but must be applied with care in circumstances where there is a significant depositional topography (e.g. fore-reef deposits may be lower than reef-crest rocks: Fig. 19.3).

Unconformities

An *unconformity* is a break in sedimentation and where there is erosion of the underlying strata this provides a clear relationship in which the beds below the unconformity are clearly older than those above it (Figs 19.4 & 19.5). All rocks which lie above the unconformity, or a surface that can be correlated with it, must be younger than those below. In cases where strata have been deformed and partly eroded prior to deposition of the younger beds, an angular unconformity is formed. A *disconformity* marks a break in sedimentation and some erosion, but without any deformation of the underlying strata.

Cross-cutting relationships

Any unit that has boundaries that cut across other strata must be younger than the rocks it cuts. This is most commonly seen with intrusive bodies such as batholiths on a larger scale and dykes on a smaller scale (Fig. 19.6). This relationship is also seen in fissure fills, sedimentary dykes (*18.1.3*) that form by younger sediments filling a crack or chasm in older rocks.

Included fragments

The fragments in a clastic rock must be made up of a rock that is older than the strata in which they are

Unconformities

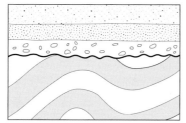
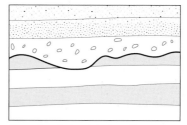

Angular unconformity: deformation and
erosion prior to deposition of younger beds

Disconformity: break in deposition and
erosion within a stratigraphic succession

Fig. 19.4 Gaps in the record are represented by unconformities: (a) angular unconformities occur when older rocks have been deformed and eroded prior to later deposition above the unconformity surface; (b) disconformities represent breaks in sedimentation that may be associated with erosion but without deformation.

Cross-cutting relationships

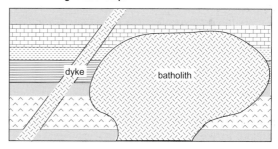

Included fragments

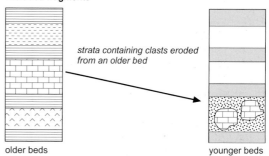

*strata containing clasts eroded
from an older bed*

older beds

younger beds

Fig. 19.5 An angular unconformity between horizontal sandstone beds above and steeply dipping shaly beds below.

Fig. 19.6 Stratigraphic relationships can be simple indicators of the relative ages of rocks: (a) cross-cutting relations show that the igneous features are younger than the sedimentary strata around them; (b) a fragment of an older rock in younger strata provides evidence of relative ages, even if they are some distance apart.

found (Fig. 19.6). The same relationship holds true for igneous rocks that contain pieces of the surrounding country rock as **xenoliths** (literally 'foreign rocks'). This relationship can be useful in determining the age relationship between rock units that are some distance apart. Pebbles of a characteristic lithology can

provide conclusive evidence that the source rock type was being eroded by the time a later unit was being deposited tens or hundreds of kilometres away.

Way-up indicators in sedimentary rocks

The folding and faulting of strata during mountain building can rotate whole successions of beds (formed as horizontal or nearly horizontal layers) through any angle, resulting in beds that may be vertical or completely overturned. In any analysis of deformed strata, it is essential to know the direction of *younging*, that is, the direction through the layers towards younger rocks. The direction of younging can be determined by small-scale features that indicate the way-up of the beds (Fig. 19.7) or by using other stratigraphic techniques to determine the order of formation.

19.3.2 Lithostratigraphic units

There is a hierarchical framework of terms used for lithostratigraphic units, and from largest to smallest these are: 'Supergroup', 'Group', 'Formation', 'Member' and 'Bed'. The basic unit of lithostratigraphic division of rocks is the *formation*, which is a body of material that can be identified by its lithological characteristics and by its stratigraphic position. It must be traceable laterally, that is, it must be mappable at the surface or in the subsurface. A formation should have some degree of lithological homogeneity and its defining characteristics may include mineralogical composition, texture, primary sedimentary structures and fossil content in addition to the lithological composition. Note that the material does not necessarily have to be lithified and that all the discussion of terminology and stratigraphic relationships applies equally to unconsolidated sediment.

A formation is not defined in terms of its age either by isotopic dating or in terms of biostratigraphy. Information about the fossil content of a mapping unit is useful in the description of a formation but the detailed taxonomy of the fossils that may define the relative age in biostratigraphic terms does not form part of the definition of a lithostratigraphic unit. A formation may be, and often is, a *diachronous unit*, that is, a deposit with the same lithological properties that was formed at different times in different places (*19.4.2*).

A formation may be divided into smaller units in order to provide more detail of the distribution of lithologies. The term *member* is used for rock units that have limited lateral extent and are consistently related to a particular formation (or, rarely, more than one formation). An example would be a formation composed mainly of sandstone but which included beds of conglomerate in some parts of the area of outcrop. A number of members may be defined

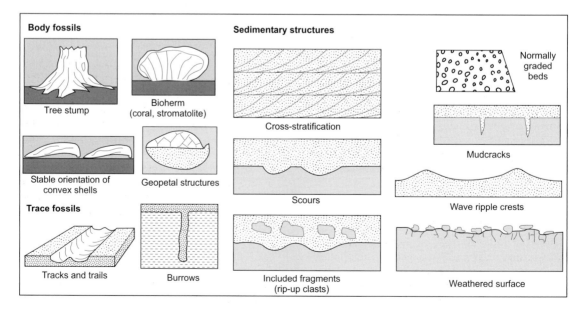

Fig. 19.7 Way-up indicators in sedimentary rocks.

within a formation (or none at all) and the formation does not have to be completely subdivided in this way: some parts of a formation may not have a member status. Individual **beds** or sets of beds may be named if they are very distinctive by virtue of their lithology or fossil content. These beds may have economic significance or be useful in correlation because of their easily recognisable characteristics across an area.

Where two or more formations are found associated with each other and share certain characteristics they are considered to form a **group**. Groups are commonly bound by unconformities which can be traced basin-wide. Unconformities that can be identified as major divisions in the stratigraphy over the area of a continent are sometimes considered to be the bounding surfaces of associations of two or more groups known as a **supergroup**.

19.3.3 Description of lithostratigraphic units

The formation is the fundamental lithostratigraphic unit and it is usual to follow a certain procedure in geological literature when describing a formation to ensure that most of the following issues are considered. Members and groups are usually described in a similar way.

Lithology and characteristics

The field characteristics of the rock, for example, an oolitic grainstone, interbedded coarse siltstone and claystone, a basaltic lithic tuff, and so on form the first part of the description. Although a formation will normally consist mainly of one lithology, combinations of two or more lithologies will often constitute a formation as interbedded or interfingering units. Sedimentary structures (ripple cross-laminations, normal grading, etc.), petrography (often determined from thin-section analysis) and fossil content (both body and trace fossils) should also be noted.

Definition of top and base

These are the criteria that are used to distinguish beds of this unit from those of underlying and overlying units; this is most commonly a change in lithology from, say, calcareous mudstone to coral boundstone. Where the boundary is not a sharp change from one formation to another, but is gradational, an arbitrary

boundary must be placed within the transition. As an example, if the lower formation consists of mainly mudstone with thin sandstone beds, and the upper is mainly sandstone with subordinate mudstone, the boundary may be placed at the point where sandstone first makes up more than 50% of beds. A common convention is for only the base of a unit to be defined at the type section: the top is taken as the defined position of the base of the overlying unit. This convention is used because at another location there may be beds at the top of the lower unit that are not present at the type locality: these can be simply added to the top without a need for redefining the formation boundaries.

Type section

A **type section** is the location where the lithological characteristics are clear and, if possible, where the lower and upper boundaries of the formation can be seen. Sometimes it is necessary for a type section to be composite within a **type area**, with different sections described from different parts of the area. The type section will normally be presented as a graphic sedimentary log and this will form the **stratotype**. It must be precisely located (grid reference and/or GPS location) to make it possible for any other geologist to visit the type section and see the boundaries and the lithological characteristics described.

Thickness and extent

The thickness is measured in the type section, but variations in the thickness seen at other localities are also noted. The limits of the geographical area over which the unit is recognised should also be determined. There are no formal upper or lower limits to thickness and extent of rock units defined as a formation (or a member or group). The variability of rock types within an area will be the main constraint on the number and thickness of lithostratigraphic units that can be described and defined. Quality and quantity of exposure also play a role, as finer subdivision is possible in areas of good exposure.

Other information

Where the age for the formation can be determined by fossil content, radiometric dating or relationships with other rock units this may be included, but note that this does not form part of the definition of the formation. A formation would not be defined as, for example, 'rocks of Burdigalian age', because an interpretation of the fossil content or isotopic dating information is required to determine the age. Information about the facies and interpretation of the environment of deposition might be included but a formation should not be defined in terms of depositional environment, for example, 'lagoonal deposits', as this is an interpretation of the lithological characteristics. It is also useful to comment on the terminology and definitions used by previous workers and how they differ from the usage proposed.

19.3.4 Lithostratigraphic nomenclature

It helps to avoid confusion if the definition and naming of stratigraphic units follows a set of rules. Formal codes have been set out in publications such as the 'North American Stratigraphic Code' (North American Commission on Stratigraphic Nomenclature 1983) and the 'International Stratigraphic Guide' (Salvador 1994). A useful summary of stratigraphic methods, which is rather more user-friendly than the formal documents, is a handbook called 'Stratigraphical Procedure' (Rawson et al. 2002).

The name of the formation, group or member must be taken from a distinct and permanent geographical feature as close as possible to the type section. The lithology is often added to give a complete name such as the Kingston Limestone Formation, but it is not essential, or necessarily desirable if the lithological characteristics are varied. The choice of geographical name should be a feature or place marked on topographic maps such as a river, hill, town or village. The rules for naming members, groups and supergroups are essentially the same as for formations, but note that it is not permissible to use a name that is already in use or to use the same name for two different ranks of lithostratigraphic unit. There are some exceptions to these rules of nomenclature that result from historical precedents, and it is less confusing to leave a well-established name as it is rather than to dogmatically revise it. Revisions to stratigraphic nomenclature may become necessary when more detailed work is carried out or more information becomes available. New work in an area may allow a formation to be subdivided and the formation may then be elevated to the rank of group and members may become formations

in their own right. For the sake of consistency the geographical name is retained when the rank of the unit is changed.

19.3.5 Lithodemic units: non-stratiform rock units

The concepts of division into stratigraphic units were developed for rock bodies that are stratiform, layered units, but many metamorphic, igneous plutonic and structurally deformed rocks are not stratiform and they do not follow the rules of superposition. Non-stratiform bodies of rock are called **lithodemic units**. The basic unit is the **lithodeme** and this is equivalent in rank to a formation and is also defined on lithological criteria. The word 'lithodeme' is itself rarely used in the name: the body of rock is normally referred to by its geographical name and lithology, such as the White River Granite or Black Hill Schist. An association of lithodemes that share lithological properties, such as a similar metamorphic grade, is referred to as a **suite**: the term **complex** is also used as the equivalent to a group for volcanic or tectonically deformed rocks.

19.4 APPLICATIONS OF LITHOSTRATIGRAPHY

19.4.1 Lithostratigraphy and geological maps

Part of the definition of a formation is that it should be a 'mappable unit', and in practice this usually means that the unit can be represented on a map of a scale of 1:50,000, or 1:100,000. Maps at this scale therefore show the distribution of formations and may also show where members and named beds occur. The stratigraphic order and, where appropriate, lateral relationships between the different lithostratigraphic units are normally shown in a stratigraphic key at the side of the map. In regions of metamorphic, intrusive igneous and highly deformed rocks the mapped units are lithodemes. There are no established rules for the colours used for different lithostratigraphic and lithodemic units on these maps, but each national geological survey usually has its own scheme. Geological maps that cover larger areas, such as a whole country or a continent, are different: they usually show the distribution of rocks in terms of chronostratigraphic units, that is, on the basis of their age, not lithology.

19.4.2 Lithostratigraphy and environments

It is clear from the earlier chapters on the processes and products of sedimentation that the environment of deposition has a fundamental control on the lithological characteristics of a rock unit. A formation, defined by its lithological characteristics, is therefore likely to be composed of strata deposited in a particular sedimentary environment. This has two important consequences for any correlation of formations in any chronostratigraphic (time) framework.

First, in any modern environment it is obvious that fluvial sedimentation can be occurring on land at the same time as deposition is happening on a beach, on a shelf and in deeper water. In each environment the characteristics of the sediments will be different and hence they would be considered to be different formations if they are preserved as sedimentary rocks. It inevitably follows that formations have a limited lateral extent, determined by the area of the depositional environment in which they formed and that two or more different formations can be deposited at the same time.

Second, depositional environments do not remain fixed in position through time. Consider a coastline (Fig. 19.8), where a sandy beach (foreshore) lies between a vegetated coastal plain and a shoreface succession of mudstones coarsening up to sandstones. The foreshore is a spatially restricted depositional environment: it may extend for long distances along a coast, but seawards it passes into the shallow marine, shoreface environment and landwards into continental conditions. The width of deposit produced in a beach and foreshore environment may therefore be only a few tens or hundreds of metres. However, a foreshore deposit will end up covering a much larger area if there is a gradual rise or fall of sea level relative to the land. If sea level slowly rises the shoreline will move landwards and through time the place where sands are being deposited on a beach would have moved several hundreds of metres (Fig. 19.8). These depositional environments (the coastal plain, the sandy foreshore and the shoreface) will each have distinct lithological characteristics that would allow them to be distinguished as mappable formations. The foreshore deposits could therefore constitute a forma-

tion, but it is also clear that the beach deposits were formed earlier in one place (at the seaward extent) than another (at the landward extent). The same would be true of formations representing the deposits of the coastal plain and shoreface environments: through time the positions of the depositional environments migrate in space. From this example, it is evident that the body of rock that constitutes a formation would be diachronous and both the upper and lower boundaries of the formation are diachronous surfaces.

There is also a relationship between environments of deposition and the hierarchy of lithostratigraphic units. In the case of a desert environment there may be three main types of deposits (Fig. 8.12): aeolian sands, alluvial fan gravels and muddy evaporites deposited in an ephemeral lake. Each type of deposit would have distinctive lithological characteristics that would allow them to be distinguished as three separate formations, but the association of the three could usefully be placed into a group. A distinct change in environment, caused, perhaps, by sea-level rise and marine flooding of the desert area, would lead to a different association of deposits, which in lithostratigraphic terms would form a separate group. Subdivision of the formations formed in this desert environment may be possible if scree deposits around the edge of the basin occur as small patches amongst the other facies. When lithified the scree would form a sedimentary breccia, recognisable as a separate member within the other formations, but not sufficiently widespread to be considered a separate formation.

19.4.3 Lithostratigraphy and correlation

Correlation in stratigraphy is usually concerned with considering rocks in a *temporal framework*, that is, we want to know the time relationships between different rock units – which ones are older, which are younger and which are the same age. Correlation on the basis of lithostratigraphy alone is difficult because, as discussed in the previous section, lithostratigraphic units are likely to be diachronous. In the example of the lithofacies deposited in a beach environment during a period of rising sea level (Fig. 19.8) the lithofacies has different ages in different places. Therefore the upper and lower boundaries of this lithofacies will cross *time-lines* (imaginary lines drawn across

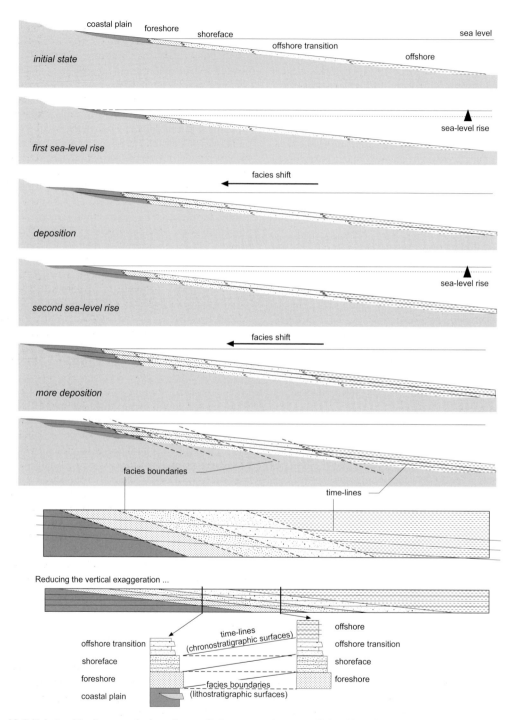

Fig. 19.8 Relationships between the boundaries of lithostratigraphic units (defined by lithological characteristics resulting from the depositional environment) and time-lines in a succession of strata formed during gradual sea-level rise (transgression).

and between bodies of rock which represent a moment in time).

If we can draw a time-line across our rock units, or, more usefully, a time-plane through an area of different strata, we would be able to reconstruct the distribution of palaeoenvironments at that time across that area. To carry out this exercise of making a palaeogeographic reconstruction we need to have some means of chronostratigraphic correlation, a means of determining the relative age of rock units which is not dependent on their lithostratigraphic characteristics.

Radiometric dating techniques (*21.2*) provide an absolute time scale but are not easy to apply because only certain rock types can be usefully dated. Biostratigraphy provides the most widely used time framework, a relative dating technique that can be related to an absolute time scale, but it often lacks the precision required for reconstructing environments and in some depositional settings appropriate fossils may be partly or totally absent (in deserts, for example). Palaeomagnetic reversal stratigraphy provides time-lines, events when the Earth's magnetism changed polarity, and may be applied in certain circumstances. The concept of sequence stratigraphy provides an approach to analysing successions of sedimentary rocks in a temporal framework. In practice a number of different correlation techniques (Chapters 20–23) are used in developing a temporal framework for rock units.

19.4.4 Lithostratigraphy and time: gaps in the record

One of the most difficult questions to answer in sedimentology and stratigraphy is 'how long did it take to form that succession of rocks?'. From our observations of sedimentary processes we can sometimes estimate the time taken to deposit a single bed: a debris-flow deposit on an alluvial fan may be formed over a few minutes to hours and a turbidite in deep water may have been accumulated over hours to days. However, we cannot simply add up the time it takes to deposit one bed in a succession and multiply it by the number of beds. We know from records of modern alluvial fans and deep seas that most of the time there is no sediment accumulating and that the time between depositional events is much longer than the duration of each event: in the case of the alluvial fan deposits and turbidites there may be hundreds or thousands of years between events. If we consider a succession of beds in terms of the passage of time, most of the time is represented by the surfaces that separate the beds: for example, if a debris flow event lasting one hour occurs every 100 years the time represented by the surfaces between beds is about a million times longer than the time taken to deposit the conglomerate. This is not a particularly extreme example: in many environments the time periods between events are much longer than events themselves – floods in the overbank areas of rivers and delta tops, storm deposits on shelves, volcanic ash accumulations, and so on. The exceptions are those places where material is gradually accumulating due to biogenic activity, such as a coral reef boundstone.

A bedding plane therefore represents a gap in the record, a *hiatus* in sedimentation, also sometimes referred to as a *lacuna* (plural *lacunae*). Usually we can only guess at how long the hiatus lasted, and our estimates may be at best to the nearest order of magnitude: were alluvial fan sedimentation events occurring every 100 years or every 1000 years? – both are equally plausible guesses. There are, however, some features that provide us with clues about the relative periods of time represented by the bedding surface. In continental environments, soils form on exposed sediment surfaces and the longer the exposure, the more mature the soil: analysis of palaeosols (*9.7.2*) can therefore provide some clues and we can conclude that a very mature palaeosol profile in a succession would have formed during a long period without sedimentation. In shallow marine environments the sea floor is bioturbated by organisms, and the intensity of the bioturbation on a bedding surface can be used as an indicator of the length of time before the next depositional event. Sediment on the sea floor can also become partly or wholly lithified if left for long enough, and it may be possible to recognise firmgrounds, with associated *Glossifungites*-type ichnofauna, and hardgrounds with a *Trypanites* ichnofacies assemblage (*11.7.2*).

Unconformities represent even longer gaps in the depositional record. On continental margins a sea-level fall may expose part of the shelf area, resulting in a period of non-deposition and erosion that will last until the sea level rises again after a period of time lasting tens to hundreds of thousands or millions of years. This results in an unconformity surface within the strata that represents a time period of that order of magnitude. Plate tectonics results in vertical movements of the crust and areas that were once places of sediment accumulation may become uplifted and eroded. Later crustal movements may cause subsidence, and the erosion surface will become preserved as an unconformity as it is overlain by younger sediment. Unconformity surfaces formed in this way may represent anything from less than a million to a billion years or more.

The problems of determining how long it takes to deposit a succession of beds and the unknown periods of time represented by any lacunae, from a bedding plane to an unconformity, make it all-but impossible to gauge the passage of time from the physical characteristics of a sedimentary succession. In the 18th and 19th centuries various different estimates of the age of the Earth were made by geologists and these were all wildly different from the 4.5 Ga we now know to be the case because they did not have any way of judging the period of time represented by the rocks in the stratigraphic record. Radiometric dating now provides us with a time frame that we can measure in years. This has made it possible to calibrate the stratigraphic chart that had already been developed for the Phanerozoic based on the occurrences of fossils.

FURTHER READING

Blatt, H., Berry, W.B. & Brande, S. (1991) *Principles of Stratigraphic Analysis*. Blackwell Scientific Publications, Oxford.

Boggs, S. (2006) *Principles of Sedimentology and Stratigraphy* (4th edition). Pearson Prentice Hall, Upper Saddle River, NJ.

Doyle, P., Bennett, M.R. & Baxter, A.N. (1994) *The Key to Earth History: an Introduction to Stratigraphy*. John Wiley and Sons, Chichester.

Friedman, G.M., Sanders, J.E. & Kopaska-Merkel, D.C. (1992) *Principles of Sedimentary Deposits: Stratigraphy and Sedimentology*. Macmillan, New York.

Gradstein, F. & Ogg, J. (2004) Geologic Time Scale 2004 — why, how, and where next! *Lethaia*, **37**, 175–181.

Rawson, P.F., Allen, P.M., Brenchley, P.J., Cope, J.C.W., Gales, A.S., Evans, J.A., Gibbard, P.L., Gregory, F.J., Hailwood, E.A., Hesselbo, S.P., Knox, R.W. O'B., Marshall, J.E.A., Oates, M., Riley, N.J., Smith, A.G., Trewin, N. & Zalasiewicz, J.A. (2002). *Stratigraphical Procedure*. Professional Handbook, Geological Society Publishing House, Bath, 58 pp.

Salvador, A. (1994) *International Stratigraphic Guide. A Guide to Stratigraphic Classification, Terminology and Procedure* (2nd edition). The International Union of Geological Sciences and the Geological Society of America.

Zalasiewicz, J., Smith, A., Brenchley, P., Evans, J., Knox, R., Riley, N., Gale, A., Gregory, F.J., Rushton, A., Gibbard, P., Hesselbo, S., Marshall, J., Oates, M. Rawson, P. & Trewin, N. (2004) Simplifying the stratigraphy of time. *Geology*, **32**, 1–4.

20

Biostratigraphy

The occurrence of fossils in beds of sedimentary rocks provided the basis for correlation of strata and the concept of a stratigraphic column when the science of geology was still young. The fundamental importance of biostratigraphy has not diminished through time, but has merely been complemented by other stratigraphic techniques discussed in preceding and following chapters. The evolution of organisms through time and the formation of new species provide the basis for the recognition of periods in the history of the Earth on the basis of the fossils that are contained within strata. In this way Earth history can be divided up into major units that are now known to represent hundreds of millions of years, some of which are familiarly known as 'the age of fish', 'the age of reptiles' and so on, because of the types of fossils found. Fossils also provide high-resolution stratigraphic tools that allow recognition of time slices of only tens to hundreds of thousands of years that are important for building up a detailed picture of events through time. Correlation between biostratigraphic units and the geological time scale therefore provides the temporal framework for the analysis of successions of sedimentary rocks.

20.1 FOSSILS AND STRATIGRAPHY

The importance of fossils as indicators of processes and environments of deposition has been mentioned in previous chapters, but the study of fossils has also provided fundamental information about the evolution of life on Earth. Skeletons and shells of animals or pieces of plant that are found as fossils are clear evidence of the fact that the nature of organisms living on the planet has changed through time. Some of these fossils resemble plants or animals living today and are evidently related to modern lifeforms, whereas others are unlike anything we are familiar with. The more spectacular of these fossils tend to capture the imagination with visions of times in the past when, for example, dinosaurs occupied ecological niches on land, in the sea and even in the air. Even casual fossil hunting reveals the remains of aquatic animals such as ammonites and fragments of plants that are unlike anything we see living around us now.

Cataloguing the fossils found in sedimentary rocks carried out in the 18th and 19th centuries provided the first clues about the passage of geological time. Early scientists and naturalists observed that different rock units contained either similar fossil remains or assemblages of fossils that were quite different from one unit to another. Moreover, the units that contained the same fossils could sometimes be traced laterally and shown to be part of the same layer. Those with different fossils could be shown by general stratigraphic principles to be either younger or older. The rocks that contained a particular fossil type were often the same lithology, but, crucially for the development of stratigraphy, sometimes the same fossil type was found in a different rock type.

With advances in the science of palaeontology it became evident that there were patterns in the distribution of fossils. Certain types of organism were found to be dominant in particular groups of strata. This led to the erection of the scheme of systems that were initially grouped into deposits formed in three eras of geological time (*19.1.1*): 'ancient life', the Palaeozoic, 'middle life', the Mesozoic and 'recent life', the Cenozoic. The actual time periods that these represented were pure speculation when these concepts were first introduced in the 19th century and the numerical ages for these eras were not known until techniques for radiometric dating were developed. The occurrence of certain types of fossils in particular stratigraphic units was simply an observation at this stage: an explanation for the distribution of the fossils in the stratigraphic record came once ideas of the evolution of life were developed.

20.2 CLASSIFICATION OF ORGANISMS

20.2.1 Species

The concept of *species*, originally defined as groups of interbreeding organisms that are reproductively isolated from other such groups, is fundamental to the classification of organisms. Modern biological analyses provide additional information that also allows the genetic characteristics of organisms to be considered when defining a species: similarities or differences in genetic make-up make it possible to rigorously define species and determine the relationships between them. Genetics therefore provide a basis for a hierarchical system of classification, although in fact such a classification system, the ***Linnaean System***, existed long before the nature of genes was understood. In the Linnaean scheme, closely related species belong to the same genus, similar genera belong to a family, and so on up to the largest unit of classification, the Kingdom (Fig. 20.1). The general term for any one of the ranks defined by the Linnaean System is a ***taxon*** (plural ***taxa***) and the fundamental taxon rank is the species.

The system was developed for living organisms but the same nomenclature and classification scheme is also used in palaeontology. However, applying the definitions of a species to fossils is problematic because it is not usually possible to demonstrate a capacity to interbreed and genetic material is usually only extractable from relatively recent fossil material: in the vast majority of cases the DNA material is too degraded in fossils. Palaeontologists therefore have to work on similarities or differences of morphology to define species and because the soft parts of organisms are only preserved in extraordinary circumstances, it is the hard parts that are principally used. The hard-part morphology is not necessarily a reliable way of defining species: there are many examples of similar-looking organisms that are genetically distinct (especially amongst birds), and at the same time some species show considerable variations in form (such as dogs). Therefore there is always an element of doubt about whether a similarity of skeletal form in a fossil is sufficient basis to assume membership of the same species.

In the definition of a fossil species it has been the practice to establish a ***holotype***, that is, a single representative specimen against which other potential representatives of the species can be compared. Additional information that is used to define a species may now include statistics about the shape and size of the organism, the ***morphometrics***, along with infor-

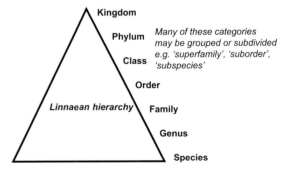

Fig. 20.1 The Linnaean hierarchical system for the taxonomy of organisms.

mation about associated fauna and the palaeoenvironmental habitat.

20.2.2 Other ranks in taxonomy

Subspecies and **races** are distinct sets which show common characteristics that set them apart from others, but which can still be considered to be part of the same species. The variations are often due to geographical separation of the sets leading to the development of different characteristics. The concept of subspecies is used in palaeontology, although the genetic basis for this cannot usually be established.

A **genus** (plural **genera**) is a group of species that are closely related, and when an organism is named it is given a genus as well as a species: for example *Homo sapiens* is the Linnaean classification name for the human species. In palaeontology species level identification is normally only required for biostratigraphic purposes, otherwise it is common to identify and classify a fossil only to generic level. For example, if fossil oysters are found in a limestone, they may be simply referred to as *Ostraea* as identification to this level

provides sufficient palaeoenvironmental information without the need to identify the particular species of *Ostraea*. (Note the conventions used in referring to species and genera: the first letter of the genus name is always capitalised, while the species is always in lower case, and *italics* are used in printed text.)

The higher ranks in the hierarchy are family, order, class, phylum and kingdom in order of scale (Fig. 20.1). The major phyla (Mollusca, Arthropoda, etc.: Fig. 20.2) have existed through the Phanerozoic and it is possible to compare fossils to modern representatives of these subsets of the main kingdoms (animal and plant). However, some classes, many orders and a large number of families have been identified as fossils but have no modern equivalents. The ammonites, for example, formed a very large and diverse order from Ordovician to Cretaceous times, but there are no modern equivalents, only organisms such as nautiloids that belong to the same class, the Cephalopoda, in the phylum Mollusca. The graptolites, which are commonly found in Palaeozoic rocks, form a class of which there are no modern representatives. As the similarities to modern organisms become fewer, the problems of classification become greater as the

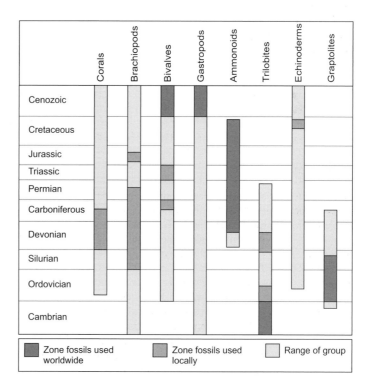

Fig. 20.2 Major groups of organisms preserved as macrofossils in the stratigraphic record and their age ranges.

significance of morphological differences is less well understood. The classification of fossils in Linnaean hierarchy is therefore in a constant state of flux as new fossil discoveries are made that shed light on the probable relationships between fossil organisms.

20.3 EVOLUTIONARY TRENDS

There is a general trend of increasing complexity and sophistication of lifeforms starting from those that occur in older rocks and finishing with the biosphere around us today. Along with this trend there is evidence of the emergence and diversification of organisms as well as signs of decline of some groups from abundance to insignificance or even extinction. Looking at particular groups we see that many of them display trends in morphology through the stratigraphic record and these trends are attributed to the evolutionary development of that group of organisms. It is one of the fundamental precepts of evolutionary theory that these changes are a one-way process: after a particular type of organism has developed a new feature to become more 'advanced', later changes do not result in a return to the more 'primitive' form. The concept of evolutionary trends therefore provides us with a way of interpreting the fossil content of rocks in terms of biological changes through time. This provides a means of correlating rocks and determining their relative ages by the fossils that they contain.

20.3.1 Population fragmentation and phyletic transformation

Many modern species of plants and animals show regional variations in their form. Charles Darwin demonstrated that geographical isolation of a part of a population can lead to changes in characteristics that make them distinctly different from the rest of the species (Darwin 1859). These changes are the result of mutations that are advantageous to the isolated population because they make them better adapted to the local habitat. Examples might be adaptations to make the organisms better suited to higher or lower water salinity, different temperatures, different sediment characteristics or less susceptibility to local predators. This can lead to the development of an isolated group with sufficiently different characteristics from the rest of the population for it to be considered a new species, a process known as *speciation*. This *popula-*

tion fragmentation process results in an increase in the number of species and would lead to an explosion in the number of species through time were it not generally balanced by the **extinction**. A species becomes extinct when its population is no longer well adapted to changing environmental conditions and/or competition from other organisms leads to a terminal decline in numbers.

An alternative process by which a new species may arise is the change through time of the whole population. The ancestral and descendant species are separated by time and therefore the 'interbreeding test' is hypothetical, but they have distinct morphological and, where it can be tested, genetic characteristics. This is called *phyletic transformation* because it is a change in form between the ancestral and descendant members of the evolutionary lineage (or *phyologeny*) and does not lead to an increase in the number of species. The ancestral species disappears, but this is not due to the death of the whole population, so the 'extinction' in this case is considered to be *phyletic extinction* or '*pseudoextinction*'. However, 'real' extinction may still occur and lead to the complete end of a branch of a particular lineage.

20.3.2 Phyletic gradualism and punctuated equilibrium

In its simplest form the theory of evolution implies that genetic changes occur in organisms in response to environmental factors and that these changes eventually result in an organism sufficiently different to be considered a separate species. This would indicate that evolution is a steady, gradual process, a *phyletic gradualism*. An alternative suggestion is that a lineage does not evolve gradually but in an episodic way with periods of *stasis,* when a species does not change genetic make-up or form, followed by short periods of rapid change, when new species develop. This process is called a *punctuated equilibria* pattern of evolution (Eldredge & Gould 1972).

It may be hoped that the stratigraphic record would provide the answer to which of these processes has been dominant. Theoretically, a continuous succession of strata would contain fossils that would reveal either a pattern of gradual change in morphology through time, up the succession, or the form would be the same through the beds and abruptly change to a different morphology at a certain horizon, if the punctuated equilibria idea is correct. In practice, the stratigraphic record does not provide clear answers (Blatt et al. 1991). First, with the exception of some deep marine and lake environments, sedimentation is not continuous and there may be significant periods of non-deposition, and hence no fossils. Furthermore, parts of the sedimentary record may be removed by erosion also leaving a gap in the record. Second, a change in environment may cause populations of a species to move from one location to another and then perhaps recolonise the original site, but only after speciation has occurred: it will then appear that the speciation occurred suddenly at that site, whereas the population itself changed gradually. Third, only a very small proportion of a population is ever fossilised and the stratigraphic record preserves only a tiny fraction of the number of organisms that existed.

20.3.3 Speciation and biostratigraphy

It is the recognition of different species at different stratigraphic horizons that underpins biostratigraphy. If evolution is a punctuated process with periods of stasis followed by rapid change then strata can be defined in terms of distinctly different fossils: speciation would occur too quickly for intermediate forms to be preserved and a zonation scheme can be devised based on the appearance and disappearance of species (Doyle et al. 1994; Doyle & Bennett 1998). A gradual evolution of morphology would mean that a new species would be defined at an arbitrary point in the lineage and hence the zonation would be based on such points. In practice, speciation appears to be a geologically sudden event in many instances because there is an absence of intermediate forms, even if the actual process was, in fact, gradual. Where there does appear to be a phyletic gradualism, it becomes necessary to define species by using statistical treatment of morphological variability, such as the ratio of the length and width of a shell.

20.4 BIOZONES AND ZONE FOSSILS

A biostratigraphic unit is a body of rock defined by its fossil content. It is therefore fundamentally different from a lithostratigraphic unit that is defined by the lithological properties of the rock. The fundamental unit of biostratigraphy is the *biozone*. Biozones are units of stratigraphy that are defined by the *zone fossils* (usually species or subspecies) that they con-

tain. In theory they are independent of lithology, although environmental factors often have to be taken into consideration in the definition and interpretation of biozones. In the same way that formations in lithostratigraphy must be defined from a type section, there must also be a type section designated as a stratotype and described for each biozone. They are named from the characteristic or common taxon (or occasionally taxa) that defines the biozone. There are several different ways in which biozones can be designated in terms of the zone fossils that they contain (Fig. 20.3).

Interval biozones These are defined by the occurrences within a succession of one or two taxa. Where the first appearance and the disappearance of a single taxon is used as the definition, this is referred to as a *taxon-range biozone*. A second type is a *concurrent range biozone*, which uses two taxa with overlapping ranges, with the base defined by the appearance of one taxon and the top by the disappearance of the second one. A third possibility is a *partial range biozone*, which is based on two taxa that do not have overlapping ranges: once again, the base is defined by the appearance of one taxon and the top by the disappearance of a second. Where a taxon can be recognised as having followed another and preceding a third as part of a phyletic lineage the biozone defined by this taxon is called a *lineage biozone* (also called a *consecutive range biozone*).

Assemblage biozones In this case the biozone is defined by at least three different taxa that may or may not be related. The presence and absence, appearance and disappearance of these taxa are all used to define a stratigraphic interval. Assemblage biozones are used in instances where there are no suitable taxa to define interval biozones and they may represent shorter time periods than those based on one or two taxa.

Acme biozones The abundance of a particular taxon may vary through time, in which case an interval containing a statistically high proportion of this taxon may be used to define a biozone. This approach can be unreliable because the relative abundance is due to local environmental factors.

The ideal zone fossil would be an organism that lived in all depositional environments all over the world and was abundant; it would have easily preserved hard parts and would be part of an evolutionary lineage

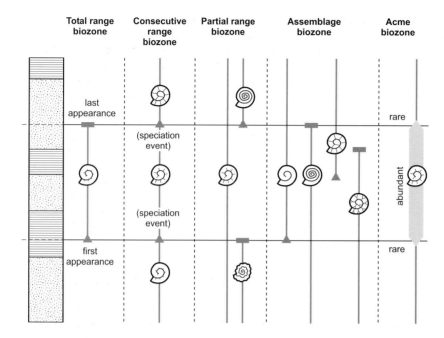

Fig. 20.3 Zonation schemes used in biostratigraphic correlation. (Adapted from North American Commission on Stratigraphic Nomenclature 1983.)

that frequently developed new, distinct species. Not surprisingly, no such fossil taxon has ever existed and the choice of fossils used in biostratigraphy has been determined by a number of factors that are considered in the following sections.

20.4.1 Rate of speciation

The frequency with which new species evolve and replace former species in the same lineage determines the resolution that can be applied in biostratigraphy. Some organisms seem to have hardly evolved at all: the brachiopod *Lingula* seems to look exactly the same today as the fossils found in Lower Palaeozoic rocks and hence is of little biostratigraphic value. The groups that appear to display the highest rates of speciation are vertebrates, with mammals, reptiles and fish developing new species every 1 to 3 million years on average (Stanley 1985). However, the stratigraphic record of vertebrates is poor compared with marine molluscs, which are much more abundant as fossils, but have slower average speciation rates (around 10 million years). There are some groups that appear to have developed new forms regularly and at frequent intervals: new species of ammonites appear to have evolved every million years or so during the Jurassic and Cretaceous and in parts of the

Cambrian some trilobite lineages appear to have developed new species at intervals of about a million years (Stanley 1985). By using more than one species to define them, biozones can commonly be established for time periods of about a million years, with higher resolution possible in certain parts of the stratigraphic record, especially in younger strata.

20.4.2 Depositional environment controls

The conditions vary so much between different depositional environments that no single species, genus or family can be expected to live in all of them. The adaptations required to live in a desert compared with a swamp, or a sandy coastline compared with a deep ocean, demand that the organisms that live in these environments are different. There is a strong environmental control on the distribution of taxa today and it is reasonable to assume that the nature of the environment strongly influenced the distribution of fossil groups as well. Some environments are more favourable to the preservation of body fossils than others: for example, preservation potential is lower on a high-energy beach than in a low-energy lagoon. There is a fundamental problem with correlation between continental and marine environments because very few animals or plants are found in both settings. In the marine environment the most widespread organisms are those that are **planktonic** (free floating) or animals that are **nektonic** (free-swimming lifestyle). Those that live on the sea bed, the **benthonic** or **benthic** creatures and plants, are normally found only in a certain water depth range and are hence not quite so useful.

The rates of sedimentation in different depositional environments are also a factor in the preservation and distribution of stratigraphically useful fossils. Slow sedimentation rates commonly result in poor preservation because the remains of organism are left exposed on the land surface or sea floor where they are subject to biogenic degradation. On the other hand, with a slower rate of accumulation in a setting where organic material has a higher chance of preservation (e.g. in an anoxic environment), the higher concentration of fossils resulting from the reduced sediment supply can make the collecting of biostratigraphically useful material easier. It is also more likely that a first or last appearance datum will be identifiable in a single outcrop section because if sediment accumulation rates are high, hundreds of metres of strata may lie within a single biozone.

20.4.3 Mobility of organisms

The lifestyle of an organism not only determines its distribution in depositional environments, it also affects the rate at which an organism migrates from one area to another. If a new species evolves in one geographical location its value as a zone fossil in a regional or worldwide sense will depend on how quickly it migrates to occupy ecological niches elsewhere. Again, planktonic and nektonic organisms tend to be most useful in biostratigraphy because they move around relatively quickly. Some benthic organisms have a larval stage that is free-swimming and may therefore be spread around oceans relatively quickly. Organisms that do not move much (a **sessile** lifestyle) generally make poor fossils for biostratigraphic purposes.

20.4.4 Geographical distribution of organisms

Two environments may be almost identical in terms of physical conditions but if they are on opposite sides of the world they may be inhabited by quite different sets of animals and plants. The contrasts are greatest in continental environments where geographical isolation of communities due to tectonic plate movements has resulted in quite different families and orders. The mammal fauna of Australia are a striking example of geographical isolation resulting in the evolution of a group of animals that are quite distinct from animals living in similar environments in Europe or Asia. This geographical isolation of groups of organisms is called **provincialism** and it also occurs in marine organisms, particularly benthic forms, which cannot easily travel across oceans. Present or past oceans have been sufficiently separate to develop localised communities even though the depositional environments may have been similar. This faunal provincialism makes it necessary to develop different biostratigraphic schemes in different parts of the world.

20.4.5 Abundance and size of fossils

To be useful as a zone fossil a species must be sufficiently abundant to be found readily in sedimentary rocks. It must be possible for the geologist to be able to find representatives of the appropriate taxon without having to spend an undue amount of time looking. There is also a play-off between size and abundance. In general, smaller organisms are more numerous and hence the fossils of small organisms tend to be the most abundant. The problem with very small fossils is that they may be difficult to find and identify. The need for biostratigraphic schemes to be applicable to subsurface data from boreholes has led to an increased use of **microfossils**, fossils that are too small to be recognised in hand specimen, but which may be abundant and readily identified under the microscope (or electron microscope in some cases). Schemes based on microfossils have been developed in parallel to macrofossil schemes. Although a scheme based on ammonites may work very well in the field, the chances of finding a whole ammonite in the core of a borehole are remote. Microfossils are the only viable material for use in biostratigraphy where drilling does not recover core but only brings up pieces of the lithologies in the drilling mud (*22.3*).

20.4.6 Preservation potential

It is impossible to determine how many species or individuals have lived on Earth through geological time because very few are ever preserved as fossils.

The fossil record represents a very small fraction of the biological history of the planet for a variety of reasons. First, some organisms do not possess the hard parts that can survive burial in sediments: we therefore have no idea how many types of worm may have existed in the past. Sites where there is exceptional preservation of the soft parts of fossils (***lagerstätten***) provide tantalising clues to the diversity of lifeforms that we know next to nothing about (Whittington & Conway-Morris 1985; Clarkson 1993). Second, the depositional environment may not be favourable to the preservation of remains: only the most resistant pieces of bone survive in the dry, oxidising setting of deserts and almost all other material is destroyed. All organisms are part of a food chain and this means that their bodies are normally consumed, either by a predator or a scavenger. Preservation is therefore the exception for most animals and plants. Finally, the stratigraphic record is very incomplete, with only a fraction of the environmental niches that have existed preserved in sedimentary rocks. The low preservation potential severely limits the material available for biostratigraphic purposes, restricting it to those taxa that had hard parts and existed in appropriate depositional environments.

20.5 TAXA USED IN BIOSTRATIGRAPHY

No single group of organisms fulfils all the criteria for the ideal zone fossil and a number of different groups of taxa have been used for defining biozones through the stratigraphic record (Clarkson 1993). Some, such as the graptolites in the Ordovician and Silurian, are used for worldwide correlation; others are restricted in use to certain facies in a particular succession, for example corals in the Carboniferous of northwest Europe. Some examples of taxonomic groups used in biostratigraphy are outlined below.

20.5.1 Marine macrofossils

The hard parts of invertebrates are common in sedimentary rocks deposited in marine environments throughout the Phanerozoic (Figs 20.2 & 20.4) (Clarkson 1993). These fossils formed the basis for the divisions of the stratigraphic column into Systems, Series and Stages (Fig. 19.1) in the 18th and 19th centuries. The fossils of organisms such as molluscs,

Fig. 20.4 Shelly fossils in a limestone bed.

arthropods, echinoderms, etc., are relatively easy to identify in hand specimen, and provide the field geologist with a means for establishing the age of rocks to the right period or possibly epoch. Expert palaeontological analysis of marine macrofossils provides a division of the rocks into stages based on these fossils.

Trilobites

These Palaeozoic arthropods are the main group used in the zonation of the Cambrian. Most trilobites are thought to have been benthic forms living on and in the sediment of shallow marine waters. They show a wide variety of morphologies and appear to have evolved quite rapidly into taxa with distinct and recognisable characteristics. They are only locally abundant as fossils.

Graptolites

These exotic and somewhat enigmatic organisms are interpreted as being colonial groups of individuals connected by a skeletal structure. They appear to have had a planktonic habit and are widespread in Ordovician and Silurian mudrocks. Preservation is normally as a thin film of flattened organic material on the bedding planes of fine-grained sedimentary rocks. The shapes of the skeletons and the 'teeth' where individuals in the colony were located are distinctive when examined with a hand lens or under a microscope. Lineages have been traced which indicate rapid evolution and have allowed a high-resolution biostratigraphy to be developed for the Ordovician and Silurian systems. The main drawback in the use of graptolites is the poor preservation in coarser grained rocks such as sandstones.

Brachiopods

Shelly, sessile organisms such as brachiopods generally make poor zone fossils but in shallow marine, high-energy environments where graptolites were not preserved, brachiopods are used for regional correlation purposes in Silurian rocks and in later Palaeozoic strata.

Ammonoids

This taxonomic group of cephalopods (phylum Mollusca) includes *goniatites* from Palaeozoic rocks as well as the more familiar *ammonites* of the Mesozoic. The nautiloids are the most closely related living group. The large size and free-swimming habit of these cephalopods made them an excellent group for biostratigraphic purposes. Fossils are widespread, found in many fully marine environments, and they are relatively robust. Morphological changes through time were to the external shape of the organisms and to the 'suture line', the relic of the bounding walls between the chambers of the coiled cephalopod. Goniatites have been used in correlation of Devonian and Carboniferous rocks, whereas ammonites and other ammonoids are the main zone fossils in Mesozoic rocks. Ammonoids became extinct at the end of the Cretaceous.

Gastropods

These also belong to the Mollusca and as marine 'snails' they are abundant as fossils in Cenozoic rocks. They are very common in the deposits of almost all shallow marine environments. Distinctive shapes and ornamentation on the calcareous shells make identification relatively straightforward and there are a wide variety of taxa within this group.

Echinoderms

This phylum includes *crinoids* (sea lilies) and *echinoids* (sea urchins). Most crinoids probably lived attached to substrate and this sessile characteristic makes them rather poor zone fossils, despite their abundance in some Palaeozoic limestones. Echinoids are benthic, living on or in soft sediment: their relatively robust form and subtle but distinctive changes in their morphology have made them useful for regional and worldwide correlation in parts of the Cretaceous.

Corals

The extensive outcrops of shallow marine limestones in Devonian and Lower Carboniferous (Mississippian) rocks in some parts of the world contain abundant corals. This group is therefore used for zonation and correlation within these strata, despite the fact that they are not generally suitable for biostratigraphic purposes because of the very restricted depositional environments they occur in.

20.5.2 Marine microfossils

Microfossils are taxa that leave fossil remains that are too small to be clearly seen with the naked eye or hand lens. They are normally examined using an optical microscope although some forms can be analysed in detail only using a scanning electron microscope. The three main groups that are used in biostratigraphy are the foraminifers, radiolaria and calcareous algae (nanofossils): other microfossils used in biostratigraphy are ostracods, diatoms and conodonts.

Foraminifera

'Forams' (the common abbreviation of foraminifers) are single-celled marine organisms that belong to the Protozoa Subkingdom. They have been found as fossils in strata as old as the Cambrian, although forms with hard calcareous shells, or 'tests', did not become well established until the Devonian. Calcareous forams generally became more abundant through the Phanerozoic and are abundant in many Mesozoic and Cenozoic marine strata. The calcareous tests of planktonic forams are typically a millimetre or less across, although during some periods, particularly the Paleogene, larger benthic forms also occur and can be more than a centimetre in diameter. Planktonic forams make very good zone fossils as they are abundant, widespread in marine strata and appear to have evolved rapidly. Schemes using forams for correlation in the Mesozoic and Cenozoic are widely used in the hydrocarbon industry because microfossils are readily recovered from boreholes and both regional and worldwide zonation schemes are used.

Radiolaria

These organisms form a subclass of planktonic protozoans and are found as fossils in deep marine strata throughout the Phanerozoic. Radiolaria commonly have silica skeletons and are roughly spherical, often spiny organisms less than a millimetre across. They are important in the dating of deep-marine deposits because the skeletons survive in siliceous oozes deposited at depths below the CCD (*16.5.2*). These deposits are preserved in the stratigraphic record as radiolarian cherts and the fossil assemblages found in them typically contain large numbers of taxa making it possible to use quite high resolution biozonation schemes. Their stratigraphic range is also greater than the forams, making them important for the dating of Palaeozoic strata.

Calcareous nanofossils

Fossils that cannot be seen with the naked eye and are only just discernible using a high-power optical microscope are referred to as **nanofossils**. They are microns to tens of microns across and are best examined using a scanning electron microscope. The most common nanofossils are **coccoliths**, the spherical calcareous cysts of marine algae. Coccoliths may occur in huge quantities in some sediments and are the main constituent of some fine-grained limestones such as the Chalk of the Upper Cretaceous in northwest Europe. They are found in fine-grained marine sediments deposited on the shelf or any depths above the CCD below which they are not normally preserved. They are used biostratigraphically in Mesozoic and Cenozoic strata.

Other microfossils

Ostracods are crustaceans with a two-valve calcareous carapace and their closest relatives are crabs and lobsters. They occur in a very wide range of depositional environments, both freshwater and marine, and they have a long history, although their abundance and distribution are sporadic. Zonation using ostracods is applied only locally in both marine and non-marine environments. **Diatoms** are chrysophyte algae with a siliceous frustule (skeleton) that can occur in large quantities in both shallow-marine and freshwater settings. The diatom frustules are less than a millimetre across and in some lacustrine settings may make up most of the sediment, forming a **diatomite** deposit. They are only rarely used in biostratigraphy. **Conodonts** are somewhat enigmatic tooth-

like structures made of phosphate and they occur in Palaeozoic strata. Despite uncertainty about the origins, they are useful stratigraphic microfossils in the older Phanerozoic rocks, which generally contain few other microfossils. **Acritarchs** are microscopic spiny structures made of organic material that occur in Proterozoic and Palaeozoic rocks. Their occurrences in Precambrian strata make them useful as a biostratigraphic tool in rocks of this age. They are of uncertain affinity, although are probably the cysts of planktonic algae, and may therefore be related to **dinoflagellates**, which are primitive organisms found from the Phanerozoic through to the present day and also produce microscopic cysts (**dinocysts**). Zonation based on dinoflagellates is locally very important, especially in non-calcareous strata of Mesozoic and Cenozoic ages: the schemes used are generally geographically local and have limited stratigraphic ranges.

20.5.3 Terrestrial fossil groups used in biostratigraphy

Correlation in the deposits of continental environments is always more difficult because of the poorer preservation potential of most materials in a subaerial setting. Only the most resistant materials survive to be fossilised in most continental deposits, and these include the organophosphates that vertebrate teeth are made of and the coatings of pollen, spores and seeds of plants. Stratigraphic schemes have been set up using the teeth of small mammals and reptiles for correlation of continental deposits of Neogene age. Pollen, spores and seeds (collectively **palynomorphs**) are much more commonly used. They are made up of organic material that is highly resistant to chemical attack and can be dissolved out of siliceous sedimentary rocks using hydrofluoric acid. Airborne particles such as pollen, spores and some seeds may be widely dispersed and the occurrence of these aeolian palynomorphs within marine strata allows for correlation between marine and continental successions. However, although palynomorphs can be used as zone fossils, they rarely provide such a high resolution as marine fossils. Identification is carried out with an optical microscope or an electron microscope after the palynomorphs have been chemically separated from the host sediment using strong acids.

20.6 BIOSTRATIGRAPHIC CORRELATION

Biostratigraphy can provide a high-resolution basis for the division of strata and hence a means of correlating between different successions. Certain conditions are, however, required for the approach to be successful. The first and most obvious is that the rocks must contain the appropriate fossils: this will be largely dependent upon the environment of deposition because it may not have been suitable for the critical taxa. The diagenetic history is also relevant because the fossil material may be altered or completely removed by chemical processes such as mineral replacement or dissolution, or physical processes such as compaction. A second major factor is the relative rate of sedimentation in the successions and the frequency of speciation events: rapid sediment accumulation and infrequent speciation result in a situation where two thick successions may be shown to lie within the same biozone, but no further subdivision and correlation is possible.

20.6.1 Correlating different environments

It is commonly the case that the rocks being studied contain fossils that have biostratigraphic value, but do not contain representatives of the taxa used in the worldwide biostratigraphic zonation scheme for that particular part of the stratigraphic record. This may be because of provincialism, the tendency for populations to occur only in a limited geographical area, or due to the depositional environment. The fossils found in the deposits of contrasting environments such as muds deposited in an offshore setting compared with a sandy foreshore are likely to be different, or on a larger scale, due to different climatic conditions at different latitudes. Differences in fossil content due to provincialism are not related to the environment, but are a result of geographical isolation of evolutionary lineages. Under these circumstances a more roundabout method of correlating using fossils may be required in which a local or regional zonation scheme is set up using the taxa that are found in the area. The strata containing the fauna or flora of the local scheme must then be correlated with the global scheme by finding a succession elsewhere in which taxa from both the local and global schemes are preserved.

The appearance or disappearance of a zone fossil may be due to changes in environment rather than be of stratigraphic importance. If the depositional environment has remained the same, the appearance of a taxon may be due to a speciation event and this will therefore have stratigraphic significance. However, an alternative explanation may be that the species had already existed for a period of time in a different geographical location before migrating to the area of the studied section. The disappearance of a species from the stratigraphic succession is likely to represent an extinction event if the depositional environment has not changed: a population is unlikely to move away from a favourable setting. Relative sea level is one of the factors that affects depositional environment and hence fossil content: appearance and disappearance of taxa within a succession may therefore be due to sea-level changes rather than to speciation and extinction events.

Organisms that are tolerant of different conditions have the widest application and most value as zone fossils. Taxa that are very sensitive to environmental conditions, such as corals, are only useful in circumstances where the environment of deposition has been constant.

20.6.2 Graphical correlation schemes

The thickness of a biostratigraphic unit at any place is determined by the rate of sediment accumulation during the time period represented by the biozone. A succession that is considered to have been a site of continuous, steady sedimentation is chosen as a reference section and the positions of biozone markers (appearance and disappearance of taxa) are noted within it. Another vertical succession of strata containing the same biozone markers can then be compared with this reference section (Carney & Pierce 1995). Tie-points are established using the biostratigraphic information and intermediate levels can be correlated graphically (Fig. 20.5). This approach is particularly effective at identifying changes in rates of sedimentation and recognising the presence of a **hiatus** (period of erosion or non-deposition) in a succession (Fig. 20.5). The recognition of depositional hiatuses is important in sequence stratigraphic analysis of successions (Chapter 23) and has been used extensively in subsurface correlation (Chapter 22).

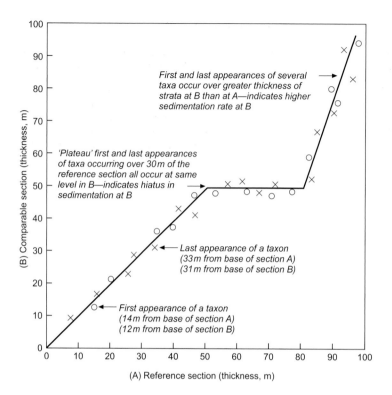

(A) Reference section (thickness, m)

Fig. 20.5 Graphical correlation methods are used to identify changes in rates of sedimentation or a hiatus in deposition. (Adapted from Shaw 1964.)

20.7 BIOSTRATIGRAPHY IN RELATION TO OTHER STRATIGRAPHIC TECHNIQUES

Correlation of strata on the basis of lithology (lithostratigraphy) has inherent limitations for the reasons outlined in *19.4.3*. Most importantly, it cannot provide any basis for correlation over large distances, especially between continents. However, if fossils are used as a correlation tool, the principles of evolutionary biology dictate that if fossils from different places are of the same species then the rocks that they are found in must both be strata deposited during the time when that species was extant. Biostratigraphy is therefore largely independent of lithostratigraphy, although because depositional environment controls the facies of the sediment and also influences the types of organisms that may be present, there are some connecting factors.

It may be argued that biozones are chronostratigraphic units (*19.2*) if a speciation event takes place rapidly enough to be considered to be an 'instant' in geological time. If the new species is dispersed very quickly (again geologically instantaneously) then the base of a biozone can be regarded as an *isochronous horizon*, that is, a surface representing a certain point in time. Hence if the concept of punctuated equilibria is accepted and biozones are defined by the first occurrences of free-swimming or floating organisms then biozones can be considered to be chronostratigraphic units within the resolution available. It may also be the case that certain taxa or groups of taxa become extinct in a geologically short period of time, so upper boundaries defined by these events also can be considered to approximate to isochronous surfaces.

Biostratigraphy plays an important role in subsurface analysis, although analysis is almost exclusively based on microfossils. Forams, radiolarians and calcareous nanofossils recovered from drill cuttings and core provide the basic means for determining the age of the strata at different levels in a borehole, and the basis on which strata can be correlated with other boreholes and with successions exposed at the surface.

The approaches used in the sequence stratigraphic analysis of successions (Chapter 22) have developed relatively recently when compared with the much older biostratigraphic approaches. However, although the recognition of depositional sequences provides a new correlation technique, it does not replace biostratigraphy because the latter is still required to provide a broad temporal framework for the sequence stratigraphic analysis. In addition, graphical correlation schemes provide information about rates of sedimentation and evidence for hiatuses, both of which are very important elements of sequence stratigraphic analysis.

FURTHER READING

Blatt, H., Berry, W.B. & Brande, S. (1991) *Principles of Stratigraphic Analysis*. Blackwell Scientific Publications, Oxford.

Clarkson, E.N.K. (1993) *Invertebrate Palaeontology and Evolution* (3rd edition). Allen and Unwin, London.

Doyle, P. (1996) *Understanding Fossils: an Introduction to Invertebrate Paleontology*. Wiley, Chichester.

McGowran, B. (2005) *Biostratigraphy: Microfossils and Geological Time*. Cambridge University Press, Cambridge.

21

Dating and Correlation Techniques

Radiometric dating techniques have been available to geologists since the discovery of radioactivity in the early part of the 20th century and they have provided an absolute scale of millions and billions of years for events in Earth history. Several different radioactive decay series are used, but because they all provide an age for the formation of a mineral they are primarily used for dating igneous rocks. Dating a grain in a sandstone bed usually provides the age when the mineral originally formed in, say, a granite, and does not provide much information about the age of the sedimentary rock. The magnetic and chemical properties of rocks can be used to carry out magnetostratigraphic and chemostratigraphic correlation, techniques that are mainly used in combination with other methods. In practice the whole process of dating and correlating rocks relies on the integration of information from a number of different sources and techniques. Dating in the Quaternary is a specialist area using a different range of techniques, including carbon-14 dating, which can be used to date organic material formed in the past few tens of thousands of years.

21.1 DATING AND CORRELATION TECHNIQUES

Lithostratigraphic techniques (19.3) do not involve any determination of the age of the rock except in a relative sense of indicating which rock units are younger or older, and correlation on the basis of lithological characteristics does not provide a temporal (time) framework for stratigraphy. The use of fossils to carry out biostratigraphic correlation (20.6) provides a means of ordering strata on both a large and small scale and can be used for correlation both

locally and regionally. These are the oldest techniques in stratigraphy, and still the most widely used, but they have now been supplemented with other approaches that utilise technological advances over the past 100 years. Of these radiometric dating is the most important because it has provided a temporal framework that is measured in years. New techniques using different isotopic systems are being developed all the time, and new instrumentation makes it possible to make measurements with higher precision on smaller samples of material. The coverage of these techniques in this chapter provides only the simplest introduction to expanding and increasingly sophisticated approaches to radiometric dating. Magnetostratigraphy and chemostratigraphy are also techniques that have benefited from technological advances, with more sensitive magnetometers for measuring the magnetism in rocks being developed and chemical analysis being carried out to higher precision.

21.2 RADIOMETRIC DATING

The discovery of radioactivity and the radiogenic decay of isotopes in the early part of the 20th century opened the way for dating rocks by an absolute, rather than relative, method. Up to this time estimates of the age of the Earth had been based on assumptions about rates of evolution, rates of deposition, the thermal behaviour of the Earth and the Sun or interpretation of religious scriptures (Eicher 1976). **Radiometric dating** uses the decay of isotopes of elements present in minerals as a measure of the age of the rock: to do this, the rate of decay must be known, the proportion of different isotopes present when the mineral formed has to be assumed, and the proportions of different isotopes present today must be measured. This dating method is principally used for determining the age of formation of igneous rocks, including volcanic units that occur within sedimentary strata. It is also possible to use it on authigenic minerals, such as glauconite (2.3.2), in some sedimentary rocks. Radiometric dating of minerals in metamorphic rocks usually indicates the age of the metamorphism.

21.2.1 Radioactive decay series

A number of elements have **isotopes** (forms of the element that have different atomic masses) that are unstable and change by radioactive decay to the isotope of a different element. Each radioactive decay series (Fig. 21.1) takes a characteristic length of time known as the **radioactive half-life**, which is the time taken for half of the original (**parent**) isotope to decay to the new (**daughter**) isotope. The decay series of most interest to geologists are those with half-lives of tens, hundreds or thousands of millions of years. If the proportions of parent and daughter isotopes of these decay series can be measured, periods of geological time in millions to thousands of millions of years can be calculated.

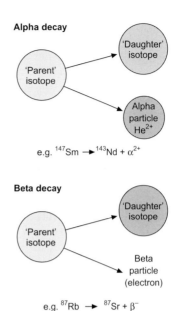

Alpha decay

e.g. $^{147}Sm \rightarrow {}^{143}Nd + \alpha^{2+}$

Beta decay

e.g. $^{87}Rb \rightarrow {}^{87}Sr + \beta^{-}$

Fig. 21.1 Radioactive decay results in the formation of a new 'daughter' isotope from the 'parent' isotope.

To calculate the age of a rock it is necessary to know the half-life of the radioactive decay series, the amount of the parent and daughter isotopes present in the rock when it formed, and the present proportions of these isotopes. The relationship can be expressed in an equation as

$$N = N_0\, e^{-lt}$$

in which 'N_0' is number of parent atoms at the start, 'N' the number of daughter atoms after a period of time ('t') and 'l' is the rate at which parent decays to daughter (the **decay constant**); 'e' is 2.718, the base of the natural logarithm (ln). This equation can be rearranged as follows:

$$t = 1/l \times \ln(N_0/N + 1)$$

It is normally assumed that when a mineral crystallises out of a magma to form part of an igneous rock there is only the parent isotope present. Similarly it is assumed that only the parent isotope is present when a mineral is precipitated chemically in sediment or forms by solid-state recrystallisation in a metamorphic rock. The radiometric 'clock' starts as the mineral is formed.

It must also be assumed that all the daughter isotope measured in the rock today formed as a result of decay of the parent. This may not always be the case because addition or loss of isotopes can occur during weathering, diagenesis and metamorphism and this will lead to errors in the calculation of the age. It is therefore important to try to ensure that decay has taken place in a 'closed system', with no loss or addition of isotopes, by using only unweathered and unaltered material in analyses.

The radiometric decay series commonly used in radiometric dating of rocks are detailed in the following sections. The choice of method of determination of the age of the rock is governed by its age and the abundance of the appropriate elements in minerals. Further details of these dating methods and their applications are found in texts such as Faure & Mensing (2004) and Dickin (2004).

21.2.2 Practical radiometric dating

The samples of rock collected for radiometric dating are generally quite large (several kilograms) to eliminate inhomogeneities in the rock. The samples are crushed to sand and granule size, thoroughly mixed to homogenise the material and a smaller subsample selected. In cases where particular minerals are to be dated, these are separated from the other minerals by using heavy liquids (liquids with densities similar to that of the minerals) in which some minerals will float and others sink, or magnetic separation using the different magnetic properties of minerals. The mineral concentrate may then be dissolved for isotopic or elemental analysis, except for argon isotope analysis, in which case the mineral grains are heated in a vacuum and the composition of the argon gas driven off is measured directly.

Measurement of the concentrations of different isotopes is carried out with a **_mass spectrometer_**. In these instruments a small amount (micrograms) of the sample is heated in a vacuum to ionise the isotopes and these charged particles are then accelerated along a tube in a vacuum by a potential difference. Part-way along the tube a magnetic field induced by an electromagnet deflects the charged particles. The amount of deflection will depend upon the atomic mass of the particles so different isotopes are separated by their different masses. Detectors at the end of the tube record the number of charged particles of a particular atomic mass and provide a ratio of the isotopes present in a sample.

21.2.2 Potassium–argon and argon–argon dating

This is the most widely used system for radiometric dating of sedimentary strata, because it can be used to date the potassium-rich authigenic mineral glauconite (*2.3.2*) and volcanic rocks (lavas and tuffs) that contain potassium in minerals such as some feldspars and micas (Fig. 21.2). One of the isotopes of potassium, ^{40}K, decays partly by electron capture (a proton becomes a neutron) to an isotope of the gaseous element argon, ^{40}Ar, the other product being an isotope of calcium, ^{40}Ca. The half-life of this decay is 11.93 billion years. Potassium is a very common element in the Earth's crust and its concentration in rocks is easily measured. However, the proportion of potassium present as ^{40}K is very small at only 0.012%, and most of this decays to ^{40}Ca, with only 11% forming ^{40}Ar. Argon is an inert rare gas and the isotopes of very small quantities of argon can be measured by a

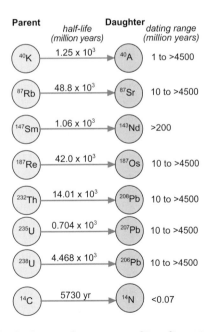

Parent	half-life (million years)	Daughter	dating range (million years)
^{40}K	1.25×10^3	^{40}A	1 to >4500
^{87}Rb	48.8×10^3	^{87}Sr	10 to >4500
^{147}Sm	1.06×10^3	^{143}Nd	>200
^{187}Re	42.0×10^3	^{187}Os	10 to >4500
^{232}Th	14.01×10^3	^{208}Pb	10 to >4500
^{235}U	0.704×10^3	^{207}Pb	10 to >4500
^{238}U	4.468×10^3	^{206}Pb	10 to >4500
^{14}C	5730 yr	^{14}N	<0.07

Fig. 21.2 The main decay series used in radiometric dating of rocks: the K–Ar, Rb–Sr and U–Pb systems are the ones most commonly used – ^{14}C dating is mainly used for dating archaeological materials.

mass spectrometer by driving the gas out of the minerals. K–Ar dating has therefore been widely used in dating rocks but there is a significant problem with the method, which is that the daughter isotope can escape from the rock by diffusion because it is a gas. The amount of argon measured is therefore commonly less than the total amount produced by the radioactive decay of potassium. This results in an underestimate of the age of the rock.

The problems of argon loss can be overcome by using the argon–argon method. The first step in this technique is the irradiation of the sample by neutron bombardment to form ^{39}Ar from ^{39}K occurring in the rock. The ratio of ^{39}K to ^{40}K is a known constant so if the amount of ^{39}Ar produced from ^{39}K can be measured, this provides an indirect method of calculating the ^{40}K present in the rock. Measurement of the ^{39}Ar produced by bombardment is made by mass spectrometer at the same time as measuring the amount of ^{40}Ar present. Before an age can be calculated from the proportions of ^{39}Ar and ^{40}Ar present it is necessary to find out the proportion of ^{39}K that has been converted to ^{39}Ar by the neutron bombardment. This can be achieved by bombarding a sample of known age (a 'standard') along with the samples to be measured and comparing the results of the isotope analysis. The principle of the Ar–Ar method is therefore the use of ^{39}Ar as a proxy for ^{40}K.

Although a more difficult and expensive method, Ar–Ar is now preferred to K–Ar. The effects of alteration can be eliminated by step-heating the sample during determination of the amounts of ^{39}Ar and ^{40}Ar present by mass spectrometer. Alteration (and hence ^{40}Ar loss) occurs at lower temperatures than the original crystallisation so the isotope ratios measured at different temperatures will be different. The sample is heated until there is no change in ratio with increase in temperature (a 'plateau' is reached): this ratio is then used to calculate the age. If no 'plateau' is achieved and the ratio changes with each temperature step the sample is known to be too altered to provide a reliable date.

21.2.3 Other radiometric dating systems

Rubidium–strontium dating

This is a widely used method for dating igneous rocks because the parent element, rubidium, is common as a trace element in many silicate minerals. The isotope ^{87}Rb decays by shedding an electron (**beta decay**) to ^{87}Sr with a half-life of 48 billion years (Fig. 21.2). The proportions of two of the isotopes of strontium, ^{86}Sr and ^{87}Sr, are measured and the ratio of ^{86}Sr to ^{87}Sr will depend on two factors. First, this ratio will depend on the proportions in the original magma: this will be constant for a particular magma body but will vary between different bodies. Second, the amount of ^{87}Sr present will vary according to the amount produced by the decay of ^{87}Rb: this depends on the amount of rubidium present in the rock and the age. The rubidium and strontium concentrations in the rock can be measured by geochemical analytical techniques such as XRF (X-ray fluorescence). Two unknowns remain: the original ^{86}Sr/^{87}Sr ratio and the ^{87}Sr formed by decay of ^{87}Rb (which provides the information needed to determine the age). The principle of solving simultaneous equations can be used to resolve these two unknowns. If the determination of the ratios of ^{86}Sr/^{87}Sr and Rb/Sr is carried out for two different minerals (e.g. orthoclase and muscovite), each will start with different proportions of strontium and rubidium because they are chemically different. An alternative method is **whole-rock dating**, in which samples from different parts of an igneous body are taken, which, if they have crystallised at different times, will contain different amounts of rubidium and strontium present. This is more straightforward than dating individual minerals as it does not require the separation of these minerals.

Uranium–lead dating

Isotopes of uranium are all unstable and decay to daughter elements that include thorium, radon and lead. Two decays are important in radiometric dating: ^{238}U to ^{206}Pb with a half-life of 4.47 billion years and ^{235}U to ^{207}Pb with a half-life of 704 million years (Fig. 21.2). The naturally occurring proportions of ^{238}U and ^{235}U are constant, with the former the most abundant at 99% and the latter 0.7%. By measuring the proportions of the parent and daughter isotopes in the two decay series it is possible to determine the amount of lead in a mineral produced by radioactive decay and hence calculate the age of the mineral. Trace amounts of uranium are to be found in minerals such as zircon, monazite, sphene and apatite: these occur as accessory minerals in igneous rocks and as heavy minerals in sediments. Dating of zircon grains using uranium–lead dating provides information about provenance of the sediment (Carter & Moss 1999). Dating of zircons has been used to establish the age of the oldest rocks in the world. Other parts of the uranium decay series are used in dating in the Quaternary (21.5.2).

Samarium–neodymium dating

These two rare earth elements in this decay series are normally only present in parts per million in rocks. The parent isotope is ^{147}Sm and this decays by alpha particle emission to ^{143}Nd with a half-life of 106 billion years (Fig. 21.2). The slow generation of ^{143}Nd means that this technique is best suited to older rocks as the effects of analytical errors are less significant. The advantage of using this decay series is that the two elements behave almost identically in geochemical reactions and any alteration of the rock is likely to affect the two isotopes to equal degrees. This eliminates some of the problems encountered with Rb–Sr caused by the different reactivity and mobility of the two elements in the decay series. This dating technique has been used on sediments to provide information about the age of the rocks that the sediment was derived from: different provenance areas, for example continental cratons of different ages, can be distinguished by analysis of mud and mudstones.

Rhenium–osmium dating

Rhenium occurs in low concentrations in most rocks, but its most abundant naturally occurring isotope ^{187}Re undergoes beta decay to an isotope of osmium ^{187}Os with a half-life of 42 Ga. This dating technique has been used mainly on sulphide ore bodies and basalts, but there have also been some successful attempts to date the depositional age of mudrocks with a high organic content (Dickin 2004). Osmium isotopes in seawater have also been shown to have varied through time and measurement of the ratio ^{187}Os/^{188}Os has been used in the same way as the strontium isotope curve (21.3.1).

21.2.4 Applications of radiometric dating

Radiometric dating is the only technique that can provide absolute ages of rocks through the strati-

graphic record, but it is limited in application by the types of rocks which can be dated. The age of formation of minerals is determined by this method, so if orthoclase feldspar grains in a sandstone are dated radiometrically, the date obtained would be that of the granite the grains were eroded from. It is therefore not possible to date the formation of rocks made up from detrital grains and this excludes most sandstones, mudrocks and conglomerates. Limestones are formed largely from the remains of organisms with calcium carbonate hard parts, and the minerals aragonite and calcite cannot be dated radiometrically on a geological time scale. Hence almost all sedimentary rocks are excluded from this method of dating and correlation. An exception to this is the mineral glauconite, an authigenic mineral that forms in shallow marine environments (11.5.1): glauconite contains potassium and may be dated by K–Ar or Ar–Ar methods, but the mineral is readily altered and limited in occurrence.

The formation of igneous rocks usually can be dated successfully provided that they have not been severely altered or metamorphosed. Intrusive bodies, including dykes and sills, and the products of volcanic activity (lavas and tuff) may be dated and these dates used to constrain the ages of the rocks around them by the laws of stratigraphic relationships (19.3.1). Dates from metamorphic rocks may provide the age of metamorphism, although complications can arise if the degree of metamorphism has not been high enough to reset the radiometric 'clock', or if there have been multiple phases of metamorphism.

General stratigraphic relations and isotopic ages are the principal means of correlating intrusive igneous bodies. Geographically separate units of igneous rock can be shown to be part of the same igneous suite or complex by determining the isotopic ages of the rocks at each locality. Radiometric dating can also be very useful for demonstrating correspondence between extrusive igneous bodies. The main drawbacks of correlation by this method are the limited range of lithologies that can be dated and problems of precision of the results, particularly with older rocks. For example, if two lava beds were formed only a million years apart and there is a margin of error in the dating methods of one million years, correlation of a lava bed of unknown affinity to one or the other cannot be certain.

21.3 OTHER ISOTOPIC AND CHEMICAL TECHNIQUES

21.3.1 Strontium isotopes

Strontium is chemically similar to calcium and is found in small quantities in many limestones. There are two common isotopes of strontium, ^{86}Sr and ^{87}Sr, and analysis of carbonates through the Phanerozoic has indicated that the ratio between these two isotopes in seawater has changed through time (De Paolo & Ingram 1985; Hess et al. 1986). A strontium isotope curve has been constructed using information from carbonate minerals formed from seawater and which have not been subsequently altered or recrystallised. By comparing the $^{86}Sr/^{87}Sr$ ratio in calcite from a sample of unknown age with the established curve, it is possible to determine the age of the sample. Note that although this approach involves isotopes, it is not an absolute dating technique and should be distinguished from absolute rubidium–strontium dating.

There are two important factors to be considered when dating using strontium isotope ratios. First, a particular $^{86}Sr/^{87}Sr$ ratio is not unique to a date as that same ratio may have existed a number of times: some other control on the age of the rock is required in order to constrain the part of the curve to be used for comparison (Fig. 21.3). Second, only carbonate minerals that have been formed from seawater and which have not been subsequently recrystallised can be used. Many organisms make shells of aragonite, but these cannot be used because aragonite recrystallises to calcite through time. Only organisms that precipitate calcite in their shells or skeletons can be used and these must be free from alteration.

21.3.2 Thermochronological techniques

Thermochronology is a process of determining the age at which a rock was at a particular temperature and is most widely used as a means of calculating the uplift and denudation history of a body of rock (6.8). The K–Ar and Ar–Ar dating techniques can be used in this way because the decay series 'clock' is reset at temperatures that vary between different minerals from about 350°C to 700°C. Fission track analysis (6.8) can be used on zircon grains to indicate when the rock was at 300°C or more, and on apatite grains to

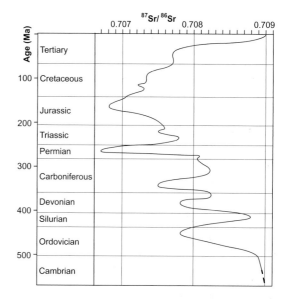

Fig. 21.3 The strontium isotope curve: these changes in the ratio of the isotopes ^{86}Sr and ^{87}Sr through geological time can be used to determine the age of some rocks, but the same ratio can occur at different ages. (Data from Faure 1986.)

indicate temperatures of over 110°C. Even lower temperatures can be determined using the U/Th--He (uranium/thorium–helium) technique of isotopic analysis. For further details of thermochronological techniques see specialist texts such as Braun et al. (2006) and Dickin (2004).

21.3.3 Chemostratigraphy

The composition of sediments is variable, so differences in the mineralogical and chemical character of sedimentary rocks are potentially a way of discriminating and correlating deposits. This approach, known as **_chemostratigraphy_**, will work in circumstances where there is variation in the mineralogy, and hence chemistry, of sediment being supplied to the area of sedimentation, that is, there are changes in provenance (Ravnås & Furnes 1995). Detrital grains in clastic sediments are derived from source areas of high ground around the basin of sedimentation that are being eroded. This is the provenance of the detrital material. Through time different rock units in the source areas are exposed by erosion and the types of clastic material supplied to the basin change. The appearance of new clast types in the succession of sediments will mark the time when that new source area lithology was exposed and eroded. A change in clast types will result in a change in the bulk chemistry of the deposit.

Three main techniques are used: (a) bulk chemical analysis, which is rapid and simple to carry out, but the data not always easy to interpret (Preston et al. 1998); (b) correlation using heavy mineral assemblages is a more time-consuming approach, but can be very effective (DeCelles 1988; Mange-Rajetzky 1995); and (c) clay mineral analysis has been applied in some circumstances (Jeans 1995). The relative simplicity of carrying out chemical analyses makes chemostratigraphy an attractive option, but application of the technique turns out to be quite limited. The 'signal' of changes in provenance is often quite subtle, and there are other factors that also influence the chemistry of a sediment, such as grain size variations and diagenetic alteration. For these reasons, the results of bulk chemical analysis tend to be difficult to interpret stratigraphically, whereas variation in the assemblages of heavy minerals in sandstones usually reflects changes in provenance through time. Chemostratigraphy has been used successfully in thick packages of continental deposits that are barren of any fossils, and at a finer scale as a means of correlating sandstone beds within subsurface hydrocarbon reservoirs.

21.4 MAGNETOSTRATIGRAPHY

The Earth's magnetic field alternates between periods of **normal magnetic polarity**, which is the field orientation of the present day, and **reversed magnetic polarity**, when the field is reversed, meaning that the 'north' arrow on a magnetic compass would point towards the South Pole. Evidence from measuring the magnetic fields of the past (**palaeomagnetism**) indicates that these **magnetic polarity reversals** (Fig. 21.4) have occurred at irregular intervals during geological time. Some reversals have been relatively short, occurring as little as a few tens of thousands of years apart, although there was a period of nearly 30 Myr in the Cretaceous when the magnetic field appears to have remained the same. Through most of the Cenozoic polarity reversals have occurred every few hundred thousand to a few million years. The time taken for a reversal to occur

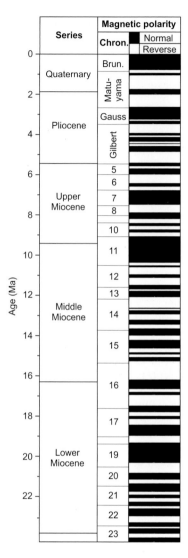

Fig. 21.4 Reversals in the polarity of the Earth's magnetic field through part of the Cenozoic. (From Haq et al. 1988.)

appears to be 'instantaneous' in the context of geological time.

21.4.1 The magnetic record in rocks

A magnetic material acquires the polarity of the ambient magnetic field as it cools through the **Curie Point**, a temperature above which the magnetic dipoles in the material are mobile and free to reorient themselves.

Once below the Curie Point, the material retains the same field when it is moved or the magnetic field changes around it. A rock may contain a number of different magnetic minerals that each have their own Curie Point temperature, and alteration of the rock may occur to create new minerals that will record the ambient magnetic field at the time of their formation. The preserved magnetisation or ***remnant magnetism*** in a rock may therefore be a complex mixture of different field orientations resident in different minerals.

The remnant magnetism in a rock sample is measured to determine the orientation of the Earth's magnetic field relative to the sample at the time of the formation of the rock (Hailwood 1989). In extrusive igneous rocks this will be recorded by the remnant magnetism in minerals such as magnetite and haematite as they cool below their Curie Point. This strong signal is relatively easily detected by a magnetometer, but of more use to the stratigrapher is a much weaker remnant magnetism preserved in sedimentary rocks. As fine magnetised particles (grains containing iron minerals such as haematite) settle out of water they tend to orient themselves parallel to the Earth's magnetic field. Clearly not all of these particles will line up perfectly parallel to the ambient field, but there will be a statistically significant pattern in their orientation that will give the sediment a remnant polarity. The effect is strongest in fine-grained sediments deposited from suspension with a high proportion of iron minerals. In coarser grained sediment the particles will be oriented by the flow that deposited them and the remnant magnetism in sediments that have a low iron content may not be detectable. The remnant magnetism in a rock will be reset when the minerals are heated above their Curie Point during metamorphism or when the minerals are altered by diagenesis or weathering.

21.4.2 Practical magnetostratigraphy

The objective of a magnetostratigraphic study will usually be to identify periods of normal and reversed magnetic polarity recorded in a succession of strata. Field sampling is normally carried out by drilling out small cores of rock from beds in the outcrop. The orientation of the cores in three dimensions and the attitude of the bedding are measured and multiple cores are normally taken from a single bed in order to provide enough samples for a statistically significant analysis of the remnant magnetism at that single site. The vertical interval between sampling sites in the succession will depend on the rates of accumulation of the sediments and the time interval between field reversals during that period of Earth history. In successions deposited at slow rates, samples may need to be taken every few metres up the succession in order to be sure of detecting all the polarity reversals, whereas higher rates of accumulation allow a wider spacing of sample sites. Once a reversal is identified, the precise location in the succession may be determined by resampling at closer intervals between the sites that show opposite field directions.

The remnant magnetism in the samples is determined in the laboratory by a ***magnetometer***. Modern instruments are capable of detecting and measuring magnetic fields in the samples that are several orders of magnitude weaker than the Earth's magnetic field. The effects of the present-day magnetic field are removed by putting the sample in a space shielded from the present-day field and either heating it up or subjecting it to the field of an alternating current. The orientation of the remaining remnant magnetism will be relict from an earlier stage in its history, hopefully the time at which the rock was formed. The remnant magnetism recorded in separate samples at the same site is compared to ensure statistical significance of the result.

21.4.3 Magnetostratigraphic correlation

By making measurements of the remnant palaeomagnetism through a succession of beds it is possible to construct a record of the periods of normal and reversed stratigraphy. These are conventionally shown as intervals marked in black for normal polarity and white for reversed polarity (Fig. 21.4). The pattern of reversals in the Earth's magnetic field through time has been established for much of the Phanerozoic and many reversal events are well dated. In order to tie the pattern measured in an individual succession to the established polarity stratigraphy it is essential to have some sort of tie-point to the geological time scale. This may be provided by absolute dating of a unit, such as a lava, within the successions, or biostratigraphic information that can be used to relate a point in the succession to the time scale (Fig. 21.5). An important proviso is that any time gaps in the record provided by the succession are recognised

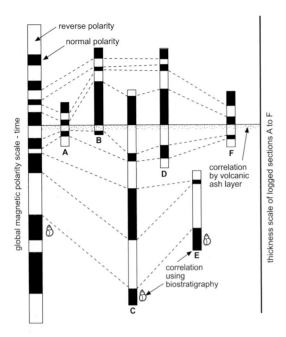

Fig. 21.5 An illustration of how different successions can be correlated using a combination of magnetic reversals, marker beds and biostratigraphic data.

and accounted for: a period of normal or reversed polarity may not be represented if there is no sedimentation or if the deposits of that period are removed by erosion. Once a reversal stratigraphy has been established in part of a sedimentary basin, correlation within the basin is possible by matching the reversal patterns at other localities, again taking any evidence for breaks in the sedimentary record into account (Hailwood 1989; Talling & Burbank 1993). The technique is normally only used when other (biostratigraphic) methods cannot be used or a high-resolution stratigraphy is required: magnetostratigraphy is often employed in continental successions that lack age-diagnostic flora or fauna and which cannot be dated biostratigraphically.

21.5 DATING IN THE QUATERNARY

A number of additional techniques are used in dating and correlating Quaternary deposits. These methods are restricted to use in the last few tens or hundreds of thousands of years and cannot be applied to older stratigraphic units.

21.5.1 Carbon-14 dating

Carbon-14 or *radiocarbon dating* is probably the best known of all radiometric dating methods because it is widely used in archaeology for the dating of materials such as bone and charcoal. The principle of this technique is that living organisms continually take up carbon from the environment which includes the isotope ^{14}C. This radioactive isotope is continually produced by cosmic bombardment of ^{14}N in the atmosphere. When the organism dies the ^{14}C in it radioactively decays at a rate determined by its half-life of 5730 years (Fig. 21.2). By measuring the proportion of ^{14}C present in a sample of once-living tissue the time elapsed since it died and stopped taking up ^{14}C from the atmosphere can be determined.

From a geological point of view, the main limitation of this technique is that with such a short half-life the levels of ^{14}C present in a sample start to become too low to be detected with accuracy in materials older than 50,000 years, although modern analytical techniques are stretching this back to 70,000 years or more (Lutgens & Tarbuck 2006). There is also an error introduced by the assumption that the levels of ^{14}C in the environment have been constant through this period; it is now known that the ^{14}C levels have varied through time and some corrections have to be made to the dates obtained.

21.5.2 Uranium-series dating

The age limitations for dating the Pleistocene of carbon-14 dating can be partly resolved by using uranium-series dating techniques. One of the products in the radioactive decay series of ^{238}U is another isotope of uranium, ^{234}U (Edwards et al. 1986). Both isotopes are present in seawater, but differences in the reactivity of the two isotopes mean that some fractionation occurs and ^{234}U is present in higher proportions than would be expected in the course of the decay. If the fractionated uranium is taken up from the seawater into a mineral, such as the carbonate of a marine organism, the proportions of the two isotopes gradually return to normal as the ^{234}U decays. This occurs over a period of about a million years, making it possible to date marine carbonates that are hundreds of thousands of years old by measuring the proportion of ^{234}U and ^{238}U. This *uranium series dating* technique works best in corals and cannot be used for carbonates formed in fresh water because the fractionation is variable in non-marine waters.

21.5.3 Oxygen isotope stratigraphy

The commonest isotope of oxygen is ^{16}O, but about 0.2% of naturally occurring oxygen is ^{18}O: these are both **stable isotopes**, that is, they do not undergo radioactive decay. The lighter isotope preferentially evaporates into the atmosphere (the process of **isotopic fractionation**) and hence atmospheric water contains relatively more ^{16}O. During glacial periods a higher proportion of the globe's water is held in ice caps, which are fed by water from the atmosphere in the form of snow. As a consequence, during glacial periods when there is more low-^{18}O ice, the oceans are relatively enriched in ^{18}O. This variation of the $^{18}O/^{16}O$ ratio in the ocean waters with temperature provides the basis for an **oxygen isotope stratigraphy** (Shackleton 1977), which was established for global ocean waters back to around 750 ka and now extends to the whole of the Quaternary (Shackleton 1997). This has been created by measuring the $^{18}O/^{16}O$ ratio in carbonate minerals formed directly from seawater by organisms and dating the deposits in which they occur by other means. The oxygen isotope ratio in a sample is expressed in terms of a comparison with a standard as $\delta^{18}O$, with negative values depleted in ^{18}O and positive values enriched in ^{18}O. The marine record of Quaternary deposition is commonly expressed in terms of 'oxygen isotope stages' or 'marine isotope stages', which reflect periods of relatively warm and cool climate.

21.5.4 Luminescence and electron spin resonance dating

Sedimentary materials are exposed to a flux of ionising radiation that originates from naturally occurring radioactivity from elements such as potassium, thorium and uranium. The radiation redistributes the electrical charge within mineral crystals and although most of this displaced charge quickly reverts to its original state, some is trapped in lattice imperfections at higher energy states. The amount of extra energy retained by the crystal depends on the length of time of the dose. This energy can be released by heat and it appears in the form of light, causing the material to luminesce: this effect is called **thermoluminescence** (TL) (Bøtter-Jensen 1997). An alternative to heating is to expose the sample to a burst of light, a procedure known as **optical stimulating luminescence** (OSL) (Bøtter-Jensen 1997).

Sediment at the surface is exposed to sunlight that causes a release of the stored energy ('bleaching'), so the build-up of the energy to produce luminescence only starts once the material is buried. Measurement of the amount of luminescence produced by heating or optically stimulating a sample can therefore be used to determine how long the sediment has been buried. The techniques only work for materials that have been thoroughly bleached when deposited, such as aeolian and fluvial sediment that has been deposited slowly. The TL and OSL techniques can be used to date the time of burial of sediment back to 150 ka with an accuracy of about 10%. They can also be used on cave stalagmites with a similar accuracy, but to about double the age range.

Electron spin resonance (ESR) dating is similar to the luminescence methods in that it is the build-up of energy within crystal lattices after burial which underlies the technique. The displacement of electrons within the lattice creates a change in the magnetic field (or spin) of the atoms. The change in magnetic field occurs progressively with time and hence by measuring the electron spin resonance this effect can be used for dating sediments. Unlike TL and OSL dating the sample is not destroyed with the ESR method, so samples can be dated more than once. Electron spin resonance can be used to date quartz sand and biogenic calcium carbonate (Rink 1997).

21.5.5 Cosmogenic isotopes

Bombardment of material on the surface by cosmic rays (mainly high-energy protons) results in the formation of cosmogenic isotopes as they react with elements in minerals exposed at the surface. The abundance of these unstable isotopes produced by this bombardment can therefore be used as a measure of how long a rock has been exposed. Three isotopes are used for this technique: ^{26}Al, ^{10}Be and ^{36}Cl. The aluminium and beryllium isotopes are usually used in combination in quartz, in which the ^{10}Be is derived from oxygen while ^{26}Al is produced from silicon: most of the production occurs in the top 50 cm of the rock surface and periods of exposure of hundreds of thousands of years can be measured using this system. The chlorine isotope, ^{36}Cl, is a product of bombardment of isotopes of calcium and potassium, both of which are common in many rock types. Periods of exposure of rocks of tens of thousands of years can be determined by using measurements of ^{36}Cl. Cosmogenic isotope techniques are particularly useful for dating features such as river terraces and other depositional surfaces in Quaternary continental successions (Bierman 1994). The results provide information about events such as uplift and incision by rivers and hence constrain the geomorphological history of a land surface.

21.5.6 Amino-acid racemisation

All living organisms contain amino-acids and these compounds exist in two geometric forms, 'L' and 'D'. In living organisms the 'L' form is dominant but when the tissue dies the process of *racemisation* occurs, converting the 'L' form into the 'D' form until they are in equilibrium (Bada 1985). The rate at which this occurs depends on the nature of the organism and is temperature sensitive. In hot conditions racemisation may take a few thousand years, but in colder conditions it may take a hundred to a thousand times longer. This technique can be used only on material where the rate of racemisation is known and the temperature can be determined.

21.5.7 Annual cycles in nature

Two techniques fall into this category: tree rings and glacial lake varves. Seasonal variations in the rate of growth of trees produce rings in the wood, the thickness of which depends on the length of the growing season. Climatic variations can be picked out in the pattern of rings, and by matching the ring patterns in very old living trees, a tree ring chronology reflecting climate fluctuations has been extended several thousand years back into the Holocene. These have been useful in calibrating the carbon-14 dating method. Varves (*10.2.3*) are millimetre-scale laminae in the deposits of glacial lakes that are caused by the seasonal influx of sediment during the summer melt. Counting these laminae back from the present in a lake deposit provides an indication of the age of the deposits.

FURTHER READING

Blatt, H., Berry, W.B. & Brande, S. (1991) *Principles of Stratigraphic Analysis*. Blackwell Scientific Publications, Oxford.

Braun, J., van der Beek, P. & Batt, G. (2006) *Quantitative Thermochronology: Numerical Methods for the Interpretation of Thermochronological Data*. Cambridge University Press, Cambridge.

Dickin, A.P. (2004) *Radiogenic Isotope Geology* (2nd edition). Cambridge University Press, Cambridge.

Faure, G. & Mensing, T.M. (2004) *Isotopes: Principles and Applications*, 3rd Edition. John Wiley, New York.

Hailwood, E.A. (1989) *Magnetostratigraphy*. Special Report 19, Geological Society of London.

Tarling, D.H. & Turner, P. (1999) *Paleomagnetism and Diagenesis in Sediments*. Special Publication 151, Geological Society Publishing House, Bath.

22

Subsurface Stratigraphy and Sedimentology

Techniques for the investigation of geology below the surface have mainly been developed to satisfy the needs of the hydrocarbon industries. Exploration for coal, oil and gas has resulted in the development of a branch of geology concerned with the analysis of stratigraphy, sedimentology and structure in the subsurface. The methods principally involve geophysical techniques such as creating seismic reflection profiles and the measurement of the properties of layers in the subsurface using instruments lowered down boreholes. Core and drill cuttings are also used to sample the rocks that have been drilled. Subsurface exploration has provided a wealth of information in some areas by oil companies and has led to a better understanding of the stratigraphy of sedimentary basins. In particular knowledge of the geology of offshore areas on continental shelves has been greatly increased as a result of these activities. The concepts and application of sequence stratigraphy grew from subsurface studies and were later transferred to outcrop geology.

22.1 INTRODUCTION TO SUBSURFACE STRATIGRAPHY AND SEDIMENTOLOGY

Geologists usually learn the principles of sedimentology and stratigraphy from outcrop relationships in the field, but many will work with subsurface data if they are employed as professional geoscientists. The exploitation of mineral resources started with miners finding layers of coal or beds rich in minerals at the surface and then following them underground by tunnelling. Modern exploration, particularly for hydrocarbons, involves using a range of techniques

for finding out what is below the surface. In some cases this will be direct sampling of what is down below by drilling a hole and bringing pieces of rock back to the surface, but most exploration uses less direct means of investigating the strata hundreds or thousands of metres below ground. These approaches involve making measurements of the physical properties of the rocks and are hence referred to as *geophysical techniques*.

Surveys of the regional variations in the Earth's magnetic field and measurements of gravity, which varies with the density of the rock below ground, are sometimes used as very general indicators of the nature of the subsurface. However, the first detailed approach in subsurface exploration is usually to create seismic reflection profiles across an area. These provide information about stratigraphic and structural relationships in the strata and also give some indication of the lithologies present. Analysis of these data helps to target locations where boreholes are drilled to take cores or make further geophysical measurements of the properties of the strata. The objective is to build up a picture of the subsurface geology, including an indication of the distribution of different facies and the large-scale stratal relationships. The principles of sedimentology and stratigraphy discussed in previous chapters of this book are applied in the same way, but using mainly geophysical data instead of the outcrop studies described in Chapter 5.

22.2 SEISMIC REFLECTION DATA

The underlying principle behind this very widely used technique in subsurface analysis is that there are variations in the acoustic properties of rocks that can be picked up by generating a series of artificial shock waves and then recording the returning waves. A sound wave is partially reflected when it encounters a boundary between two materials of different density and **sonic velocity** (the speed of sound in the material). The product of the density and sonic velocity of a material is the **acoustic impedance** of that material. A strong reflection of sound waves occurs when there is a strong contrast between the acoustic impedance of one material and another. In geological terms there is a strong reflection of the sound waves at the contact between two rocks that have different acoustic properties, such as a limestone and a mudstone. In general, crystalline or well-cemented rocks have a higher sonic velocity than clay-rich or porous lithologies.

The time taken for a sound wave to reach a reflector and return to the surface can be recorded: this is called the **two-way time** (TWT) and it can then be related to depth of the reflector at that point. The strength of the reflection is governed by the contrast in the acoustic properties at the boundary between the two rock units. By recording multiple sound waves reaching multiple reflectors across an area an image of the subsurface can be generated and

subsequently interpreted in terms of geological structures and stratigraphy.

22.2.1 Acquisition of seismic reflection data

Seismic reflection profiling can be carried out on land or at sea. Marine surveys are generally more straightforward because the ship can follow a course optimised for the data collection, whereas land-based surveys are restricted by topography, access and land use. The source of the energy at the surface is provided by a number of different mechanisms. On land, explosives may be used but it is now more common to use a **vibraseis** set-up, a vehicle or group of vehicles that vibrate at the surface at an appropriate frequency to generate shock waves. At sea the sound energy is provided by an **airgun**, a device that builds up and releases compressed air with explosive force: it is usual to have multiple airguns forming an array, releasing energy every 10 to 20 seconds. The horizontal spacing of the points where the energy is released (the **shot points**) is usually 12.5 or 25 m.

The returning sound waves are detected by receivers: these are essentially microphones that are referred to as **geophones** on land and **hydrophones** at sea. The pattern of these receivers depends on whether the survey is two-dimensional, a **2-D survey**, or three-dimensional, a **3-D survey**. For 2-D surveys a single string of receivers is spread out along a line spaced 12.5 to 25 m apart: in marine surveys this is called a **streamer** and it may be 3 to 12 km long. The returning sound waves are recorded along one vertical plane, producing a single profile that may be many tens of kilometres in length. For 3-D surveys a series of 6 to 12 parallel streamers, each about 100 m apart, are towed behind the ship to create an array of receivers arranged in a grid pattern (Fig. 22.1). These record the reflected sound waves in a 3-D volume of rock in the subsurface and a 3-D survey may cover tens of square kilometres in a series of parallel swathes.

In the initial stages of exploration in an area a series of widely-spaced 2-D survey lines are shot to provide a general picture of the structure and stratigraphy of the region. 3-D surveys are more expensive to acquire and are usually used in the later stages of exploration to provide more detailed information about the exploration target.

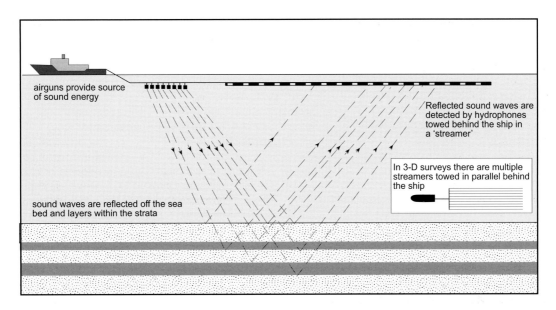

Fig. 22.1 In marine seismic reflection surveys the ship tows the energy source, the airgun, and the receivers either as a single line or in multiple lines to generate a 3-D survey.

22.2.2 Processing of seismic reflection data

The signals generated by each reflection from one burst of energy are very weak. However, each reflection point in the subsurface will generate multiple return signals recorded at many different receivers from successive shots. These signals can be merged in a process called *stacking*, which greatly enhances the signal strength. Another processing technique is also used to allow for the fact that the reflected sound waves do not come back to the surface along a vertical pathway. *Migration* of the data is a process of adjusting the time taken for the return from each reflection point to take account of the longer, oblique pathway the sound wave has taken on its journey. An important component of the processing involves converting the vertical scale of the data from two-way time to depth in metres. This *depth conversion* requires information about the acoustic characteristics (sonic velocity) of all the stratigraphic units from the surface down to the chosen limits of interpretation of the profile. The sonic velocity of the layers varies with lithology (*22.4.1*) and depth, becoming higher as more compacted lithologies are encountered at greater depth. Values for the sonic velocity of the stratigraphic units can be obtained from measurements made in boreholes and these can be used to convert the two-way time into a true thickness for that interval. If carried out in a series of steps for each unit a pattern of reflectors can be presented scaled to depth below surface.

After the processing is carried out, the results from a seismic reflection survey can be presented as an image that appears to be a series of dark lines on a white background when presented as a 2-D profile (Fig. 22.2) (colours are often used in profiles generated from 3-D surveys). These images are built up of a series of closely spaced vertical traces, each of which is a record of the acoustic impedance contrasts that generated reflections. The peaks on the right-hand side of each trace representing high contrasts are filled in black, and when these traces are put next to each other, lines of strong impedance contrast, *reflectors*, show as black lines on the profile. The data from 3-D surveys are also combined into images built up from closely spaced vertical traces in a 3-D volume of rock.

The data collected in the course of a single 3-D seismic reflection survey run to hundreds of gigabytes, and the processing of the raw data into a form that can be readily interpreted in terms of the subsurface geology requires significant amounts of computer processing power. An important factor in the

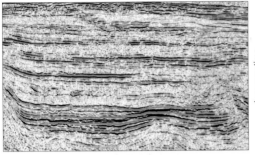

distance

two-way time

Fig. 22.2 Example of a seismic reflection profile: the horizontal scale is distance (in this case several kilometres across), but the vertical scale is in two-way travel time, that is, the time it takes for sound waves to reach a subsurface reflector and return to the surface. If the acoustic properties of the rock are known (these vary with the bulk density) this can be converted to depth.

development of more sophisticated data acquisition and processing techniques in recent years has been the availability of more powerful computers able to store, handle and rapidly process data volumes on these scales.

22.2.3 Visualisation of seismic reflection data

2-D profiles are presented as black and white paper copy, typically rolls of paper a metre or more wide and many metres long. These will show a horizontal scale in metres and kilometres, marked with the shot points of the survey. The vertical scale will be in milliseconds of two-way time (TWT ms) unless a depth conversion has been carried out prior to printing. The patterns of reflectors can be visually assessed and interpreted in terms of structures and stratigraphy as described below. If a series of lines has been shot to form a grid pattern, cross-cutting lines are matched up and a correlation between all of the lines in the grid is carried out.

The scope for visualisation of data from a 3-D survey is much greater and has expanded as computing technology has advanced. 2-D profiles can be extracted from the data and presented on-screen in any orientation, vertically, obliquely or horizontally. It is also possible to create three-dimensional images that can be perspective images on the screen or using 3-D projection technology to generate a virtual three-

dimensional effect. These latter visualisation techniques allow the interpreter to 'move' through the volume of data as if they were moving through the volume of rock and view the geology from different perspectives, angles and at different scales.

22.2.4 Interpretation of seismic reflection data

At a first glance there is a lot in common between a seismic reflection profile and a cross-section compiled from surface outcrop data. Layers looking like beds of rock may be seen on the profile, unconformities, folds and faults may be picked out and contrasts in the detailed pattern of the reflectors suggest that different rocks may be identified on a seismic reflection profile. Although all these features can indeed be related to stratigraphic and structural features seen in rocks, comparison and interpretation must be carried out with caution because there are important differences too.

First, there is a question of scale. In dealing with outcrop, a geologist is accustomed to looking at beds centimetres to metres thick and features tens to hundreds of metres across are considered large scale. The vertical resolution on a seismic reflection profile is related to the wavelength of the sound waves and the best resolution that can be achieved is about 15 m, so the units defined by reflectors are packages of beds, not individual beds. Sound waves reflected from deeper in the succession have lower energy so there is also a decrease in resolution with depth and detail can be much more clearly seen in shallower strata than in rocks buried a few thousand metres below ground.

Second, a contact between two rock units will not show up on a seismic profile if there is no acoustic impedance contrast between them. The boundary between a thick sandstone and a conglomerate body might be easily recognised in outcrop, but if they have the same acoustic properties the contact between the two would not be imaged as a reflector. The clearest reflectors are generated by the contacts between beds of contrasting properties, such as a mudstone and a well-cemented limestone, a basalt lava and a sandstone or a bed of halite overlain by anhydrite.

Third, processing techniques that attempt to convert the geometries imaged on the profile into the true subsurface relationships become less effective with

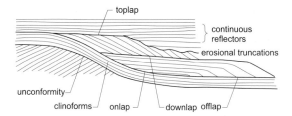

Fig. 22.3 Reflector patterns and reflector relationships on seismic reflection profiles.

increasing depth. The relative horizontal positions of reflectors are distorted such that the true location is not correctly shown, and the angular relationships are also not accurate. The interpretation of both stratal and structural geometries on seismic reflection profiles must therefore be carried out with care and an awareness of these potential distortions.

22.2.5 Stratigraphic relationships on seismic profiles

By tracing reflectors across profiles it is possible to recognise stratal relationships (Fig. 22.3) that are on the scale of hundreds of metres to kilometres. When traced and marked these form a framework for the interpretation of the whole succession of rocks imaged on the profile.

Continuous reflectors

A well-defined reflector marks a boundary between two layers of different acoustic impedance and for this to be continuous over kilometres it must mark a change in lithological characteristics of the same extent. Changes in lithology in a sedimentary succession result from changes in depositional environment and a widespread change in depositional environment can result from events such as a change in sea level or sediment supply. For example, a sea-level rise may cause sandy, shallow-water deposits to be replaced by muddy, deeper-water sediments over a wide area. A similar widespread change may occur when a carbonate shelf environment receives an influx of mud and the lithology deposited changes from limestone to mudstone. In deeper water the progradation of a sandy submarine fan lobe over muddier turbidites may also mark a change in depositional style over a

wide area. Continuous reflectors therefore may be seen as markers that indicate a significant, widespread change in deposition in the basin. For this reason, prominent reflectors are often considered to represent time-lines, isochronous surfaces, within a basin-fill succession, although care should be exercised in making this assumption where there are complex stratigraphic relationships or where reflectors merge. Changes in depositional environment usually occur over a period of time because events such as transgressions that result in retrogradation of facies (*23.1.6*) do not occur instantaneously.

Clinoforms

Inclined surfaces bounding stratal packages on seismic reflection profiles are referred to as ***clinoforms*** (Mitchum et al. 1977) and they form a pattern that indicates a progradational geometry of packages of sediment building out into deeper water. Depositional slopes of a few degrees occur at delta fronts (especially sandy or gravelly deltas: *12.4.4*), on the edges of clastic shelves and in carbonate environments, a fore-reef slope may be 25° or more. The angle of the clinoform seen on a seismic reflection profile may not always represent the true depositional geometry, and the angle may be enhanced by compaction in some instances. Sandstone has a much lower initial porosity than mudstone and therefore compacts to a lesser degree on burial, so units that grade from sandstone to mudstone would tend to taper distally upon compaction, resulting in inclined surfaces on a large scale.

Unconformities

An unconformity surface will not be represented by a reflector unless there is a consistent change in lithology across it to create an acoustic impedance contrast. In many cases, an unconformity may be identified on a seismic reflection profile by the presence of ***reflector terminations***, the points at which relatively continuous reflectors end (Mitchum et al. 1977). Some terminations are not related to unconformities (see below) but result from the shapes of the stratal packages. The breaks in the sedimentary record represented by unconformities are also often considered to be time-lines within the stratigraphy, but an unconformity may actually represent a series of events over a period of time. There may be a long time period between the erosion and subsequent deposition above the erosion surface and deposition may not occur across the whole unconformity at one time (see 'onlap' below).

Erosional truncation

If the surface of truncation is at a high angle to the orientation of the layers it intersects, erosional truncations are relatively easy to recognise (Fig. 22.3). They are assumed to result from the removal of packages of beds by subaerial or submarine erosion and are most distinct where the underlying layers have been uplifted and tilted prior to erosion. A truncation surface caused by the incision of a river valley into shelf strata following a sea-level fall may also be recognised, but only if the incision is several tens of metres and therefore enough to be resolved on seismic profiles. Low-angle erosional truncations may be difficult to identify.

Onlap

This relationship forms where there is a clear topography at the edge of or within the basin. Reflectors indicate that stratal packages are banked up against this topography, with the younger layers successively covering more of the underlying unit and sometimes covering it completely. Geometries of this type may form by the drowning of topography. Onlap relationships are an example of an unconformity representing multiple events through time: erosion may still be continuing at the upper part of the underlying unit, while deposition occurs further down dip on the surface, and deposition above this unconformity is clearly later at the top than at the bottom.

Downlap

This term is used to describe inclined surfaces that terminate downwards against a horizontal surface. This geometrical relationship is rarely seen in the smaller scale of outcrop because steeply inclined bedding surfaces are uncommon, although fore-reef slopes (*15.3.2*) and Gilbert-type deltas (*12.4.4*) are notable exceptions. Downlap surfaces seen on some seismic reflection profiles may be due to a merging of reflectors at the base of a clinoform slope where thicker sandstone beds pass distally into thinner mudrock units.

Toplap

Inclined reflectors that have upper surfaces that terminate against a horizontal surface create a pattern that is described as toplap (Fig. 22.3). This relationship occurs where there is a succession of packages of sediment that prograde basinwards, without any aggradation.

Offlap

This relationship refers to a pattern of reflectors, rather than a reflector termination. Offlap is a pattern of stratal packages that build upwards and outwards into the basin (Fig. 22.3).

22.2.6 Structural features on seismic reflection profiles

A fault surface is not often seen on a seismic line as a distinct reflector. Even if there is an acoustic impedance contrast across the fault, steeply dipping structures are poorly imaged by conventional seismic surveys because the reflected sound waves return to the surface at a high angle and are not picked up by the recording array. Faults are normally recognised by the displacement of continuous reflectors. If distinctive individual reflectors can be recognised on both sides of the fault, the direction and amount of displacement can be determined. Folds can be identified on seismic profiles although steep limbs are poorly imaged for the same reasons as discussed for steep fault surfaces. The angles of bedding or faults imaged on seismic reflection profiles are not always the true geometries and should be interpreted with caution.

22.2.7 Seismic facies

The character of patterns of reflectors on seismic reflection profiles can be used to make a preliminary interpretation of rock type and depositional facies (Mitchum et al. 1977; Friedman et al. 1992). For example, continuous reflectors suggest an environment that is relatively stable with periodic changes, such as a shelf affected by sea-level changes or a deep basin with periodic progradation of submarine fan lobes. In continental environments lateral facies patterns tend to be complex as rivers change course and widespread surfaces are less common so a discontinuous reflector pattern results. Some lithologies are characterised by an absence of parallel reflectors. For example, salt and other evaporites tend to have a 'chaotic' pattern (random reflectors) or 'transparent' pattern (lacking internal reflectors). A basement of metamorphic or igneous rocks generally lacks regular reflectors. The geometry of units bounded by reflectors can also give an indication of the depositional setting. Estuarine or fluvial deposits may be underlain by an erosional truncation and confined to a valley fill. Large reefs may be picked out by their morphology and chaotic to transparent internal reflectors.

The character of some units on a profile may give some indication of the lithology and facies but interpretation of the layers in terms of a stratigraphy of rock units can be carried out with any confidence only if the succession imaged has been drilled. The seismic facies can then be related to the rock units encountered in the borehole.

22.2.8 Interpretation of three-dimensional data

Cubes of 3-D seismic reflection data and the computing power to manipulate and analyse these data have made it possible to take interpretations much further than is possible using 2-D profiles alone. For example, horizontal slicing techniques have made it possible to recognise and determine the shape of erosional features such as fluvial and estuarine palaeovalleys, and positive features such as reefs. Similarly, the depth of the basement can be shown as a map if the contact between the basement and the basin fill has been identified across the area. Variations in the thickness of a particular unit can also be shown as a map from information about the position of the top and bottom of that unit within the data cube.

In addition to providing information about geometrical relationships, 3-D data can be used to provide information about spatial variations of the rock or fluid properties. One example of this is that a single reflector can be traced through the cube and its intensity mapped: variations in the intensity can be related to lithological changes, such as a sandstone bed being more muddy in one part of the area and hence showing less of a contrast with an overlying mudrock unit. An assessment of the fluid present can also be made because the acoustic properties of a bed depend on both the

lithology and the fluid present in pore spaces: areas where gas fills the pore spaces can be distinguished from oil- or water-bearing rocks using this approach.

The possibilities offered by the manipulation of 3-D seismic data cubes are considerable, but the interpretations ultimately require corroboration by lithological data from boreholes (see below). However, these techniques make it possible to consider stratal units in three-dimensions in a way that is rarely, if ever, possible from outcrop data alone. This has greatly improved the understanding of large-scale stratigraphy and structure of sedimentary basins.

22.3 BOREHOLE STRATIGRAPHY AND SEDIMENTOLOGY

The interpretation of seismic reflection profiles provides a model for the stratigraphic and structural relationships that may exist in the subsurface. Data from these sources can provide some indicators of the lithologies in the subsurface, but a full geological picture can be obtained only by the addition of information on lithology and facies. This can be provided by drilling boreholes through the succession and either taking samples of the rocks and/or using geophysical tools to take detailed measurements of the rock properties. When a borehole is drilled there are a number of ways of collecting information from the subsurface, and these are briefly described below.

22.3.1 Borehole cuttings

In the course of drilling a deep borehole, a fluid is pumped down to the drill bit to lubricate it, remove the rock that has been cut (cuttings) and to counteract formation fluid pressures in the subsurface. Due to the weight of rocks above, fluids (water, oil and gas) trapped in porous and permeable strata will be under pressure, and without something to counteract that pressure they would rush to the surface up the borehole. The drilling fluid is therefore usually a 'mud', made up of a mixture of water or oil and powdered material, which gives the fluid a higher density: powdered barite ($BaSO_4$) is often used because this mineral has a density of 4.48. The density of the drilling mud is varied to balance the pressure in the formations in the subsurface.

The drilling mud is recirculated by being pumped down the inside of the ***drill string*** (pipe) and returning up the outside: because it is a dense, viscous fluid, it will bring the cuttings with it as it reaches the surface. The cuttings are filtered from the mud with a sieve and washed to provide a record of the strata that have been drilled. These cuttings are typically 1–5 mm in diameter and are sieved out of the drilling mud at the surface. Recording the lithology of these drill chips (**mud-logging**) provides information about the rock types of the strata that have been penetrated by the borehole, but details such as sedimentary structures are not preserved. Microfossils such as foraminifera, nanofossils and palynomorphs (*20.5.3*) can be recovered from cuttings and used in biostratigraphic analysis. There is usually a degree of mixing of material from different layers as the fluid returns up the borehole, so it is the depth at which a lithology or fossil first appears that is most significant.

22.3.2 Core

A drill bit can be designed such that it cuts an annulus of rock away leaving a cylinder in the centre, a **core**, that can be brought up to the surface. Where coring is being carried out the drilling is halted and the section of core is brought up to the surface in a sleeve inside the hollow drill string. As each section of core is brought to the surface it is placed in a box, which is labelled to show the depth interval it was recovered from. Recovery is often incomplete, with only part of the succession drilled preserved, and the core may be broken up during drilling. The core is then usually cut vertically to provide a smooth-surfaced slab of rock that is typically 90 mm to 150 mm across, depending on the width of the borehole being drilled. Cores cut in this way provide a considerable amount of detail of the lithologies present, the small-scale sedimentary structures, body and trace fossils.

In exploration for oil and gas and in the development of fields for hydrocarbon production, cores are cut through '**target horizons**', that is, parts of the succession that have been identified from the interpretation of seismic interpretation as likely source rocks, or, more importantly, reservoir bodies. Core is usually only cut and recovered through these parts of the stratigraphy: the rest of the succession has to be interpreted on the basis of geophysical wireline logs (*22.4*). However, continuous cores may be cut through successions that cannot be interpreted satisfactorily using geophysical information alone, as

can occur when the properties of the rock units do not allow differentiation between different lithologies using wireline logging tools.

In contrast to oil and gas exploration, coal and mineral exploration normally involves taking a complete core through the section drilled. The width of the core that is cut is smaller, often just 40 mm, and the core is not split vertically (Fig. 22.4). The small size

Fig. 22.4 Cores cut by a drill bit and brought to the surface provide information about subsurface strata.

and the curved surface of the core may make it more difficult to recognise sedimentary structures than in the conventional, larger, split core used in oil and gas exploration, but the continuous core provides good vertical coverage of the drilled succession.

22.2.3 Core logging

The procedure for recording the details of the sedimentary rocks in a core is very similar to making a graphic sedimentary log of a succession exposed in the field. Core logging sheets are similar in format to field logging sheets (Fig. 5.3), and the same types of information are recorded (lithology, bed thickness, bed boundaries, sedimentary structures, biogenic structures, and so on). The scale is usually 1:20 or 1:50. In some ways recording information about strata from core is easier than field description. If the core recovery is good then there will be an almost complete record of the succession, including the finer grained lithologies. Weathering of mudrocks in the field usually means that they are less well preserved than the coarser beds, but in core this tends to be less of a problem, although weaker, finer grained beds will often break up more during the drilling. The main limitations are those imposed by the width of the core. It is not possible to see the lateral geometry of the beds and recognise features such as channels easily, and only parts of larger scale sedimentary structures are preserved. On the other hand, the details of ripple-scale features may be more easily seen on the smooth, cut surface of a core. Palaeocurrent data can be recorded from sedimentary structures only if the orientation of the core has been recorded during the drilling process, and this is not always possible. The other, not insignificant, difference between core and outcrop is that the geologist can carry out the recording of data in the relative comfort of a core store, although it is unlikely to be such an interesting environment to work in as a field location in an exotic place.

Not all cores pass through the strata at right angles to the bedding. If the strata are tilted then a vertical drill core will cut through the beds at an angle, so all bed boundaries and sedimentary structures observed in the core will be inclined. During the development phase of oil and gas extraction, drilling is often directed along pathways (directional drilling) that can be at any angle, including horizontal. Interpretation of

inclined and near-horizontal cores therefore requires information about the angle of the well.

22.4 GEOPHYSICAL LOGGING

There is a wide range of instruments, *geophysical logging tools*, that are lowered down a borehole to record the physical and chemical properties of the rocks. These instruments are mounted on a device called a *sonde* that is lowered down the drill hole (on a *wireline*) once the drill string has been removed. Data from these instruments are recorded at the surface as the sonde passes up through the formations (Fig. 22.5). An alternative technique is to fix a sonde mounted with logging instruments behind the drill bit and record data as drilling proceeds.

The tools can be broadly divided into those that are concerned with the *petrophysics* of the formations, that is, the physical properties of the rocks and the fluids that they contain, and geological tools that provide sedimentological information. The interpretation of all the data is usually referred to as *formation evaluation* – the determination of the nature and properties of formations in the subsurface. A brief introduction to some of the tools is provided below (see also Fig. 22.6), while further details are provided

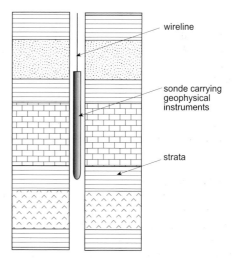

Fig. 22.5 Geophysical instruments are normally mounted on a sonde that passes through formations on the end of a wireline.

(a)

(b)

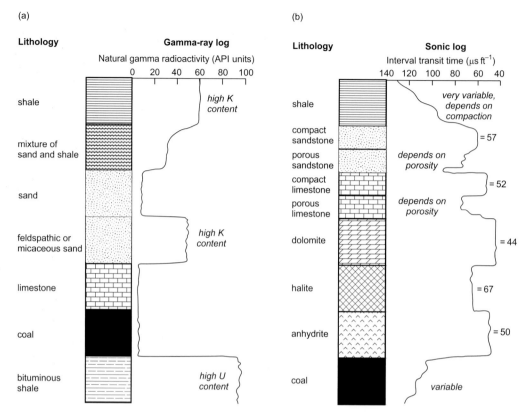

Fig. 22.6 (a) Determination of lithology using information provided by a gamma-ray logging tool. (b) Determination of lithology and porosity using information provided by a sonic logging tool. (From Rider 2002.)

in specialist texts such as Rider (2002). Many of these tools are now used in combinations and provide an integrated output that indicates parameters such as sand:mud ratio, porosity, permeability and hydrocarbon saturation.

22.4.1 Petrophysical logging tools

Caliper log

The width of the borehole is initially determined by the size of the drill bit used, but it can vary depending on the nature of the lithology and the permeability of the formation (Fig. 22.7). The borehole wall may cave in where there are less indurated lithologies such as mudrocks, and this can be seen as an anomalously

wide interval of the hole. The caliper log can also detect parts of the borehole where the diameter is reduced by the accumulation of a **mud cake** on the inside: mud cakes are made up of the solid suspension in the drilling mud and form where there is a porous and permeable bed that allows the drilling fluid to penetrate, leaving the mud filtered out on the borehole wall.

Gamma-ray log

This records the natural gamma radioactivity in the rocks that comes from the decay of isotopes of potassium, uranium and thorium. The main use of this tool is to distinguish between mudrocks, which generally have a high potassium content and hence high natural radioactivity, and sandstone and limestone, both

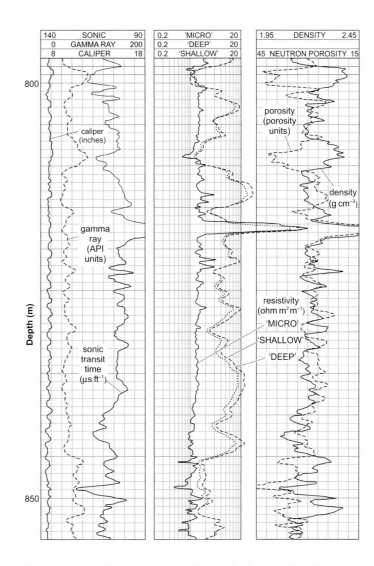

Fig. 22.7 Wireline logging traces produced by geophysical logging tools.

of which normally have a lower natural radioactivity. The gamma-ray log is often used to determine the '*sand: shale ratio*' in a clastic succession (note that for petrophysical purposes, all mudrocks are called 'shales'). However, it should be noted that mica, feldspar, glauconite and some heavy minerals are also radioactive, and sandstones rich in any of these cannot always be distinguished from mudstones using this tool. Organic-rich rocks can also be detected with this tool because uranium is often naturally associated with organic matter. Mudrocks with high organic contents are sometimes referred to as '*hot*

shales' because of their high natural radioactivity. The *spectral gamma-ray log* records the radioactivity due to potassium, thorium and uranium separately, allowing the signal due to clay minerals to be separated from radioactivity associated with organic matter.

Resistivity logs

Resistivity logging tools are a range of instruments that are used to measure the electrical conductivity of the rocks and their pore fluids by passing an electrical current from one part of the sonde, through the rocks of the borehole wall measuring the current at another part of the sonde. Most minerals are poor conductors, with the exception of clay minerals that have charged ions in their structures (*2.4.3*). The resistivity measurements provide information about the composition of the pore fluids because hydrocarbons and fresh water are poor electrical conductors but saline groundwater is a good conductor of electricity. Resistivity logging tools are usually configured so that they are able to measure the resistivity at different distances into the formation away from the borehole wall. A **microresistivity** tool records the properties at the borehole wall, a 'shallow' log measures a short distance into the formation and a 'deep' log records the current that has passed through the formation well away from the borehole (these are sometimes called **laterologs**). Comparison of readings at different distances from the borehole wall can provide an indication of how far the drilling mud has penetrated into the formation and this gives a measure of the formation permeability. **Induction logs** are resistivity tools that indirectly generate and measure the electrical properties by the process of induction of a current.

Sonic log

The velocity of sound waves in the formation is determined by using a tool that comprises a pulsing sound source and receiver microphone that records how long it has taken for the sound to pass through the rock near the borehole. The sonic velocity is dependent upon two factors. First, lithologies composed of high-density material transmit sound faster than low-density rocks: for example, coal is a low-density material, basalt is high-density, and sandstones and limestones have intermediate densities. Second, if the rock is porous, the bulk density of the formation will be reduced, and hence the sonic velocity, so if the lithology is known, the porosity can be calculated, or vice versa. The velocities determined by this tool can be used for depth conversion of seismic reflection profiles.

Density logs

These tools operate by emitting gamma radiation and detecting the proportion of the radiation that returns to detectors on the tool. The amount of radiation returned is proportional to the electron density of the material bombarded and this is in turn proportional to the overall density of the formation. If the lithology is known, the porosity can be calculated as density decreases with increased porosity. The application of this tool is therefore very similar to that of the sonic logging tool.

Neutron logs

In this instance the tool has a source that emits neutrons and a detector that measures the energy of returning neutrons. Neutrons lose energy by colliding with a particle of similar mass, a hydrogen nucleus, so this logging tool effectively measures the hydrogen concentration of the formation. Hydrogen is mostly present in the pore spaces in the rock filled by formation fluids, oil or water (which have approximately the same hydrogen ion concentration) so the neutron log provides a measure of the porosity of the formation. However, clay minerals contain hydrogen ions as part of the mineral structure, so this tool does not provide a reliable indicator of the porosity in mudrocks or muddy sandstones or limestones.

Electromagnetic propagation log

The dielectric properties of the formation fluids are measured with this tool. It consists of microwave transmitters that propagate a pulse of electromagnetic energy through the formation and measures the attenuation of the wave with receivers. The measurements are related to the dielectric constant of the formation, which is in turn determined by the amount of water present. The tool therefore can be used to distinguish between oil and water in porous formations.

Nuclear magnetic resonance logs

Conventional porosity determination techniques do not provide information about the size of the pore spaces or how easily the fluid can be removed from those pores. Fluids that are bound to the surface of grains by capillary action cannot easily be removed and are therefore not producible fluids, and if pore spaces are small more fluid will be bound into the formation. The nuclear magnetic resonance (NMR) tool works by producing a strong magnetic field that polarises hydrogen nuclei in water and hydrocarbons.

When the field is switched off the hydrogen nuclei relax to their previous state, but the rate at which they do so, the relaxation time, increases if they interact with grain surfaces. Measurement of the electromagnetic 'echo' produced during the relaxation period can thus be used as a measure of how much of the fluid is 'free' and how much of it is close to, and bound on to, grain surfaces. The tool operates by producing a pulsed magnetic field and measuring the echo many times a second.

22.4.2 Geological logging tools

Dipmeter log

The sonde for this tool has four or six separate devices for measuring the resistivity at the borehole wall. They are arranged around the sonde so that if there is a difference in the resistivity on different sides of the borehole, this will be detected. If the layering in the formations is inclined due to a tectonic tilt or cross-stratification it is possible to detect the degree and direction of the tilt by comparing the readings of the different, horizontal resistivity devices. Hence this tool has the potential to measure the sedimentary or tectonic dip of layering.

Microimaging tools

These tools, often called **borehole scanners**, are also resistivity devices and use a large number of small receiving devices to provide an image of the resistivity of the whole borehole wall. If there are fine-scale contrasts in electrical properties, for instance where there are fine alternations of clay and sand, it is possible to image sedimentary structures as well as fractures in the rock. The images generated superficially resemble a photograph of the borehole wall, but is in fact a 'map' of variations in the resistivity.

Ultrasonic imaging logs

High-resolution measurements of the acoustic properties of the formations in the borehole walls are made by a rotating transmitter that emits an ultrasonic pulse and then records the reflected pulse with a receiver. The main use of this tool is to detect how uneven the borehole wall is, and this can be related to both lithology and the presence of fractures.

22.4.3 Sedimentological interpretation of wireline logs

It is common for the interpretation of subsurface formations to be based very largely on wireline log data, with only a limited amount of core information being available. Modern systems often provide a large amount of 'automatic' interpretation of the data, but there is nevertheless a requirement for sedimentological interpretation based on an understanding of sedimentary processes and facies analysis.

Certain lithologies have very distinctive log responses that allow them to be readily distinguished in a stratigraphic succession. Coal, for example, has a low density that makes it easily recognisable in a succession of higher density sandstones and mudstones (Fig. 22.6). A bed of halite may also be picked out from a succession of other evaporite deposits and limestones because it is also relatively low density. Igneous rocks such as basalt lavas have markedly higher densities than other strata. Organic-rich mudrocks have high natural gamma radioactivity that allows them to be distinguished from other beds, especially if a spectral gamma-ray tool is used to pick out the high uranium content. However, many common lithologies cannot easily be separated from each other using these tools, including quartz sandstone and limestone, which have similar densities, natural radioactivity and electrical properties. Information from cuttings and core is therefore often an essential component of any lithological analysis.

The gamma-ray log is the most useful tool for subsurface facies analysis as it can be used to pick out trends in lithologies (Fig. 22.8). An increase in gamma value upwards suggests that the formation is becoming more clay-rich upwards, and this may be interpreted as a fining-up trend, such as a channel fill in a fluvial, tidal or submarine fan environment. A coarsening-up pattern, as seen in prograding clastic shorelines, shoaling carbonate successions and submarine fan lobes may be recorded as a decrease in natural gamma radiation upwards. A drawback of using these trends is that they are not unique to particular depositional settings and other information will be required to identify individual environments. Borehole imaging tools (scanners) provide centimetre-scale detail of the beds in the borehole and can allow sedimentary structures such as cross-bedding, horizontal laminae, wave and ripple lamination to be recognised. Detailed facies analysis can therefore be

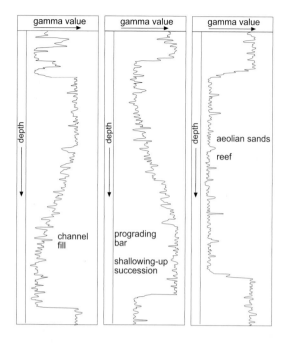

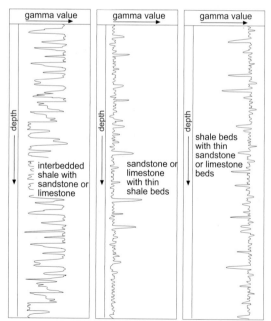

Fig. 22.8 Trends in gamma-ray traces can be interpreted in terms of depositional environment provided that there is sufficient corroborative evidence from cuttings and cores. (From Cant 1992.)

carried out using these tools, although patterns are not always easy to interpret and the most reliable interpretations can be made if there is also some core with which to make comparisons.

22.5 SUBSURFACE FACIES AND BASIN ANALYSIS

From the foregoing it should be apparent that a combination of information from the interpretation of seismic reflection data, core, cuttings and wireline logging tools can be used to carry out a full stratigraphic and sedimentary analysis of a subsurface succession of rocks. The vast majority of oil and gas reserves are to be found in the subsurface, so, from an economic point of view, the techniques described in this chapter are fundamental to exploiting those reserves. The principles of interpretation of sedimentary facies and the application of stratigraphic principles are the same whether the data are collected from below ground or from outcrops. Seismic reflection data provide information about large-scale structural and stratigraphic relationships which, with the advent of 3-D seismic cubes of data, offer a more complete image of the geology than scattered outcrops at the surface. Although data from boreholes may be scattered and one-dimensional, they provide a record of the sedimentary history that is more complete than is available from surface exposures, and seismic reflection data can be used to help correlate. Further description of these techniques is beyond the scope of this book and the reader is referred to the books and articles listed below for more detailed information.

FURTHER READING

Bacon, M., Simm, R. & Redshaw, T. (2003). *3-D Seismic Interpretation.* Cambridge University Press, Cambridge.

Blackbourn, G.A. (2005). *Cores and Core Logging for Geologists.* Whittles Publishing, Caithness.

Brown, A.R. (2005). *Interpretation of 3-D Seismic Data* (6th Edition). Memoir 42, American Association of Petroleum Geologists, Tulsa, OK.

Ellis, D.V. & Singer, J.M. (2007). *Well Logging for Earth Scientists* (2nd edition). Wiley, Chichester, 692 pp.

Emery, D. & Myers, K.J. (Eds) (1996) *Sequence Stratigraphy.* Blackwell Science, Oxford.

Rider, M.H. (2002). *The Geological Interpretation of Well Logs* (2nd edition). Whittles Publishing, Caithness.

23

Sequence Stratigraphy and Sea-level Changes

Every now and then a new theory arrives on the scene that has a revolutionary effect on science. In earth sciences, the development of the theory of plate tectonics in the mid-1960s had a profound effect and completely changed the way that geologists, and others, looked at the world around us. It is a truly global, unifying theory of the behaviour of the planet that now underpins all modern geology. The emergence of the concept of sequence stratigraphy a decade or more later has had nothing like the widespread impact of plate tectonics, but within the field of sedimentology and stratigraphy it has resulted in a fundamentally different way of thinking about successions of sedimentary rocks. The underlying tenet of sequence stratigraphy is that a change in base level (usually relative sea level) results in a change in the patterns of sedimentation in almost all depositional environments. If we can recognise the pattern in the sediments produced by, say, a rise in sea level in a coastal plain, a beach, a shallow shelf and the deep sea, then it may be possible to correlate using these patterns. In this chapter the principles that underlie sequence stratigraphy are presented and the application of the methodology explained. To some extent the concepts are based on long established ideas from traditional stratigraphy, but a plethora of new terms has been introduced as part of the new approach and these require some explanation. Once the jargon has been pushed aside the sequence stratigraphy approach can be seen to be a commonsense way of looking at successions of sedimentary rocks.

23.1 SEA-LEVEL CHANGES AND SEDIMENTATION

There is evidence from around the world today that the position of the shoreline is not constant, even in the geologically short time-span of historical records: harbours built hundreds or thousands of years ago are in some places drowned, in others left high and dry away from the shoreline. The first obvious cause is tectonic activity that moves the crust vertically, as well as the horizontal movements due to plate motion. This movement of the crust itself up or down relative to the sea level may affect the crust within a few kilometres of a single fault, or may be large-scale '**thermo-tectonic**' activity that has an effect on whole continental margins. Second, there can be changes in the volume of water in the world's oceans: this is called '**eustatic sea-level change**' (**eustasy**) and is caused by melting and freezing of continental ice caps, among other things. Debates about the effects of global warming on the level of the sea worldwide have brought this phenomenon to the attention of most people. Third, there is the effect of sedimentation: sand, gravel and mud piled up at the shoreline can result in the shoreline moving away from its former position.

These three factors – tectonic uplift/subsidence, eustatic sea-level rise/fall, and sedimentation – and how they occur, where they occur, their rates and how they interact are fundamental to sedimentology and stratigraphy. The character of sediment deposited in environments ranging from rivers and floodplains to shorelines, shelves and even the deep seas is in some way influenced by these three factors. The study of the relationships between sea-level changes and sedimentation is often referred to as '**sequence stratigraphy**'. In the following sections the principles underlying the basic concepts are considered and then there is an explanation of some of the terminology that has evolved to describe the relationships between strata under conditions of changing sea level. The causes of sea-level fluctuations and the use of a sea-level curve as a correlative tool are also discussed.

23.1.1 Changes to a shoreline

If the three variables are considered in isolation of each other, five different scenarios can be considered (Fig. 23.1). Consider what will happen to a palm tree growing on a beach and a crab sitting on the sea floor a few hundred metres away.
1 Eustatic sea-level rise: the palm tree is drowned and the crab will find itself in deeper water.
2 Eustatic sea-level fall: the palm tree will end up growing some distance from the shoreline, and the crab is now in shallower water.
3 Uplift of the crust: the palm tree will end up growing some distance from the shoreline, and the crab is now in shallower water.
4 Subsidence of the crust: the palm tree is drowned and the crab will find itself in deeper water.

5 Addition of sediment at the shoreline: the palm tree will end up growing some distance from the shoreline, and the crab is now in shallower water (providing it is not engulfed by sediment, but instead moves up to the new sea floor).

It is important to note that scenarios 1 and 4 are exactly the same, and viewed from just one point on the Earth's surface it is not possible to distinguish between these two possible causes. The same is true of scenarios 2 and 3, which are indistinguishable at a local scale, and often the difference between either of these and scenario 5 can be subtle. The controls on sea-level fluctuations are considered in section 23.8, but because it is difficult to distinguish between uplift and eustatic sea-level fall on the one hand and subsidence and eustatic sea-level rise on the other, it is usually best to refer to changes in '**relative sea level**' or '**relative base level**' when looking at strata in one place. The drowning of a palm tree may therefore be considered to be evidence of a 'relative sea-level rise', without any implication of the cause, and in the same way, the crab is now in relatively deeper water. The impact that these relative changes have on processes and products of sedimentation is something that will be considered in the next section, along with the importance of the third factor, sedimentation.

23.1.2 Sea level and sedimentation

Although we may find palm trees fossilised in sedimentary rocks, the evidence for the position of the shoreline and the relative depth of water comes mainly from the character of the sediments themselves. In Chapters 12 to 16 the characteristic facies of sediment deposited at different positions relative to the shoreline and in different depths of seawater were considered. If, therefore, we can establish the water depth/position relative to shoreline by examining the sedimentary facies, we can also recognise relative changes in the shoreline/water depth from changes in those facies. In fact, the analysis of strata in terms of relative sea-level changes can be carried out only if a facies analysis is carried out first. Once all the beds in a succession have been analysed and classified according to environment of deposition using the approaches described in earlier chapters in this book, the effects of sea-level changes on their deposition can then be considered.

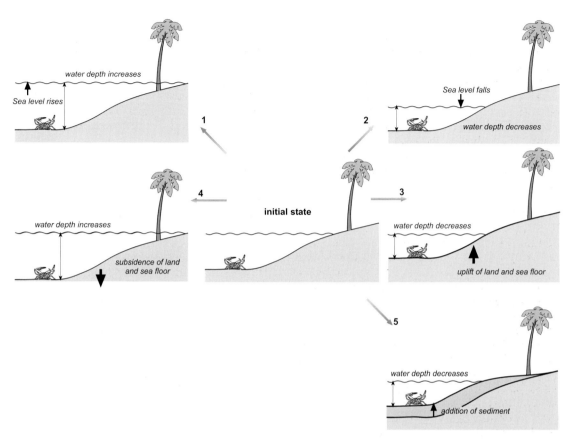

Fig. 23.1 Relative sea-level change (the change in water depth at a point) may be due to uplift or subsidence of the crust, increase or decrease in the amount of water, or addition of sediment to the sea floor. It is often not possible to determine which mechanism is responsibie if there is information from only one place.

23.1.3 Transgression, regression and forced regression

If there is a relative sea-level rise the shoreline will move landward: this is referred to as ***transgression*** (Fig. 23.2a). Movement of the shoreline seawards as a result of sedimentation occurring at the coast is called a ***regression*** (Fig. 23.2b), but if it is due to a relative sea-level fall it is known as a ***forced regression*** (Fig. 23.2c). The sedimentary response to these changes in shoreline can be preserved in strata as changes in facies going up through a succession, changes that reflect either a landward movement of the shoreline, transgression, or a seaward movement of the shoreline, regression (forced or otherwise).

Under conditions of transgression, the shoreline will move to a place that used to be land, and the coastal plain deposits are overlain by beach deposits. Similarly, beach (foreshore) deposits will be overlain by shoreface deposits because the former beach is now under shallow water. The same pattern of changes in facies from shallower to deeper will be seen all the way across the shelf (Fig. 23.2a). It is therefore possible to recognise the signature of a transgression in a succession of beds by the tendency for the environment of deposition, as indicated by the facies, to become deeper upwards. If there is a regression, the pattern seen in vertical succession will be the opposite: as the sea becomes shallower, either due to a relative sea-level fall (forced regression) or addition of more sediment (regression), the facies will reflect this: shoreface facies will be overlain by foreshore deposits, offshore transition sediments by shoreface deposits, and so on (Fig. 23.2b & c). In some circumstances,

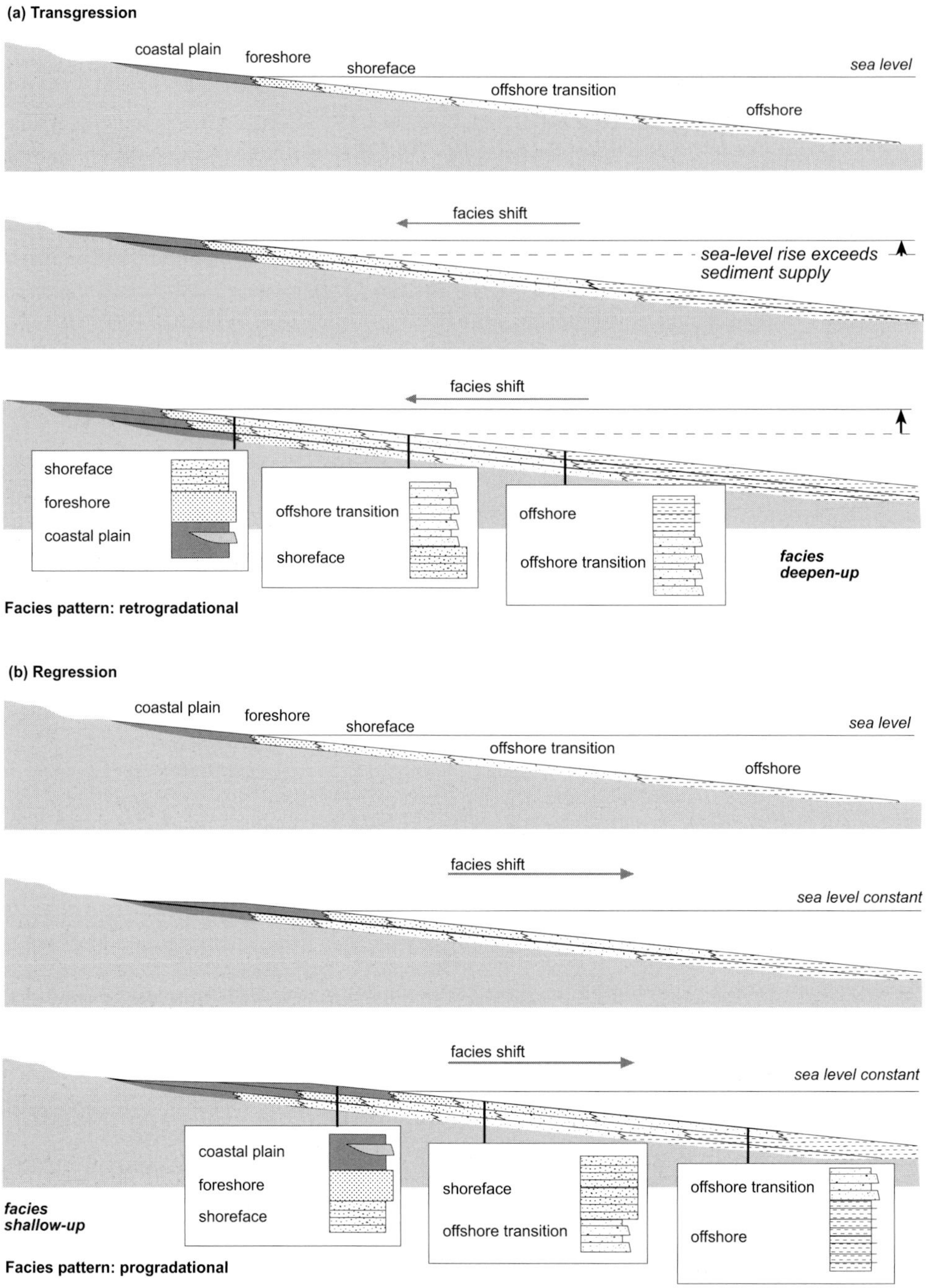

Fig. 23.2 (a) If sea level rises faster than sediment is supplied the coastline shifts landward: this is known as **transgression** and the pattern in the sediments is **retrogradational**. (b) If sediment is supplied to a coast where there is no (or relatively slow) sea-level rise the coastline moves seaward: this is **regression** and the sediment pattern is **progradational**. (c) Sea-level fall results in a **forced regression** and the sediment pattern is **retogradational**, and may include erosion surfaces. (d) A situation where the coastline stays in the same position for long periods of time is relatively unusual and requires a balance between relative sea-level rise and sediment supply producing a pattern of **aggradation** in the sediments.

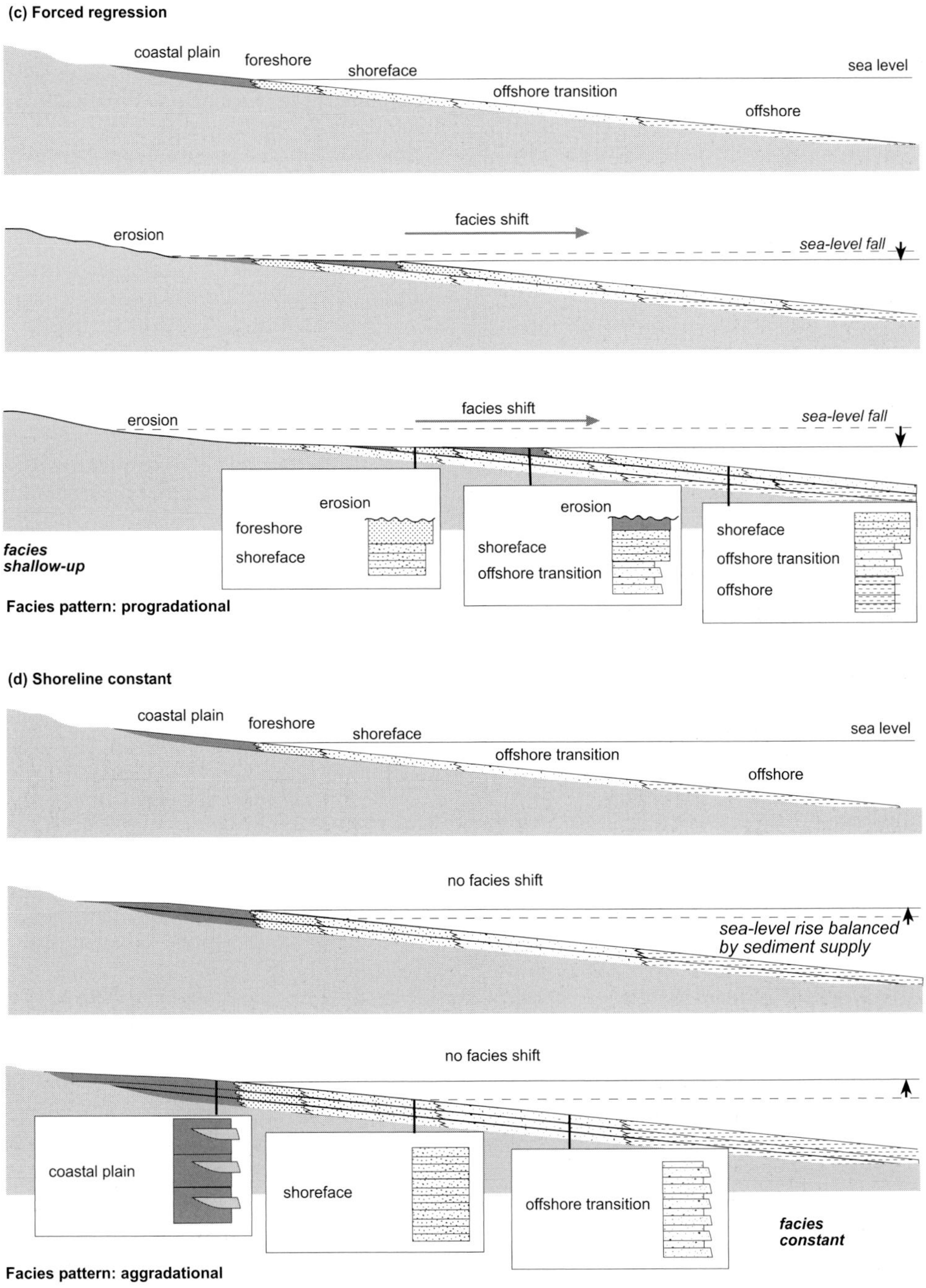

(c) Forced regression

coastal plain   foreshore   shoreface   offshore transition   offshore   sea level

erosion   facies shift   sea-level fall

erosion   facies shift   sea-level fall

facies shallow-up

erosion
foreshore
shoreface

erosion
shoreface
offshore transition

shoreface
offshore transition
offshore

Facies pattern: progradational

(d) Shoreline constant

coastal plain   foreshore   shoreface   offshore transition   offshore   sea level

no facies shift   sea-level rise balanced by sediment supply

no facies shift

coastal plain

shoreface

offshore transition

facies constant

Facies pattern: aggradational

Fig. 23.2 *(cont'd)*

a forced regression may be distinguished from a simple regression by evidence of erosion in the coastal and shallowest marine deposits: as sea level falls, the river may have to erode the older coastal deposits as it cuts a new path to the shoreline. However, this may not always happen, and depends on rates of sediment supply and the slope of the foreshore/shoreface.

23.1.4 The concept of accommodation

Sediment will be deposited in places where there is space available to accumulate material: this is the concept of **accommodation** (or **accommodation space**) and its availability is determined by changes in relative sea level (Muto & Steel 2000). In shallow marine environments an increase in relative sea level creates accommodation that is then filled up with sediment until an equilibrium profile is reached. The **equilibrium profile** is a notional surface of deposition relative to sea level and sedimentation occurs on any point in the shallow marine environment until this surface is reached: any material deposited above the surface is reworked by processes such as waves and tidal currents. The equilibrium profile is at different positions relative to sea level in different environments: in the foreshore it is at sea level, in the shoreface a few metres below sea level and then progressively deeper through further offshore.

Accommodation in shallow marine environments is created by any mechanism that results in a relative rise in sea level, including eustatic sea-level rise, tectonic subsidence and compaction of sea-floor sediments. Accommodation is reduced by the addition of sediment to fill the space or by tectonic or eustatic mechanisms that lower the relative sea level. The rate of change of accommodation is determined by the relative rates of relative sea-level change and sediment supply. Deposits in places where there has been a relative sea-level fall will often be eroded, and this can be considered to be a condition where there is negative accommodation.

The ideas of accommodation and equilibrium profiles can also be applied to fluvial environments. A mature river will erode in its upper tracts and deposit in the downstream parts until it develops an equilibrium profile, whereby the main channel is neither eroding nor depositing. Under these conditions erosion still continues in the hillslopes above the main channel valley, but sediment is carried through the river down to the sea. This profile may be disturbed by a fall in sea

level that creates negative accommodation along part of the profile, resulting in erosion, or by sea-level rise generating accommodation that allows sediment to accumulate in the channels and overbank areas until it returns to the equilibrium position.

The concept can also be applied to non-marine systems such as lakes and river systems feeding them, where it is the level of the water in the lake that determines the amount of accommodation available. In the following discussion, accommodation is considered in terms of relative sea level, and depositional systems described are either marine or have marine connections. The same principles can be applied to lacustrine systems and the deposits of large lakes can be considered in terms of relative changes in the lake level. Global eustasy does not directly control the level of water in lakes, but climatic controls are important because the balance between precipitation/run-off and evaporation determines the amount of water in the lake and hence its level.

23.1.5 Rates of sea-level change and sediment supply

In a previous section it was stated that if there is a relative sea-level rise, the shoreline will move landwards. In fact, this is not necessarily the case: if the rate at which sediment is supplied is greater than the rate at which the sea level is rising, then the shoreline will still move seawards. Similarly, if the rate of sediment supply and the rate of sea-level rise are in balance, the shoreline position does not change (Fig. 23.2d). Several different situations can be envisaged when the sea level is rising (either due to subsidence or eustatic sea-level rise) which give rise to different stratal geometries (Fig. 23.3).

I If the rate of sediment supply is very low then the shoreline will move landward without deposition occurring and with the possibility of erosion.

II With moderate sedimentation rates, but high rates of sea-level rise, deposition will occur as the shoreline moves landward.

III If it is a higher sedimentation rate, then as fast as the sea level rises the space is filled up with sediment and the shoreline stays in the same place.

IV At high sedimentation rates, the shoreline will still move seawards, even though the sea level is rising.

V During periods when the sea level is static the addition of sediment causes the shoreline to shift seawards.

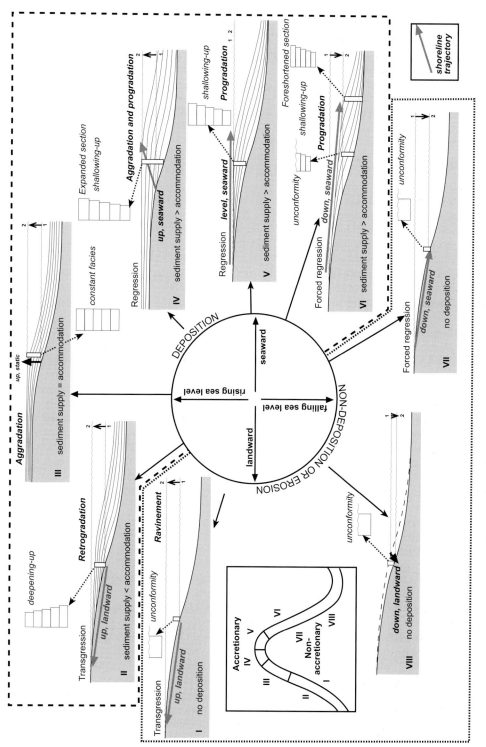

Fig. 23.3 The various possible patterns of sedimentation that can result from different relative amounts of sediment supply and relative sea-level change are summarised in this diagram. The responses to the different combinations are expressed in terms of vertical sedimentary successions, as seen in successions of strata in outcrop or boreholes, or as geometries seen in regional cross-sections or seismic reflection profiles expressed in terms of shoreline trajectories. Eight main scenarios (I–VIII) are recognised.

VI At low rates of sea-level fall and/or high rates of sediment supply deposition occurs as the shoreline moves seawards.

VII If the rate of sea-level fall is relatively high and the rate of sedimentation is low, there is no sedimentation, and there may be erosion.

VIII A coast undergoing rapid erosion during sea-level fall could theoretically fall into this category.

Shoreline trajectory

The different relationships between relative sea-level rise/fall and sediment supply shown in Fig. 23.3 can be divided into situations where there is transgression (I and II), regression (IV and V) and forced regression (VI and VII). These terms refer to changes in the position of the shoreline, so none of them apply to case III in which the shoreline remains fixed. Another way of considering these different relationships is in terms of '*shoreline trajectory*' (Helland-Hansen & Martinsen 1996). The arrows on Fig. 23.3 indicate the trajectory of the shoreline relative to a fixed horizontal datum, and in the scheme suggested by these authors, a shoreline trajectory of $0°$ is no change in vertical position (geometry V), negative values, round to $-90°$ are cases where the shoreline has changed its relative position to a lower elevation (geometries VI and VII), and positive values ($+1°$ to $+179°$) are instances where the shoreline has moved to a higher elevation. Values between $-89°$ and $+89°$ represent scenarios where the shoreline has moved seawards (regression) and values over $+90°$ are cases where the shoreline has moved landwards (transgression). This scheme can be readily applied to seismic reflection profiles (*22.2.5*) and can be used to help interpret the subsurface stratigraphy in terms of relative sea-level changes.

Depositional slope, onshore and offshore

In all of the above scenarios it is the relative rates of sea-level rise/fall and sediment supply that are important. A further factor that should be considered is the *physiography* of the margin, that is, the slope of both the land onshore and the sea-floor offshore. If the onshore slope is a low angle, then the shoreline will move much further landward during sea-level rise. Similarly, a gently sloping sea floor will result in the shoreline shifting further seaward during sea-level fall because the accommodation will be filled

more quickly by the available sediment. Other aspects of the relationship between sea-floor bathymetry and the distribution of sediments during cycles of sea-level rise and fall are considered in *23.2*.

23.1.6 Progradation, aggradation and retrogradation

The concepts of transgression, regression and forced regression refer to the change in the position of the shoreline. Another way of looking at it is to consider the arrangement of the strata deposited during periods of sea-level rise, standstill or fall and to look at the relative positions in time and space of the facies within those strata. If the rate of creation of accommodation is exactly balanced by the rate of sediment supply (Fig. 23.3, geometry III) the sediment in all environments along the profile will simply build up without any variation of character: foreshore deposits will be overlain by more foreshore deposits, shoreface sediments by shoreface sediments, and so on. This may be referred to as *aggradation* of the sedimentary succession (Fig. 23.3). Shorelines that receive sediment at a higher rate than accommodation is created build out through time, with foreshore deposits on top of shoreface sediments, shoreface deposits overlying offshore transition facies and so on: this pattern of shallowing up of the facies through the succession is called *progradation* and it is the signature in the sediments of geometries IV, V and VI (Fig. 23.3). A distinction can be drawn between the progradational pattern in IV, which is building up as well as out, and the pattern in VI, which is stepping down as it progrades. Where there is a high rate of creation of accommodation relative to the sediment supply (scenario II) the pattern is one of deeper deposits progressively though the succession, as foreshore facies will be overlain by shoreface and so on: this is referred to as a *retrogradation* within the succession and is characteristic of transgression (Fig. 23.3).

The fundamental approach to considering succession of strata using the sequence stratigraphic approach is to look for these patterns in the beds. A retrogradational pattern in the succession indicates an increase in accommodation due to transgression. A progradational pattern is indicative of a reduction in the rate of creation of accommodation relative to sediment supply and may be interpreted as occurring during sea-level fall, sea-level stasis or relatively slow

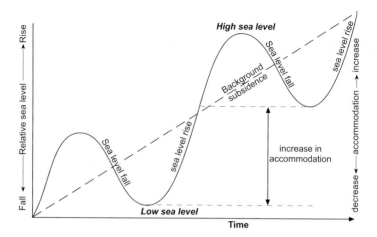

Fig. 23.4 A sinusoidal curve of sea-level variation combined with a long-term increase in relative sea-level results in periods when there is relative sea-level rise (and hence transgression) and periods of relative sea-level fall (resulting in regression).

sea-level rise. Aggradational patterns are relatively uncommon because they require the specific condition of a balance between the creation of accommodation and sediment supply. Recognition of these patterns within a succession of strata makes it possible to divide the beds up into groups on the basis of changes in relative sea level.

23.1.7 Cycles of sea-level change

Analysis of strata of different ages throughout the world has revealed that in many instances there is evidence for cycles of sea-level change. The periods of alternating sea-level rise and fall can be represented as a sinusoidal curve (Fig. 23.4) that shows the change in accommodation through time. Assuming that sediment supply is constant (which may or may not be the case, but this assumption provides the simplest scenario), the patterns of progradation, aggradation and retrogradation can be matched to different sections of the curve. The different stratal geometries shown in Fig. 23.3 can also be related to the curve in Figure 23.4. Two cases can be considered, depending on whether the sediment supply is relatively high or relatively low compared with the rate of change of accommodation.

1 With a high sediment supply rate, deposition may occur throughout the cycle, with stratal geometry II deposited during transgression, geometries III and IV forming as the rate of sea-level rise slows down and

stops V followed by geometry VI as relative sea level falls.

2 Under conditions of low sediment supply and/or rapid changes in sea level, erosion may occur during sea-level fall (geometry VII) and there can also be situations where erosion occurs during transgression (geometry I).

The shape of the curve in Fig. 23.4 signifies that the base of the second cycle is at a higher level than the base of the first. This means that there is an overall increase in the amount of accommodation through time and this is required in order that there may be a net accumulation of sediment. If the sea level at the bottom of the second cycle fell to the same point as the first, and if there was erosion during sea-level fall, then all the sediment deposited earlier in the second cycle might be completely removed. A condition of net accommodation creation though time is therefore required in order to preserve a cycle of sedimentation.

23.2 DEPOSITIONAL SEQUENCES AND SYSTEMS TRACTS

In the discussion thus far the terminology used predates the advent of sequence stratigraphy: terms such as transgression and regression to describe sea-level changes have been in use for a long time and the idea that strata are sometimes arranged into repeating cycles of lithologies was established in the 19th century. The development of the concepts of sequence

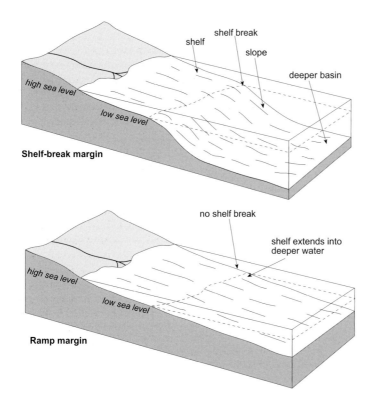

Fig. 23.5 Two main types of continental margin are recognised, each resulting in different stratigraphic patterns when there are sea-level fluctuations: (a) shelf-break margins have a shallow shelf area bordered by a steeper slope down to the deeper basin floor; (b) ramp margins do not have a distinct change in slope at a shelf edge.

stratigraphy has resulted in the introduction of a large number of new terms into sedimentary geology and although at first this new terminology seems to be quite daunting, it is generally quite logical and easy to relate to the concepts. To illustrate and introduce the application of sequence stratigraphic terminology, and to show how facies distributions can relate to changes in relative sea level, two types of continental margin are considered, each with certain conditions regarding rates of sea-level change and sediment supply. These two case studies are the ones most often illustrated in texts on sequence stratigraphy (e.g. Coe 2003; Cateneaunu 2006), but variations on these two are not only possible, but common, depending on the rates of sea-level rise and fall, changes in sediment supply, and the shape of the bathymetry of the shelf.

1 A continental margin with a distinct shelf break: sediment supply is assumed to be constant, and the sea level falls to below the edge of the shelf so that there is erosion on the shelf during sea-level fall (Fig. 23.5). Beyond the edge of the shelf lies a slope and a deeper basin area that receives sediment during certain stages of the sea-level cycle.

2 A continental shelf that is a sloping ramp with no distinct change in slope: the sediment supply is again considered to be constant and is relatively high such that deposition occurs throughout the cycle (Fig. 23.5). A deeper basin area may exist, but is not strongly influenced by sea-level fluctuations on the ramp.

23.2.1 Shelf-break margin depositional sequence (Fig. 23.6)

Highstand

The ***highstand*** is the period of high sea level during the cycle and the beds deposited during this period are called the ***highstand systems tract*** (**HST**). (A '***systems tract***' is the term used in sequence stratigraphy for strata deposited during a stage of a depositional sequence.) The beds show either an aggradational or a progradational pattern as the shoreline shifts seawards across the shelf. Sediment is supplied by rivers from the hinterland and most of the accumulation occurs on the shelf with little sediment reaching the deeper basin.

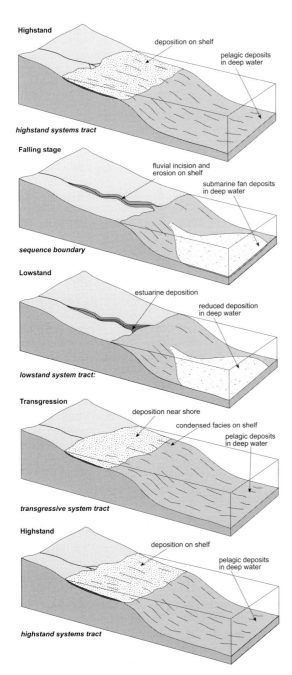

Fig. 23.6 Variations in sea level that follow the pattern in Fig. 23.4 affecting a shelf-break margin result in a series of systems tracts formed at different stages: the lowstand is characterised by deep-basin turbidites if the sea level falls to the shelf edge.

Sequence boundary

During sea-level fall erosion of the shelf occurs as rivers erode into the sediment deposited during the previous cycle: where erosion is localised the rivers cut *incised valleys*. This erosion creates an unconformity, which in this context is also called a *sequence boundary (SB)*. It marks the end of the previous depositional sequence and the start of a new one: depositional sequences are defined as the packages of beds that lie between successive sequence boundaries. If the sea level falls to the level of the shelf edge then both the detritus from the hinterland being carried by the rivers and the material eroded from the shelf are carried beyond the edge of the shelf. This sediment forms a succession of turbidites on the basin floor that are deposited during the period of falling sea level, forming a basin floor submarine fan. There is no unconformity within the basin floor succession to mark the sequence boundary, so the start of the next sequence is marked by a *correlative conformity*, a surface that is laterally equivalent to the unconformity that forms the sequence boundary on the shelf.

Lowstand

The interval of low sea level is called a *lowstand* and the deposits of this period are called the *lowstand systems tract (LST)*. The relative sea level is rising slowly but the rate of sediment supply is relatively high. The rivers cease to erode on the shelf but the shelf continues to be by-passed: sedimentation continues to occur on the *basin-floor fan* as turbidites (also referred to as *lowstand fan* deposits). Sediments also start to build up above the fan at the base of the slope to form a *lowstand wedge* (not shown in Fig. 23.6). The pattern of beds in these deposits is initially progradational, becoming aggradational in the lowstand wedge as the rate of sea level increases.

Transgressive surface

The point at which the rate of creation of accommodation due to relative sea-level rise exceeds the rate of sediment supply to fill the space is called the *transgressive surface (TS)*. It marks the start of retrogradational patterns within the sedimentary succession as accommodation outpaces sediment supply. If sediment supply is relatively low the transgressive surface may be erosional: surfaces of erosion formed during transgression are called **ravinement surfaces** and they form because high wave energy in the shallow water that floods over the land surface can result in erosion.

Transgressive systems tract

Deposits on the shelf formed during a period of relative sea level rising faster than the rate of sediment supply are referred to as the **transgressive systems tract (TST)**. They show a retrogradational pattern within the beds as the shoreline moves landwards. Sediment is no longer supplied to the basin floor because there is now sufficient accommodation on the shelf. The relative sea-level rise results in the formation of estuaries as the incised valleys are flooded with seawater: estuarine sedimentation (*13.6*) is characteristic of the transgressive systems tract. The rise in base level further upstream creates accommodation for the accumulation of fluvial deposits within the valleys.

Maximum flooding surface

As the rate of sea-level rise slows down the depositional system reaches the point where the accommodation is balanced by sediment supply: when this happens transgression ceases and the shoreline initially remains static and then starts to move seawards. This point of furthest landward extent of the shoreline is called the **maximum flooding surface (MFS)**: it should be noted that it does not represent the highest sea level in the cycle, which occurs later in the highstand systems tract. As the point of the maximum flooding surface is approached the outer part of the shelf is starved of sediment because there is abundant accommodation near the shoreline: very low sedimentation rates on the shelf can be recognised by a number of features including concentrations of authigenic glauconite and phosphorites (*11.5.1, 3.4*), condensed beds rich in fossils (*11.3.2*) and evidence of sea-floor cementation from hardgrounds and firmgrounds (*11.7.4*).

Highstand

A return to aggradational and progradational patterns of shelf sedimentation marks the onset of the highstand systems tract above the maximum flooding surface. Continued relative sea-level rise creates

accommodation within the continental realm: fluvial deposition is no longer confined within incised valleys (cf. transgressive systems tract) resulting in deposition in rivers and on overbank areas over wide areas of the coastal plain.

23.2.2 Ramp margin depositional sequence (Fig. 23.7)

Highstand

Highstand deposition on a ramp margin is essentially the same as for the shelf-break example, with an aggradational to progradational pattern of deposition on the inner part of the margin.

Sequence boundary

The sequence boundary in the ramp succession is placed on the surface on which there is the first evidence of erosion caused by sea-level fall (Coe 2003). Erosion starts at the landward end, but further seaward there will be continued sedimentation, with the geometry of the strata changing from building up to stepping down (i.e. from geometry IV to geometry VI on Fig. 23.3). This change in stratal geometry may not always be easy to recognise in practice. It is worth noting that in the original schemes for shelf-break margins summarised in publications such as Van Wagoner et al. (1990) the highstand was considered to continue into the initial stages of sea-level fall and the sequence boundary placed at the point when erosion was widespread across the shelf: if this approach is applied to a ramp margin it would be placed at some point within the succession deposited during sea-level fall. Hunt & Tucker (1992) suggested an alternative scheme in which the sequence boundary is placed above all the strata deposited during the period of falling sea level, but this creates the situation where the sequence boundary lies within the package of strata deposited in the basin in the shelf-break setting.

Falling stage systems tract

Sediments deposited during the period from the onset of the relative fall in sea level until the point where it stops falling are considered to form the **falling stage systems tract (FSST)** (Plint & Nummedal 2000). The

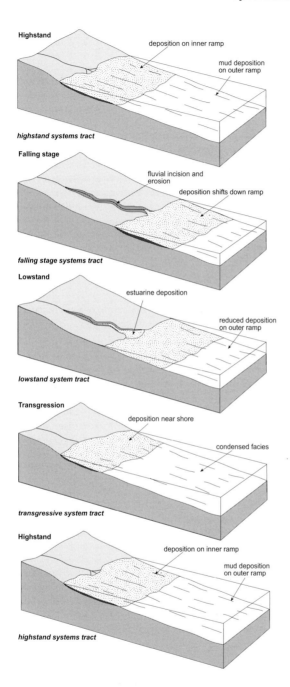

Fig. 23.7 Variations in sea level that follow the pattern in Fig. 23.4 affecting a ramp margin result in a series of systems tracts formed at different stages: during lowstand the deposition is shifted down the ramp.

sediment is supplied by rivers from the hinterland and also as a result of erosion of the landward part of the ramp. Erosion by rivers may create incised valleys between the landward part of the ramp and the shore-line. The shoreline moves seawards and steps down as the relative sea level falls and hence the geometry of the strata is progradational and down-stepping. Note that under the alternative scheme suggested by Hunt

& Tucker (1992), these deposits would be considered to be below the sequence boundary and they refer to them as the ***forced regressive wedge systems tract (FRWST)***.

Lowstand

The lowstand systems tract is deposited during the early stages of sea-level rise. The inner part of the ramp is no longer erosional and the water level starts to rise in the incised valleys. The geometry of the strata in the outer part of the ramp is progradational, becoming aggradational as the rate of sea-level rise increases and the shoreline stops moving seawards and becomes stationary.

Transgressive surface, transgressive systems tract, maximum flooding surface, highstand

The processes and patterns of sedimentation during these stages of rising sea level are essentially the same as for the shelf-break margin depositional sequence.

23.3 PARASEQUENCES: COMPONENTS OF SYSTEMS TRACTS

23.3.1 Parasequences

Each of the systems tracts in Figs 23.6 and 23.7 is shown as consisting of several separate packages of strata. In detail each package is itself a cycle of beds showing a progradational pattern and these depositional cycles are known as ***parasequences*** in sequence stratigraphy terminology. They form because the actual variation in sea level is not the smooth sinusoidal curve shown in Fig. 23.4: sea level does not rise or fall at steadily varying rates as the smooth curve indicates, but occurs in a series of shorter stages. A curve shape generated by superimposing a short-term sinusoidal pattern onto the longer term trend can account for the variations in accommodation indicated by the presence of parasequences (Fig. 23.8) During the

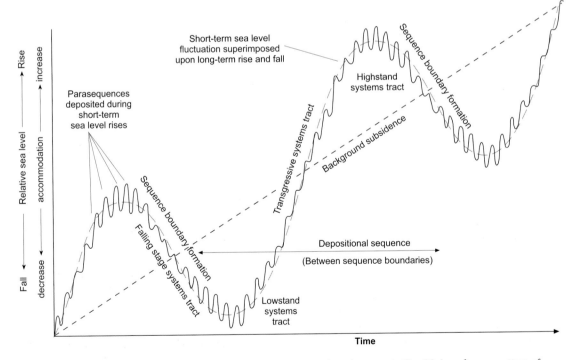

Fig. 23.8 A higher frequency sea-level fluctuation curve superimposed on the curve in Fig. 23.4 produces a pattern of short-term rises and falls in sea level within the general trends of transgression and regression. These short-term fluctuations result in creation of small amounts of accommodation being created, even during the falling stage.

rising limb of the depositional sequence curve the superimposed curve results in short periods of accelerated relative sea-level rise separated by periods when sea level is rising more slowly or not at all. Superimposition of curves on the falling limb creates short periods when the relative sea-level rise is enough to create some accommodation.

Parasequence boundaries

Parasequences are hence packages of beds deposited as a consequence of the creation of a small amount of accommodation that is then filled up by sediment. Within a succession, the boundary of a parasequence is marked by a facies shift from shallower to deeper water deposition (e.g. from foreshore to offshore transition facies), which signals a sudden increase in relative sea level. This surface marking the parasequence boundary is called a ***flooding surface*** (not to be confused with a 'maximum flooding surface').

Parasequence thickness

The thickness of a parasequence measured between flooding surfaces is determined by the relative sea-level rise: it is usually in the range of metres to tens of metres (Fig. 23.9). The accommodation created is filled by sediment until the next flooding surface occurs, and there are three different possibilities for the succession of beds within the parasequence. If the sea level is static between flooding events the thickness of the beds representing each of the facies belts will be equivalent to the depth ranges of those facies belts: for example, if the foreshore covers the range from 0 to 2 m depth, the foreshore deposits will be 2 m, and if the shoreface is from 2 to 15 m deep, the shoreface facies will be 13 m thick, and so on. However, if the sea level is continuing to slowly rise during the period of deposition of the parasequence, as could happen in a transgressive systems tract, each of these facies units will be thicker: this can be considered to be an ***expanded succession*** within a parasequence. On the other hand, falling sea level during parasequence formation would result in thinner units, perhaps only 1 m of foreshore facies and 6 m of shoreface deposits: such a ***foreshortened succession*** within a parasequence would be characteristic of falling stage systems tract. The facies present in a parasequence and the thickness of beds will also be determined by the position on the shelf where deposition is occur-

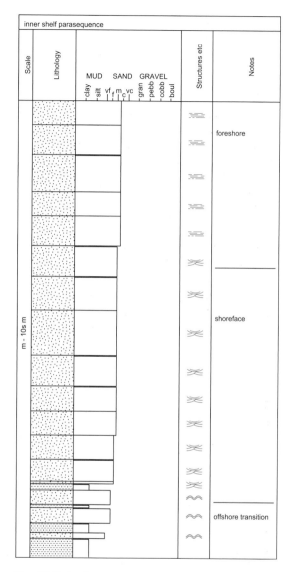

Fig. 23.9 A schematic graphic sedimentary log of a parasequence in an inner shelf shallow-marine setting.

ring: in a more distal setting the parasequence will be made up of finer grained facies (Fig. 23.10).

Parasequence sets

Although an individual parasequence shows an internal progradational bed pattern, they may be arranged into ***parasequence sets*** that may overall show progradational, aggradational or retrogradational patterns (Fig. 23.10). In a progradational parasequence set each successive parasequence going up through the set shows shallower water facies, that is, an overall shallowing-up trend. Retrogradational parasequence sets are made up of parasequences that show a trend of becoming deeper up through the succession, that is, the top bed of the higher parasequence is a deeper water facies than the beds that form the top of the parasequence below. In aggradational parasequence sets all the individual parasequences show the same cycle of facies with no overall trend of shallowing or deepening.

Parasequence sets and systems tracts

The trends within parasequence sets are indicative of the systems tract that they make up (Fig. 23.11): a highstand systems tract will have parasequence sets with aggradational to progradational trends, a transgressive systems tract is characterised by parasequence sets with a retrogradational trend, and a lowstand by progradational to aggradational trends within parasequence sets. A systems tract may be made up of one or more parasequence sets, but all sets in the same tract will show the same trends. In falling stage systems tracts the parasequences are not stacked vertically but are instead arranged laterally (Fig. 23.12). The relationship between parasequences within a fallings stage systems tract depends on the rate of sea-level fall in relation to the rate of sediment supply: with a high sedimentation rate or slow sea-level fall the individual parasequences are stacked against each other to form an ***attached falling stage systems tract***, whereas faster sea-level fall and/or lower sediment supply results in a ***detached falling stage systems tract*** (Fig. 23.12).

23.3.2 Sequences and parasequences: scales and variations

Building up from parasequences, which are metres to tens of metres thick, to parasequence sets and systems tracts, which are several tens of metres thick, depositional sequences can be expected to be in the order of a hundred metres or more in thickness. In practice, depositional sequences vary from a few metres to hundreds of metres thick depending on the amount of accommodation that is created. The greatest amount of accommodation occurs in places where there is a large amount of tectonic subsidence that occurs in certain types of sedimentary basin (Chapter 24). Accommodation tends to be much less in regions that are tectonically stable and where the sea-level variations are due to global eustatic fluctuations. The mechanisms and magnitudes of relative sea-level change are discussed in section 23.8.

The two conceptual models for the arrangement of strata in depositional sequences present in section 23.2 are idealised and merely represent possible

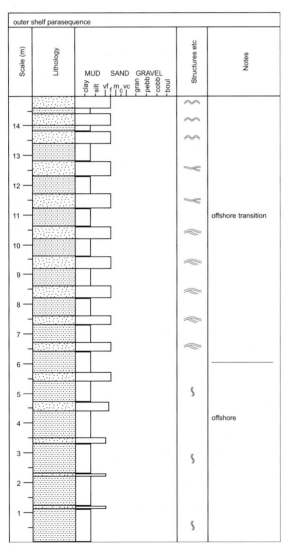

Fig. 23.10 A schematic graphic sedimentary log of a parasequence in an offshore shelf shallow-marine setting.

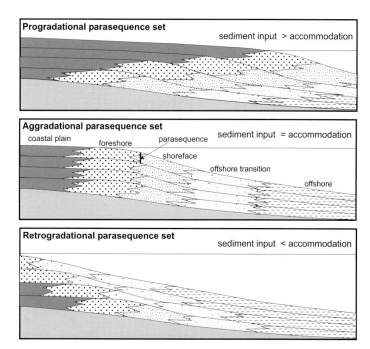

Fig. 23.11 The stacking patterns of parasequences to form parasequence sets are characteristic of different systems tracts.

scenarios. In addition to variations in the scales, there is huge scope for variations in the character of the deposits and the way that the strata are arranged depending on the variables that control sequence development. These variables are (a) the magnitude of relative sea-level change, (b) the rate of relative sea-level change, (c) the supply of sediment and (d) the physiography of the margin. Different combinations of these variables can generate any number of stratal patterns, but there are several points to consider.

1 Rapid changes in relative sea level can lead to the complete omission of systems tracts. For example, a rapid sea-level rise can result in a succession of a lowstand systems tract directly overlain by highstand deposits, with no deposition during transgression. Also, highstand deposits may be absent if relative sea level falls abruptly after transgression and similarly the lowstand systems tract may not be represented if there is a quick turnaround from fall to rise of relative sea level.

2 The physiography of the margin and the magnitude of the relative sea-level changes determine the proportion of the deposition that occurs in shallow marine environments. Margins with a very wide shelf with a shelf break below the lowest point of sea level experience predominantly ramp-type geometries and

most of the sedimentation is accommodated on the shelf, with little reaching the basin. Conversely, narrow shelves do not accommodate very much sediment and it may be that only highstand deposits are present, with the bulk of the deposition occurring on the basin floor.

3 Sedimentation rates are important because if they are low then accommodation will not be filled up and progradation will be subdued, but at high sedimentation rates progradational geometries will tend to dominate the succession.

23.4 CARBONATE SEQUENCE STRATIGRAPHY

A fundamental difference between the behaviour of carbonate depositional systems and terrigenous clastic settings is the control on the supply of sediment. In clastic depositional systems the sediment supply is mainly detritus coming from a hinterland area supplemented at times by the erosion and reworking of material from the margin. In small clastic depositional systems, such as alluvial-fan supplied fan-deltas at active basin margins, there may be a close link between relief, sediment supply and relative sea

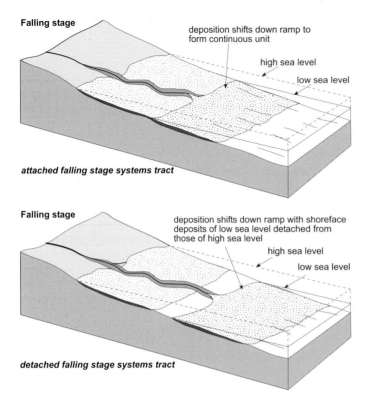

Falling stage

deposition shifts down ramp to
form continuous unit

high sea level

low sea level

attached falling stage systems tract

Falling stage

deposition shifts down ramp with shoreface
deposits of low sea level detached from
those of high sea level

high sea level

low sea level

detached falling stage systems tract

Fig. 23.12 A falling-stage systems tract
(FSST) on a ramp margin can show different patterns of deposition: if the sea
level falls relatively slowly with respect to
sediment supply deposition forms a continuous succession across the ramp as an
attached FSST; relatively fast sea-level
fall results in a **detached FSST**.

level. However, although there are some links
between climate, sea level and rates of erosion, the
sediment supply through larger scale, continent-wide
fluvial systems, however, is largely independent of
relative sea-level fluctuations. In contrast, supply of
material in carbonate settings is governed mainly by
the factors that control biogenic productivity
(15.1.1), such as water temperature, salinity, nutrient supply, suspended sediment content, water depth
and the area of the shelf that is available for production (Tucker & Wright 1990). Fluctuations in relative
sea level determine both the water depth and the shelf
area and so there is a direct link between relative sea
level and sediment supply in carbonate depositional
systems. There is a lot of production of sediment if
there is a large area of shallow water, but if this area
is reduced, the amount of sediment supply also drops.

The pattern of strata in a carbonate rimmed shelf
during a cycle of relative sea-level rise and fall is
shown in Fig. 23.13. The character of the systems
tracts will depend on the relative rates of sea-level rise
and carbonate production. If the sea-level rise is rapid
the shelf floods and there is a thin retrogradational

package of beds making up the transgressive systems
tract: the outer shelf area may be starved of sediment
allowing the development of condensed beds and
hardgrounds that will mark the maximum flooding
surface. The following highstand systems tract will be
aggradational to progradational. Under conditions of
slow relative sea-level rise the carbonate production
keeps pace and the pattern of beds is aggradational to
progradational: the transgressive systems tract will
pass without a maximum flooding surface into a prograding highstand systems tract. A fall in sea level
that results in the exposure of the shelf or ramp
carbonate sediment can normally be recognised by
the presence of solution, karstification, soil development and/or subaerial evaporite formation. Smaller
falls in sea level can result in a restriction of circulation to the shelf lagoon, leading to hypersaline conditions in arid regions and reduced salinity in humid
climates. Either way, the change in water chemistry
will be reflected in the organisms in the lagoon as
biodiversity is reduced under these conditions. Fall of
the sea level below the shelf edge dramatically reduces
the area of the carbonate factory to a small patch on

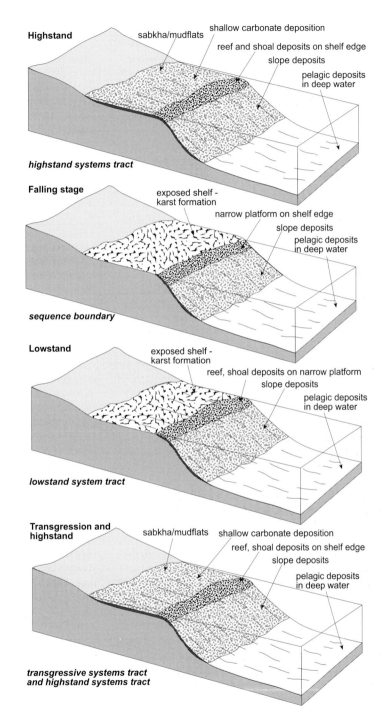

Highstand

sabkha/mudflats

shallow carbonate deposition

reef and shoal deposits on shelf edge

slope deposits

pelagic deposits in deep water

highstand systems tract

Falling stage

exposed shelf - karst formation

narrow platform on shelf edge

slope deposits

pelagic deposits in deep water

sequence boundary

Lowstand

exposed shelf - karst formation

reef, shoal deposits on narrow platform

slope deposits

pelagic deposits in deep water

lowstand system tract

Transgression and highstand

sabkha/mudflats

shallow carbonate deposition

reef, shoal deposits on shelf edge

slope deposits

pelagic deposits in deep water

transgressive systems tract and highstand systems tract

Fig. 23.13 The responses of a carbonate rimmed shelf to changes in sea level. An important difference with clastic systems tracts is that the carbonate productivity varies because most carbonate material is biogenic and forms in shallow water. During highstand and transgressive systems tracts wide areas of shallow water allow more sediment to be formed, whereas at falling stage and lowstand production of carbonate sediment is much lower.

the slope: consequently there may be little sedimentation associated with falling stage and lowstand except for some redeposited material at the base of the slope.

Carbonate ramps are generally areas of lower carbonate productivity (Bosence & Wilson 2003). The productivity may be unable to keep up with sea-level rise during transgression and the transgressive systems tract will consist of retogradational parasequence sets. A maximum flooding surface marked by condensed beds and/or hardgrounds is likely to form. With a reduced rate of sea-level rise in the highstand the parasequence sets will be aggradational to progradational. In contrast to rimmed shelves, sedimentation continues during deposition of the falling stage and lowstand systems tracts: the parasequences show a progradational downstepping geometry, while exposure on the inner part of the ramp results in solution and karst formation.

Parasequences in carbonate depositional systems normally show a shallowing-up character. They typically consist of beds deposited in the lower subtidal zone comprising wackestones that coarsen up into packstones and then to grainstones deposited in the higher energy wave-reworked zone of shallower water. Parasequences that form part of a retrogradational or aggradational package tend to be thicker with higher proportions of finer grained facies, whereas parasequences in progradational parasequence sets tend to become thinner upwards with more shallow water grainstone facies present.

23.5 SEQUENCE STRATIGRAPHY IN NON-MARINE BASINS

In basins that are not connected to the oceans the relative sea level does not act as a control on sedimentation. The deposition in a basin that has a permanent central lake is affected by the water level in that lake in a manner that is similar to a relative sea-level control. Climate directly controls the volume of water in lakes. A shift to a more arid climate causes a reduction in water supply and an increase in evaporation and the result is a fall in the lake level. Wetter climatic conditions mean that rivers supply more water, evaporation is reduced and the lake level consequently rises. These base-level fluctuations can be of greater magnitude than global eustasy, and accommodation in the fluvial and lacustrine depositional systems within the basin is also determined by

tectonic subsidence. In areas of accumulation of wind-blown sand accommodation is controlled by the level of the water table as it limits the extent of wind deflation (Kocurek 1996).

23.6 ALTERNATIVE SCHEMES IN SEQUENCE STRATIGRAPHY

It is perhaps a consequence of the relatively recent development of the concepts involved in sequence stratigraphy that there is not general agreement on definitions and terminology amongst those applying it to field and subsurface geology. A summary of the history and development of models, conceptual approaches and methods in sequence stratigraphy is provided in Nystuen (1998). Schemes that place the sequence boundary in a completely different part of the cycle, such as the 'genetic sequence' approach of Galloway (1989) in which the 'sequence boundary' is placed at the maximum flooding surface, have fallen out of favour. The approach presented in this chapter is based partly on the original 'Exxon model' developed as a tool for analysis of subsurface stratigraphy seen on seismic reflection profiles (Payton 1977; Vail et al. 1977; Jervey 1988) and later extended to sedimentary successions (Van Wagoner et al. 1988, 1990). Revisions to these models presented by Coe (2003) and Cateneaunu (2006) have been incorporated, and some of the original ideas from the Exxon scheme, such as the concept of 'type 1 and type 2' sequences, depending on whether there was erosion at the sequence boundary (type 1) or not (type 2), have not been explored. As noted in section *23.2.2* there are various approaches used in the analysis of events during sea-level fall, particularly the problem of where to place a sequence boundary in a setting where there is continuous deposition on the margin (e.g. Hunt & Tucker 1990; Posamentier & Morris 2000). The shoreline trajectory concept, which is summarised in Helland-Hansen & Martinsen (1996), provides an alternative way of treating stratal relationships (particularly on seismic reflection profiles) that is based on considering the relative rates of creation/reduction of accommodation and sediment supply along with the physiography of the margin. Different approaches are best suited to particular situations, depending on the scale of investigation and the geological setting, and no single scheme can be considered to be ideal for all circumstances.

23.7 APPLICATIONS OF SEQUENCE STRATIGRAPHY

One of the advantages of using the sequence stratigraphic approach for the analysis of sedimentary successions is that it can be applied to different types of data and be used to help combine information from different sources. It can be applied in the field using graphic sedimentary logs of the strata, it can be used in the subsurface in a similar way using borehole core and wireline log data (22.4.3) and it can be applied to seismic reflection profiles that provide images of subsurface stratigraphic relationships (22.2.5). Patterns can be related to a general model (e.g. Fig. 23.14) and predictions made about likely trends both laterally and vertically.

23.7.1 Sequence stratigraphic analysis of seismic sections

The sequence stratigraphic approach first became well established as a means of analysing seismic reflection profiles (Payton 1977; Vail et al. 1977; Jervey 1988). Patterns in the stratal geometries can be readily related to changes in relative sea level: onlap relationships (22.2.5) on the shelf are characteristic of transgression, a widespread unconformity can be recognised and interpreted as a sequence boundary, prograding clinoforms form during highstand, and so on. The scale of the resolution of seismic reflection profiles (22.2.4) means that individual parasequences can rarely be recognised, so interpretation is at the scale of depositional sequences. Analysis is carried out by identifying key stratigraphic surfaces, such as sequence boundaries identifiable as unconformities on the shelf and maximum flooding surfaces that can be recognised by the change in stratal geometries from retrogradational to progradational. The vertical interval between maximum flooding surfaces can be used as a means of estimating the magnitude of relative sea-level change, provided that post-depositional compaction of the sediment (18.2.1) is taken into account. Shoreline trajectories determined from the geometry of the strata between the stratigraphic surfaces can be used to infer information about rates of relative sea-level change and sedimentation.

23.7.2 Sequence stratigraphic analysis of graphic sedimentary logs

The procedure for carrying out an analysis of a succession of sedimentary rocks in terms of changes in relative sea level must start with facies analysis of the entire sections. In outcrop and cores this is achieved by examining the lithology, sedimentary structures and biogenic features within each bed and using them to determine the environment of deposition of each bed. The principles of relating depositional facies and facies associations to environment of deposition that have been considered in earlier chapters in this book are used. Trends in facies

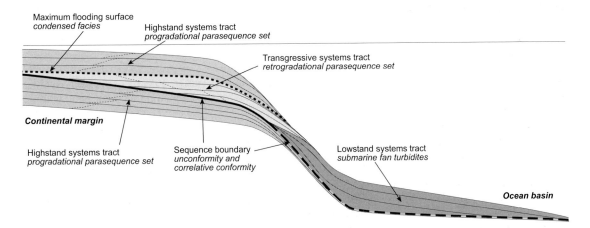

Fig. 23.14 A summary of the stratal geometries and the patterns within systems tracts in a depositional sequence.

and their associations can then be used to identify parasequences: the beds in a parasequence can be expected to show a shallowing-up pattern, that is, going up through the succession the facies indicate shallower water deposition. Flooding surfaces can be recognised by relatively abrupt changes from shallower to deeper water facies. Once parasequences have been identified they can be grouped into parasequence sets and the trends within the sets determined. A progradational trend is indicated if the higher parasequences indicate generally shallower water than the lower ones in the set; a deepening-up trend through the parasequence set is the signature of a retrogradational trend and if all the parasequences in the set show the same range of facies the trend is aggradational. Trends within parasequence sets can then be used to establish the systems tracts and hence the whole succession can be divided up into systems tracts and depositional sequences (Fig. 23.15).

The recognition of key surfaces such as sequence boundaries and maximum flooding surfaces is an important step in the analysis of the sedimentary succession (Figs 23.14 & 23.15). The nature of the sequence boundary depends on the relative landward/seaward location on the margin: on the middle to outer regions the surface may mark an abrupt change in facies from offshore or offshore transition facies of the preceding highstand systems tract to estuarine deposits of the transgressive systems tract above the sequence boundary. The presence of estuarine facies is generally a very useful indicator in a sequence stratigraphic analysis because estuaries form by flooding a river valley (13.6) and are therefore characteristic of transgression. Further landward the sequence boundary may be entirely within fluvial deposits and recognition can be difficult: the best indicator is the identification of an incised valley with fluvial channel deposits concentrated within in it during the transgression. There is no erosional surface within deep basin deposits and the sequence boundary is identified as a correlative conformity: below this sedimentation is dominantly fine-grained and pelagic but at the boundary the onset of turbidites deposited on a basin floor fan marks the onset of erosion and by-pass on the shelf.

Within carbonate successions the principles of analysing all the strata in terms of water depth and then identifying parasequences, patterns within parasequence sets and systems tracts also apply. On ramps and shelves the sequence boundary can be picked out by evidence of subaerial exposure in the form of karstic surfaces. In the deeper water environment the correlative conformity may not be easy to identify because the reduced sediment supply during lowstand means that there is not necessarily a significant increase in sedimentation at the sequence boundary. The maximum flooding surface may be marked by condensed beds and/or hardgrounds if sea-level rise is relatively rapid, but in other cases it may not be possible to define a distinct surface between the transgressive and highstand systems tracts.

23.7.3 Sequence stratigraphic analysis of geophysical logs

The most useful petrophysical logging tool in sequence stratigraphic analysis is the gamma-ray log (22.4.1). This tool is used to assess the relative proportions of sand and mud within the succession and, in general terms, sandier successions indicate shallower marine deposition than muddy sediments. A trend of decreasing value upwards on a gamma-ray log can therefore be taken to indicate increase in sand content and hence interpreted as a shallowing-up succession (Fig. 23.16). An abrupt increase in the reading means a higher mud content and this can be used as evidence for a flooding surface (i.e. a parasequence boundary). Trends in sand and mud content on gamma logs within parasequence sets can be used to identify patterns of progradation, retrogradation and aggradation (Fig. 23.16) through the succession. Recognition of channel-fill successions, characterised by sharp increases in sand content (low gamma log value) at the base followed by a trend over several metres to become more mud-rich (increasing gamma log values), can be important to help identify fluvial and estuarine facies which, in turn, can be indicators of sequence boundaries. Maximum flooding surfaces may also be picked up on spectral gamma logs (22.4.1) by a characteristic increase in the potassium content indicating a concentration of glauconite. Other geophysical logs, such as imaging tools, provide more information that helps to interpret the environment of deposition and hence aids in stratigraphic analysis.

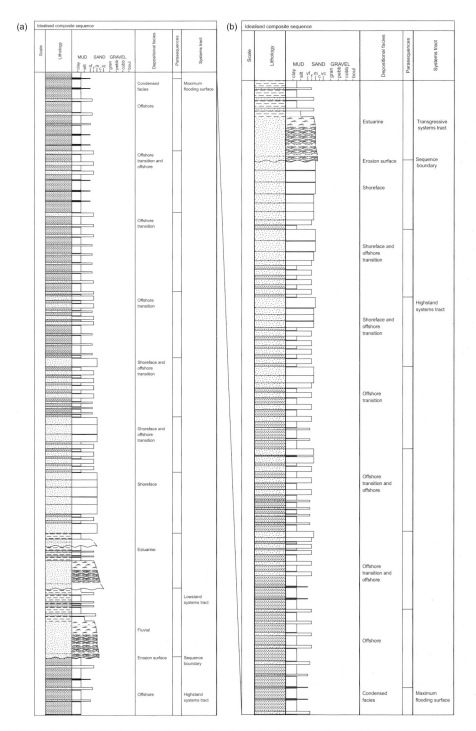

Fig. 23.15 Schematic sedimentary log through an idealised depositional sequence: in practice, the succession seen in outcrop or in the subsurface will often include only parts of this whole pattern, with considerable variations in thicknesses of the systems tracts. (a) Lower part of the succession. (b) Continuation into the upper part of the succession.

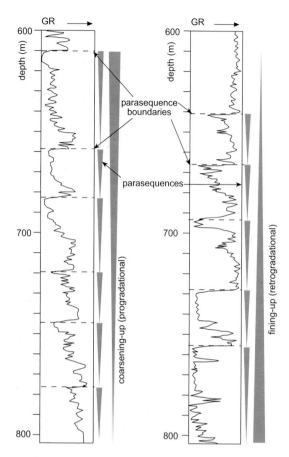

Fig. 23.16 Interpretation of gamma-ray logs in terms of parasequences and systems tracts.

23.7.4 Correlation of sections using sequence stratigraphic principles

The benefits of using a sequence stratigraphic approach become apparent when it comes to correlating outcrop sections or boreholes kilometres to tens of kilometres apart. In Chapter 19 the limitations of lithostratigraphic analysis were discussed: it is clear that this approach does not provide a time framework and hence is of limited value. Biostratigraphic techniques (Chapter 20) provide essential information for correlating strata on the basis of their age, but the resolution available may mean that hundreds of metres of strata fall within the same biozone and also correlation between terrestrial, shallow marine and deep marine environments may not always be easy because different fossils are

found in beds deposited in these different environments. Radiometric dating is of even more limited value in correlation because of the problems in finding material that can be used to provide a depositional age (*21.2*). By using changes in relative sea level as the template for analysis the sequence stratigraphic approach overcomes some of the problems inherent in other techniques.

Changes in relative sea level usually occur over relatively large areas, and can even be global (see *23.8* for further discussion): they affect sedimentation in environments ranging from rivers on coastal plains, to coastlines, shelves and the areas of the deep seas adjacent to continental margins. If there is evidence of a sea-level rise or fall in one of these environments, then there should also be a signature in the others as well. Correlation can be carried out on the basis of comparing patterns in different sections: a progradational pattern seen in coastal facies will be matched by a progradational pattern in the offshore and offshore transition zone and the two may therefore be correlated, even though the lithologies are quite different and they may not contain the same fossils. Sequence boundaries, seen as erosional unconformities on the shelf or as correlative conformities in deeper water, may be traced over large areas. Maximum flooding surfaces may be traced by identifying evidence for sediment starvation on the outer parts of the shelf and a change from retrogradational to progradational patterns in parasequence sets nearer shore. A theoretical example of the process of correlation using sequence stratigraphy is shown in Fig. 23.17.

Correlation of sections using the recognition of patterns and key surfaces can be undertaken only if there is some additional information to help the process. Biostratigraphic information is needed in many cases to provide an overall temporal framework: it is necessary to know if the sections being correlated are approximately the same age, but higher resolution can be provided by looking at the sequences because it may be possible to identify several depositional sequences within a single biozone. In the subsurface a general correlation between boreholes can be achieved by locating them on seismic reflection profiles and then tracing reflector surfaces between them.

One of the most useful applications of sequence stratigraphy is that it can be used as a predictive tool. Consider the four sections in Fig. 23.17: even if only the three shelf successions are present, it is still possible to make a prediction that a succession similar to that seen in the deep basin section will be present because the presence of a sequence boundary unconformity indicates non-deposition on the shelf and so accumulation of turbidites in the deep water can be expected. Similarly, using the same logic, if any of the other sections are missing a prediction can be made of what might be expected to occur there. This predictive facility is especially valuable in subsurface stratigraphic analysis. A seismic reflection survey provides information about the general patterns of the strata, but even a single borehole can provide enough information to make an informed guess about the distribution of facies across the area using sequence stratigraphic principles in both the sedimentary section and the seismic profile. With more data from boreholes the depositional model for the area becomes more sophisticated and, hopefully, more accurate.

23.8 CAUSES OF SEA-LEVEL FLUCTUATIONS

In the introduction section to this chapter two main causes of changes of relative sea level were identified: tectonics and global eustasy. Uplift and subsidence are never on a global scale and are always localised to some extent, affecting just part of a continent at its broadest scale. There are, however, aspects of plate tectonic processes that affect the sea level on a global scale and these also have to be considered. Eustasy is a global phenomenon involving changes in the volume of water in the world's oceans, so every shoreline will experience the same amount of sea-level rise or fall at the same time. The different mechanisms resulting in sea-level fluctuations are considered in the following sections.

23.8.1 Local changes in sea level

Tectonic forces and related thermal effects acting on the margins of continents result in the land mass being raised or lowered relative to sea level (Fig. 23.18). In rifted basins, blocks move down in the rift, but along the flanks blocks may be uplifted: this can give rise to either relative rises or fall in sea level depending on which faulted block the coast is situated. Along passive margins at the edges of ocean basins the continental crust cools and contracts

through time resulting in a relative rise in sea level along these margins. Relative rises and falls in sea level may occur also at active continental margins where ocean crust is being consumed at a subduction zone. In all these cases, the sea-level changes are localised to the region affected by the thermal or tectonic event. In the case of rift basins this may be a region only a few kilometres across, but on passive margins the subsidence may affect the whole margin of the continent. The influence of these localised events does not extend to other coastlines around the world.

23.8.2 Glacio-eustasy

Melting of continental ice caps at the poles can release large quantities of water to the oceans that can potentially raise the sea levels around the world by several tens of metres. At the South Pole the Antarctic continent hosts most of the world's fresh water as ice sheets and ice caps, which are thousands of metres thick in places. In the northern hemisphere there is much less continental ice on Greenland, with the remainder of the polar ice being floating sea ice that does not change the level of the sea when it melts because the mass above water level is compensated by the reduction in density when the mass of ice changes to water.

There is abundant evidence that the ice sheets at the poles expanded and contracted in volume during the Quaternary, resulting in worldwide changes in sea level of tens of metres. Glacial deposits from the Pleistocene are evidence of times when there were areas of northern Eurasia and the North American continent covered by ice sheets. Elsewhere morphological features such as raised beaches testify to periods of higher sea level in interglacial periods. The fluctuations between glacial and interglacial conditions are attributed to periodic cooling and warming of the global climate during the Pleistocene. The connection between climate change, glacial accretion/wastage and global sea-level changes is well established as a *glacio-eustatic mechanism* in the Quaternary (Chappell & Shackleton 1986; Matthews 1986).

Glacio-eustasy is therefore a mechanism that can explain global sea-level changes for periods of Earth history when there were ice caps at the poles to store water during cooler climate periods. The volumes of water involved are sufficient to cause a rise and fall in

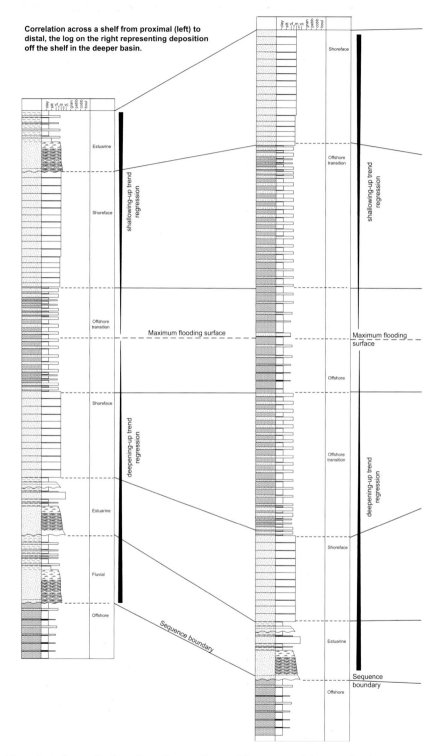

Fig. 23.17 A hypothetical example of correlation between logs in different parts of the coastal and marine environments using sequence stratigraphic principles. Note that correlation is on the basis that in different places on the shelf and in the basin there will be different facies deposited at the same time.

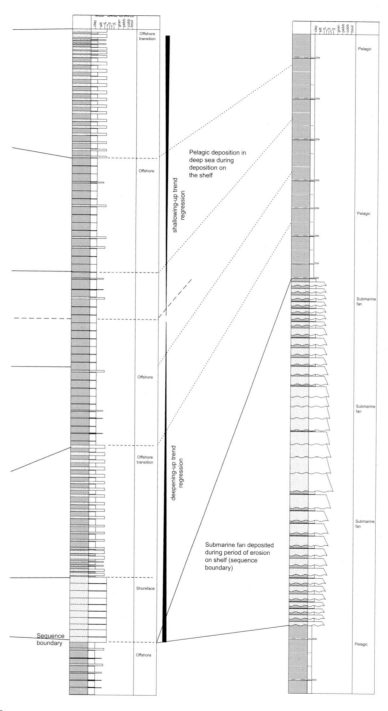

Fig. 23.17 (*cont'd*)

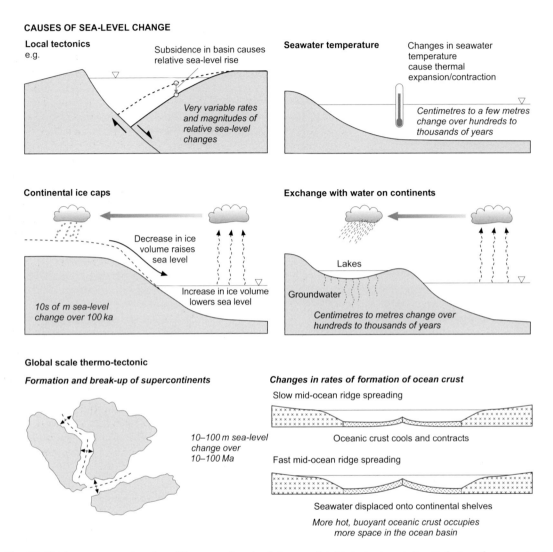

CAUSES OF SEA-LEVEL CHANGE

Local tectonics
e.g.

Subsidence in basin causes
relative sea-level rise

*Very variable rates
and magnitudes of
relative sea-level
changes*

Seawater temperature

Changes in seawater
temperature
cause thermal
expansion/contraction

*Centimetres to a few metres
change over hundreds to
thousands of years*

Continental ice caps

Decrease in ice
volume raises
sea level

Increase in ice volume
lowers sea level

*10s of m sea-level
change over 100 ka*

Exchange with water on continents

Lakes

Groundwater

*Centimetres to metres change over
hundreds to thousands of years*

Global scale thermo-tectonic

Formation and break-up of supercontinents

*10–100 m sea-level
change over
10–100 Ma*

Changes in rates of formation of ocean crust

Slow mid-ocean ridge spreading

Oceanic crust cools and contracts

Fast mid-ocean ridge spreading

Seawater displaced onto continental shelves

*More hot, buoyant oceanic crust occupies
more space in the ocean basin*

Fig. 23.18 There are a number of possible causes of sea-level change related to tectonic and climatic factors; the approximate magnitudes of change and the rates at which it will occur are indicated in each case.

sea level of over a hundred metres. The time period over which glacio-eustasy operates is also significant because climate changes apparently occur very quickly (Plint et al. 1992). The resulting sea-level change may take place over a few thousand years and the interval between glacial and interglacial periods in the Pleistocene is in the order of tens to hundreds of thousands of years. Climatically driven glacio-eustasy can therefore provide quick and frequent global sea-level fluctuations.

23.8.3 Thermo-tectonic causes of sea-level change

The configuration of the tectonic plates around the globe is constantly changing with long-term patterns of amalgamation to form supercontinents and subsequent dispersal as the continental mass breaks up into smaller units. During supercontinent break-up new oceanic spreading centres develop: this young oceanic crust is hot and relatively buoyant and mid-ocean ridges are at about 2000 to 2500 m water depth. It therefore takes up more space in the ocean basins than older, colder ocean crust that sinks to a lower level allowing 4000 to 5000 m of water above it. If the total length of spreading ridges in the world's oceans is great, more of the space in the ocean basins will be taken up by the ocean crust and this will cause the ocean water to spill further onto the continental margins (Fig. 23.18). These will therefore be periods of higher sea level. Conversely during periods of supercontinent formation the total length of spreading ridges will be reduced and result in more capacity in the ocean basins for the water and the sea level falls. Changes in the rate of sea-floor spreading also result in eustatic sea-level changes for much the same reasons. When spreading rates are high there is more hot crust in the ocean basins and sea level rises: slower spreading rates result in a fall in sea level.

It is estimated that sea-level changes of a hundred metres or more could be produced by these tectono-eustatic mechanisms (Plint et al. 1992). However, these changes in global tectonics take place slowly. The cycle of supercontinent amalgamation and break-up takes hundreds of millions of years and the changes in spreading rates probably occur over tens of millions of years. The rates of rise or fall in sea level generated by these mechanisms would therefore be slow.

23.8.4 Other causes of global sea-level change

Glacio-eustasy is climatically controlled and in addition to changing the proportion of ice on polar ice caps there is a second effect of changes in global temperature on the world's oceans. When water warms up, the volume increases by thermal expansion. An increase in global atmospheric temperature would result in a warming of the oceans although a deep water circulation is required to affect all the ocean waters. However, a rise in temperature of several degrees Celsius would only result in a few metres eustatic sea-level rise, so these thermal volume changes in the oceans would have a very limited effect. A similarly small change of sea level is predicted from changing the proportion of the world's water (hydrosphere) which is resident on the continents in rivers, lakes and groundwater (Plint et al. 1992).

23.8.5 Cyclicity in changes in sea level

An analysis of data from seismic reflection profiles, boreholes and outcrop sections around the world carried out by geoscientists in Exxon led to the publication of a 'global sea-level curve' (Vail et al. 1977; Haq et al. 1987, 1988). There are disputes about aspects of these curves regarding the timing and evidence for global synchroneity (e.g. Miall 1992, 1997), but they appear to show that there is a hierarchy of cycles of sea-level fluctuations through the Phanerozoic. Other authors have also remarked upon the evidence for cycles of sea-level rise and fall (Hallam 1963; Pitman 1978; Worsley et al. 1984) and proposed mechanisms for generating these cycles. If the shorter term variations are subtracted from the longer term trends a number of orders of cyclicity can be recognised (Fig. 23.19).

First-order cycles

The smoothed global sea level for the whole of the Phanerozoic (Fig. 21.19) shows the trend of a rise during the Cambrian to a peak in the early Ordovician, followed by a steady decline through to the end of the Palaeozoic (Boggs 2006). A slow rise in the Jurassic followed by a steep rise during the Cretaceous culminated in a peak in global sea level in the Late Cretaceous, which has been followed by a steady fall to present-day levels. There is a strong correlation between this curve and the patterns of continental dispersal and amalgamation through the Phanerozoic (Vail et al. 1977; Worsley et al. 1984). Supercontinent break-up occurred in the early Palaeozoic and was followed by a long period of dispersal of continents prior to amalgamation of the supercontinent of Pangea in the Permian. Sea level rose in the Cambrian as the new spreading centres formed between the continental masses and then fell again as the continents regrouped during the Late Palaeozoic. Break-up of Pangea and in particular the dispersal of the fragments that made up Gondwana led to the formation of new active ocean ridges between the continents of South America, Africa, Antarctica, Australia and India. Sea level rose sharply during this period until continents started to amalgamate again in the Late Cretaceous with the collision between India and Eurasia and further west the closure of the Tethys Ocean to form the Alpine mountain belt.

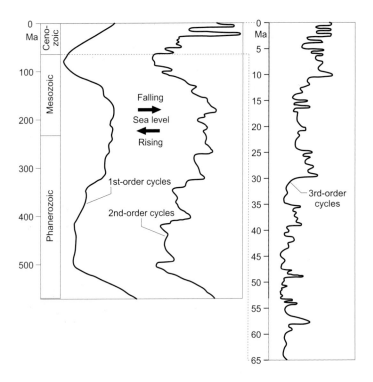

Fig. 23.19 First-, second- and third-order sea-level cycles are considered to be global signatures due to tectonic and climatic controls outlined in Fig. 23.18. (Modified from Vail et al. 1977.)

Second-order cycles

Superimposed on the first-order cycles, which show a duration of hundreds of millions of years, is a pattern of rises and falls with durations of tens of millions of years. The causes of second-order sea-level changes are thought to be changes in the rates of spreading at mid-ocean ridges, which may account for changes in ocean water levels on a scale of tens of millions of years (Hallam 1963; Pitman 1978). The shorter term global cycles in the Neogene may be accounted for by long-term trends in glaciation and deglaciation.

Third-order cycles

Rises and falls of sea level with a magnitude of several tens of metres and a periodicity of one to ten million years are recognised throughout the Phanerozoic stratigraphic record (Fig. 23.19). There is no general agreement on the mechanisms that may cause sea-level fluctuations at this third order of cyclicity (Plint et al. 1992) and it may be that more than one mechanism is responsible. Mechanisms that are thought to be less likely include changing the lengths

and volumes of material in mid-ocean spreading centres because this operates too slowly; thermal expansion and contraction of the world's seawater and changes in the volume of water in and on the continents cannot generate the magnitude of change indicated. More likely is glacio-eustasy because it can generate the appropriate magnitude of sea-level change in a short period of time (Vail et al. 1977; Haq et al. 1988). However, this mechanism requires the existence of ice caps on continents in polar regions, and although there is abundant evidence for this during periods of major glaciations in the Ordovician–Silurian, the Carboniferous–Permian and the Neogene–Quaternary, there is some doubt about other periods. In particular, the mid-Cretaceous was one of the warmest periods of the Phanerozoic and there is doubt over whether there were ice caps present at this time, although there do seem to be signs of global sea-level fluctuations. Other mechanisms that may generate the frequency and magnitude implied by the third-order cycle curve are applicable only to individual basins. Changes in the regional tectonic stresses acting on a basin may result in basin-wide subsidence or uplift (Cloetingh 1988) although the rates at which these changes occur are not well known.

The third-order cycles are of particular relevance to sequence stratigraphy because they are of the appropriate magnitude and period to be responsible for the cycles of sea-level rise and fall that generate depositional sequences. It must be emphasised though, that not all depositional sequences in all sedimentary basins formed as a response to global sea-level changes. Local tectonic activity can also generate the relative sea-level fluctuations required to form depositional sequences, and in many cases it seems likely that a combination of global eustatic and local tectonic mechanisms was responsible for the sea-level changes that are recoded in depositional sequences.

23.8.6 Short-term changes in sea level

In addition to these three orders of cyclicity, shorter term changes in sea level have also been recognised. These are fourth-order cycles of 200,000 to 500,000 years duration and fifth-order cycles lasting 10,000 to 200,000 years. The magnitude of sea-level change in these cycles ranges from a few metres to 10 or 20 m, although short-term sea-level changes in the Quaternary have much higher magnitudes. Evidence for frequent relative changes in sea level of a few metres has been found in successions throughout the stratigraphic record and in sequence stratigraphy terminology these are referred to as parasequences.

A very detailed record of sea-level changes has been established for the Quaternary and related to estimates of palaeotemperature from the oxygen isotope record (*21.5.3*). These indicate that changes in global climate caused periodic wasting and accretion of ice masses with a cyclicity of tens to hundreds of thousands of years. These global climatic variations have been related to the behaviour of the Earth in its orbit around the Sun and changes in the axis of rotation. Three orbital rhythms were recognised and their periodicity calculated by the mathematician Milankovitch, and these cycles are commonly known as ***Milankovitch cycles*** (Fig. 23.20). The longest period rhythm, approximately 100,000 years, is due to changes of the eccentricity of the Earth's orbit around the Sun, that is, the orbit is elliptical and changes its shape with time. The rota-

Changes in the eccentricity of the Earth's orbit around the Sun
100 ka cycle

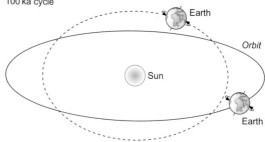

Changes in the obliquity (tilt) of the Earth's axis of rotation
41 ka cycle

Precession of the axis of rotation
22 ka cycle

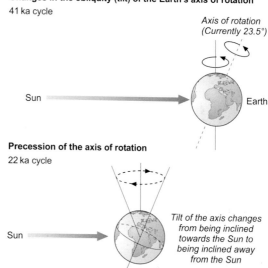

Fig. 23.20 Milankovitch cycles: the eccentricity of the Earth's orbit of the Sun, changes on the obliquity of the axis of rotation of the Earth and the precession of the axis of rotation may result in global climatic cycles on the scale of tens of thousands of years.

tion of the Earth on its axis shows two patterns of variation. The axis of rotation is oblique with respect to the plane of the Earth's rotation about the Sun and the angle of tilt changes over a period of 40,000 years between 21.5° and 24.5°. The shortest rhythm is 21,000 years caused by the precession or 'wobble' in axis of rotation analogous to the behaviour of a spinning top. These three cycles, working independently and in combinations, are believed to exert fundamental controls on the global climate leading to cycles of global warming and cooling and hence sea-level fluctuations.

23.8.7 Global synchroneity of sea-level fluctuations

The sea-level curves presented by Vail et al., (1977) and Haq et al. (1987) were originally considered to be global signatures. Dating of the strata in which they were recognised led to assertions that the curve itself could be used as a correlation tool: if there was evidence of a sea-level fall within a succession of strata of, say, early Oligocene, then the sequence boundary formed by the sea-level fall could be dated by matching it to one of the 'global' sea-level falls. There are a number of objections to using this approach to correlation. First, the sequence boundary in the section being examined may be the result of a local relative sea-level fall and not related to global eustasy. Second, some authors doubt whether all of the sea-level fluctuations shown on these published charts are actually global signatures. Third, most sedimentary successions are dated biostratigraphically, and in many cases there are known to be errors of hundreds of thousands of years when relating them to radiometric ages: the length of some of the cycles falls within the error in the dating of these cycles, so confidence in the accuracy of the dates placed on the curve is not assured (Miall 1992, 1997). The procedure of dating depositional sequences by comparison with a global sea-level chart has largely fallen out of favour, although sea-level curves for particular areas have been published more recently (Hardenbol et al. 1998). It is reasonable to construct a sea-level curve for a particular continental margin and use this as a tool in correlation across that margin, but detailed global correlation is probably not appropriate. However, there is little doubt that the signature of global eustasy is present in many parts of the stratigraphic record, and in the latter half of the Cenozoic there does seem to be a global signature.

23.9 SEQUENCE STRATIGRAPHY: SUMMARY

The sequence stratigraphy approach to the analysis of sedimentary successions has taken some time to become widely accepted and widely used. There were two issues in the earlier stages of its development which caused problems. First, the concept was initially linked with the idea that a global eustatic sea-level curve could be established and used as a means of correlating strata: doubts about the curve led to doubts about the methodology, but in fact the principles underlying sequence stratigraphy are sound and valid whether there is a global sea-level curve or not (Posamentier & James 1993). A second barrier to acceptance was the plethora of new terms that were introduced as the concepts were developed. To some extent these served to make the subject appear more complicated than it actually is, because sequence stratigraphy is really quite a straightforward and elegant approach to the analysis of sedimentary successions. It is now applied very extensively, especially in the hydrocarbon exploration industry, and several texts are now available that provide comprehensive accounts of the principles and applications of sequence stratigraphy.

FURTHER READING

Catuneanu, O. (2006) *Principles of Sequence Stratigraphy.* Elsevier, Amsterdam.

Coe, A.L. (Ed.). (2003) *The Sedimentary Record of Sea-level Change.* Cambridge University Press, Cambridge.

De Boer, P.L. & Smith, D.G. (Eds) (1994) *Orbital Forcing and Cyclic Sequences.* Special Publication 19, International Association of Sedimentologists. Blackwell Scientific Publications, Oxford.

Emery, D. & Myers, K.J. (Eds) (1996) *Sequence Stratigraphy.* Blackwell Science, Oxford.

Miall, A.D. (1997) *The Geology of Stratigraphic Sequences.* Springer-Verlag, Berlin.

Posamentier, H.W. & Allen, G.P. (1999) *Siliciclastic Sequence Stratigraphy: Concepts and Applications.* Society of Economic Paleontologists and Mineralogists, Tulsa, OK.

Van Wagoner, J.C., Mitchum, R.M., Campion, K.M. & Rahmanian, V.D. (1990) *Siliciclastic Sequence Stratigraphy in Well Logs, Cores and Outcrop: Concepts for High Resolution Correlation of Time and Facies.* Methods in Exploration Series 7, American Association of Petroleum Geologists, Tulsa, OK.

24

Sedimentary Basins

Sedimentary basins are regions where sediment accumulates into successions hundreds to thousands of metres in thickness over areas of thousands to millions of square kilometres. The underlying control on the formation of sedimentary basins is plate tectonics and hence basins are normally classified in terms of their position in relation to plate tectonic setting and tectonic processes. Each basin type has distinctive features, and the characteristics of sedimentation and the stratigraphic succession that develops in a rift valley can be seen to be distinctly different from those of an ocean trench. A stratigraphic succession can therefore be interpreted in terms of plate tectonics and places the study of sedimentary rocks into a larger context. The sedimentary rocks in a basin provide a record of the tectonic history of the area. They also provide the record of the effects of other controls on deposition, such as climate, base level and sediment supply.

24.1 CONTROLS ON SEDIMENT ACCUMULATION

The issues of how and where sediment is preserved could perhaps have been considered before a discussion of environments of deposition, because not every river, lake, delta, estuary or so on is necessarily a place where sediments will accumulate and form a succession of strata. In fact, the preservation of deposits that will eventually form part of the sedimentary record is actually the exception, rather than the rule. The transitory nature of deposition is most obvious in upland areas. The deposits left by glaciers retreating a few thousand years ago may be familiar as lateral and terminal moraines in some of our modern landscapes, but they occur in areas that are undergoing erosion, and will not be preserved as glacial features in the stratigraphic record. Similarly, the sediment that we currently see in rivers, estuaries, deltas and coasts is mostly only passing through on its way to the open seas where they may be preserved on the shelf or in the deep seas.

The concept of 'the present being the key to the past', introduced in Chapter 1, can be difficult to apply, because most of what we see happening today in modern environments of deposition is not necessarily representative of events that will lead to the formation of sedimentary rocks. For example, tidal currents may form bars of cross-bedded sands in an estuary, but those sands may be washed back and forth by the tide for millennia, with some material added by the river, and some moved out to sea. To create a set of strata from these processes, something else has to happen, usually some form of change in the environment. At a small scale this may be the change in the position of a river due to avulsion leading to abandonment of the old river course, or the shift in a lake shoreline due to a change in climate covering the old lake margin deposits with water and more sediment. However, at a larger scale it is *tectonic subsidence*, local and regional changes in the vertical position of the crust, that ultimately allows sediment to become preserved as strata.

24.1.1 Tectonics of sedimentary basins

The importance of tectonic subsidence as a mechanism for creating accommodation space has already been considered in the context of sequence stratigraphy (*23.1.4*), but there is a broader implication of this process. Put simply, without tectonics creating areas that are 'lows' on the Earth's surface, there would be no long-term accumulation of sediment, no sedimentary rocks and no stratigraphy as we know it. Places where sediment accumulates are known as *sedimentary basins* and they can range in size from a few kilometres across to ocean basins covering half the planet. A 'basin' can also be a geomorphological

feature, a bowl-shaped depression on the land surface that may or may not be a place where sediment is accumulating – in geology we are really only concerned with basins that preserve strata, and provide us with our record of depositional environments through Earth history.

Distinct areas of sediment accumulation were recognised by geologists in the late 19th and early 20th centuries. They were then referred to as '*geosynclines*' and defined as broad down-folds in the crust where successions of strata were first preserved and then subsequently deformed. With the advent of plate tectonic theory the geosynclinal concept became redundant and it is now conventional to categorise sedimentary basins in terms of their plate tectonic setting (Ingersoll 1988; Busby & Ingersoll 1995).

24.1.2 Climate, sediment supply and base-level controls

The role of tectonics in creating the accommodation for sediment to accumulate is fundamental to sedimentology and stratigraphy, but there are a number of other factors that control the volume, type and distribution of sediment. These are summarised in Fig. 24.1. Climate, tectonics, bedrock geology and

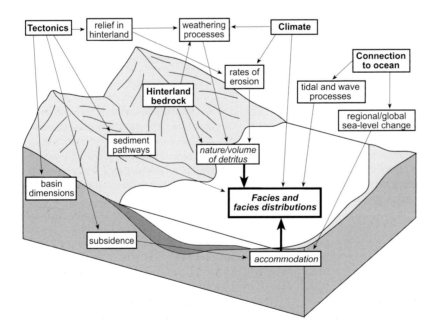

Fig. 24.1 The facies of deposits in sedimentary basins and their distributions in three dimensions are controlled by climatic and tectonic factors, the nature of the hinterland bedrock and the connection with the oceans.

ocean connection/base level all interact within and around all types of sedimentary basin to govern the character of the basin-fill succession.

Connection to oceans and sea-level changes

The role of relative changes in sea level has been considered during the discussion of sequence stratigraphy in Chapter 23. In shallow marine environments the sea level directly determines the amount of accommodation available for sediment to accumulate, but it also influences fluvial deposition and deep-sea sedimentation. Sea-level changes do not necessarily affect all basins because some are wholly within continental landmasses and have no link or direct exchange of water with the oceans. These basins of internal drainage (or '*endorheic*' basins) can form in a variety of tectonic settings, principally as rifts, foreland basins and strike-slip basins (see below). They may be dominated by lacustrine conditions, but in more arid climates fluvial and aeolian processes dominate (Nichols 2005).

Climatic effects of weathering, transport and deposition

The significance of climate as a control on processes has been considered in the context of a number of different surface processes and depositional environments. To start with, weathering processes are determined by the availability of water and the temperature (6.4): under warm, humid conditions more clay minerals and ions in suspension are generated, whereas colder environments form more coarse clastic material. The transport of sediment by water, ice or wind is also climatically controlled, both in terms of the amount of water available and the temperature. Depositional processes in all continental environments and many coastal settings are sensitive to the climate: a comparison of clastic lagoons formed in a temperate or tropical setting (13.3.2) and an evaporite lagoon formed in an arid environment (15.2.2) makes clear the importance of climate in determining depositional facies.

Bedrock and topography controls on sediment supply

Sediment supply is a further important factor, both in terms of character and volume of material. It is obvious that a delta cannot be a site of deposition of sand if no sand is supplied by the river, and similarly a beach cannot form a foreshore body of rounded pebbles if there is no gravel available to be deposited there. A deposit derived from the weathering and erosion of basaltic rock will have a very different character to one derived from a limestone terrain. The nature of all facies in all depositional environments is ultimately determined by the grain size of the sediment available, the mineralogical and petrological character of the detritus and the chemistry of the water. The relationship between sediment supply and accommodation was considered in the context of relative sea-level changes in Chapter 23, but the volume of sediment supply has an impact on the nature of the whole basin fill. The availability of sediment is principally determined by tectonic controls on uplift in the hinterland, but climate and bedrock character also play a role. If the rate of sediment supply exceeds the rate of tectonic subsidence, the basin fills up (is **overfilled**) and the facies will be shallow marine or continental. A low supply compared with subsidence rate results in a basin that is **underfilled** or **starved**: in a marine setting these basins will accumulate mainly deep-water facies. Continental basins that are underfilled may end up below sea level (e.g. the Dead Sea, Jordan, and Death Valley, USA).

24.1.3 Tectonic setting classification of sedimentary basins

The movement of tectonic plates results in mountain belts where two areas of continental crust collide, subduction zones with associated volcanic arcs where oceanic crust is consumed at plate margins, oceans form at places where plates are moving apart and major fault zones where plates move past each other. All these different tectonic settings are also areas where sediment can accumulate, and at a simple level three main settings of basin formation can be recognised:

1 basins associated with regional extension within and between plates;
2 basins related to convergent plate boundaries;
3 basins associated with strike-slip plate boundaries.

  In the following discussion the main basin types and the transitions between them are considered in terms of the plate tectonic setting. An elementary knowledge of plate tectonic processes and the nature of continental and oceanic lithosphere is assumed, as is an understanding of the basic terminology of structural geology. A more detailed consideration of the tectonic setting of both modern and ancient basins (Ingersoll 1988; Busby & Ingersoll 1995) indicates that at least 20 types can be recognised. In addition hybrid forms exist because of the complexities of plate tectonic processes, for example where crustal extension is oblique and resulting basins have characteristics of both a rift and a strike-slip setting.

24.2 BASINS RELATED TO LITHOSPHERIC EXTENSION

The motion of tectonic plates results in some areas where lithosphere is under extension and other places where it is under compression. Horizontal stress within continental crust causes brittle fracture in the surface layers while the stretching is accommodated by ductile flow in the lower part of the lithosphere. In the early stages of this extension, rifts form and are typically sites of continental sedimentation. If the stretching continues, the continental lithosphere may rupture completely and the injection of basaltic magmas results in the formation of new oceanic crust within the zone of extension. This stage is known as a 'proto-oceanic trough' and is the first stage in the initiation of an ocean basin: the remnant flanks of the rift become the passive margins of the ocean basin as it develops. However, not all crustal extension follows the same path: continental rift basins may exist for long periods without making the **rift to drift transition** of forming an ocean basin, especially if the driving force for the extension fades.

One tectonic setting where lithospheric extension occurs is associated with a 'hot spot', an area of increased heat flow in the crust generated by thermal plumes in the mantle. Rupture of the continental lithosphere over a plume creates three branches along which extension occurs, a triple junction of plates that can be seen today centred on the Afar Triangle where the East African Rift valley, the Red Sea and the Gulf of Aden meet. These three extensional regimes are in different stages of development – continental rift, proto-oceanic trough and young ocean basin respectively. On the other side of Africa, an older triple junction now centred on the Niger Delta had two arms forming the South Atlantic, while the third arm, the Benue Trough, was a 'failed

rift' that subsequently became an area of intracratonic subsidence.

Not all lithospheric extension is related to hot spots and the formation of new ocean basins. Areas of thickened crust and high heat flow due to asthenospheric upwelling, such as the Basin and Range Province in western USA, are also regions of widespread rift basin development as the upper layer of the crust responds to the doming. Furthermore, in arc–trench systems (24.3) local tectonic forces lead to the rifting of the crust and the formation of intra-arc and subsequently backarc basins due to extension.

24.2.1 Rift basins (Fig. 24.2)

In regions of extension continental crust fractures to produce **rifts**, which are structural valleys (Fig. 24.3) bound by extensional (normal) faults (Leeder 1995). The axis of the rift lies more-or-less perpendicular to the direction of the stress. The down-faulted blocks are referred to as **graben** and the up-faulted areas as **horsts**. The bounding faults may be planar or listric, and if the displacement is greater on one side they form asymmetric valleys referred to as **half-graben**. The structural weakness in the crust and high heat flow associated with rifting may result in volcanic activity. Uplift on the flanks of rifts due to regional high heat flow and the effect of relative movements on the rift-bounding faults creates local sediment sources for rift valleys.

The controls on sedimentation in rift valleys are a combination of tectonic factors that determine the rift flank relief and hence availability of material, as well as the pathways of sediment into the basin, and climate, which influences weathering, water availability for transport and facies in the rift basin

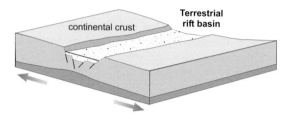

Fig. 24.2 Rift basins form by extension in continental crust: sediment is supplied from the rift flanks or may also be brought in by rivers flowing along the axis of the rift.

Fig. 24.3 Rift valleys are characterised by steep sides formed by the extensional faults that form the basin (East African Rift Valley).

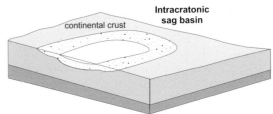

Fig. 24.4 Intracratonic basins are broad regions of subsidence within continental crust: they are typically broad and shallow basins.

(Nichols & Uttamo 2005). Connection to oceans is also important. Death Valley, California, is a **terrestrial rift valley**, isolated from the sea and has an arid climate, such that alluvial fan, desert dune and evaporative lake environments are dominant. In contrast, the Gulf of Corinth, Greece, is a **maritime rift** and is the site of fan-delta and deeper marine clastic deposits (Leeder & Gawthorpe 1987). Extensional basins with low clastic supply may be sites of carbonate deposition. The patterns of sedimentation in rifts evolve as the basins deepen, separate basins combine and links to the marine realm become established (Gawthorpe & Leeder 2000).

24.2.2 Intracratonic basins (Fig. 24.4)

Areas of broad subsidence within a continental block (**craton**) away from plate margins or regions of orogeny are known as **intracratonic basins** (Klein 1995). The cratonic crust is typically ancient, and with low relief: the area may be very large, but the amount of subsidence is low and the rate is very slow. The mechanism of subsidence varies between different basins, as some are apparently related to antecedent rifting episodes, whereas others are not. After the cessation of rifting within continental crust there is a change in the thermal regime of the area. When continental crust is extended it is thinned and this brings hotter mantle material closer to the surface. Rifts are therefore areas of high heat flow, a high **geothermal gradient** (rate of change of temperature

with depth). When rifting stops the geothermal gradient is reduced and the crust in the region of the rift starts to cool, contract and sink resulting in **thermal subsidence**. Intracratonic basins that apparently have no precursor rift history may also be a product of thermal subsidence. Irregularities in the temperature distribution within the mantle associated with cold crustal slabs relict from long-extinct subduction zones create areas where there is downward movement. Cratonic areas above these zones may be subject to subsidence and the formation of a broad, shallow basin. Long-wavelength lithospheric buckling has also been suggested as a mechanism for forming intracratonic basins.

Fluvial and lacustrine sediments are commonly encountered in intracratonic basins, although flooding from an adjacent ocean may result in a broad epicontinental sea. Intracratonic basins in wholly continental settings are very sensitive to climate fluctuations as increased temperature may raise rates of evaporation in lakes and reduce the water level over a wide area.

24.2.3 Proto-oceanic troughs: the transition from rift to ocean (Fig. 24.5)

Continued extension within continental crust leads to thinning and eventual complete rupture. Basaltic magmas rise to the surface in the axis of the rift and start to form new oceanic crust. Where there is a thin strip of basaltic crust in between two halves of a rift system the basin is called a **proto-oceanic trough** (Leeder 1995). The basin will be wholly or partly flooded by seawater by the time this amount of extension has occurred and the trough has the form of a narrow seaway between continental blocks. Sediment supply to this seaway comes from the flanks of the

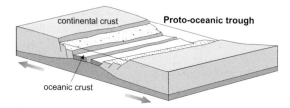

Fig. 24.5 With continued extension in a rift, the lithosphere thins and oceanic crust starts to form in a proto-oceanic trough where sedimentation occurs in a marine setting.

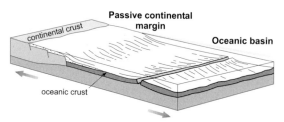

Fig. 24.6 An ocean basin is flanked by thinned continental crust, which subsides to form passive margins to the ocean basin.

trough, which will still be relatively uplifted. Rivers will feed sediment to shelf areas and out into deeper water in the axis of the trough as turbidity currents. Connection to the open ocean may be intermittent during the early stage of basin formation and in arid areas with high evaporation rates the basin may periodically desiccate. Evaporites may form part of the succession in these circumstances and this phase of basin development may be recognised by beds of gypsum or halite in the lower part of a passive margin succession.

24.2.4 Passive margins (Fig. 24.6)

The regions of continental crust and the transition to oceanic crust along the edges of spreading oceans basins are known as **passive margins**. The term 'passive' is used in this sense as the opposite to the 'active' margins between oceans and continents where subduction is occurring. The continental crust is commonly thinned in this region and there may be a zone of transitional crust before fully oceanic crust of the ocean basin is encountered. **Transitional crust** forms by basaltic magmas injecting into continental crust in a diffuse zone as a proto-oceanic trough develops. Subsidence of the passive margin is due mainly to continued cooling of the lithosphere as the heat source of the spreading centre becomes further away, augmented by the load on the crust due to the pile of sediment that accumulates (Einsele 2000).

Morphologically the passive margin is the continental shelf and slope (Fig. 11.1) and the clastic sediment supply is largely from the adjacent continental land area. The climate, topography and drainage pattern on the continent therefore determines the nature and volume of material supplied to the shelf. Adjacent to desert areas the clastic supply is low, and the margin will be a **starved margin**, experiencing a low clastic sedimentation rate. In contrast, a large river system may carry large amounts of detritus and build out a large deltaic wedge of sediment onto the margin. In the absence of terrigenous detrital supply, the shelf may be the site of accumulation of large amounts of biogenic carbonate sediment, although the volume and character of the material will be determined by the local climate.

Passive margins are important areas of accumulation of both carbonate and clastic sediment: they may extend over tens to hundreds of thousands of square kilometres and develop thicknesses of many thousands of metres. They are also areas that are sensitive to the effects of eustatic changes in sea level because most of the deposition occurs in water depths of up to 100 m. Sea-level fluctuations of tens of metres result in significant shifts in the patterns of sedimentation on passive margins and the effects of a sea-level rise or fall can be correlated over large distances in a passive margin setting.

24.2.5 Ocean basins (Fig. 24.6)

Basaltic crust formed at mid-oceanic ridges is hot and relatively buoyant. As the basin grows in size by new magmas created along the spreading ridges, older crust moves away from the hot mid-ocean ridge. Cooling of the crust increases its density and decreases relative buoyancy, so as crust moves away from the ridges, it sinks. Mid-ocean ridges are typically at depths of around 2500 m. The depth of the ocean basin increases away from the ridges to between 4000 and 5000 m where the basaltic crust is old and cool.

The ocean floor is not a flat surface. Spreading ridges tend to be irregular, offset by transform faults that create some areas of local topography. Isolated volcanoes and linear chains of volcanic activity related to hotspots (mantle plumes) such as the Hawaiian Islands form submerged seamounts or exposed islands. In addition to the formation of volcanic rocks in these areas, the shallow water environment may be a site of carbonate production and the formation of reefs. In the deeper parts of the ocean basins sedimentation is mainly pelagic, consisting of fine-grained biogenic detritus and clays. Nearer to the edges of the basins terrigenous clastic material may be deposited as turbidites.

24.2.6 Obducted slabs

Most oceanic crust is subducted at destructive plate margins, but there are circumstances under which slabs of ocean crust are *obducted* up onto the overriding plate to lie on top of continental or other oceanic crust. Outcrops of oceanic crust preserved in these situations are known as *ophiolites* (Gass 1982). Ophiolites may represent the stratigraphic succession formed in an ocean basin or the fill of a backarc basin. Until drilling in the deep oceans became possible ophiolite complexes provided the only tangible evidence of oceanic crust and deep-sea sediments. An

ophiolite suite consists of the ultrabasic and basic intrusive rocks of the lower oceanic crust (peridotites and gabbros), a dolerite dyke swarm which represents the feeders to the basaltic pillow lavas that formed on the ocean floor. The lavas are overlain by deep-ocean sediments deposited at or close to the spreading centre. If sea-floor spreading occurred above the CCD these sediments would have been calcareous oozes, preserved as fine-grained pelagic limestones. Red clays and siliceous oozes deposited below the CCD are lithified to form red mudstones and cherts. Concentrations of metalliferous ores are common, formed as hydrothermal deposits close to the volcanic vents.

24.3 BASINS RELATED TO SUBDUCTION

At convergent plate margins involving oceanic lithosphere subduction occurs (Fig. 24.7). The downgoing ocean plate descends into the mantle beneath the overriding plate, which may be either another piece of oceanic lithosphere or a continental margin. As the downgoing plate bends to enter the subduction zone a trough is created at the contact between the two plates: this is the *ocean trench*. The descending slab is heated as it goes down and partially melts. The magmas generated rise to the surface through the overriding plate to create a line of volcanoes,

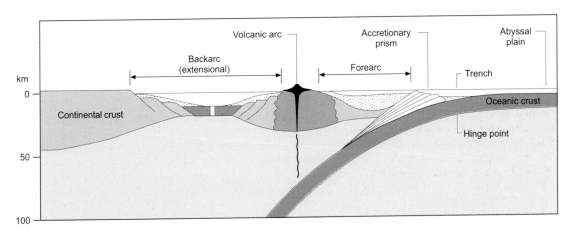

Fig. 24.7 An arc–trench system forms where oceanic crust is subducted at an ocean trench and the downgoing plate releases magma at depth, which rises to form a volcanic arc. Sediment may accumulate in the trench, in the forearc basin between the trench and the arc and in the region behind the arc called a backarc basin if there is subsidence due to extension (see also retroarc basins, Fig. 24.13).

Fig. 24.8 Arc–trench systems include chains of volcanoes that form the arc.

a *volcanic arc* (Fig. 24.8). The magmas start to form when the downgoing slab reaches 90 to 150 km depth. The *arc–trench gap* (distance between the axis of the ocean trench and the line of the volcanic arc) will depend on the angle of subduction: at steep angles the distance will be as little as 50 km and where subduction is at a shallow angle it may be over 200 km.

Arc–trench systems are regions of plate convergence, however, the upper plate of an active arc must be in extension in order for magmas to reach the surface and generate volcanic activity. The amount of extension is governed by the relative rates of plate convergence and subduction and this is in turn influenced by the angle of subduction. If the angle of subduction is steep then convergence is slower than subduction at the trench, the upper plate is in net extension and an *extensional backarc basin* forms (Dickinson 1980). Steep subduction occurs if the downgoing plate consists of old, cold

crust. However, not all backarc areas are under extension: some are 'neutral' and others are sites of the formation of a flexural basin due to thrust movements at the margins of the arc massif (retroarc basins).

24.3.1 Trenches (Fig. 24.9)

Ocean trenches are elongate, gently curving troughs that form where an oceanic plate bends as it enters a subduction zone. The inner margin of the trench is formed by the leading edge of the overriding plate of the arc–trench system. The bottoms of modern trenches are up to 10,000 m below sea level, twice as deep as the average bathymetry of the ocean floors. They are also narrow, sometimes as little as 5 km across, although they may be thousands of kilometres long. Trenches formed along margins flanked by continental crust tend to be filled with sediment derived from the adjacent land areas. Intra-oceanic trenches are often starved of sediment because the only sources of material apart from pelagic deposits are the islands of the volcanic arc. Transport of coarse material into trenches is by mass flows, especially turbidity currents that may flow for long distances along the axis of the trench (Underwood & Moore 1995).

24.3.2 Accretionary complexes

The strata accumulated on the ocean crust and in a trench are not necessarily subducted along with the crust at a destructive plate boundary. The sediments may be wholly or partly scraped off the downgoing

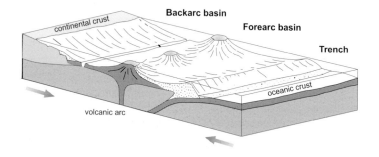

Fig. 24.9 Forearc basins, trenches and extensional backarc basins are supplied by volcaniclastic material from the adjacent arc and may also receive continentally derived detritus if the overriding plate is continental crust.

plate and accrete on the leading edge of the overriding plate to form an ***accretionary complex*** or ***accretionary prism***. These prisms or wedges of oceanic and trench sediments are best developed where there are thick successions of sediment in the trench (Einsele 2000). A subducting plate can be thought of as a conveyor belt bringing ocean basin deposits, mainly pelagic sediments and turbidites, to the edge of the overriding plate. In some places this sediment is carried down the subduction zone, but in others it is sliced off as a package of strata that is then accreted on to the overriding plate (Fig. 24.10).

24.3.3 Forearc basins (Fig. 24.9)

The inner margin of a forearc basin is the edge of the volcanic arc and the outer limit the accretionary complex formed on the leading edge of the upper plate. The width of a forearc basin will therefore be determined by the dimensions of the arc–trench gap, which is in turn determined by the angle of subduction. The basin may be underlain by either oceanic crust or a continental margin (Dickinson 1995). The thickness of sediments that can accumulate in a forearc setting is partly controlled by the height of the accretionary complex: if this is close to sea level the forearc basin may also fill to that level. Subsidence in the forearc region is due only to sedimentary loading. The main source of sediment to the basin is the volcanic arc and, if the arc lies in continental crust, the hinterland of continental rocks. Intraoceanic arcs are commonly starved of sediment because the island-arc volcanic chain is the only source of detritus apart from pelagic sediment. Given sufficient supply of detritus a forearc basin succession will consist of deep-

water deposits at the base, shallowing up to shallow-marine, deltaic and fluvial sediments at the top (Macdonald & Butterworth 1990). Volcaniclastic debris is likely to be present in almost all cases.

24.3.4 Backarc basins (Fig. 24.9)

Extensional backarc basins form where the angle of subduction of the downgoing slab is steep and the rate of subduction is greater than the rate of plate convergence. Rifting occurs in the region of the volcanic arc where the crust is hotter and weaker. At this stage an 'intra-arc basin' forms, a transient extensional basin that is bound on both sides by active volcanoes and is the site of accumulation of mainly volcanically derived sediment. With further extension the arc completely splits into two parts, an active arc with continued volcanism closer to the subduction zone and a remnant arc. As divergence between the remnant and active arcs continues a new spreading centre is formed to generate basaltic crust between the two. This backarc basin continues to grow by spreading until renewed rifting in the active arc leads to the formation of a new line of extension closer to the trench. Once a new backarc basin is formed the older one is abandoned. The lifespan of these basins is relatively short: in the Western Pacific Cenozoic backarc basins have existed for around 20 Myr between formation and abandonment. Extensional backarc basins can form in either oceanic or continental plates (Marsaglia 1995). The principal source of sediment in a backarc basin formed in an oceanic plate will be the active volcanic arc. Once the remnant arc is eroded down to sea level it contributes little further detritus. More abundant supplies are

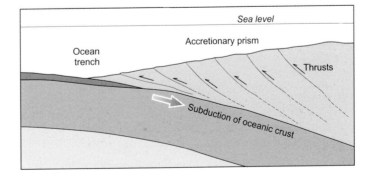

Fig. 24.10 Sediment deposited in an ocean trench includes both material derived from the overriding plate and pelagic material. As subduction proceeds sediment is scraped off the downgoing plate to form an accretionary prism of deformed sedimentary material.

available if there is continental crust on either or both sides of the basin. Backarc basins are typically under-filled, containing mainly deep-water sediment of vol-caniclastic and pelagic origin.

24.4 BASINS RELATED TO CRUSTAL LOADING

When an ocean basin completely closes with the total elimination of oceanic crust by subduction the two continental margins eventually converge. Where two continental plates converge subduction does not occur because the thick, low-density conti-nental lithosphere is too buoyant to be subducted. Collision of plates involves a thickening of the litho-sphere and the creation of an *orogenic belt*, a moun-tain belt formed by collision of plates. The Alps have formed by the closure of the Tethys Ocean as Africa has moved northwards relative to Europe, and the Himalayas (Fig. 24.11) are the result of a series of collisions related to the northward movement of India. The edges of the two continental margins that collide are likely to be thinned, passive margins.

Fig. 24.11 The major mountainous areas of the world occur in areas of plate collision where an orogenic belt forms.

Shortening initially increases the lithosphere thick-ness up to 'normal' values before it overthickens. As the crust thickens it undergoes deformation, with metamorphism occurring in the lower parts of the crust and movement of material outwards from the core of the orogenic belt along major fault planes. In the shallower levels of the mountain belt, low-angle faults (thrusts) also move rock outwards, away from the centre of the belt. This combination of movement by *thick-skinned tectonics* (which involves faults that extend deep into the crust) and *thin-skinned tectonics* (superficial thrust faults) transfers mass lat-erally and results in a *loading* of the crust adjacent to the mountain belt. This crustal loading results in the formation of a 'peripheral' foreland basin.

Crustal loading can also occur in settings other than the collision between two blocks of continental crust. At ocean–continent convergence settings, shortening in the overriding continental plate and subduction-related magmatism can also create a mountain belt. The Andes, along the western margin of South Amer-ica, have been uplifted by crustal thickening and the intrusion of magma associated with subduction to the west. Thrust belts on the landward side of mountain chains in these settings result in loading and the for-mation of a 'retroarc' foreland basin.

24.4.1 Peripheral foreland basins (Fig. 24.12)

Loading of the foreland crust either side of the orogenic belt causes the crust to flex in the same way that adding a mass to the unsupported end of a beam will cause it to bend downwards. The crust is not wholly unsupported, but the mantle/asthenosphere below the lithosphere is mobile and allows a flexural deformation of the loaded crust to form a *peripheral foreland basin*. The width of the basin will depend on the amount of load and the flexural rigidity of the foreland lithosphere, the ease with which it bends when a load is added to one end (Beaumont 1981). Rigid (typically older, thicker) lithosphere will respond to form a wide, shallow basin, whereas younger, thinner lithosphere flexes more easily to create a narrower, deeper trough. Increasing the load increases the basin depth.

In the initial stages of foreland basin formation the collision will have only proceeded to the extent of thickening the crust (which was formerly thinned at a passive margin) up to 'normal' crustal thickness. Although this results in a load on the foreland and

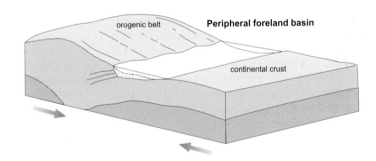

Fig. 24.12 Collision between two continental plates results in the formation of an orogenic belt where there is thickening of the crust: this results in an additional load being placed on the crust either side and causes a downward flexure of the crust to form peripheral foreland basins.

lithospheric flexure, the orogenic belt itself will not be high above sea level at this stage and little detritus will be supplied by erosion of the orogenic belt. Early foreland basin sediments will therefore occur in a deep-water basin, with the rate of subsidence exceeding the rate of supply. Turbidites are typical of this stage. When the orogenic belt is more mature and has built up a mountain chain there is an increase in the rate of sediment supply to the foreland basin. Although the load on the foreland will have increased, the sediment supply normally exceeds the rate of flexural subsidence. Foreland-basin stratigraphy typically shallows up from deep water to shallow marine and then continental sedimentation, which dominates the later stages of foreland-basin sedimentation (Miall 1995).

The stratigraphy of a foreland-basin fill is complicated by the fact that thrusting, and hence loading, at the margin of the basin continues as the basin evolves. The basin will tend to become larger with time as more load is added, and the later deformation at the margin will include some of the earlier basin deposits. Erosion and reworking of older basin strata into the younger deposits are common. As a consequence, the full succession of basin deposits will not be exposed in a single profile. Sometimes thrusting is not restricted to the basin margin, and may subdivide the basin to form **piggy-back basins** (Ori & Friend 1984) that lie on

top of the thrust sheets and which are separate from the **foredeep**, the basin in front of all the thrusts.

24.4.2 Retroarc foreland basins (Fig. 24.13)

Thickening of the crust in the continental magmatic arc results in the landward movement of masses of rock along thrusts. The loading of the crust on the opposite side of the arc to the trench results in flexure, and the formation of a basin: these basins are called **retroarc foreland basins** because of their position behind the arc. The continental crust will be close to sea level at the time the loading commences so most of the sedimentation occurs in fluvial, coastal and shallow marine environments. Continued subsidence occurs due to further loading of the basin margin by thrusted masses from the mountain belt, augmented by the sedimentary load. The main source of detritus is the mountain belt and volcanic arc.

24.5 BASINS RELATED TO STRIKE-SLIP TECTONICS

If a plate boundary is a straight line and the relative plate motion purely parallel to that line there would be neither uplift nor basin formation along strike-slip

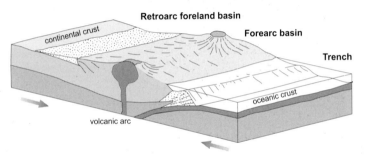

Fig. 24.13 The thickness of the crust increases due to emplacement of magma in a volcanic arc at a continental margin, resulting in flexure of the crust behind the arc to form a retroarc foreland basin.

plate boundaries. However, such plate boundaries are not straight, the motion is not purely parallel and they consist not of a single fault strand but of a network of branching and overlapping individual faults. Zones of localised subsidence and uplift create topographic depressions for sediment to accumulate and the source areas to supply them (Christie-Blick & Biddle 1985).

24.5.1 Strike-slip basins (Fig. 24.14)

Most basins in strike-slip belts are generally termed **transtensional basins** and are formed by three main mechanisms (Reading 1980). First, the overlap of two separate faults can create regions of extension between them known as **pull-apart basins**. Such basins are typically rectangular or rhombic in plan with widths and lengths of only a few kilometres or tens of kilometres. They are unusually deep, especially compared with rift basins. Second, where there is a branching of faults a zone of extension exists between the two branches forming a basin. Third, the curvature of a single fault strand results in bends that are either **restraining bends** (locally compressive) or **releasing bends** (locally extensional): releasing bends form elliptical zones of subsidence. Most strike-slip basins are bounded by faults that extend deep into the crust; 'thin-skinned' strike-slip basins are an exception, as the faulting affects only the upper part of the crust.

Strike-slip basins bounded by deep faults are relatively small, usually in the range of a hundred to a thousand square kilometres, and often contain thicker successions than basins of similar size formed by other mechanisms. Subsidence is usually rapid and several kilometres of strata can accumulate in a few million years (Allen & Allen 2005). Typically the margins are sites of deposition of coarse facies (alluvial fans and fan deltas) and these pass laterally over very short distances to lacustrine sediments in continental settings or marine deposits. In the stratigraphic record, facies are very varied and show lateral facies changes over short distances. 'Thin-skinned' strike-slip basins are broader and relatively shallower (Royden 1985).

24.6 COMPLEX AND HYBRID BASINS

Not all basins fall into the simple categories outlined above because they are the product of the interaction of more than one tectonic regime. This most commonly occurs where there is a strike-slip component to the motion at a convergent or divergent plate boundary. A basin may therefore partly show the characteristics of, say, a peripheral foreland but also contain strong indicators of strike-slip movement. Such situations exist because plate motions are commonly not simply orthogonal or parallel and examples of both oblique convergence and oblique extension between plates are common.

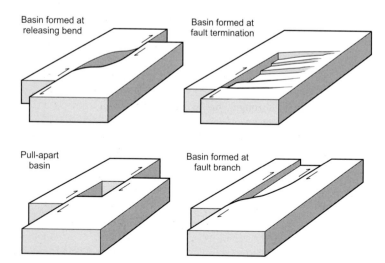

Basin formed at releasing bend

Basin formed at fault termination

Pull-apart basin

Basin formed at fault branch

Fig. 24.14 Basins may form by a variety of mechanisms in strike-slip settings: (a) a releasing bend, (b) a fault termination, (c) a fault offset (usually referred to as a pull-apart basin) and (d) at a junction between faults. Note that if the relative motion of the faults were reversed in each case the result would be uplift instead of subsidence.

24.7 THE RECORD OF TECTONICS IN STRATIGRAPHY

Tectonic forces act slowly on a human time scale but in the context of geological time the surface of the planet is in a continuous state of flux. Rift basins form and evolve into proto-oceanic troughs and eventually into ocean basins bordered by passive margins. After a period of tens to hundreds of millions of years the ocean basin starts to close with subduction zones around the margins consuming oceanic crust. Final closure of the ocean results in continental collision and the formation of an orogenic belt. These patterns of plate movement through time are known as the **Wilson Cycle** (Fig. 24.15) (Wilson 1966). The whole cycle starts again as the continent breaks up by renewed rifting. This relatively straightforward sequence of events may become complicated by oblique and strike-slip plate motion and over hundreds of millions of years regions of the crust may experience a succession of different tectonic settings, particularly those areas adjacent to plate margins.

The record of changing tectonic setting is contained within stratigraphy. For example, within the Wilson Cycle, the rift basin deposits may be recognised by river and lake deposits overlying the basement, evaporites may mark the proto-oceanic trough stage, and a thick succession of shallow-marine carbonate and clastic deposits will record passive margin deposition. If this passive margin subsequently becomes a site of subduction, arc-related volcanics will occur as the margin is transformed into a forearc region of shallow-marine, arc-derived sedimentation. Upon complete closure of the ocean basin, loading by the orogenic belt may then result in foreland flexure of this same area of the crust, and the environment of deposition will become one of deeper water facies. As the mountain belt rises, more sediment will be shed into the foreland basin and the stratigraphy will show a shallowing-up pattern.

The same principles of using the character of the association of sediments to determine the tectonic setting of deposition can be applied to any strata of any age. An objective of sedimentary and stratigraphic analysis of a succession of rocks is therefore to determine the type of basin that they were deposited in, and then use changes in the sedimentary character as an indicator of changing tectonic setting. In this way, a history of plate movements through geological history can be built up by combining the sedimentary and stratigraphic analysis with data from palaeomagnetic studies, which provide information about relative plate motions through time, and palaeobiogeographical information, which tells us about the distribution of plants and animals. The geological history of an area is now typically divided into stages that reflect different phases in the regional tectonic development: for example, in northeast America and northwest Europe, Palaeozoic strata are divided into a succession of marine deposits that formed within and on the margins of the Iapetus Ocean, sedimentary successions deposited in trenches and arc-related basins as this ocean closed, and, following the Caledonian orogeny, a thick sequence of Devonian red beds deposited mainly in extensional and strike-slip related basins within a supercontinent land mass.

The frequency with which the tectonic setting may change varies according to the position of a region with respect to plate margins. It is only in the centre of a stable continental area that the tectonic setting is unchanging over long periods of geological time. For example, the central part of the Australian continent has not experienced the tectonic forces of plate margins for 400 million years and in the latter part of that time a broad intracratonic basin, the Lake Eyre Basin, has formed by very slow subsidence. In regions closer to plate margins basins typically have a lifespan of a few tens of millions of years. The backarc basins in the West Pacific appear to be active for 20 million years or so. In contrast the passive margins of the Atlantic have been sites of sedimentation at the edges of the continents for over 200 million years.

24.8 SEDIMENTARY BASIN ANALYSIS

A succession of sedimentary rocks can be considered first in terms of the depositional environment of individual beds or associations of beds (Chapters 7–10 and 12–17), and second in the context of changes through time by the application of a time scale and means of correlation of strata (Chapters 19–23). The spatial distribution of depositional facies and variations in the environment of deposition through time will depend upon the tectonic setting (see above), so a comprehensive analysis of the sedimentology and stratigraphy of an area must take place in the context of the basin setting. **Sedimentary basin analysis** is the aspect of geology that considers all the controls on the accumulation of a succession of sedimentary rocks to develop a model for the evolution of the

Rift basin

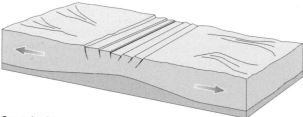

Ocean basin

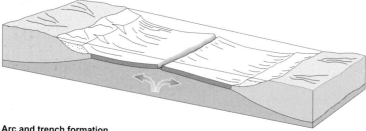

Arc and trench formation

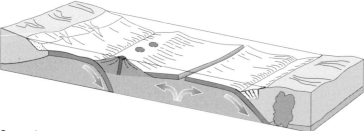

Ocean closure

Mountain belt

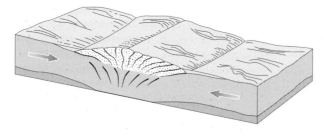

Fig. 24.15 The Wilson Cycle of extension to form a rift basin and ocean basin followed by basin closure and formation of an orogenic belt. (Adapted from Wilson 1966.)

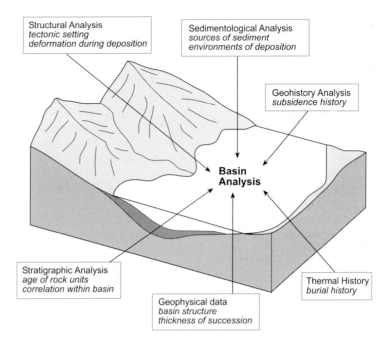

Structural Analysis
tectonic setting
deformation during deposition

Sedimentological Analysis
sources of sediment
environments of deposition

Geohistory Analysis
subsidence history

**Basin
Analysis**

Stratigraphic Analysis
age of rock units
correlation within basin

Geophysical data
basin structure
thickness of succession

Thermal History
burial history

Fig. 24.16 Basin analysis techniques.

sedimentary basin as a whole (Fig. 24.16). A comprehensive summary of basin analysis is provided in texts such as Allen & Allen (2005) and a brief introduction is outlined below.

24.8.1 Structural analysis

The patterns of deformation within a sedimentary succession provide information about the crustal stresses that existed in the area during and after deposition. **Synsedimentary faults** and folds are evidence of tectonic activity during deposition: a layer of strata may show structures (such as normal faults) that indicate extension during sedimentation or evidence of compressive forces (reverse faults or folds) acting as the strata were accumulating. These **growth structures** can be identified by the fact that their occurrence is limited to a particular stratigraphic unit within the succession. In general, extensional settings such as rift basins can be distinguished from basins formed under compressional regimes (such as foreland basins) on the basis of the recognition of these syndepositional structures, although local variations in stress commonly occur. Structural features that affect the whole succession are evidence

of tectonic forces occurring after deposition in the basin, but nevertheless provide information about the plate tectonics history of the area.

24.8.2 Geophysical data

Information from seismic reflection surveys (22.2) provides subsurface structural data that can be used in the structural analysis of a succession. Other geophysical data are in the form of information about the magnetic properties of the rocks below the surface and variations in the strength of the gravitational field in the region of the basin. **Magnetic surveys** over an area can indicate the nature of the rocks that lie at depth below the sedimentary succession: in general terms, oceanic crust retains a higher remnant magnetism than continental crust, allowing the crustal substrate of a basin to be determined. The strength of the Earth's magnetic field at any point on the surface depends on the density of the rocks below the surface at that point. **Gravity surveys** can therefore provide an indication of the thickness of sedimentary strata present, as they are of lower density than igneous or metamorphic rocks. Geophysical surveys are therefore a useful way of distinguishing between different basin types (e.g. extensional backarc basins are floored by oceanic crust) and the amount of subsidence that has occurred (basins in strike-slip setting commonly have very thick sedimentary successions).

24.8.3 Thermal history

Burial of sediments results in diagenetic changes (*18.2*) that include the effects of increased temperature with burial depth. The temperature that a body of sediment has been subjected to can be determined by fission-track analysis (*6.8*) and by studying the vitrinite reflectance characteristics of organic matter present in the sediment (*3.6.2*). The burial history of a body of sediment can therefore be reconstructed using these palaeothermometers, as they record the maximum temperature that the material has reached, and this can be used to infer the depth to which it has been buried. Combining these burial history data with the age of the strata can provide information on the history of subsidence in the basin and this can be related to the basin setting.

24.8.4 Stratigraphic analysis

The relative or absolute dating of the strata in the basin can be carried out using techniques described in Chapters 19 to 23. These provide a time framework for the basin history, indicating when the basin first started to form (the age of the rocks that lie at the bottom of the basin), and when sedimentation ceased (the youngest strata preserved), as well as events in between. The rate of sediment accumulation, that is, the thickness of strata deposited between two datable horizons, can be a characteristic indicator of basin setting: for example, rift basin sediments will commonly accumulate at a faster rate than passive margin deposits. On a shorter time scale, changes in sediment accumulation rate may reflect the relative sea level (Chapter 23).

24.8.5 Sedimentological analysis

The nature and the distribution of sediment present in a succession will reflect the basin setting. Three main aspects of the sedimentological analysis of basins can be considered: provenance studies, the distribution of facies and palaeoenvironments, and the changes in these through time during the basin evolution.

Provenance studies (*5.4.1*) are a key element of the analysis of a basin, providing information about the tectonic setting. Arc-related basins, such as backarc and forearc basins, are most likely to contain volcanic material derived from the magmatic arc. Rift basins in continental crust contain material derived from the surrounding cratonic area and are likely to include clasts of plutonic igneous or metamorphic origin. Peripheral foreland basins normally contain a high proportion of reworked sedimentary rocks that have been uplifted and subsequently eroded as part of the mountain-building process. Changes in clast composition through time can be used as an indicator of depth of erosion in the hinterland source area and hence provide a record of the uplift and unroofing history of an orogenic belt.

Once a stratigraphic framework for the basin succession has been established (24.8.4), the basin succession can be divided up into packages of strata, each deposited during an interval of time. The distribution of facies within an individual package provides a picture of the distribution of palaeoenvironments for that time interval, and hence a palaeogeography can be established. Different basin types can be expected to show different patterns of sedimentation: for example, a rift or strike-slip basin may be expected to have coarse facies such as alluvial fans or fan-deltas at its margins, a backarc basin would have an apron of volcaniclastic deposits at one margin, and a passive margin succession would be dominated by shallow-marine clastic or carbonate facies.

The tectonic setting of the basin is a major factor controlling the changes in the facies distributions and palaeogeographical patterns through time. Peripheral foreland basins and forearc basins both typically comprise deep-water facies in the lower part of the basin succession, shallowing up to shallow-marine or continental deposits. In contrast, rifts and backarc basins commonly show a progressive change from continental deposits formed in the early stage of rifting, followed by shallow-marine and sometimes deeper-marine facies. Changing palaeogeography within a basin therefore reflects the tectonic evolution of both the basin and the surrounding area.

24.8.6 Geohistory analysis

The quantitative study of the history of subsidence and sedimentation in a basin is known as ***geohistory*** *analysis* (Allen & Allen 2005). As sediment accumulates, the material in the lower part of the succession undergoes compaction (*18.2.1*), so the thickness of each stratal unit decreases through time. The compaction effect varies considerably with different lithologies, with clay-rich sediments decreasing in volume by up to 80% through time, whereas sandstones typically lose between 10 and 20% of their porosity as a result of compaction. The history of subsidence in a basin can be calculated by '*decompacting*' the sedimentary succession in a series of stages and by taking into account the **palaeobathymetry**, the water depth at which the sedimentation occurred determined from facies and palaeontological studies. Further information about the burial history can also be obtained from fission-track analysis (*6.8*) and vitrinite reflectance studies (*3.6.2 & 18.7.2*), which provide a measure of the thermal history of the strata. Geohistory analysis is important in hydrocarbon exploration because it provides information on the porosity and permeability changes through time, and also the thermal history of any part of the succession, which is critical to the generation of oil and gas (*18.7.3*).

24.9 THE SEDIMENTARY RECORD

Sedimentology equips us with the tools to interpret rocks in terms of processes and environments. These depositional environments are considered through geological time in the context of stratigraphic principles and analytical techniques. Sedimentological and stratigraphic analysis of the rock record using the material exposed at the surface, drilled and surveyed in the subsurface provides us with the means to reconstruct the history of the surface of the Earth as far as the data allow.

FURTHER READING

Allen, P.A. & Allen, J.R. (2005) *Basin Analysis: Principles and Applications* (2nd edition). Blackwell Science, Oxford.

Busby, C. & Ingersoll, R.V. (Eds) (1995) *Tectonics of Sedimentary Basins*. Blackwell Science, Oxford.

Einsele, G. (2000) *Sedimentary Basins, Evolution, Facies and Sediment Budget* (2nd edition). Springer-Verlag, Berlin.

Ingersoll, R.V. (1988) Tectonics of sedimentary basins. *Geological Association of America Bulletin*, **100**, 1704–1719.

Miall, A.D. (1999) *Principles of Sedimentary Basin Analysis*, 3rd edition. Springer-Verlag, Berlin.

References

Adams, A.E. & Mackenzie, W.S. (1998) *A Colour Atlas of Carbonate Sediments and Rocks under the Microscope.* Manson Publishing, London.

Adams, A., Mackenzie, W. & Guilford, C. (1984) *Atlas of Sedimentary Rocks under the Microscope.* Wiley, Chichester.

Aigner, T. (1985) *Storm Depositional Systems.* Lecture Notes in Earth Sciences 3, Springer-Verlag, Berlin.

Allen, J.R.L. (1965) The sedimentation and palaeogeography of the Old Red Sandstone of Anglesey, North Wales. *Proceedings of the Yorkshire Geological Society*, **35**, 139–185.

Allen, J.R.L. (1972) A theoretical and experimental study of climbing ripple cross lamination with a field application to the Uppsala esker. *Geografiska Annaler*, **53A**, 157–187.

Allen, J.R.L (1974) Studies in fluviatile sedimentation: implications of pedogenic carbonate units, Lower Old Red Sandstone, Anglo-Welsh outcrop. *Geological Journal*, **9**, 181–208.

Allen, J.R.L. (1982) *Sedimentary Structures: their Character and Physical Basis*, Vol. 1. Developments in Sedimentology. Elsevier, Amsterdam.

Allen, J.R.L. (1994) Fundamental properties of fluids and their relation to sediment transport processes. In: *Sediment Transport and Depositional Processes* (Ed. Pye, K.). Blackwell Science, Oxford; 25–60.

Allen, P.A. (1997) *Earth Surface Processes.* Blackwell Science, Oxford, 404 pp.

Allen, P.A. & Allen, J.R. (2005) *Basin Analysis: Principles and Applications* (2nd edition). Blackwell Science, Oxford.

Alsop, G.I., Blundell, D.J. & Davison, I. (Eds) (1996). *Salt Tectonics.* Special Publications 100, Geological Society Publishing House, Bath.

Amorosi, A. (2003) Glaucony and verdine. In: *Encyclopedia of Sediments and Sedimentary Rocks* (Ed. Middleton, G.V.). Kluwer Academic Publishers, Dordrecht; 331–333.

Anderton, R. (1985) Clastic facies models and facies analysis. In: *Sedimentology, Recent Developments and Applied Aspects* (Eds Brenchley, P.J & Williams, B.P.J). Blackwell Scientific Publications, Oxford; 31–47.

Anderton, R. (1995) Sequences, cycles and other nonsense: are submarine fan models any use in reservoir geology? In: *Characterization of Deep Marine Clastic Systems* (Eds Hartley, A.J. & Prosser, D.J.). Special Publication 94, Geological Society Publishing House, Bath; 5–11.

Astin, T.R. (1991) Subaqueous shrinkage or syneresis cracks in the Devonian of Scotland reinterpreted. *Journal of Sedimentary Petrology*, **61**, 850–859.

Augustinus, P.G. (1989) Cheniers and chenier plains: a general introduction. *Marine Geology*, **90**, 219–229 .

Baas, J.H. (1994) A flume study on the development and equilibrium morphology of small-scale bedforms in very fine sand. *Sedimentology*, **41**, 185–209.

Baas, J.H. (1999) An empirical model for the development and equilibrium morphology of current ripples in fine sand. *Sedimentology*, **46**, 123–138.

Bada, J.L. (1985) Amino acid racemisation dating of fossil bones. *Annual Reviews of Earth and Planetary Science*, **13**, 241–268.

Basu, A. (2003) Provenance. In: *Encyclopedia of Sediments and Sedimentary Rocks* (Ed. Middleton, G.V.). Kluwer Academic Publishers, Dordrecht; 544–549.

Bathurst, R.G.C. (1987) Diagenetically enhanced bedding in argillaceous platform limestones: stratified cementation and selective compaction: *Sedimentology*, **34**, 749–778.

Beaumont, C. (1981) Foreland basins. *Geophysical Journal of the Royal Astronomical Society*, **65**, 291–329.

Belknap, D.F. (2003) Salt marshes. In: *Encyclopedia of Sediments and Sedimentary Rocks* (Ed. Middleton, G.V.). Kluwer Academic Publishers, Dordrecht; 586–588.

Benn, D.I. & Evans, D.J.A. (1998) *Glaciers and Glaciation.* Arnold, London.

Bernoulli, D. & Jenkyns, H.C. (1974) Alpine, Mediterranean and Central Atlantic Mesozoic facies in relation to the early evolution of the Tethys. In: *Modern and Ancient Geosynclinal Sedimentation* (Eds Dott, R.H. & Shaver, R.H.). Special Publication 22, Society of Economic Paleontologists and Mineralogists, Tulsa, OK; 129–160.

Bhattacharya, J.P. (2003) Deltas and estuaries. In: *Encyclopedia of Sediments and Sedimentary Rocks* (Ed. Middleton, G.V.). Kluwer Academic Publishers, Dordrecht; 195–203.

Bhattacharya, J.P. & Davies, R.K. (2001) Growth faults at the pro-delta to delta front transition, Cretaceous

Ferron Sandstone, Utah. *Marine and Petroleum Geology*, **18**, 525–534.

Bhattacharya, J.P. & Giosan, L. (2003) Wave-influenced deltas: geomorphological implications for facies reconstruction. *Sedimentology*, **50**, 187–210.

Bhattacharya, J.P. & Walker, R.G. (1992) Deltas. In: *Facies Models: Response to Sea Level Change* (Eds Walker, R.G. & James, N.P.). Geological Association of Canada, St Johns, Newfoundland; 157–178.

Bierman, P.R. (1994) Using in situ produced cosmogenic isotopes to estimate rates of landscape evolution: A review from the geomorphic perspective. *Journal of Geophysical Research*, **99**, 13,885–13,896.

Blair, T.C. (2000a) Cause of dominance by sheetflood vs. debris-flow processes on two adjoining fans, Death Valley, California. *Sedimentology*, **46**, 1015–1028.

Blair, T.C. (2000b) Sedimentary processes and facies of the waterlaid Anvil Spring Canyon alluvial fan, Death Valley, California. *Sedimentology*, **46**, 913–940.

Blair, T.C. & McPherson, J.G. (1994) Alluvial fans and their natural distinction from rivers based on morphology, hydraulic processes, sedimentary processes and facies assemblages. *Journal of Sedimentary Research*, **A64**, 450–589.

Blatt, H. (1985) Provenance studies and mudrocks. *Journal of Sedimentary Petrology*, **55**, 69–75.

Blatt, H., Middleton, G.V. & Murray, R.C. (1980) *Origin of Sedimentary Rocks* (2nd edition). Prentice-Hall, Englewood Cliffs, New Jersey.

Blatt, H., Berry, W.B. & Brande, S. (1991) *Principles of Stratigraphic Analysis*. Blackwell Scientific Publications, Oxford.

Boggs, S. (2006) *Principles of Sedimentology and Stratigraphy* (4th edition). Pearson Prentice Hall, Upper Saddle River, NJ.

Bohacs, K.M. & Suter, J. (1997) Sequence stratigraphic distribution of coaly rocks; fundamental controls and paralic examples. *American Association of Petroleum Geologists Bulletin*, **81**, 1612–1639.

Bohacs, K.M., Carroll, A.R., Neal, J.E. & Mankiewicz, P.J. (2000) Lake-basin type, source potential, and hydrocarbon character: an integrated-sequence-stratigraphic–geochemical framework. In *Lake Basins through Space and Time* (Eds Gierlowski-Kordesch, E.H. & Kelts, K.R.). Studies in Geology 46, American Association of Petroleum Geologists, Tulsa, OK; 3–34.

Bohacs, K.M., Carroll, A.R. & Neal, J.E. (2003) Lessons from large lake systems – thresholds, nonlinearity, and strange attractors. In: *Extreme Depositional Environments: Mega End Members in Geologic Time* (Eds Chan, M.A. & Archer, A.W.). Geological Society of America Special Paper 370, Boulder, CO; 75–90.

Boothroyd, J.C. (1985) Tidal inlets and tidal deltas. In: *Coastal Sedimentary Environments* (Ed. Davis, R.A.) Springer-Verlag, Berlin; 445–533.

Boothroyd, J.C. & Ashley, G.M. (1975) Process, bar morphology and sedimentary structures on braided outwash fans, northeastern Gulf of Alaska. In: *Glaciofluvial and Glaciomarine Sedimentation* (Eds Jopling, A.V. & McDonald, B.C.). Special Publication 23, Society of Economic Paleontologists and Mineralogists, Tulsa, OK; 193–222.

Boothroyd, J.C. & Nummedal, D. (1978) Proglacial braided outwash: a model for humid alluvial fan deposits. In: *Fluvial Sedimentology* (Ed. Miall, A.D.). Memoir 5, Canadian Society of Petroleum Geologists, Calgary; 641–668.

Bosence, D.W.J. (2005) A genetic classification of carbonate platforms based on their basinal and tectonic settings in the Cenozoic. *Sedimentary Geology*, **175**, 49–72.

Bosence, D.W.J. & Wilson, R.C.L. (2003) Carbonate depositional systems. In: *The Sedimentary Record of Sea-Level Change* (Ed. Coe, A.L.). Cambridge University Press, Cambridge; 209–233.

Bosscher, H. & Schlager, W. (1992) Computer simulation of reef growth. *Sedimentology*, **39**, 503–512.

Bouma, A.H. (1962) *Sedimentology of some Flysch Deposits. A Graphic Approach to Facies Interpretation*. Elsevier, Amsterdam.

Bovis, M.J. (2003) Avalanche and rock fall. In: *Encyclopedia of Sediments and Sedimentary Rocks* (Ed. Middleton, G.V.). Kluwer Academic Publishers, Dordrecht; 31–34.

Bown, T.M. & Kraus, M.J. (1987) Integration of channel and overbank suites, I. Depositional sequences and lateral relations of alluvial paleosols. *Journal of Sedimentary Petrology*, **57**, 587–601.

Bøtter-Jensen, L. (1997) Luminescence techniques: instrumentation and methods. *Radiation Measurements*, **27**, 749–768.

Braithwaite, C. (2005) *Carbonate Sediments and Rocks*. Whittles Publishing, Dunbeath.

Braun, J., van der Beek, P. & Batt, G. (2006) *Quantitative Thermochronology: Numerical Methods for the Interpretation of Thermochronological Data*. Cambridge University Press, Cambridge.

Bridge, J.S. (1978) Origin of horizontal lamination under turbulent boundary layers. *Sedimentary Geology*, **20**, 1–16.

Bridge, J.S (2003) *Rivers and Floodplains: Forms, Processes, and Sedimentary Record*. Blackwell Science, Oxford.

Bridge, J.S. & Leeder, M.R. (1979) A simulation model of alluvial stratigraphy. *Sedimentology*, **26**, 617–644.

Briere, P.R. (2000) Playa, playa lake, sabkha: proposed definitions for old terms. *Journal of Arid Environments*, **45**, 1–7.

Brookfield, M.E. (1992) Eolian systems. In: *Facies Models: Response to Sea Level Change* (Eds Walker, R.G. & James, N.P.). Geological Association of Canada, St Johns, Newfoundland; 143–156.

Burbank, D.W. & Pinter, N. (1999) Landscape evolution: the interactions of tectonics and surface processes. *Basin Research*, **11**, 1–6.

Burley, S.D., Kantorowicz, J.D. & Waugh, B. (1985) Clastic diagenesis. In: *Sedimentology, Recent Developments and Applied Aspects* (Eds Brenchley, P.J & Williams, B.P.J). Blackwell Scientific Publications, Oxford; 189–228.

Burns, S.J., Mackenzie, J.A. & Vasconcelos, C. (2000) Dolomite formation and biochemical cycles in the Phanerozoic. *Sedimentology*, **47**, 49–61.

Busby, C. & Ingersoll, R.V. (Eds) (1995) *Tectonics of Sedimentary Basins*. Blackwell Science, Oxford.

Bustin, R.M. & Wüst, R.A.J. (2003) Maturation, organic. In: *Encyclopedia of Sediments and Sedimentary Rocks* (Ed. Middleton, G.V.). Kluwer Academic Publishers, Dordrecht; 425–429.

Calvert, S.E. (2003) Iron-manganese nodules. In: *Encyclopedia of Sediments and Sedimentary Rocks* (Ed. Middleton, G.V.). Kluwer Academic Publishers, Dordrecht; 376–379.

Cant, D.J. (1982) Fluvial facies models. In: *Sandstone Depositional Environments* (Eds Scholle, P.A. & Spearing, D.). *American Association of Petroleum Geologists Memoir*, **31**, 115–138.

Carey, S.N. (1991) Transport and deposition by pyroclastic flows and surges. In: *Sedimentation in Volcanic Settings* (Eds R.V. Fisher & G.A. Smith). Special Publication 45, SEPM (Society for Sedimentary Geology), Tulsa, OK; 39–57.

Carney, J.L. & Pierce, R.W. (1995) Graphic correlation and composite standard databases as tools for the exploration biostratigrapher. In: *Graphic Correlation* (Eds Mann, K.O. & Lane, H.R.). Special Publication 53, SEPM (Society for Sedimentary Geology), Tulsa, OK; 23–43.

Carson, M.A. (1977) Angles of repose, angles of shearing resistance and angles of talus slopes. *Earth Surface Processes and Landforms*, **2**, 363–380.

Carter, A. & Moss, S.J. (1999) Combined detrital-zircon fission-track and U–Pb dating: a new approach to understanding hinterland evolution. *Geology*, **27**, 235–238

Cas, R.A.F, & Wright, J.V. (1987) *Volcanic Successions: Modern and Ancient*. Unwin Hyman, London.

Catuneanu, O. (2006) *Principles of Sequence Stratigraphy*. Elsevier, Amsterdam.

Chappell, J. & Shackleton, N.J. (1986) Oxygen isotopes and sea level. *Nature*, **324**, 137–140.

Cheel, R.J. & Leckie, D.A. (1993) Hummocky cross-stratification. In: *Sedimentology Review 1* (Ed. Wright, V.P.). Blackwell Scientific Publications, Oxford; 103–122.

Christie-Blick, N. & Biddle, K.T. (1985) Deformation and basin formation along strike-slip faults. In: *Strike-slip Deformation, Basin Formation and Sedimentation* (Eds Biddle, K.T. & Christie-Blick, N.). Special Publication 37, Society of Economic Paleontologists and Mineralogists, Tulsa, OK; 1–34.

Church, M. & Jones, D. (1982) Channel bars in gravel-bed rivers. In: *Gravel-bed Rivers* (Eds Hey, R.D., Bathurst, J.C. & Thorne, C.R.). Wiley, Chichester; 291–324.

Clarkson, E.N.K. (1993) *Invertebrate Palaeontology and Evolution* (3rd edition). Allen and Unwin, London.

Clifton, H.E. (2003) Coastal sedimentary facies. In: *Encyclopedia of Sediments and Sedimentary Rocks* (Ed. Middleton, G.V.). Kluwer Academic Publishers, Dordrecht; 149–157.

Clifton, H.E. (2006) A re-examination of facies models for clastic shorelines. In: *Facies Models Revisited* (Eds Walker, R.G. & Posamentier, H.). Special Publication 84, Society of Economic Paleontologists and Mineralogists, Tulsa, OK; 293–337.

Cloetingh, S. (1988) Intraplate stresses: a tectonic cause for third-order cycles in apparent sea level? In: *Sea Level Changes: an Integrated Approach* (Eds C.K. Wilgus, B.S. Hastings, C.G.St.C. Kendall, H.W. Posamentier, C.A. Ross and J.C. Van Wagoner). Special Publication 42, Society of Economic Paleontologists and Mineralogists, Tulsa, OK; 19–29.

Coe, A.L. (Ed.). (2003) *The Sedimentary Record of Sea-level Change*. Cambridge University Press, Cambridge.

Collinson, J.D. (1969) The sedimentology of the Grindslow Shales and the Kinderscout Grit: a deltaic complex in the Namurian of northern England. *Journal of Sedimentary Petrology*, **39**, 194–221.

Collinson, J.D. (1986) Alluvial sediments. In: *Sedimentary Environments and Facies* (Ed. Reading, H.G.). Blackwell Scientific Publications, Oxford; 20–62.

Collinson, J.D. (2003) Deformation structures and growth faults. In: *Encyclopedia of Sediments and Sedimentary Rocks* (Ed. Middleton, G.V.). Kluwer Academic Publishers, Dordrecht; 193–195.

Collinson, J.D., Martinsen, O.J., Bakken, B. & Kloster, A. (1991) Early fill of the Western Irish Namurian Basin: A complex relationship between turbidites and deltas. *Basin Research*, **3**, 223–242.

Collinson, J.D., Mountney, N. & Thompson, D. (2006) *Sedimentary Structures*. Terra Publishing, London.

Conway-Morris, S. (1992) The early evolution of life. In: *Understanding the Earth, a New Synthesis* (Eds Brown, G.C., Hawkesworth, C.J. & Wilson, R.C.L.). Cambridge University Press, Cambridge; 436–457.

Dalrymple, R.W. (1984) Morphology and internal structure of sand waves in the Bay of Fundy. *Sedimentology*, **31**, 365–382.

Dalrymple, R.W. (1992) Tidal depositional systems. In: *Facies Models: Response to Sea Level Change* (Eds Walker, R.G. & James, N.P.). Geological Association of Canada, St Johns, Newfoundland; 195–218.

Dalrymple, R.W. & Choi, K. (2007) Morphologic and facies trends through the fluvial–marine transition in tide-dominated depositional systems: a schematic framework for environmental and sequence-stratigraphic interpretation. *Earth-Science Reviews*, **81**, 135–174.

Dalrymple, R.W., Zaitlin, B.A. & Boyd, R. (1992) Estuarine facies models: conceptual basis and stratigraphic implications. *Journal of Sedimentary Petrology*, **62**, 1130–1146.

Darwin, C. (1859) *The Origin of Species by Means of Natural Selection*. John Murray, London.

Davis, R.A. Jr & Fitzgerald, D.M. (2004) *Beaches and Coasts*. Blackwell Science, Oxford.

DeCelles, P.G. (1988) Lithologic provenance modelling applied to the Late Cretaceous synorogenic Echo Canyon Conglomerate, Utah: a case of multiple source areas. *Geology*, **16**, 1039–1043.

De Paolo, D.J. & Ingram, B.L. (1985) High resolution stratigraphy with strontium isotopes. *Science*, **217**, 938–941.

Dickin, A.P. (2004) *Radiogenic Isotope Geology* (2nd edition). Cambridge University Press, Cambridge.

Dickinson, W.R. (1980) Plate tectonics and key petrologic associations. In: *The Continental Crust and its Mineral Deposits* (Ed. Strangway, D.W.). Special Paper 20, Geological Association of Canada, St Johns, Newfoundland; 341–360.

Dickinson, W.R. (1995) Forearc basins. In: *Tectonics of Sedimentary Basins* (Eds Ingersoll, R.V. & Busby, C.J.). Blackwell Science, Oxford; 221–262.

Dickinson, W.R. & Suczek, C.A. (1979) Plate tectonics and sandstone composition, *American Association of Petroleum Geologists Bulletin*, **63**, 2164–2182.

Dott, R.H. Jr & Bourgeois, J. (1982) Hummocky cross stratification: significance of its variable bedding sequences. *Geological Society of America Bulletin*, **93**, 663–680.

Douglas, R.G. (2003) Oceanic sediments, In: *Encyclopedia of Sediments and Sedimentary Rocks* (Ed. Middleton, G.V.). Kluwer Academic Publishers, Dordrecht; 481–492.

Doyle, P. & Bennett, M.R. (1998) *Unlocking the Stratigraphical Record. Advances in Modern Stratigraphy*. Wiley, Chichester.

Doyle, P., Bennett, M.R. & Baxter, A.N. (1994) *The Key to Earth History: an Introduction to Stratigraphy*. Wiley, Chichester.

Drewry, D.D. (1986) *Glacial Geologic Processes*. Arnold, London.

Droser, M.L. & Bottjer, D.J. (1986) A semiquantitative field classification of ichnofabric. *Journal of Sedimentary Research*, **56**, 558–559.

Dunham, R.J. (1962) Classification of carbonate rocks according to depositional texture. In: *Classification of Carbonate Rocks* (Ed. Ham, W.E.). Memoir 1, American Association of Petroleum Geologists, Tulsa, OK; 108–121.

Edwards, L.R., Chen, J.H. & Wasserburg, G.J. (1986) ^{238}U–^{234}U–^{230}Th–^{232}Th systematics and the precise measurement of time over the past 500 000 years. *Earth and Planetary Science Letters*, **81**, 175–192.

Eicher, D.L. (1976) *Geologic Time* (2nd edition). Prentice-Hall, Englewood Cliffs, New Jersey.

Einsele, G. (2000) *Sedimentary Basins, Evolution, Facies and Sediment Budget* (2nd edition). Springer-Verlag, Berlin.

Ekdale, A.A. & Bromley, R.G. (1991) Analysis of composite ichnofabrics: an example in uppermost Cretaceous Chalk of Denmark. *Palaios*, 6, 232–249.

Ekdale, A.A., Bromley, R.G. & Pemberton, S.G. (1984) *Ichnology: the Use of Trace Fossils in Sedimentology and Stratigraphy*. Short Course Notes 15, Society of Economic Paleontologists and Mineralogists, Tulsa, OK.

Eldredge, N. & Gould, S.J. (1972) Punctuated equilibria: an alternative to phyletic gradualism. In: *Models in Paleobiology* (Ed. Schopf, T.J.M) Freeman, Cooper, San Francisco; 82–115.

Elliott, T. (1986) Deltas. In: *Sedimentary Environments and Facies* (Ed. Reading, H.G.). Blackwell Scientific Publications, Oxford; 113–154.

Embry, A.F. & Klovan, J.E. (1971) A late Devonian reef tract on north-eastern Banks Island, Northwest Territories. *Bulletin of Canadian Petroleum Geology*, **19**, 730–781.

Emery, D. & Myers, K.J. (Eds) (1996) *Sequence Stratigraphy*. Blackwell Science, Oxford.

Eugster, H.P. (1985) Oil shales, evaporites and ore deposits. *Geochimica Cosmochimica Acta*, **49**, 619–635.

Eugster, E.P. & Hardie, L.A. (1978) Saline lakes. In: *Lakes: Chemistry, Geology, Physics* (Ed. Lerman, A.). Springer-Verlag, Berlin; 237–293.

Faure, G. (1986) *Principles of Isotope Geology* (2nd edition). Wiley, New York, 589 pp.

Faure, G. & Mensing, T.M. (2004) *Isotopes: Principles and Applications* (3rd edition). Wiley, New York.

Fine, I.V., Rabinovich, A.B., Bornhold, B.D., Thomson, R.E. & Kulikov, E.A. (2005) The Grand Banks landslide-generated tsunami of November 18, 1929: preliminary analysis and numerical modelling. *Marine Geology*, **215**, 45–57.

FitzGerald, D.M. & Buynevich, I.V. (2003) Barrier Islands. In: *Encyclopedia of Sediments and Sedimentary Rocks* (Ed. Middleton, G.V.). Kluwer Academic Publishers, Dordrecht; 43–48.

Folk, R.L. (1974) *Petrology of Sedimentary Rocks*. Hemphill, Austin, Texas.

Folk, R.L. & Lynch, F.L. (2001) Organic matter, putative nannobacteria and the formation of ooids and hardgrounds. *Sedimentology*, **48**, 215–229.

Fowler, C.M.R. (2005) *The Solid Earth: An Introduction to Global Geophysics*. Cambridge University Press, Cambridge.

Fralick, P. & Barrett, T.J. (1995) Depositional controls on iron formation associations in Canada. In: *Sedimentary Facies Analysis: a Tribute to the Research and Teaching of Harold G. Reading* (Ed. Plint, A.G.). Special Publication 22, International Association of Sedimentologists. Blackwell Science, Oxford; 137–156.

Friedman, G.M., Sanders, J.E. & Kopaska-Merkel, D.C. (1992) *Principles of Sedimentary Deposits: Stratigraphy and Sedimentology*. Macmillan, New York.

Friend, P.F. (1978) Distinctive features of some ancient river systems. In: *Fluvial Sedimentology* (Ed. Miall, A.D.). Memoir 5, Canadian Society of Petroleum Geologists, Calgary; 531–542.

Galloway, W.E. (1989) Genetic stratigraphic sequences in basin analysis 1: Architecture and genesis of flooding-bounded depositional units. *American Association of Petroleum Geologists Bulletin*, **73**, 125–142.

Gass, I.G. (1982) Ophiolites. *Scientific American*, **247**, 122–131.

Gawthorpe, R.L. & Leeder, M.R. (2000) Tectono-sedimentary evolution of active extensional basins. *Basin Research*, **12**, 195–218.

Gischler, E. & Lomando, A.J. (1997) Holocene cemented beach deposits in Belize. *Sedimentary Geology*, **110**, 277–297.

Glenn, C.R. & Garrison, R.E. (2003) Phosphorites. In: *Encyclopedia of Sediments and Sedimentary Rocks* (Ed. Middleton, G.V.). Kluwer Academic Publishers, Dordrecht; 519–526.

Gradstein, F. & Ogg, J. (2004) Geologic Time Scale 2004 – why, how, and where next! *Lethaia*, **37**, 175–181.

Gradstein F.M., Ogg J.G., Smith A.G., Agterberg F.P., Bleeker W., Cooper R.A., Davydov V., Gibbard P., Hinnov L.A., House, M.R., Lourens, L., Luterbacher H-P., McArthur J., Melchin M.J., Robb L.J., Shergold J., Villeneuve M., Wardlaw B.R., Ali J., Brinkhuis H., Hilgen F.J., Hooker J., Howarth R.J., Knoll A.H., Laskar J., Monechi S., Powell J., Plumb K.A., Raffi I., Röhl U., Sanfilippo A., Schmitz B., Shackleton N.J., Shields G.A., Strauss H., Van Dam J., Veizer J., van Kolfschoten, Th., & Wilson, D. (2004) *A Geologic Time Scale 2004*. Cambridge University Press, Cambridge.

Gribble, C.D. & Hall, A.J. (1999) Optical *Mineralogy, Principles and Practice*. Routledge, London.

Hailwood, E.A. (1989) *Magnetostratigraphy*. Special Report 19, Geological Society of London.

Hallam, A. (1963) Major epirogenic and eustatic changes since the Cretaceous and their possible relationship to crustal structure. *American Journal of Science*, **261**, 164–177.

Hambrey, M.J. (1994). *Glacial Enviroments*. UCL Press, London.

Hambrey, M.J. & Glasser, N.F. (2003) Glacial sediments: processes, environments and facies. In: *Encyclopedia of Sediments and Sedimentary Rocks* (Ed. Middleton, G.V.). Kluwer Academic Publishers, Dordrecht; 316–330.

Haq, B.U., Hardenbol, J., & Vail, P.R. (1987) Chronology of fluctuating sea levels since the Triassic. *Science*, **235**, 1156–1167.

Haq, B.U., Hardenbol, J., & Vail, P.R. (1988) Mesozoic and Cenozoic chronostratigraphy and cycles of sea level change. In: *Sea Level Changes: an Integrated Approach* (Eds Wilgus, C.K., Hastings, B.S., Kendall, C.G.St.C., Posamentier H.W., Ross C.A. and Van Wagoner J.C.). Special Publication 42, Society of Economic Paleontologists and Mineralogists, Tulsa, OK; 71–108.

Hardenbol, J., Thierry, J., Farley, M.B., Jacquin, T., de Graciansky, P.C. & Vail, P.R. (1998) Mesozoic and Cenozoic sequence chronostratigraphic framework of European basins In: *Mesozoic and Cenozoic Sequence Stratigraphy of European Basins* (Eds de Graciansky, P.C., Hardenbol, J., Jacquin, T. & Vail, P.R.). Special Publication 60, SEPM (Society for Sedimentary Geology), Tulsa, OK; 3–13.

Hardie, L.A. (1996) Secular variation in seawater chemistry: an explanation for the coupled secular variation in the mineralogies of marine limestones and potash evaporites over the past 600 m.y. *Geology*, **24**, 279–283.

Hart, B.S. & Plint, A.G. (1995) Gravelly shoreface and beach deposits. In: *Sedimentary Facies Analysis: a Tribute to the Research and Teaching of Harold G. Reading* (Ed. Plint, A.G.). Special Publication 22, International Association of Sedimentologists. Blackwell Science, Oxford; 75–90.

Harvey, A.M., Mather, A.E. & Stokes, M. (Eds) (2005) *Alluvial Fans: Geomorphology, Sedimentology, Dynamics*. Special Publication 251, Geological Society Publishing House, Bath.

Hasiotis, S.T. (2002) *Continental Trace Fossils*. Short Course Notes 51, Society of Economic Paleontologists and Mineralogists, Tulsa, OK.

Hazeldine, R.S. (1989) Coal reviewed: depositional controls, modern analogues and ancient climates. In: *Deltas: Sites and Traps for Fossil Fuels* (Eds Whateley, M.K., & Pickering, K.T.). Special Publication 41, Geological Society Publishing House, Bath; 289–308.

Helland-Hansen, W. & Martinsen, O. (1996) Shoreline trajectories and sequences: a description of variable depositional-dip scenarios. *Journal of Sedimentary Research*, **66**, 670–688.

Hess, J., Bender, M.L. & Schilling, J-G. (1986) Evolution of the ratio of strontium-87 to strontium-86 in seawater from the Cretaceous to the present. *Science*, **231**, 979–984.

Heward, A.P. (1978) Alluvial fan sequence and megasequence models: with examples from Westphalian D–Stephanian B coalfields, northern Spain. In: *Fluvial Sedimentology* (Ed. Miall, A.D.). Memoir 5, Canadian Society of Petroleum Geologists, Calgary; 669–702.

Hiscott, R.N. (2003) Latest Quaternary Baram Prodelta, Northwestern Borneo. In: *Tropical Deltas of Southeat Asia – Sedimentology, Stratigraphy and Petroleum Geology* (Eds Sidi, F.H., Nummedal, D., Imbert, P., Darman, H. & Posamentier, H.W.). Special Publication 76, SEPM (Society for Sedimentary Geology), Tulsa, OK; 89–107.

Horton, B.K. & DeCelles, P.G. (2001) Modern and ancient fluvial megafans in the foreland basin system of the central Andes, southern Bolivia: implications for drainage network evolution in fold–thrust belts, *Basin Research*, **13**, 43–63.

Hounslow, M.W. (1997) Significance of localised pore pressures to the genesis of septarian concretions. *Sedimentology*, **44**, 1133–1147.

Hoyt, J.H. (1967) Barrier Island Formation. *Geological Society of America Bulletin*, **78**, 1123–1136.

Hsü, K.J. (1972) The origin of saline giants: a critical review after the discovery of the Mediterranean evaporite. *Earth-Science Reviews*, **8**, 371–396.

Hughes, D.A. & Lewin, J. (1982) A small-scale flood plain, *Sedimentology*, **29**, 891–895.

Humphrey, J.D. & Quinn, T.M. (1989) Coastal mixing zone dolomite, forward modeling and massive dolomitization of platform-margin carbonates. *Journal of Sedimentary Petrology*, **59**, 438–454.

Hunt, D. & Tucker, M.E. (1992) Stranded parasequences and the forced regressive wedge systems tract: deposition during base-level fall. *Sedimentary Geology*, **81**, 1–9.

Ingersoll, R.V. (1988) Tectonics of sedimentary basins. *Geological Association of America Bulletin*, **100**, 1704–1719.

James, N.P. (2003) Neritic carbonate depositional environments. In: *Encyclopedia of Sediments and Sedimentary Rocks* (Ed. Middleton, G.V.). Kluwer Academic Publishers, Dordrecht; 464–475.

James, N.P. & Bourque, P-A. (1992) Reefs and mounds. In: *Facies Models: Response to Sea Level Change* (Eds Walker, R. G. & James, N.P.). Geological Association of Canada, St Johns, Newfoundland; 323–348.

Jeans, C.V. (1995) *Clay Mineral Stratigraphy in Palaeozoic and Mesozoic Red Bed Facies Onshore and Offshore UK*. Special Publication 89, Geological Society, London; 31–55.

Jervey, M.T. (1988) Quantitative geological modelling of siliciclastic rock sequences and their seismic expressions. In: *Sea Level Changes: an Integrated Approach* (Eds Wilgus C.K., Hastings B.S., Kendall C.G.St.C., Posamentier H.W., Ross C.A. and Van Wagoner J.C.). Special Publication 42, Society of Economic Paleontologists and Mineralogists, Tulsa, OK; 47–69.

Johns, D.R., Mutti, E., Rosell, J. & Seguret, M. (1981) Origin of a thick, redeposited carbonate bed in Eocene turbidites of the Hecho Group, south-central Pyrenees, Spain, *Geology*, **9**, 161–164.

Johnson, H.D. & Baldwin, C.T. (1996) Shallow clastic seas. In: *Sedimentary Environments: Processes, Facies and Stratigraphy* (Ed. Reading, H.G.). Blackwell Science, Oxford; 232–280.

Jones, B. & Desrochers, A. (1992) Shallow platform carbonates. In: *Facies Models: Response to Sea Level Change* (Eds Walker, R.G. & James, N.P.). Geological Association of Canada, St Johns, Newfoundland; 277–302.

Kearey, P. & Vine, F.J. (1996) *Global Tectonics* (2nd edition), Blackwell Science, Oxford, 333 pp.

Kendall, A.C. (1992) Evaporites. In: *Facies Models: Response to Sea Level Change* (Eds Walker, R.G. & James, N.P.). Geological Association of Canada, St Johns, Newfoundland; 375–409.

Kendall, A.C. & Harwood, G.M. (1996) Marine evaporites: arid shorelines and basins. In: *Sedimentary Environments: Processes, Facies and Stratigraphy* (Ed. Reading, H.G.). Blackwell Science, Oxford; 281–324.

Kiessling, W. (2003) Reefs. In: *Encyclopedia of Sediments and Sedimentary Rocks* (Ed. Middleton, G.V.). Kluwer Academic Publishers, Dordrecht; 557–560.

Klein, G.D. (1995) Intracratonic basins. In: *Tectonics of Sedimentary Basins* (Eds Ingersoll, R.V. & Busby, C.J.). Blackwell Science, Oxford; 459–478.

Knauth, L.P. (1979) A model for the origin of chert in limestone. *Geology*, **7**, 274–277.

Knauth, L.P. (1994) Petrogenesis of chert. *Reviews of Mineralogy*, **29**, 233–258.

Kocurek, G.A. (1996) Desert aeolian systems. In: *Sedimentary Environments: Processes, Facies and Stratigraphy* (Ed. Reading, H.G.). Blackwell Science, Oxford; 125–153.

Krauskopf, K.B. (1979) *Introduction to Geochemistry*. McGraw-Hill, New York, 617 pp.

Krumbein, W.C. & Pettijohn, F.J. (1938) *Manual of Sedimentary Petrography*. Appleton-Century-Crofts, New York. (Reprinted by Society of Economic Palaeontologists and Mineralogists, Reprint Series 18, Tulsa, OK.)

Krumbein, W.C. & Sloss, L.L. (1951) *Stratigraphy and Sedimentation*. Freeman, San Francisco, 497 pp.

Leeder, M.R. (1975) Pedogenic carbonate and floodplain accretion rates: a quantitative model for alluvial, arid-zone lithofacies. *Geological Magazine*, **112**, 257–270.

Leeder, M.R. (1995) Continental rifts and proto-oceanic troughs. In: *Tectonics of Sedimentary Basins* (Eds Ingersoll, R.V. & Busby, C.J.). Blackwell Science, Oxford; 119–148.

Leeder, M.R. (1999) *Sedimentology and Sedimentary Basins: from Turbulence to Tectonics*. Blackwell Science, Oxford.

Leeder, M.R. & Gawthorpe, R.L. (1987) Sedimentary models for extensional tilt-block/half graben basins. In: *Continental Extensional Tectonics* (Eds Coward, M.P. & Dewey, J.F). Special Publication 28, Geological Society, London; 271–284.

Lees, A. (1975) Possible influences of salinity and temperature on modern shelf carbonate sedimentation. *Marine Geology*, **19**, 159–198.

Lewis, D.G. & McConchie, D. (1994) *Analytical Sedimentology*. Chapman and Hall, New York, London.

Lowe, D.R. (1982) Sediment gravity flows II: depositional models with special reference to the deposits of high density turbidity currents. *Journal of Sedimentary Petrology*, **52**, 279–297.

Lowenstein, T.K. & Hardie, L.A. (1985) Criteria for the recognition of salt pan evaporates. Sedimentology, **32**, 627–644.

Lutgens, F.K. & Tarbuck, E.J. (2006) *Essentials of Geology* (9th edition). Pearson Prentice Hall, Upper Saddle River, NJ.

Macdonald, D.I.M. & Butterworth, P.J. (1990) The stratigraphy, setting and hydrocarbon potential of the Mesozoic Sedimentary Basins of the Antarctic Peninsula. In: *Antarctic as an Exploration Frontier* (Ed. St. John, B.). Studies in Geology 31, American Association of Petroleum Geologists, Tulsa, OK; 1001–1036.

Machel, H.G. (2003) Dolomites and dolomitization. In: *Encyclopedia of Sediments and Sedimentary Rocks* (Ed. Middleton, G.V.). Kluwer Academic Publishers, Dordrecht; 235–243.

Mack, G.H., James, W.C. & Monger, H.C. (1993) Classification of paleosols, *Bulletin of the Geological Society of America*, **105**, 129–136.

Mackenzie, F.T. (2003) Carbonate Mineralogy and Geochemistry. In: *Encyclopedia of Sediments and Sedimentary Rocks* (Ed. Middleton, G.V.). Kluwer Academic Publishers, Dordrecht; 91–100.

Maizels, J.K., (1989) Sedimentology, palaeoflow dynamics and flood history of jökulhlaup deposits: palaeohydrology of Holocene sediment sequences in southern Iceland sandur deposits. *Journal of Sedimentary Petrology*, **59**, 204–223.

Maizels, J. (1993) Lithofacies variations within sandur deposits: the role of runoff regime, flow dynamics and sediment supply characteristics. In: Current research in fluvial sedimentology (ed C.R. Fielding). *Sedimentary Geology*, **85**, 299–325.

Major, J.J. (2003) Debris flow. In: *Encyclopedia of Sediments and Sedimentary Rocks* (Ed. Middleton, G.V.). Kluwer Academic Publishers, Dordrecht; 186–188.

Makaske, B. (2001) Anastomosing rivers: a review of their classification, origin and sedimentary products. *Earth-Science Reviews*, **53**, 149–196.

Mange, M.A. & Maurer, H.F.W. (1992) *Heavy Minerals in Colour*. Chapman and Hall, London, 147 pp.

Mange-Rajetzky, M.A. (1995) Subdivision and correlation of monotonous sandstone sequences using high-resolution heavy mineral analysis, a case study: the Triassic of the Central Graben. In: *Non-biostratigraphical Methods of Dating and Correlation* (Eds Dunay, R.E. & Hailwood, E.A.). Special Publication 89, Geological Society Publishing House, Bath; 23–30.

Marsgalia, K.M., (1995) Interarc and backarc basins. In: *Tectonics of Sedimentary Basins* (Eds Ingersoll, R.V. & Busby, C.J.). Blackwell Science, Oxford; 299–330.

Masson, D.G. (1994) Late Quaternary turbidite current pathways to the Madeira Abyssal Plain and some constraints on turbidity current mechanisms. *Basin Research*, **6**, 17–33.

Masson, D.G., Kidd, R.B., Gardner, J.V., Huggett, Q.J. & Weaver, P.P.E (1992) Saharan continental rise: facies distribution and sediment slides. In: *Geologic Evolution of Atlantic Continental Rises* (Eds Poag, C.W. & de Graciansky, P.C.). Van Nostrand Reinhold, New York; 327–343.

Matthews, R.K. (1986) Oxygen isotope record of ice-volume history: 100 million years of glacio-eustatic sea level fluctuation. In: *Inter-regional Unconformities and Hydrocarbon Accumulation* (Ed. Schlee, J.S.). Memoir 36, American Association of Petroleum Geologists, Tulsa, OK; 97–107.

Mazullo, S.J. (2000) Organogenic dolomitization in peritidal to deepsea sediments. *Journal of Sedimentary Research*, **70**, 10–23.

McCabe, P.J. (1984) Depositional environments of coal and coal-bearing strata. In: *Sedimentology of Coal and Coal-bearing Sequences* (Eds Rahmani, R.A. & Flores, R.M.). Special Publication 7, International Association of Sedimentologists. Blackwell Scientific Publications, Oxford; 13–42.

McCave, I.N. (1984) Erosion, transport and deposition of fine-grained marine sediments. In: *Fine-grained Sediments: Deep-water Processes and Facies* (Eds Stow D.A.V. & Piper D.W.J.). Special Publication 15, Geological Society, London; 35–69.

McKee, E.D. (1979) An introduction to the study of global sand seas. In: *Global Sand Seas* (Ed. McKee, E.D.). *United States Geological Survey Professional Paper*, **1052**, 1–19.

McKee, E.D. & Ward, W.C. (1983) Eolian environment. In: *Carbonate Depositional Environments* (Eds Scholle, P.A.,

Bebout, D.G. & Moore, C.H.). Memoir 33, American Association of Petroleum Geologists, Tulsa, OK; 132–170.

Miall, A.D. (1978) Lithofacies types and vertical profile models of braided river deposits, a summary. In: *Fluvial Sedimentology* (Ed. Miall, A.D.). Memoir 5, Canadian Society of Petroleum Geologists, Calgary; 597–604.

Miall, A.D. (1992) Exxon global cycle chart: an event for every occasion. *Geology*, **20**, 787–790.

Miall, A.D. (1995) Collision-related foreland basins. In: *Tectonics of Sedimentary Basins* (Eds Ingersoll, R.V. & Busby, C.J.). Blackwell Science, Oxford; 393–424.

Miall, A.D. (1997) *The Geology of Stratigraphic Sequences*. Springer-Verlag, Berlin.

Miall, A.D. (1999) *Principles of Sedimentary Basin Analysis* (3rd edition). Springer-Verlag, Berlin.

Middleton, G.V. (1973) Johannes Walther's law of correlation of facies. *Bulletin of the Geological Society of America*, **84**, 979–988.

Middleton, G.V. & Hampton, M.A. (1973) Sediment gravity flows: mechanics of flow and deposition. In: *Turbidites and Deep Water Sedimentation*. Short Course Notes, American Geophysical Institute–Society of Economic Palaeontologists and Mineralogists.

Miller, J.M.G. (1996) Glacial sediments. In: *Sedimentary Environments: Processes, Facies and Stratigraphy* (Ed. Reading, H.G.). Blackwell Scientific Publications, Oxford; 454–484.

Miller, R.M.C., McCave, I.N. & Komar, P.D. (1977) Threshold of sediment motion under unidirectional currents. *Sedimentology*, **24**, 507–527.

Milliken, K.L. (2003) Diagenesis. In: *Encyclopedia of Sediments and Sedimentary Rocks* (Ed. Middleton, G.V.). Kluwer Academic Publishers, Dordrecht; 214–219.

Mitchum, R.M., Vail, P.R. & Thompson, S. (1977) Seismic stratigraphy and global changes in sea level part 2: the depositional sequence as a basic unit for stratigraphic analysis. In: *Seismic Stratigraphy – Applications to Hydrocarbon Exploration* (Ed. Payton, C.E.). Memoir 26, American Association of Petroleum Geologists, Tulsa, OK; 117–134.

Monty, C.L.V., Bosence, D.W.J., Bridges, P.H. & Pratt, B.R. (Eds) (1995) *Carbonate Mud-mounds, Their Origin and Evolution*. Special Publication 23, International Association of Sedimentologists. Blackwell Science, Oxford.

Morrow, D.W. (1999) Regional subsurface dolomitization: models and constraints. *Geoscience Canada*, **25**, 57–70.

Morton, A.C. (2003) Heavy minerals. In: *Encyclopedia of Sediments and Sedimentary Rocks* (Ed. Middleton, G.V.). Kluwer Academic Publishers, Dordrecht; 356–358.

Morton, A.C. & Hallsworth, C.R. (1994) Identifying provenance-specific features of detrital heavy mineral assemblages in sandstones. *Sedimentary Geology*, **90**, 241–256.

Morton, A.C., Todd, S.P. & Haughton, P.D.W. (Eds) (1991) *Developments in Sedimentary Provenance Studies*. Special Publication 57, Geological Society Publishing House, Bath.

Mozley, P. (2003) Diagenetic structures. In: *Encyclopedia of Sediments and Sedimentary Rocks* (Ed. Middleton, G.V.). Kluwer Academic Publishers, Dordrecht; 219–225.

Mullins, H.T. & Cook, H.E. (1986) Carbonate apron models: alternatives to the submarine fan model for palaeoenvironmental analysis and hydrocarbon exploration. *Sedimentary Geology*, **48**, 37–79.

Muto, T. & Steel, R.J. (2000) The accommodation concept in sequence stratigraphy: some dimensional problems and possible redefinition. *Sedimentary Geology*, **130**, 1–10.

Mutti, E. (1992) *Turbidite Sandstones*. Agip Instituto di Geologia, Universita di Parma, Milano.

Nemec, W. (1990a) Deltas – remarks on terminology and classification. In: *Coarse-grained Deltas* (Eds Colella, A. & Prior, D.B.). Special Publication 10, International Association of Sedimentologists. Blackwell Scientific Publications, Oxford; 3–12.

Nemec, W. (1990b) Aspects of sediment movement on steep delta slopes. In: *Coarse-grained Deltas* (Eds Colella, A. & Prior, D.B.). Special Publication 10, International Association of Sedimentologists. Blackwell Scientific Publications, Oxford; 29–73.

Nesse, W.D. (2004) *Introduction to Optical Mineralogy*. Oxford University Press, Oxford.

Nichols, G.J. (1987) Syntectonic alluvial fan sedimentation, southern Pyrenees. *Geological Magazine*, **124**, 121–133.

Nichols, G.J. (2005) Sedimentary evolution of the Lower Clair Group, Devonian, west of Shetland: climate and sediment supply controls on fluvial, aeolian and lacustrine deposition. In: *Petroleum Geology: North West Europe and Global Perspectives – Proceedings of the 6th Petroleum Geology Conference* (Eds Doré, A.G. & Vining, B.A.); 957–967.

Nichols, G.J. & Fisher, J.A. (2007) Processes, facies and architecture of fluvial distributary system deposits. *Sedimentary Geology*, **195**, 75–90.

Nichols, G.J. & Thompson, B. (2005) Bedrock lithology control on contemporaneous alluvial fan facies, Oligo-Miocene, southern Pyrenees, Spain. *Sedimentology*, **52**, 571–585.

Nichols, G.J. & Uttamo, W. (2005) Sedimentation in a humid, interior, rift basin: the Cenozoic Li Basin, northern Thailand. *Journal of the Geological Society, London*, **162**, 333–348.

Nickling, W.G. (1994) Aeolian sediment transport and deposition. In: *Sediment Transport and Depositional Processes* (Ed. Pye, K.). Blackwell Science, Oxford; 293–350.

Ninkovich, D., Shackleton, N.J., Abdel-Monem, A.A., Obradovich, J.D. & Izett, G. (1978). K–Ar age of the late Pleistocene eruption of Toba, North Sumatra. *Nature*, **276**, 574–577.

North American Commission on Stratigraphic Nomenclature (1983) North American Stratigraphic Code, *American Association of Petroleum Geologists Bulletin*, **67**, 841–875.

Nurmi, R.D. & Friedman, G.M. (1977) Sedimentology and depositional environments of basin centre evaporites,

Lower Salina Group (Upper Silurian) Michigan Basin. In: *Reefs and Evaporites – Concepts and Depositional Models* (Ed. Fisher, J.H.). Studies in Geology 5, American Association of Petroleum Geologists, Tulsa, OK; 23–52.

Nystuen J.P. (1998) History and development of sequence stratigraphy. In: *Sequence Stratigraphy – Concepts and Applications* (Eds Gradstein, F.M. Sandvik, K.O. & Milton, N.J.). Special Publication 8, Norwegian Petroleum Society, Oslo; 31–116.

Oberhänsli, R. & Stoffers, P. (Eds) (1988) Hydrothermal activity and metaliferous sediments on the ocean floor. *Marine Geology* (Special Issue), **84**, 145–284.

O'Brien, P.E. & Wells, A.T. (1986) A small, alluvial crevasse splay. *Journal of Sedimentary Petrology*, **56**, 876–879.

Ori, G.G. & Friend, P.F. (1984) Sedimentary basins formed and carried piggy-back on active thrust sheets. *Geology*, **12**, 475–478.

Orton, G.J. (1996) Volcanic environments. In: *Sedimentary Environments: Processes, Facies and Stratigraphy* (Ed. Reading, H.G.). Blackwell Science, Oxford; 485–567.

Orton, G.J. & Reading, H.G. (1993) Variability of deltaic processes in terms of sediment supply, with particular emphasis on grain size. *Sedimentology*, **40**, 475–512.

Oudin, E., & Constantinou, G. (1984) Black smoker chimney fragments in Cyprus sulphide deposits. *Nature*, **308**, 349–353.

Owen, G. (2003) Ball-and-pillow (pillow) structures. In: *Encyclopedia of Sediments and Sedimentary Rocks* (Ed. Middleton, G.V.). Kluwer Academic Publishers, Dordrecht; 39–40.

Patterson, R.J. & Kinsman, D.J.J. (1982) Formation of diagenetic dolomite in coastal sabkha along the Arabian (Persian) Gulf. *American Association of Petroleum Geologists Bulletin*, **66**, 28–43.

Pauley, J.C. (1995) Sandstone megabeds from the Tertiary of the North Sea. In: *Characterization of Deep Marine Clastic Systems* (Eds Hartley, A.J. & Prosser, D.J.). Special Publication 94, Geological Society Publishing House, Bath; 103–114.

Payton, C.E. (Ed.) (1977) *Seismic Stratigraphy – Applications to Hydrocarbon Exploration*. Memoir 26, American Association of Petroleum Geologists, Tulsa, OK.

Pejrup, M. (2003) Flocculation. In: *Encyclopedia of Sediments and Sedimentary Rocks* (Ed. Middleton, G.V.). Kluwer Academic Publishers, Dordrecht; 284–285.

Pemberton, S.G. & MacEachern, J.A. (1995) The sequence stratigraphic significance of trace fossils: examples from the Cretaceous foreland basin of Alberta, Canada. In: *Sequence Stratigraphy of Foreland Basin Deposits* (Eds Van Wagoner, J.C. & Bertram, G.T.). Memoir 64, American Association of Petroleum Geologists, Tulsa, OK; 429–476.

Pemberton, S.G., MacEachern, J.A. & Frey, R.W. (1992) Trace fossil facies models: environmental and allostratigraphic significance. In: *Facies Models: Response to*

Sea Level Change (Eds Walker, R.G. & James, N.P.). Geological Association of Canada, St Johns, Newfoundland; 47–72.

Percival, C.J. (1986) Paleosols containing an albic horizon: examples from the Upper Carboniferous of northern England. In: *Paleosols: their Recognition and Interpretation* (Ed. Wright, V.P.). Blackwell Scientific Publications, Oxford; 87–111.

Pettijohn, F.J. (1975) *Sedimentary Rocks* (3rd edition). Harper and Row, New York.

Pettijohn, F.J., Potter, P.E. & Siever, R. (1987) *Sand and Sandstone*. Springer-Verlag, New York.

Pickering, K.T., Hiscott, R.N. & Hein, F.J. (1989) *Deep Marine Environments: Clastic Sedimentation and Tectonics*. Unwin Hyman, London.

Pilkey, O.H. (1988) Basin plains: giant sedimentation events. In: *Sedimentologic Consequences of Convulsive Geologic Events* (Ed. Clifton, H.E.). *Geological Society of America Special Paper*, **229**, 93–99.

Pitman, W.C. (1978) Relationship between eustasy and stratigraphic sequences of passive margins. *Geological Society of America Bulletin*, **89**, 1387–1403.

Plint, A.G. (1988) Sharp-based shoreface sequences and off-shore bars of the Cardium Formation of Alberta: their relationship to relative changes in sea-level. In: *Sea Level Changes: an Integrated Approach* (Eds C.K. Wilgus, B.S. Hastings, C.G.St.C. Kendall, H.W. Posamentier, C.A. Ross and J.C. Van Wagoner). Special Publication 42, Society of Economic Paleontologists and Mineralogists, Tulsa, OK; 357–370.

Plint, A.G. & Nummedal, D. (2000) The falling stage systems tract: recognition and importance in sequence stratigraphic analysis. In: *Sedimentary Responses to Forced Regressions* (Eds Hunt, D. & Gawthorpe, R.L.). Special Publication 172, Geological Society Publishing House, Bath; 1–17.

Plint, A.G., Eyles, N., Eyles, C.H. & Walker, R.G. (1992) Control of sea level change. In: *Facies Models: Response to Sea Level Change* (Eds Walker, R.G. & James, N.P.). Geological Association of Canada, St Johns, Newfoundland; 15–26.

Porebski, S.J. & Gradzinski, R. (1990) Lava-fed Gilbert-type delta in the Plonez Cove Formation (Lower Oligocene), King George Island, West Antarctica. In: *Coarse-grained Deltas* (Eds Colella, A. & Prior, D.B.). Special Publication 10, International Association of Sedimentologists. Blackwell Scientific Publications, Oxford; 335–351.

Posamentier, H.W. & James, N.P. (1993) An overview of sequence-stratigraphic concepts: uses and abuses. In: *Sequence Stratigraphy and Facies Associations* (Eds Posamentier, H.W., Summerhayes, C.P., Haq, B.U. & Allen, G.P.). Special Publication 18, International Association of Sedimentologists. Blackwell Science, Oxford; 3–18.

Posamentier, H.W. & Morris, W.R. (2000) Aspects of the stratal architecture of forced regressive deposits. In: *Sedimentary Responses to Forced Regressions* (Eds Hunt, D. & Gawthorpe, R.L.). Special Publication 172, Geological Society Publishing House, Bath; 19–46.

Pratt, B.R., James, N.P. & Cowan, C.A. (1992) Peritidal carbonates. In: *Facies Models: Response to Sea Level Change* (Eds Walker, R.G. & James, N.P.). Geological Association of Canada, St Johns, Newfoundland; 303–322.

Press, F. & Siever, R. (1986) *Earth* (2nd edition). W.H. Freeman, New York, 649 pp.

Preston, J., Hartley, A., Hole, M., Buck, S., Bond, J., Mange, M. & Still, J. (1998) Integrated whole-rock trace element geochemistry and heavy mineral chemistry studies: aids to correlation of continental red-bed reservoirs in the Beryl Field, UK North Sea. *Petroleum Geoscience*, **4**, 7–16.

Purser, B., Tucker, M.E. & Zenger, D. (Eds) (1994) *Dolomites – a Volume in Honour of Dolomieu*. Special Publication 21, International Association of Sedimentologists. Blackwell Science, Oxford.

Pye, K. (1987) *Aeolian Dust and Dust Deposits*. Academic Press, London.

Qing, H. & Mountjoy, E.W. (1992) Large-scale fluid flow in the Middle Devonian Presqu'ile barrier, Western Canada Sedimentary Basin. *Geology*, **20**, 903–906.

Ravnås, R. & Furnes, H. (1995) The use of geochemical data in determining the provenance and tectonic setting of ancient sedimentary successions: the Kalvåg Melange, western Norwegian Caledonides. In: *Sedimentary Facies Analysis: a Tribute to the Research and Teaching of Harold G. Reading* (Ed. Plint, A.G.). Special Publication 22, International Association of Sedimentologists. Blackwell Science, Oxford; 237–264.

Rawson, P.F., Allen, P.M., Brenchley, P.J., Cope, J.C.W., Gales, A.S., Evans, J.A., Gibbard, P.L., Gregory, F.J., Hailwood, E.A., Hesselbo, S.P., Knox, R.W. O'B., Marshall, J.E.A., Oates, M., Riley, N.J., Smith, A.G., Trewin, N. & Zalasiewicz, J.A. (2002). *Stratigraphical Procedure*. Professional Handbook, Geological Society Publishing House, Bath, 58 pp.

Raymo, M.E., & Ruddiman, W.F. (1992) Tectonic forcing of late Cenozoic climate: *Nature*, **359**, 117–122.

Read, J.F. (1985) Carbonate platform facies models. *American Association of Petroleum Geologists Bulletin*, **69**, 1–21.

Reading, H.G. (1980) Characteristics and recognition of strike-slip systems. In: *Sedimentation in Oblique-slip Mobile Zones* (Eds Ballance, P.F. & Reading, H.G.). Special Publication 4, International Association of Sedimentologists. Blackwell Scientific Publications, Oxford; 7–26.

Reading, H.G. (Ed.) (1996) *Sedimentary Environments: Processes, Facies and Stratigraphy* (3rd edition). Blackwell Science, Oxford.

Reading, H.G. & Collinson, J.D. (1996) Clastic coasts. In: *Sedimentary Environments: Processes, Facies and Stratigraphy* (Ed. Reading, H.G.). Blackwell Science, Oxford; 154–231.

Reading, H.G. & Levell, B.K. (1996) Controls on the sedimentary record In: *Sedimentary Environments: Processes, Facies and Stratigraphy* (Ed. Reading, H.G.). Blackwell Science, Oxford; 5–36.

Reading, H.G. & Richards (1994) Turbidite systems in deepwater basin margins classified by grain size and feeder system. *American Association of Petroleum Geologists Bulletin*, **78**, 792–822.

Reid, I. & Frostick, L.E. (1985) Beach orientation, bar morphology and the concentration of metalliferous placer deposits: a case study, Lake Turkana, Kenya. *Journal of the Geological Society of London*, **142**, 837–848.

Reineck, H.E. & Singh, I.B. (1972) Genesis of laminated sand and graded rhythmites in storm-sand layers of shelf mud. *Sedimentology*, **18**, 123–128.

Reineck, H.E. & Singh, I.B. (1980) *Depositional Sedimentary Environments* (2nd edition). Springer-Verlag, Berlin.

Reinson, G.E. (1992) Transgressive barrier island and estuarine systems. In: *Facies Models: Response to Sea Level Change* (Eds Walker, R.G. & James, N.P.). Geological Association of Canada, St Johns, Newfoundland; 179–194.

Renard, F. & Dysthe, D. (2003) Pressure solution. In: *Encyclopedia of Sediments and Sedimentary Rocks* (Ed. Middleton, G.V.). Kluwer Academic Publishers, Dordrecht; 542–543.

Retallack, G.J. (2001) *Soils of the Past: an Introduction to Paleopedology* (2nd edition). Blackwell Science, Oxford.

Rider, M.H. (2002) *The Geological Interpretation of Well Logs* (2nd edition). Whittles Publishing, Caithness.

Riding, R. (2000) Microbial carbonates: the geological record of calcified bacterial–algal mats and biofilms. *Sedimentology*, **47**, 179–214.

Rink, W.J. (1997) Electron spin resonance (ESR) dating and ESR applications in Quaternary science and archaeometry. *Radiation Measurements*, **27**, 975–1025.

Ross, D.J. & Skelton, P.W. (1993) Rudist formations of the Cretaceous: a palaeoecological, sedimentological and stratigraphical review. In: *Sedimentology Review 1* (Ed. Wright, V.P.). Blackwell Science, Oxford; 73–91.

Royden, L. (1985) The Vienna Basin: a thin-skinned pull-apart basin. In: *Strike-slip Deformation, Basin Formation, and Sedimentation* (Eds Biddle, K.& Christie-Blick, N.). Special Publication 37, Society of Economic Paleontologists and Mineralogists, Tulsa, OK; 319–338.

Russell, A.J. & Knudsen, Ó. (2002) The effects of glacier outburst flood flow dynamics on ice-contact deposits: November (1996) jökulhlaup, Skeiðoarársandur, Iceland. In: *Flood and Megaflood Deposits: Recent and Ancient* (Eds Martini, I.P., Baker, V.R. & Garzón, G.). Special Publication 32, International Association of Sedimentologists. Blackwell Scientific Publications, Oxford; 67–84.

Salvador, A. (1994) *International Stratigraphic Guide. A Guide to Stratigraphic Classification, Terminology and Procedure* (2nd edition). The International Union of Geological Sciences and the Geological Society of America.

Scheffers, A. & Kelletat, D. (2003) Sedimentologic and geomorphologic tsunami imprints worldwide – a review. *Earth-Science Reviews*, **63**, 83–92.

Schminke, H.V., Fisher, R.V & Waters, A.C. (1973) Antidune and chute-and-pool structures in the base surge deposits of the Laacher See area, Germany. *Sedimentology*, **20**, 553–574.

Scholle, P.A. (1978) *A Color Illustrated Guide to Carbonate Rock Consituents, Textures, Cements and Porosities.* Memoir 27, American Association of Petroleum Geologists, Tulsa.

Scholle, P.A. & Ulmer-Scholle, D. (2003) Cements and cementation. In: *Encyclopedia of Sediments and Sedimentary Rocks* (Ed. Middleton, G.V.). Kluwer Academic Publishers, Dordrecht; 110–119.

Scholle, P.A., Arthur, M.A. & Ekdale, A.A. (1983) Pelagic environments. In: *Carbonate Depositional Environments* (Eds Scholle, P.A., Bebout, D.G. & Moore, C.H.). Memoir 33, American Association of Petroleum Geologists, Tulsa, OK; 620–691.

Schumm, S.A. (1968) Speculations concerning palaeohydraulic controls of terrestrial sedimentation. *Geological Society of America Bulletin*, **79**, 1573–1588.

Schumm, S.A. (1981) Evolution and response of the fluvial system; sedimentologic implications. In: *Recent and Ancient Nonmarine Depositional Environments* (Eds Ethridge, F.G. & Flores, R.M.). Special Publication 31, Society of Economic Paleontologists and Mineralogists, Tulsa, OK; 19–29.

Schutz, L. (1980) Long range transport of desert dust with special emphasis on the Sahara. *New York Academy of Science Annals*, **338**, 515–532.

Seilacher, A. (2007) *Trace Fossil Analysis.* Springer, Berlin.

Shackleton, N.J. (1977) The oxygen isotope stratigraphic record of the late Pleistocene. *Philosophical Transactions of the Royal Society, London*, **280**, 169–182.

Shaw, A.B. (1964) *Time in Stratigraphy.* McGraw-Hill, New York.

Simonson, B. (2003) Ironstones and iron formations. In: *Encyclopedia of Sediments and Sedimentary Rocks* (Ed. Middleton, G.V.). Kluwer Academic Publishers, Dordrecht; 379–385.

Simpson, S. (1975) Classification of trace fossils. In: *The Study of Trace Fossils* (Ed. Frey, R.W.). Springer-Verlag, Berlin; 39–45.

Smith, D.G. (1983) Anastomosed fluvial deposits: examples from western Canada. In: *Modern and Ancient Fluvial Systems* (Eds Collinson, J.D. & Lewin, J.). Special Publication 6, International Association of Sedimentologists. Blackwell Scientific Publications, Oxford; 155–168.

Smith, D.G. & Smith, N.D. (1980) Sedimentation in anastomosing river systems: examples from alluvial valleys near Banff, Alberta. *Journal of Sedimentary Petrology*, **50**, 157–164.

Smith, G.A. & Lowe, D.R. (1991) Lahars: volcano-hydrologic events and deposition in the debris flow – hyperconcentrated flow continuum. In: *Sedimentation in Volcanic Settings* (Eds Fisher, R.V. & Smith, G.A.). Special Publication 45, SEPM (Society for Sedimentary Geology), Tulsa, OK; 59–70.

Smith, N.D. (1978) Some comments on terminology for bars in shallow rivers. In: *Fluvial Sedimentology* (Ed. Miall, A.D.). Memoir 5, Canadian Society of Petroleum Geologists, Calgary; 85–92.

Smith, R.B. & Braile, L.W. (1994) The Yellowstone hotspot. *Journal of Volcanology and Geothermal Research*, **61**, 121–187.

Southard, J.B. (1991) Experimental determination of bedform stability. *Annual Review of Earth and Planetary Sciences*, **19**, 423–455.

Sparks, R.S.J. (1976) Grain size variations in ignimbrites and implications for the transport of pyroclastic flows. *Sedimentology*, **23**, 146–188.

Spears, D.A. (2003) Bentonites and tonsteins. In: *Encyclopedia of Sediments and Sedimentary Rocks* (Ed. Middleton, G.V.). Kluwer Academic Publishers, Dordrecht; 61–63.

Stanley, S.M. (1985) Rates of evolution. *Palaeobiology*, **11**, 13–26.

Stow, D.A.V. (1979) Distinguishing between fine-grained turbidites and contourites on the Nova Scotia deep water margin. *Sedimentology*, **26**, 371–387.

Stow, D.A.V. (1985) Deep-sea clastics: where are we and where are we going? In: *Sedimentology, Recent Developments and Applied Aspects* (Eds Brenchley, P.J. & Williams, B.P.J.). Blackwell Scientific Publications, Oxford; 67–94.

Stow, D.A.V. (1986) Deep clastic seas. In: *Sedimentary Environments and Facies* (Ed. Reading, H.G.). Blackwell Scientific Publications, Oxford; 400–444.

Stow, D.A.V. (1994) Deep-sea processes of sediment transport and deposition. In: *Sediment Transport and Depositional Processes* (Ed. Pye, K.). Blackwell Science, Oxford; 257–291.

Stow, D.A.V. (2005) *Sedimentary Rocks in the Field: a Colour Guide*. Manson, London.

Stow, D.A.V. & Lovell, J.P.B. (1979) Contourites; their recognition in modern and ancient sediments. *Earth-Science Reviews*, **14**, 251–291.

Stow, D.A.V., Reading, H.G. & Collinson, J.D. (1996) Deep seas. In: *Sedimentary Environments: Processes, Facies and Stratigraphy* (Ed. Reading, H.G.). Blackwell Science, Oxford; 395–453.

Stow, D.A.V., Faugères, J-C., Viana, A. & Gonthier, E. (1998) Fossil contourites: a critical review. *Sedimentary Geology*, **115**, 3–31.

Stubblefield, W.L., McGrail, D.W. & Kersey, D.G. (1984) Recognition of transgressive and post-transgressive sand ridges on the New Jersey continental shelf. In: *Siliciclastic Shelf Sediments* (Eds Tillman, R.W. & Siemers, C.T.). Special Publication 34, SEPM (Society for Sedimentary Geology), Tulsa, OK; 1–23.

Sturm, M. & Matter, A. (1978) Turbidites and varves in Lake Brienz (Switzerland): deposition of clastic detritus by density currents. In: *Modern and Ancient Lake Sediments* (Eds Matter, A. & Tucker, M.E.) Special Publication 2,

International Association of Sedimentologists. Blackwell Scientific Publications, Oxford; 147–168.

Swan, A.R.H. & Sandilands, M. (1995) *Introduction to Geological Data Analysis*. Blackwell Science, Oxford.

Talbot, M.R. & Allen, P.A. (1996) Lakes. In: *Sedimentary Environments: Processes, Facies and Stratigraphy* (Ed. Reading, H.G.). Blackwell Scientific Publications, Oxford; 83–124.

Talling, P.J. & Burbank, D.W. (1993) Assessment of uncertainties in magnetostratigraphic dating of sedimentary strata. In: *Applications of Palaeomagnetism to Sedimentary Geology* (Eds Aïssauoi, D.M., McNeill, D.F. & Hurley, N.F.). Special Publication 49, SEPM (Society for Sedimentary Geology), Tulsa, OK; 59–69.

Tanner, P.W.G. (2003) Syneresis. In: *Encyclopedia of Sediments and Sedimentary Rocks* (Ed. Middleton, G.V.). Kluwer Academic Publishers, Dordrecht; 718–720.

Tanner, L.H. & Hubert, J.F. (1991) Basalt breccias and conglomerates in the Lower Jurassic McCoy Brook Formation, Fundy Basin, Nova Scotia: differentiation between talus and debris-flow deposits. *Journal of Sedimentary Petrology*, **61**, 15–27.

Taylor, C.M. (1990) Late Permian – Zechstein. In: *Introduction to the Petroleum Geology of the North Sea* (Ed. Glennie, K.W.). Blackwell Scientific Publications, Oxford; 153–190.

Thomas, R.G., Smith, D.G., Wood, J.M., Visser, J., Calverley-Range, E.A. & Koster, E.H. (1987) Inclined heterolithic stratification – terminology, description, interpretation and significance. *Sedimentary Geology*, **53**, 123–179.

Trendall, A.F. (2002) The significance of iron-formation in the Precambrian stratigraphic record. In: *Precambrian Sedimentary Environments: a Modern Approach to Ancient Depositional Systems* (Eds Altermann W. & Corcoran, P.L.). Special Publication 33, International Association of Sedimentologists. Blackwell Scientific Publications, Oxford; 33–66.

Tucker, M.E. (Ed) (1988) *Techniques in Sedimentology*. Blackwell Scientific Publications, Oxford, 394 pp.

Tucker, M.E. (1991) *Sedimentary Petrology* (2nd edition). Blackwell Scientific Publications, Oxford.

Tucker, M.E. (1992) Limestones through time. In: *Understanding the Earth, a New Synthesis* (Eds Brown, G.C., Hawkesworth, C.J. & Wilson, R.C.L.). Cambridge University Press, Cambridge; 347–363.

Tucker, M.E. (1996) *Sedimentary Rocks in the Field* (2nd edition). Wiley, Chichester, 153 pp.

Tucker, M.E. (2003) *Sedimentary Rocks in the Field* (3rd edition). Wiley, Chichester.

Tucker, M.E. & Wright, V.P. (1990) *Carbonate Sedimentology*. Blackwell Scientific Publications, Oxford, 482 pp.

Turner, P. (1980) *Continental Red Beds*. Elsevier, Amsterdam.

Udden, J.A. (1914) Mechanical composition of clastic sediments. *Geological Society of America Bulletin*, **25**, 655–744.

Underwood, M.B. & Moore, G.F. (1995) Trenches and trench-slope basins. In: *Tectonics of Sedimentary Basins*

(Eds Ingersoll, R.V. & Busby, C.J.). Blackwell Science, Oxford; 179–220.

Urgeles, R., Canals, M., Baraza, J., Alonso, B. & Masson, D. (1997) The most recent megalandslides of the Canary Islands: El Golfo debris avalanche and Canary debris flow, west El Hierro Island. *Journal of Geophysical Research*, **102**, 20305–20323.

Vail, P.R., Mitchum, R.M. & Thompson, S. (1977) Seismic stratigraphy and global changes in sea level, part 4; global cycles of relative changes of sea level. In: *Seismic Stratigraphy – Applications to Hydrocarbon Exploration* (Ed. Payton, C.E.). Memoir 26, American Association of Petroleum Geologists, Tulsa, OK.

Van Wagoner, J.C., Posamentier, H.W., Mitchum, R.M., Vail, P.R., Sard, J.F., Loutit, T.S. & Hardenbol, J. (1988) An overview of sequence stratigraphy and key definitions. In: *Sea Level Changes: an Integrated Approach* (Eds C.K. Wilgus, B.S. Hastings, C.G.St.C. Kendall, H.W. Posamentier, C.A. Ross and J.C. Van Wagoner). Special Publication 42, Society of Economic Paleontologists and Mineralogists, Tulsa, OK; 39–45.

Van Wagoner, J.C., Mitchum, R.M., Campion, K.M. & Rahmanian, V.D. (1990) *Siliciclastic Sequence Stratigraphy in Well Logs, Cores and Outcrop: Concepts for High Resolution Correlation of Time and Facies*. Methods in Exploration Series 7, American Association of Petroleum Geologists, Tulsa, OK.

Wal, A. & McManus, J. (1993) Wind regime and sand transport on a coastal beach-dune complex, Tentsmuir, eastern Scotland. In: *The Dynamics and Environmental Context of Aeolian Sedimentary Systems* (Ed. Pye, K.). Special Publications 72, Geological Society Publishing House, Bath; 159–171.

Walker, R.G. (1992) Facies models. In: *Facies Models: Response to Sea Level Change* (Eds Walker, R.G. & James, N.P.). Geological Association of Canada, St Johns, Newfoundland; 1–14.

Walker, R.G. (2006) Facies models revisited: introduction. In: *Facies Models Revisited* (Eds Walker, R.G. & Posamentier, H.). Special Publication 84, Society of Economic Paleontologists and Mineralogists, Tulsa, OK; 1–17.

Walker, R.G. & James, N.P. (Eds) (1992) *Facies Models: Response to Sea Level Change*. Geological Association of Canada, St Johns, Newfoundland.

Walker, R.G. & Plint, A.G. (1992) Wave- and storm-dominated shallow marine systems. In: *Facies Models: Response to Sea Level Change* (Eds Walker, R.G. & James, N.P.). Geological Association of Canada, St Johns, Newfoundland; 219–238.

Walsh, S., Gradstein, F. & Ogg, J. (2004) History, philosophy, and application of the Global Stratotype Section and Point (GSSP). *Lethaia*, **37**, 201–218.

Waltham, D.A. (2000) *Mathematics: a Simple Tool for Geologists* (2nd edition). Blackwell Science, Oxford.

Walther, J. (1894) *Einleitung in die Geologicals Historische Wissenschaft*, Bd 3, *Lithogenesis der Gegenwart*. Fischerverlag, Jena: 535–1055.

Warren, J. (1999) *Evaporites: their Evolution and Economics*. Blackwell Science, Oxford.

Warren, J.K. & Kendall, C.G. St.C. (1985). Comparison of sequences formed in marine sabkha (subaerial) and salina (subaqueous) settings – modern and ancient. *American Association of Petroleum Geologists Bulletin*, **69**, 1013–1023.

Warren, W.P. & Ashley, G.M. (1994) Origins of the ice-contact stratified ridges (eskers) of Ireland. *Journal of Sedimentary Research*, **A64**, 433–449.

Wasson, R.J. & Hyde, R. (1983) Factors determining desert dune type. *Nature*, **304**, 37–339.

Wells, N.A. & Dorr Jr, J.A. (1987) A reconnaissance of sedimentation on the Kosi alluvial fan of India. In: *Recent Developments in Fluvial Sedimentology* (Eds Ethridge, F.G., Flores, R.M. & Harvey, M.D.). Special Publication 39, Society of Economic Paleontologists and Mineralogists, Tulsa, OK; 51–61.

Wentworth, C.K. (1922) A scale of grade and class terms for clastic sediments. *Journal of Geology*, **30**, 377–394.

Wescott, W.A. & Ethridge, F.G. (1990) Fan deltas – alluvial fans in coastal settings. In: *Alluvial Fans: a Field Approach* (Eds Rachocki, A.H. & Church, M.). Wiley, Chichester; 195–211.

Whitham, A.G. & Sparks, R.S.J. (1986) Pumice. *Bulletin of Volcanology*, **48**, 209–223.

Whittington, H.B. & Conway-Morris, S. (Eds) (1985) Extraordinary fossil biotas: their ecological and evolutionary significance. *Philosophical Transactions of the Royal Society, Series B*, **311**, 1–192.

Wignall, P.B. (1994) *Black Shales*. Oxford University Press, Oxford.

Willis, B.J., Bhattacharya, J.B., Gabel, S.L. & White, C.D. (1999) Architecture of a tide-influenced delta in the Frontier Formation of central Wyoming, U.S.A. *Sedimentology*, **46**, 667–688.

Wilson, I.G. (1972) Universal discontinuities in bedforms produced by the wind. *Journal of Sedimentary Petrology*, **42**, 667–669.

Wilson, J.T. (1966) Did the Atlantic close and then re-open? *Nature*, **211**, 676–681.

Wilson, M.E.J. (2005) Development of equatorial delta-front patch reefs during the Neogene, Borneo. *Journal of Sedimentary Research*, **75**, 114–133

Wilson, M.E.J. & Lokier, S.W. (2002) Siliciclastic and volcaniclastic influences on equatorial carbonates: insights from the Neogene of Indonesia. *Sedimentology*, **49**, 583–601

Wilson, M.E.J. & Vecsei, A. (2005) The apparent paradox of abundant foramol facies in low latitudes: their environmental significance and effect on platform development. *Earth-Science Reviews*, **69**, 133–168.

Wise, S. (2003) Calcite compensation depth. In: *Encyclopedia of Sediments and Sedimentary Rocks* (Ed. Middleton, G.V.). Kluwer Academic Publishers, Dordrecht; 88–89.

Woodroffe, C.D. (2003) *Coasts: Form, Process and Evolution.* Cambridge University Press, Cambridge.

Worsley, T.W., Nance, D. & Moody, J.B. (1984) Global tectonics and eustasy for the past 2 billion years. *Marine Geology*, **58**, 373–400.

Wright, V.P. (1986) Facies sequences on a carbonate ramp: the Carboniferous Limestone of South Wales. *Sedimentology*, **33**, 221–241.

Wright, V.P. & Burchette, T.P. (1996) Shallow-water carbonate environments. In: *Sedimentary Environments: Processes, Facies and Stratigraphy* (Ed. Reading, H.G.). Blackwell Science, Oxford; 325–394.

Wright, V.P. & Tucker, M.E. (1991) Calcretes: an introduction. In: *Calcretes* (Ed. Wright, V.P.). Reprint Series 2, International Association of Sedimentologists. Blackwell Science, Oxford; 1–22.

Wüst, R.A.J. & Bustin, R.M. (2003) Kerogen. In: *Encyclopedia of Sediments and Sedimentary Rocks* (Ed. Middleton, G.V.). Kluwer Academic Publishers, Dordrecht; 400–403.

Yang, C-S. & Nio, S-D. (1985) The estimation of palaeohydrodynamic processes from subtidal deposits using time series analysis methods. *Sedimentology*, **32**, 41–57.

Yechieli, Y. & Wood, W.W. (2002) Hydrogeologic processes in saline systems: playas, sabkhas, and saline lakes. *Earth-Science Reviews*, **58**, 343–365.

Zalasiewicz, J., Smith, A., Brenchley, P., Evans, J., Knox, R., Riley, N., Gale, A., Gregory, F.J., Rushton, A., Gibbard, P., Hesselbo, S., Marshall, J., Oates, M. Rawson, P. & Trewin, N. (2004) Simplifying the stratigraphy of time. *Geology*, **32**, 1–4.

Index